Bacterial Diversity and Systematics

FEDERATION OF EUROPEAN MICROBIOLOGICAL SOCIETIES SYMPOSIUM SERIES

Recent FEMS Symposium volumes published by Plenum Press

1991 • GENETICS AND PRODUCT FORMATION IN *STREPTOMYCES*
Edited by Simon Baumberg, Hans Krügel, and Dieter Noack
(FEMS Symposium No. 55)

1991 • THE BIOLOGY OF *ACINETOBACTER*: Taxonomy, Clinical
Importance, Molecular Biology, Physiology, Industrial Relevance
Edited by K. J. Towner. E. Bergogne-Bérézin, and C. A. Fewson
(FEMS Symposium No. 57)

1991 • MOLECULAR PATHOGENESIS OF GASTROINTESTINAL INFECTIONS
Edited by T. Wadström, P. H. Mäkelä, A.-M. Svennerholm, and
H. Wolf-Watz
(FEMS Symposium No. 58)

1992 • MOLECULAR RECOGNITION IN HOST–PARASITE INTERACTIONS
Edited by Timo K. Korhonen, Tapani Hovi, and P. Helena Mäkelä
(FEMS Symposium No. 61)

1992 • THE RELEASE OF GENETICALLY MODIFIED
MICROORGANISMS—REGEM 2
Edited by Duncan E. S. Stewart-Tull and Max Sussman
(FEMS Symposium No. 63)

1993 • RAPID DIAGNOSIS OF MYCOPLASMAS
Edited by Itzhak Kahane and Amiram Adoni
(FEMS Symposium No. 62)

1993 • BACTERIAL GROWTH AND LYSIS: Metabolism and
Structure of the Bacterial Sacculus
Edited by M. A. de Pedro, J.-V. Höltje, and W. Löffelhardt
(FEMS Symposium No. 65)

1994 • THE GENUS *ASPERGILLUS*: From Taxonomy and Genetics to
Industrial Application
Edited by Keith A. Powell, Annabel Renwick, and John F. Peberdy
(FEMS Symposium No. 69)

1994 • BACTERIAL DIVERSITY AND SYSTEMATICS
Edited by Fergus G. Priest, Alberto Ramos-Cormenzana, and
B. J. Tindall
(FEMS Symposium No. 75)

Bacterial Diversity and Systematics

Edited by

Fergus G. Priest

Heriot Watt University
Edinburgh, Scotland, United Kingdom

Alberto Ramos-Cormenzana

University of Granada
Granada, Spain

and

B. J. Tindall

Deutsche Sammlung von Mikroorganismen
und Zellkulturen GmbH
Braunschweig, Germany

SPRINGER SCIENCE+BUSINESS MEDIA, LLC

Library of Congress Cataloging-in-Publication Data

Bacterial diversity and systematics / edited by Fergus G. Priest,
Alberto Ramos-Cormenzana, and B.J. Tindall.
        p.   cm. -- (Federation of European Microbiological Societies
symposium series ; 75)
    Includes bibliographical references and index.
    ISBN 978-0-306-44832-4   ISBN 978-1-4615-1869-3 (eBook)
    DOI 10.1007/978-1-4615-1869-3
    1. Bacterial diversity. 2. Bacteriology--Classification.
I. Priest, F. G. II. Ramos-Cormenzana, Alberto. III. Tindall, B.J.
IV. Series: FEMS symposium ; no. 75.
QR73.B326 1994
589.9'0012--dc20   94-43384
                                                                  CIP

Proceedings of a symposium held under the auspices of the Federation of European Microbiological Societies, September 19–22, 1993, in Granada, Spain

ISBN 978-0-306-44832-4

© 1994 Springer Science+Business Media New York
Originally published by Plenum Press, New York in 1994

Preface

Bacterial taxonomy as a specialized discipline is practised by a minority but the applications of taxonomy are important to most, if not all microbiologists. It is the implementation of taxonomic ideas and practises which gives rise to identification and typing systems, procedures for the analysis and characterization of biodiversity, hypotheses about the evolution of micro-organisms, and improved procedures for the isolation and implementation of bacteria in biotechnological processes. Without taxonomic theory providing a sound basis to these many facets of microbiology there would be severe problems faced by many scientists working with micro-organisms.

Taxonomy comprises three sequential but independent processes; classification, nomenclature and identification. The first two stages are the prime concern of the specialist taxonomist but the third stage should result in identification schemes of value to all microbiologists. As the classification and identification of micro-organisms improves, largely due to the introduction of new technologies, so does its contribution to the subject as a whole. It therefore seemed timely to hold a conference in the autumn of 1993 devoted to microbial identification. Such a topic could not be addressed without some reference to the enabling discipline of classification, but the principal aims were to assess improvements in identification and typing and how these were benefiting microbiological topics ranging from ecological and biotechnological studies of extremophilic bacteria to the use of pyrolysis mass spectrometry in epidemiology. The meeting, which was held in Granada, Spain, was supported by FEMS (FEMS Symposium No. 75) and the British (SGM), German and Spanish Microbiology Societies. The financial support of Novo Nordisk was also greatly appreciated by the organizers and all who attended, as was the extensive hospitality provided by the state of Andalucia.

This book is derived from the Granada Conference. It is not intended as a proceedings of that conference, but each chapter has been drawn from the speakers and topics covered at the meeting. Inevitably, there has been some rearrangement and modification of material and with minor changes in emphasis it was considered that a name change was appropriate to reflect better the contents of the book. Thus 'Microbial Diversity and Systematics' was chosen, but the emphasis remains on **identification** of bacterial diversity.

The Editors are grateful to all the authors who have contributed to this book. The Granada Conference was considered a great success by those who attended, we hope that this book will similarly be found to be timely and useful.

Fergus G. Priest; Edinburgh
Alberto Ramos-Cormenzana; Granada
Brian Tindall; Braunschweig

CONTENTS

Molecular Taxonomy : Classification and Identification.. 1
  K.H. Schleifer and W. Ludwig

DNA Sequence Analysis of the Genetic Structure of Populations
     of *Salmonella enterica* and *Escherichia coli*.............................. 17
  R.K. Selander, J. Li, E.F. Boyd, F.-S. Wang, and K. Nelson

Identification and Typing of Bacteria by Protein Electrophoresis............................ 51
  K. Kersters, B. Pot, D. Dewettinck, U. Torck, M. Vancanneyt,
  L. Vauterin, and P. Vandamme

Characterisation and Identification of Micro-organisms by
     FT-IR Spectroscopy and FT-IR Microscopy.................................. 67
  D. Naumann, D. Helm, and C. Schultz

Curie Point Pyrolysis Spectrometry and Its Application to
     Bacterial Systematics.. 87
  M. Goodfellow, J. Chun, E. Atalan, and J.-J. Sanglier

New Methods for Diagnosis and Epidemiological Studies of
     Tuberculosis Based on PCR and RFLP.....................................105
  C. Martín, S. Samper, I. Otal, P. Asensio, R. Goméz-Lus, G. Torrea,
  and B. Gicquel

Typing *in situ* with Pobes.. 115
  R. Amann and W. Ludwig

The Use of Molecular Markers for the Detection and Typing of
     Bacteria in Soil.. 137
  E.M.H. Wellington, A.S. Huddleston, and P. Marsh

Phylogenetic Diversity of Methanogen Endosymbionts of
    Anaerobic Ciliates... 153
T.M. Embley and B.J. Finlay

Diversity, Dynamics and Topographic Arrangement of
    Microorganisms are Essential Parameters That Identify
    a Microbial Consortium... 161
E. Conway de Macario and A.J.L. Macario

Chemotaxonomy and the Identification of Thermophilic Bacteria........................ 173
M.S. da Costa and M.F. Nobre

Alkaliphiles : Diversity and Identification... 195
B.E. Jones, W.D. Grant, N.C. Collins, and W.E. Mwatha

Taxonomy and Phylogeny of Moderately Halphilic
    Bacteria.. 231
A. Ventosa

Chemical Analysis of Archaea and Bacteria : A Critical
    Evaluation of Its Use in Taxonomy and Identification................................... 243
B.J. Tindall

The Biotechnological Importance of Molecular Biodiversity
    Studies for Metal Bioleaching.. 259
B.M. Goebel and E. Stackebrandt

Systematics of Insect Pathogenic Bacilli : Uses in Strain
    Identification and Isolation of Novel Pathogens... 275
F.G. Priest, M. Aquino de Muro, and D.A. Kaji

Industries Requirements with Regard to Identification
    of Bacteria.. 297
H. Gürtler and L. Anker

Present Trends and Future Prospects for Rapid Methods
    and Automation in the Chemical Laboratory... 309
K.A. Feltham and M. Stevens

Contributors... 323

Index... 327

MOLECULAR TAXONOMY: CLASSIFICATION AND IDENTIFICATION

Karl-Heinz Schleifer and Wolfgang Ludwig

Lehrstuhl für Mikrobiologie, Technische Universität, D-80290 München, Germany

INTRODUCTION

Bacterial **taxonomy** or **systematics** may be defined as the scientific study of the diversity of organisms with the ultimate object of characterizing and arranging them in an orderly manner (Trüper and Schleifer, 1991). It comprises the three subdisciplines classification, nomenclature and identification. **Classification** deals with the orderly arrangements of taxonomic units (taxa) into groups on the basis of similarities or relationships. **Nomenclature** is the assignment of names to the taxonomic units according to the International Code of Nomenclature of Bacteria (Sneath, 1992). Finally, in **identification**, members of a distinct taxonomic unit are identified on the basis of common characteristic properties which distinguish them from other organisms.

Classification is often confused with identification. However, it is necessary to describe and characterize first the basic taxonomic unit (classification) before an isolate can be identified as belonging to this taxon. An isolate, .pl which cannot be identified as a member of a known taxon, has to be described and classified as a new species or even genus. Therefore, classification precedes identification.

There are two main types of classification based on phenetic (non-evolutionary) and phylogenetic (evolutionary) relationships. The aim of the former is to group organisms on the basis of their phenotypic properties. However, phenetic resemblance is often an unreliable guide to genealogy. Adaptive characters are often acquired independently by unrelated taxa (convergence and lateral gene transfer) and in other cases organisms can resemble one another merely by lacking a particular character. Evolutionary or phylogenetic classification, on the other hand, is not based on phenetic resemblances between species but on their genealogical relationships.

PHYLOGENY OF BACTERIA

The morphological diversity and fossil records of plants and higher animals allows reconstruction of the evolution of these organisms. The morphological simplicity of bacteria as well as the lack of meaningful developmental stages and fossil records discouraged microbiologists in their attempt to elucidate the phylogenetic relationships of prokaryotes. This is documented by the ready resignation of van Niel, once the leading exponent of a phylogenetic bacterial classification system, who stated "... that comparable efforts in the realm of the bacteria are doomed to failure because it does not appear likely that criteria of truly phylogenetic significance can be devised for these organisms" (van Niel, 1955). A similar pessimistic view can be found in the 1970 edition of the well known textbook "The Microbial World" by Stanier et al.. They wrote ".....for (most) major biological groups (such as bacteria), the general course of evolution will never be known, and there is simply not enough objective evidence to base their classification on phylogenetic grounds"(Stanier et al., 1970). However, the development of techniques that allowed the sequencing of genetic material revolutionized biology including the taxonomy of bacteria.

MOLECULAR CHRONOMETERS

The sequence of a macromolecule can be changed in a large number of ways without altering its function. This realization has led to the concept of the "evolutionary clock" and to the use of molecular sequences as chronometers (Zuckerkandl and Pauling, 1965). The evolution occurs on the level of the genotype and of the phenotype. The changes of the deoxyribonucleic acid sequences occur more or less continously on an evolutionary time scale. The majority of these changes are either deleterious or selectively neutral, and therefore do not alter the phenotype (Kimura, 1983). Changes at the phenotypic level, on the other hand, are relatively rare and sporadic.

The genotypic changes are the basis for inferring phylogenetic relationships. However, not all molecular chronometers are equally useful for reconstructing phylogenies. A useful chronomter for inferring the phylogeny of all living organisms has to fulfill the following properties:

1. it has to be ubiquitously distributed among organisms

2. it should show a high degree of functional constancy
3. it should be sufficiently conserved to span the full evolutionary spectrum
4. it should share a common ancestor

Large size ribosomal ribonucleic acids (rRNA) fulfill these properties. Therefore, a breakthrough in the study of the phylogenetic relationships of prokaryotes was achieved by Woese and coworkers who compared first partial, and later total sequences of 16S rRNA (Woese, 1987). For the first time, it was possible to reconstruct the phylogeny of proka- ryotes. The majority of information comes from the analysis of one molecule, the 16S rRNA. Currently about 1500 complete or almost complete 16S rRNA sequences from Bacteria are available. The fact that the whole phylogentic superstructure rests almost solely on a single molecule species is disquie- ting. Therefore, the question arises whether these data really reflect the organismal phylogeny or merely the history of a single molecule. Therefore, we decided to sequence further conserved macromolecules which fulfill the properties of a useful chronometer.

PHYLOGENTIC TREES INFERRED FROM COMPARATIVE SEQUENCE ANALYSIS OF DIFFERENT HOMOLOGOUS MACROMOLECULES

Comparative sequence analyses of genes coding for the following molecules were carried out: 23S rRNA, elongation factor Tu (EF-Tu) and ß-subunit of ATP-synthase. 23S rRNA is found in the large subunit of the ribosomes, whereas 16S rRNA is present in the small subunit. EF-Tu is one of the most abundant cellular proteins in *E. coli*. It is highly conserved among all Bacteria and plays, like rRNAs, an important role in protein biosynthesis. The ß-subunit of ATP-synthase, on the other hand, is not involved in protein biosynthesis, but is part of a membrane protein complex. It is present in all Bacteria studied so far and shows a high degree of functional constancy.

Phylogentic trees derived from comparative sequence analyses of 23S rRNA, EF-Tu and ß-subunit of ATP-synthase molecules are in very good agreement with the tree inferred from 16S rRNA sequences (Schleifer and Ludwig, 1989; Figs. 1 - 3). There are, however, some minor differences in the topology of the phylogenetic trees derived from the different chronometers. The protein molecules seem to contain less phylogenetic information than the rRNAs. One difference between the 16S rRNA based tree, as published by Woese (1987), and the phylogenetic trees derived from 23S rRNA and EF-Tu sequence data is worth considering. In the 16S rRNA tree shown by Woese (1987), the gram-positive bacteria form a monophyletic group consisting of two subgroups: the gram-positive bacteria with a high DNA G+C content and those with a low DNA G+C content. However, the unity of the two subgroups is not significant (Fig. 1-3). Moreover, the 23S rRNA and EF-Tu sequence data do not support the monophyletic origin of the gram-positive bacteria (Fig. 1 and 2). Based on the 16S and 23S rRNA trees mycoplasmas form a separate, deep branching cluster within the gram-positive bacteria with a low G+C content, whereas within the EF-Tu and ATPase based trees they appear as a distinct line of descent.

In conclusion one can state that the similar topology of the phylogenetic trees derived from comparative sequence analysis of such unrelated macromolecules as 16/23S rRNAs, EF-Tu and ß-subunit of ATP-synthase is a strong indication that the trees reflect not only the phylogeny of individual genes but also that of the organisms. The similar topology is also an evidence against lateral chromosomal gene transfer in taxonomically unrelated organisms.

GENERAL CONSIDERATIONS ON PHYLOGENETIC TREES

The correctness of the sequences, their careful alignment and analysis are prerequisites for the reconstruction of the most probable phylogeny of bacteria. The proper alignment of the sequences is the first step in phylogenetic analysis. A correct alignment ensures that only homologous characters, i.e., residues which are derived from a common position within the ancestor molecule, are compared. Invariant and conserved positions can easily be aligned, whereas the correct alignment of variable and highly variable positions is much more difficult. The extensive higher order structure of rRNA molecules, in particular the occurrence of helical structures, are often very helpful for the correct alignment of the more variable positions.

The underlying assumption of the approach is, that functionally equivalent regions of the ribosomal RNAs from phylogenetically diverse sources will have biologically equivalent structures. Therefore, secondary structure analysis allows to recognize homologous regions involved in helix or loop formation despite a low primary structure similarity.

Unfortunately, sequence information is only available from extant organisms and each individual position carries merely the information of a rather small evolutionary period. Therefore, only a minor part of the evolutionary events can be reconstructed. None of the methods for the inference of phylogenetic trees (e.g., pairwise distance, maximum parsimony, maximum likelihood; Ludwig and Schleifer, in press) is perfect and results should not be accepted without tests for significance. In cases where lineages are branching in close proximity to one another and the relative branching order does not remain stable when applying the different treeing methods, the branching positions should be specifically indicated within the trees as multifurcations.

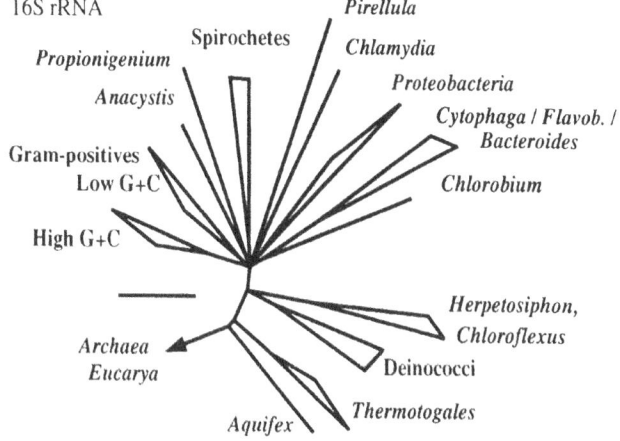

16S rRNA

Pirellula
Spirochetes
Chlamydia
Propionigenium
Anacystis
Proteobacteria
Cytophaga / Flavob. /
Bacteroides
Gram-positives
Low G+C
Chlorobium
High G+C
Herpetosiphon,
Chloroflexus
Archaea
Eucarya
Deinococci
Thermotogales
Aquifex

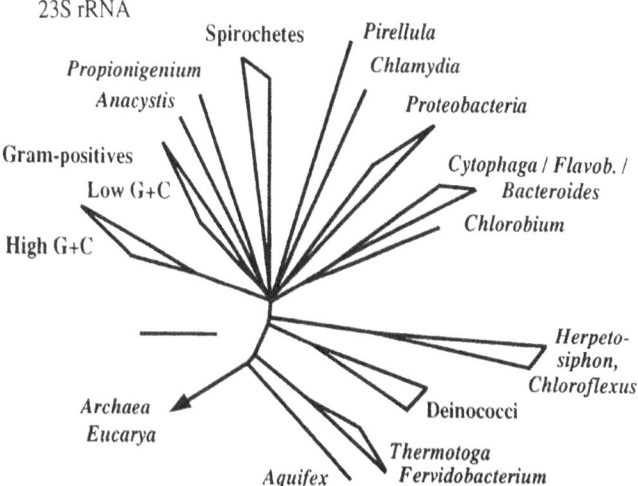

23S rRNA

Spirochetes
Pirellula
Propionigenium
Chlamydia
Anacystis
Proteobacteria
Gram-positives
Cytophaga / Flavob. /
Low G+C
Bacteroides
Chlorobium
High G+C
Herpeto-
siphon,
Chloroflexus
Archaea
Eucarya
Deinococci
Thermotoga
Fervidobacterium
Aquifex

Figure 1. Phylogenetic trees inferred from comparative sequence analyses of 16S and 23S rRNAs. The bar indicates 10% estimated changes.

5

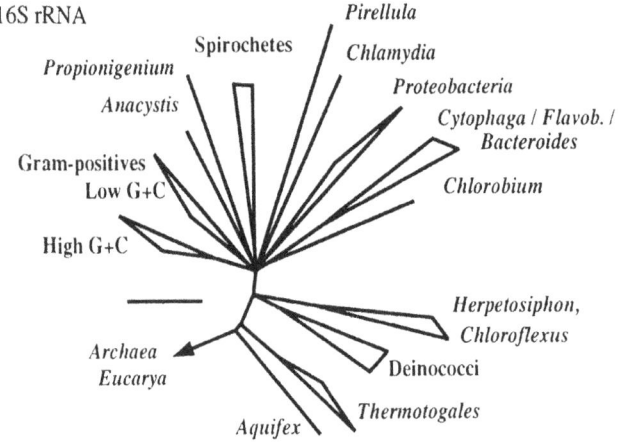

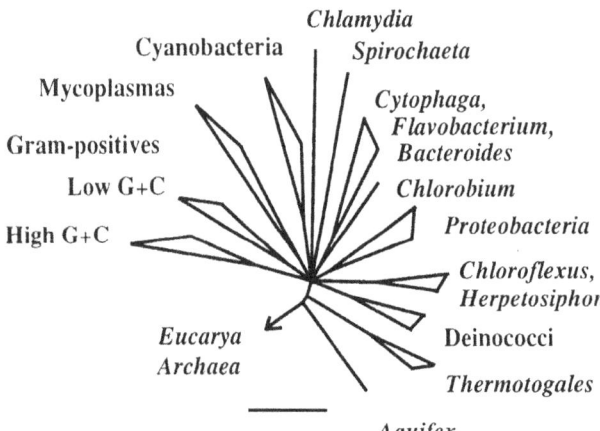

Figure 2. Phylogenetic trees inferred from comparative sequence analyses of 16S rRNA and elongation factor Tu. The bar indicates 10% estimated changes.

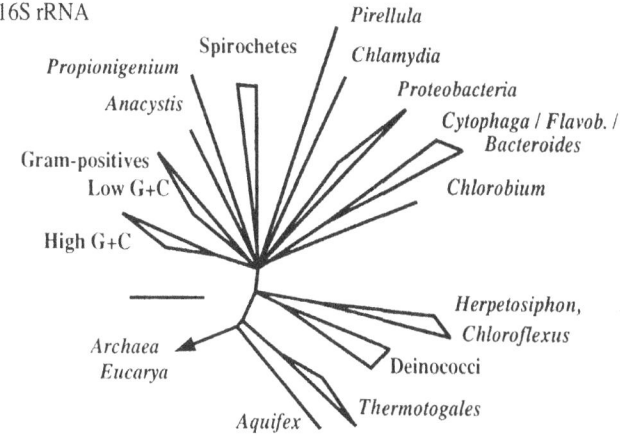

16S rRNA

Propionigenium
Spirochetes
Anacystis
Gram-positives
Low G+C
High G+C

Pirellula
Chlamydia
Proteobacteria
Cytophaga / Flavob. /
Bacteroides
Chlorobium

Archaea
Eucarya
Aquifex
Thermotogales
Deinococci
Herpetosiphon,
Chloroflexus

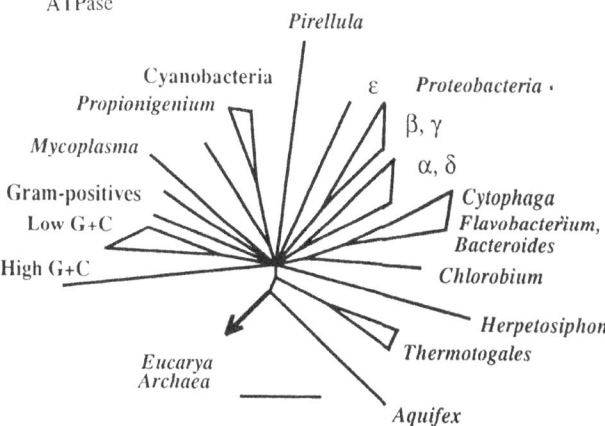

ATPase

Cyanobacteria
Propionigenium
Mycoplasma
Gram-positives
Low G+C
High G+C

Pirellula
ε   Proteobacteria ·
β, γ
α, δ
Cytophaga
Flavobacterium,
Bacteroides
Chlorobium

Eucarya
Archaea
Aquifex
Thermotogales
Herpetosiphon

Figure 3. Phylogenetic trees inferred from comparative sequence analyses of 16S rRNA and ß-subunit of ATP-synthase. The bar indicates 10% estimated changes.

PROBLEMS WITH THE DEVELOPMENT OF A PHYLOGENY-BASED CLASSIFI-
CATION SYSTEM

Despite the enormous progress which has been made in unravelling the phylogenetic relationships of bacteria in recent years, bacterial systematists have not yet reached their ultimate goal, namely the development of a stable, phylogeny-based classification system. The practical approach is still dominating in the latest edition of "Bergey's Manual of Systematic Bacteriology". Phylogenetic relationships were mentioned in this text, but the necessary taxonomic consequences were not carried out, e.g., staphyloccocci and micrococci were still combined in one family although it has been known for a long time that phylogenetically they are not closely related. The editors of the second edition of "The Prokaryotes" (Balows et al., 1991) at least tried to arrange the sections on the basis of phylogenetic relationships. But despite these efforts there is not a generally accepted phylogeny-based classification system.

What is the reason for the slow progress in developing such a classification system? The traditional classification system is based largely on phenotypic properties. Taxa were described on the basis of similar morphological, physiological and biochemical properties. However, the phylogenetic studies have clearly shown that organisms sharing similar properties are not necessarily genealogically related to each other, e.g., phototrophic or gliding bacteria are found in different lines of descent. Therefore, traditional dogmas have to be renounced, i.è. morphological or physiological characters can no longer be used as sole properties for classifying an organism. A typical example is the exclusion of organisms from the genera *Bacillus* and *Clostridium* merely on their inability to form endospores. It is no longer justifiable to define a genus solely on the basis of a unique character particularly as a deletion or even an inactivation of one gene can be sufficient to loose the ability to form endospores. Therefore, it is no wonder that non-sporeforming bacteria such as *Caryophanon* or *Filibacter* are phylogentically closely related to certain bacilli. A similar situation is found among non-sporeforming anaerobic bacteria such as certain *Eubacterium* species which are closely related to clostridia. Moreover, molecular taxonomic studies have shown that the genera *Bacillus* and *Clostridium* are phylogenetically heterogeneous and should be split into additional genera. Almost fifty years ago van Niel (1946) had recognized that morphological criteria allow no conclusions about phylogeny. This was later confirmed by Stackebrandt and Woese (1979) showing the phylogentic dissection of the family Micrococcaceae.

Lactobacilli are an example where one can find a discrepancy between common metabolic traits and phylogenetic relationships. Based on their type of fermentation, lactobacilli can be subdivided into three groups: obligately homofermentative, facultatively heterofermentative and obligately heterofermentative. Unfortunately, these subdivisions do not correlate well with the phylogenetic relatedness of lactobacilli (Schleifer and Ludwig, in press).

However, there are also genetically closely related organisms showing 100% DNA-DNA similarity which are nevertheless placed in different species or even genera. Typical examples are *Mycobacterium avium, M. intracellulare* and *M. paratuberculosis* (Hurley et al., 1988), *Yersinia pestis* and *Y. pseudotuberculosis* (Bercovier et al., 1980) or *Escherichia coli* and the different species of *Shigella* (Brenner et al., 1973). These species or genera are differentiated on the basis of phenotypic characters and for clinical and public health purposes they did not become subspecies but maintained their original taxonomic status.

To develop an applicable phylogeny-based classification system we have to rely on a polyphasic approach based on phylogenetic and phenotypic information. New informative phenotypic characters have to be found to delineate those taxa that currently can only be defined by phylogenetic data. Phenotypic characters should not be restricted to morphological and physiological properties but should also include chemotaxonomic characters and genetic properties. There are some chemotaxonomic markers which are characteristic for certain phylogenetic taxa. Archaea can be distinguished from Bacteria due to the absence of peptidoglycan, the presence of glycerol ethers instead of esters and the occurrence of a rather complex RNA polymerase. The occurrence of lipopolysaccharides is typical for Proteobacteria and other gram-negative bacteria, whereas gram-positive bacteria lack this cell wall polymer. Peptidoglycan and teichoic acid types are helpful for the differentiation of gram-positive bacteria. Corynebacteria, mycobacteria and nocardia are characterized by the occurrence of typical mycolic acids. Lipid, quinone, polyamine and fatty acid patterns are also often helpful in characterizing bacterial taxa, e.g. the different subclasses of Proteobacteria reveal distinct polyamine and quinone patterns. The ultimate goal is to delineate taxa that are phylogenetically coherent and easy to recognize by phenotypic properties. Currently, there are many phylogenetically coherent groups, in particular within the various subclasses of Proteobacteria and among gram-positive bacteria such as bacilli or clostridia, that urgently need taxonomic evaluation. However, appropriate phenotypic characters have so far not been found to delineate these taxa. The current taxonomic situation is also a challenge to the more traditionally oriented systematists to screen their data bases for phylogenetically meaningful phenetic characters and to the chemotaxonomists and biochemists to look for new features of taxonomic significance.

IMPROVED IDENTIFICATION OF BACTERIA BY USING MOLECULAR METHODS

Nucleic Acid Probe Technique

The nucleic acid probe technique is based upon the unique property of nucleic acids to form a double strand from complementary single strands. The two strands are held together mainly by hydrogen bonds between the corresponding nucleotide bases and can be dissociated by heat or high pH. On cooling down or changing the pH to neutral, the single strands will reanneal into the double stranded conformation. The reannealing process is called hybridization.

Nucleic acid probes are generally short single-stranded molecules (mostly DNA) which bind (hybridise) to complementary DNA or RNA (target nucleic acids). Probes can either be isolated from the DNA of the target organisms or synthesized as oligonucleotides in the laboratory. The hybridization is rapid, stable and highly specific. Probes will bind to their complementary sequences, even if these sequences account for only a minor fraction of the total target nucleic acid. Differences between target sequences as little as one nucleotide can be distinguished.

The nucleic acid probe technique consists of four steps (Schleifer et al., 1993). First, the probe has to be labelled with a radioactive or non-radioactive detector group. Second, the target nucleic acid has to be extracted or the cells have to be made permeable for the labelled probe. Third, the probe has to react with the target nucleic acid (hybridization) and the unbound or non-specifically bound probe has to be removed from the specific hybrids. Fourth, the specific hybrids are detected by measuring the amount of labelled probe bound to the target nucleic acid.

Nucleic acid probe assays offer several advantages over alternative identification methods. First, they can be designed to be very specific and only recognise a particular species or in certain cases even only a distinct strain. Second, they can be used for the identification of organisms prior to their cultivation. Third, probes inferred from rRNA genes reflect the phylogenetic relationship. Fourth, the genome of an organism is rather stable and, in contrast to antigenic structures or certain chemotaxonomic properties, not affected by exogeneous or endogeneous factors.

There are several different approaches to the design of nucleic acid probes (Schleifer et al., 1993) of which the most important are summarized in Table 1.

The unique properties and the ubiquitous occurrence of rRNAs render these molecules particularly suitable for rapid and specific identification of bacteria. The nucleotide sequences of 16S and 23S rRNA consist of stretches of highly, moderateley and less conserved regions. Therefore, it is possible to design different probes of a wide specificity ranging from species specificity to a universal probe which will react with any cellular life form. Moreover, using rRNA as the target nucleic acid increases considerably the sensitivity of the test, since rRNAs are present in high copy numbers (up to 10^5 copies per cell) in growing bacterial cells.

Different formats can be used for nucleic acid probe assays. For the **dot-blot** hybridization assay, small amounts of target nucleic acids are extracted from cells and immobilised onto nylon, nitrocellulose or teflon membranes. **Colony hybridization** is a convenient, rapid and simple method for analysing mixed cultures in which the colonies are transferred to membranes and lysed. Gram-negative bacteria can be lysed easily using simple alkali treatment, whereas grampositive bacteria require usually an additional heating step or treatment with cell wall lytic enzymes. For the *in situ* identification of individual **whole cells**, it is necessary to render the cells permeable for the oligonucleotide probes which react with rRNA. The whole cell hybridization technique even allows the identification and *in situ* detection of individual cells without prior cultivation using fluorescently or enzyme labelled rRNA-targeted oligonucleotide probes (Amann et al., 1992). The hybridized labelled cells can be analysed by light microscopy, whereby an epifluorescence microscope is necessary for fluorescently labelled probes, or flow cytometry (Amann et al., 1990). This approach is also

Table 1. Design and specificity of nucleic acid probes

Origin	Design	Specificity
Whole cell DNA	Labelled whole cell DNA	Cross-hybridization with closely related species
Randomly cloned or amplified DNA fragment	Isolation and cloning of non cross-hybridizing fragments	Strain to species
Gene or gene fragment	Isolation and cloning of genes or fragments of genes	Organisms containing these genes
16S/23S rRNA	Comparative sequence analysis, synthesis of oligonucleotides	Wide specificity, species to universal

helpful for the identification and detection of non-culturable bacteria (Amann et al., 1991; Spring et al., 1992). From
most ecosytems less than 10% and sometimes even less than 1%
of the actual ocurring bacteria can be cultivated. A better
knowledge of these currently non-culturable bacteria is
absolutely necessary to understand their role in the ecosystem and to monitor their distribution and population shifts.

Use of the Polymerase Chain Reaction for Direct Detection of Bacteria

The polymerase chain reaction (PCR) is a highly sensitive
method for the specific detection of bacteria. PCR requires a
set of two oligonucleotides (primers) complementary to regions flanking the DNA sequence to be amplified. By choosing
primer sequences which are specific for a characteristic gene
of the bacterial species (Bej et al., 1990) or distinct
regions of the 16S or 23S rRNA genes (e.g., Roller et al.,
1992) the organisms can be directly detected without the
need to culture them. DNA is extracted from cells or the
sample and amplified *in vitro* by the PCR. Often it is even
possible to use a crude lysate of a single colony without DNA
extraction or purification (Frothingham et al., 1991; Hynes
et al., 1992). The amplified products are fractionated by
agarose gel electrophoresis and visualized by ethidium bromide staining or hybridization with specific probes. Using a
mixture of primers against different characteristic genes
(multiplex PCR), it is even possible to differentiate certain
species (Bej et al., 1991, Way et al., 1993).
For identification of bacteria on the basis of rRNA
sequence differences, PCR-based assays are usually done by
amplifying a highly variable fragment of a 16S or 23S rRNA
gene, fixing the amplified DNA onto a series of nylon
membranes, and hybridizing each membrane with one of the
labelled oligonucleotides which are specific for distinct
taxa. However, this is a rather tedious and time-consuming
procedure. Therefore, we tried to adapt the reverse dot-blot
hybridization for the rapid identification of bacteria without prior cultivation. This method was first described by
Saiki et al. (1989) for the detection of polymorphisms.
Specific oligonucleotides, containing at the 3' end a deoxy-
ribothymidine homopolymer tail, were attached to the nylon
support and the target segment of the rRNA gene was PCR-
amplified with digoxigenin (DIG) labelled primers and then
hybridized to the bound oligonucleotides. Hybrids are detected nonradioactively with enzyme-conjugated anti-DIG F_{ab}-
fragments. Using this method it was possible to detect the
presence of certain lactic acid bacteria in food samples
without prior cultivation. DNA was extracted from the sample,
a distinct region of the 23S rRNA genes was amplified *in
vitro*, DIG-labelled and hybridized to specific, membrane-

bound oligonucleotides (Ehrmann, Ludwig and Schleifer, unpublished results).

CONCLUSIONS

The development of new molecular techniques has revolutionized bacterial taxonomy. For the first time in history, it is now possible to reconstruct the phylogenetic relationships of prokaryotes. Based on these data it should be possible to establish an evolutionary classification system. Moreover, as a spin-off of the 16S and 23S rRNA sequence studies, oligonucleotide probes can be designed which allow a rapid and reliable identification of bacteria.

Acknowledgements

Original studies were supported by grants from the Commission of European Communities (BIOT-CT91-0294) and Deutsche Forschungsgemeinschaft.

REFERENCES

Amann, R.I., Binder, B.J., Olson, R.J., Chisholm, S.W., Devereux, R. and Stahl, D. (1990). Combination of 16S rRNA-targeted oligonucleotide probes with flow cytometry for analyzing mixed microbial populations. Appl. Environ. Microbiol. 56,1919-1925.

Amann, R.I., Springer, N., Ludwig, W., Görtz, H.D. and Schleifer, K.H. (1991). Identification and phylogeny of uncultured bacterial endosymbionts. Nature 351,161-164.

Amann, R.I., Ludwig, W. and Schleifer, K.H. (1992). Identification and in situ detection of individual bacterial cells. FEMS Microbiol. Letters 100,45-50.

Balows, A., Trüper, H.G., Dworkin, M., Harder, W. and Schleifer, K.H. (Eds.) The Prokaryotes. Second Edition. Volume I-IV, Springer Verlag, New York 1991.

Bej, A.K., Steffan, R.J., DiCesare, J., Haff, L. and Atlas, R.M. (1990). Detection of colifrom bacteria in water by polymerase chain reaction and gene probes. Appl. Environ. Microbiol. 56,307-314.

Bej, A.K., McCarty, S.C. and Atlas, R.M. (1991). Detection of coliform bacteria and Escherichia coli by multiplex polymerase chain reaction: comparison with defined substrate and plating methods for water quality monitoring. Appl. Environ. Microbiol. 57, 2429-2432.

Bercovier, H.H., Mollaret, H., Alonso, J.M., Brault, J., Fanning, G.R., Steigerwalt, A.G. and Brenner, D.J. (1980. Intra- and interspecies relatedness of Yersinia pestis by DNA hybridization and its relationship to Yersinia pseudotuberculosis. Curr. Microbiol.4, 225-230.

Brenner, D.J., Fanning, G.R., Miklos, G.V. and Steigerwalt, A.G. (1973). Polynucleotide sequence relatedness among *Shigella* species. Int. J. Syst. Bacteriol. 23,1-7.

Frothingham, R., Allen, R.L. and Wilson, K.H. (1991). Rapid 16S ribosomal DNA sequencing from a single colony without DNA extraction or purification. BioTechniques 11,40-44.

Hurley, S.S., Splitter, G.A. and Walch, R.A. 1988. Deoxyribonucleic acid relatedness of *Mycobacterium pseudotuberculosis* to other members of the family Mycobacteriaceae. Int. J. Syst. Bacteriol. 38, 143-146.

Hynes, W.L., Ferretti, J.J., Gilmore, M.S. and Segarra, R.A. (1992. PCR amplification of streptococcal DNA using crude cell lysates. FEMS Microbiol. Letters 94,139-142.

Kimura, M. (1983). The Neutral Theory of Molecular Evolution. Cambridge University Press, Cambridge.

Ludwig, W. and Schleifer, K.H. (1993. Bacterial phylogeny based on 16S and 23S rRNA sequence analyses. FEMS Microbiol. Reviews, in press.

Roller, C., Ludwig, W. and Schleifer, K.H. (1992). Gram-positive bacteria with a high DNA G+C content are characterized by a common insertion within their 23S rRNA genes. J. Gen. Microbiol. 138,1167-1175.

Saiki, R.K., Walsh, P.S., Levenson, C.H. and Erlich, H.A. (1989). Genetic analysis of amplified DNA with immobilized sequence-specific oligonucleotide probes. Proc. Natl. Acad. Sci. USA 86, 6230-6234.

Schleifer, K.H. and Ludwig, W. (1989). Phylogenetic relationships among bacteria. pp.103-117. *in*: Hierarchy of Life. B. Fernholm, K. Bremer and H. Jörnvall (eds.), Elsevier Science Publishers, Amsterdam.

Schleifer, K.H. and Ludwig, W. (in press). Phylogenetic relationships of lactic acid bacteria.*in*:The Lactic Acid Bacteria. Vol. II: The Genera of Lactic Acid Bacteria. B.J.B. Wood and W.H. Holzapfel (eds.), Elsevier Applied Science Publishers Ltd., Amsterdam.

Schleifer, K.H., Ludwig, W. and Amann, R.I. (1993). Nucleic acid probes. pp. 463-510. *in*: The New Bacterial Systematics. M. Goodfellow and A.G. O'Donnell (eds.), Academic Press, New York.

Sneath, P.H.A. (1992). International Code of Nomenclature of Bacteria. American Society for Microbiology, Washington.

Spring, S., Amann, R., Ludwig, W., Schleifer, K.H. and Petersen, N. (1992). Phylogenetic diversity and identification of non-culturable magnetotactic bacteria. Syst. Appl. Microbiol. 15,116-122.

Stackebrandt, E. and Woese. C.R. (1979). A phylogenetic dissection of the family Micrococcaceae. Curr. Microbiol. 2,317-322.

Stanier, R.Y., Doudoroff, M. and Adelberg, E.A. (1970). The Microbial World. 3rd ed. Prentice-Hall, Englewood Cliffs.

Trüper, H.G. and Schleifer, K.H. (1991). Principles of characterization and identification of prokaryotes. pp. 126-148. *in*: The Prokaryotes. Second Edition. Balows,A., Trüper,H.G., Dworkin,M., Harder,W. and Schleifer,K.H. (eds.). Springer Verlag, New York.

Van Niel, C.B. (1946). The classification and natural relationships of bacteria. Cold Spring Harbor Symp. Quant. Biol. 11,285-301.

Van Niel, C.B. (1955). Classification and taxonomy of the bacteria and bluegreen algae. A Century of Progress in the Natural Sciences 1853-1953. pp.89-114. California Academy of Sciences, San Francisco.

Way, J.S., Josephson, K.L., Pillai, S.D., Abbaszadegan, M., Gerba, C.P. and Pepper, I.L. (1993). Specific detection of *Salmonella* by multiplex polymerase chain reaction. Appl. Environ. Microbiol. 59, 1473-1479.

Woese, C.R. (1987). Bacterial evolution. Microbiol. Rev. 51,221-271.

Zuckerkandl, E. and Pauling, L. (1965). Molecules as documents of evolutionary history. J. Theor. Biol. 8,357-366.

DNA SEQUENCE ANALYSIS OF THE GENETIC STRUCTURE
OF POPULATIONS OF *SALMONELLA ENTERICA*
AND *ESCHERICHIA COLI*

Robert K. Selander, Jia Li, E. Fidelma Boyd, Fu-Sheng Wang,
and Kimberlyn Nelson

Institute of Molecular Evolutionary Genetics
Pennsylvania State University
University Park, PA 16802

INTRODUCTION

The goal of bacterial population genetics is to understand the factors that determine genetic structure and mediate evolutionary change in natural populations. This broad objective transcends the practical needs of microbiologists for methods of species identification and strain discrimination for epidemiological and other purposes, but the findings of population genetics research have important implications for several branches of medical microbiology (Selander and Musser, 1990), as well as for bacterial systematics.

Work in our laboratory involves sampling the genomes of a variety of bacteria – mostly pathogenic forms – to determine the extent and organization of genetic variation as a basis for inferring the structure of natural populations and assessing the evolutionary roles of natural selection, random drift, mutation, and horizontal genetic exchange. Our primary methods of studying genetic variation are multilocus enzyme electrophoresis (MLEE) (Selander *et al.*, 1986) and comparative nucleotide sequencing of genes encoding proteins of various functional types (Nelson *et al.*, 1991*b*). Both methods yield quantitative measures of genetic similarity among isolates, which are required for the analysis of population structure and the reconstruction of evolutionary histories. The great advantage of MLEE is that many genes (20-38 loci in application in our laboratory) can be readily examined in hundreds of isolates; but the technique detects only part of the existing allelic variation, convergence to the same electromorph (allozyme) is not uncommon, and intragenic recombination events cannot be identified. The singular advantage of sequencing is, of course, that all allelic variation is revealed, but thus far the numbers of both genes and strains examined have been relatively small.

In this chapter, we review some of the major conclusions that have emerged from research on the population genetics of bacteria and provide examples from our recent work on nucleotide sequence variation in several genes of *Salmonella enterica* and its comparatively close relative *Escherichia coli*.

Bacterial Diversity and Systematics
Edited by F.G. Priest *et al.*, Plenum Press, New York, 1994

TYPES OF GENETIC STRUCTURE

Clonality Versus Free Recombination

The genetic structure of natural populations of bacteria is determined by the frequency and pattern of horizontal exchange and recombination of genetic material among cell lineages. The two theoretical extremes are clonality and free recombination (panmixia), as illustrated by the results of computer simulations shown in Figure 1, in which 20 polymorphic loci in 30 haploid genotypes were followed over several generations (Whittam, 1992).

The top panel (A) of Figure 1 illustrates clonal population structure: in the absence of recombination, new alleles arising by mutation remain nonrandomly associated in cell lineages, producing the condition of linkage disequilibrium (Whittam *et al.*, 1983a,b). There are some deep branches (the dendrogram is tree-like) and the pairwise mismatch distribution (number of allelic differences) is bimodal, with a relatively large variance. The bottom panel (B) illustrates free recombination: genes are exchanged so frequently among genotypes that their allelic variants become randomly assorted – alleles at different loci are in linkage equilibrium. The dendrogram is bush-like and the pairwise mismatch distribution is unimodal, with a relatively small variance.

The modes of evolution may be very different under these two types of population structure. Consider the occurrence of a new mutation that confers increased fitness (adaptation to the environment). Under strict clonality, the mutant allele is confined to a single cell-lineage, the members of which may then increase in frequency at the expense of those of other lineages, with a consequent purging of variation in the population as the alleles at all loci hitchhike along with the favored mutant allele – the phenomenon of periodic selection

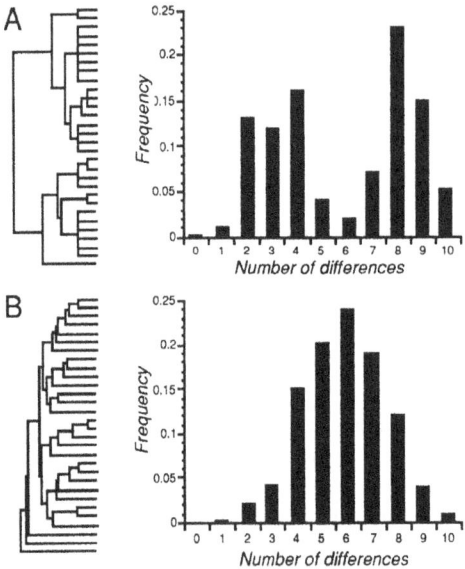

Figure 1. The two extremes of population structure. **A.** Dendrogram based on genotypes (20 loci) of 30 cell lineages under clonal population structure. The pairwise mismatch distribution at the right is bimodal with a large variance as a result of linkage disequilibrium. **B.** Dendrogram based on genotypes of lineages under free recombination. The mismatch distribution is unimodal with a smaller variance, reflecting linkage equilibrium. (After Whittam, 1992)

(Koch, 1974; Levin, 1981). But with frequent recombination, the new allele can be rapidly transferred to other genotypes, and, consequently, there may be little or no reduction in genetic variation in the population as the allele goes to fixation. It follows that the degree to which a bacterial population or species will be perceived as clonal in structure ultimately depends on the ratio of the rate of recombination to that of geographic spread (Achtman, 1994). Additionally, species of bacteria with a strongly clonal structure may be expected to show less single-locus allelic diversity and multilocus genotypic diversity than those in which recombination among lineages is frequent. Allelic diversity cannot be generated by intragenic recombination of exogenous DNA in clonal populations, and episodes of periodic selection reduce allelic diversity generated by mutation.

The Concept of Clonal Population Structure

Although the association of particular serobiotypes with independent outbreaks of specific diarrheal disease syndromes had earlier suggested the existence of clones of certain enteropathogenic *E. coli* (Ørskov *et al.*, 1976, 1977), genetic evidence of a clonal structure for bacterial populations was first provided in 1980 by Selander and Levin's (1980) demonstration, by MLEE analysis, of the recovery of strains of *E. coli* of identical or closely similar multilocus genotypes at different times and places in North America. Soon thereafter, the concept of a clonal structure for the species *E. coli* as a whole, on both local and global scales, was firmly established through a series of empirical and statistical studies, based largely on MLEE data, in our laboratory (Caugant *et al.*, 1981; Whittam *et al.*, 1983 *a,b*; Ochman *et al.*, 1983; Ochman and Selander, 1984*a*) and, independently, in that of Mark Achtman (Achtman *et al.*, 1983; Achtman and Pluschke, 1986).

Since 1980, MLEE has been extensively employed to study genetic variation and infer population structure in most of the common human pathogenic bacteria (including, in our laboratory, one or more species of *Bordetella, Haemophilus, Legionella, Mycobacterium, Neisseria, Pseudomonas, Salmonella, Shigella, Staphylococcus, Streptococcus, and Yersinia*) and in *E. coli*, most strains of which are commensals rather than pathogens, as well as in a few soil bacteria (*Rhizobium* and *Bacillus*). Particularly large and important contributions to an understanding of the population genetics of pathogenic bacteria have been made by D. A. Caugant's numerous studies of *Neisseria meningitidis*, including the tracing of the global spread of the ET-5 clone complex (Caugant *et al.*, 1986, 1987; Sacchi *et al.*, 1992*a*) and other clones (Ashton *et al.*, 1991; Sacchi *et al.*, 1992*b*; Wang *et al.*, 1992*a,b*), and by J. M. Musser's extensive work on the evolution and global molecular epidemiology of *Haemophilus influenzae* (Musser *et al.*, 1985, 1986, 1988*a,b*, 1990; Quentin *et al.*, 1990; Musser and Kapur, 1994). MLEE has also played a part in the research of Mark Achtman's group on the genetic structure of populations of *N. meningitidis* serogroup A organisms in relation to epidemics and pandemics of meningococcal disease, which is a model for the study of the evolutionary genetics of pathogenic bacteria (Achtman *et al.*, 1992; Wang *et al.*, 1992*a*; Morelli *et al.*, 1994; Achtman, 1994).

From this work, three major generalizations have emerged (Selander and Musser, 1990). First, most nominal species of bacteria are highly polymorphic, with several electrophoretically detectable alleles at the average enzyme locus. Second, *E. coli* and many pathogenic species show evidence of strong linkage disequilibrium within local populations as well as regionally, if not globally (Whittam *et al.*, 1983*a,b*; Musser *et al.*, 1990; Maynard Smith *et al.*, 1993). And third, in the case of most pathogenic species, a relatively small number cell lineages marked by distinctive multilocus enzyme genotypes – designated as electrophoretic types or ETs – cause most of the disease worldwide (which means that inter-lineage variance in pathogenicity is large), and isolates of the same ET may be recovered over periods of many decades.

These findings indicated that the genetic structure of many species of pathogenic bacteria is basically clonal in the sense that the effective (realized) recombination rates for most genes are sufficiently low to permit the mutational diversification of cell lineages in terms of host distribution, disease specificity, and virulence, and the maintenance of differentially adapted chromosomal genotypes over periods sufficiently long for them to become widely distributed geographically (Achtman and Pluschke, 1986; Musser *et al.*, 1987; Selander *et al.*, 1985, 1987; Selander and Musser, 1990).

An example from T. S. Whittam's extensive studies of *E. coli* (Whittam, 1989; Whittam and Ake, 1993) will suffice to illustrate clonal structure and the type of evolutionary insight that population genetics can provide.

Following outbreaks of a clinically distinctive syndrome of diarrheal disease — hemorrhagic colitis — in North America in 1982, the causative agents, strains of *E. coli* of serotype O157:H7 (Wells *et al.*, 1983), were recognized as a new class of enteric pathogens, designated as EHEC. In addition to being serologically distinct from the enteropathogenic *E. coli* (EPEC), they lack the invasive properties of enteroinvasive *E. coli* (EIEC), and they do not produce the classical toxins of the enterotoxigenic *E. coli* (ETEC). Instead, they express potent Shiga-like cytotoxins, carry plasmids that encode adhesins mediating attachment to intestinal cells, and have a chromosomal gene (*eae*) that is essential for the production of intimin and the attaching and effacing lesions that are characteristic of some EPEC strains.

MLEE analysis revealed that all O157:H7 strains belong to a single clonal lineage (Whittam *et al.*, 1988) that is not closely allied with those of other serotypes that produce Shiga-like cytotoxins or with other strains of serogroup O157 that are associated with enteric infections in animals (Whittam and Wilson, 1988). These findings suggested that, if the O157:H7 clone has recently evolved, a close relative might be found among strains of serogroups other than O157. Searching for a possible ancestor, Whittam *et al.* (1993) screened by MLEE (20 loci) 1,300 strains of five serogroups representing 16 O:H serotypes that are associated with diarrhea and other types of intestinal disease in humans. There was substantial genotypic variation among the isolates of each serotype, but 70% of the strains belonged to only 15 of the 191 distinctive multilocus enzyme genotypes (ETs) that were identified. The O157:H7 clone was shown to be closely related (95% identity in MLEE genotype) to a clone of serotype O55:H7 strains that has long been associated with worldwide outbreaks of infantile diarrhea. The clear inference is that the O157:H7 clone, a new pathogen causing a new disease, recently emerged when an O55:H7-like ancestor acquired secondary virulence factors (Shiga-like cytotoxins and plasmid-encoded adhesins) via horizontal transfer and recombination.

The concept of clonal population structure has had major impact on the fields of bacterial epidemiology and pathogenesis and has recently been applied to certain parasitic protozoans as well (Tibayrenc *et al.*, 1990, 1993). Indeed, it has been widely regarded as a paradigm for all bacteria (Istock *et al.*, 1992; Maynard Smith *et al.*, 1993), although we have noted that linkage disequilibrium analyses of MLEE data do not support a clonal structure for the pathogenic species *Neisseria gonorrhoeae* (see O'Rourke and Stevens, 1993) and *Pseudomonas aeruginosa* (Selander and Musser, 1990; see Denamur *et al.*, 1993) or for major phylogenetic divisions of the soil bacterium *Rhizobium meliloti* (Eardly *et al.*, 1990). It is unlikely that any bacterial species is strictly clonal, and certainly none consists of freely recombining cell lineages. The critical problem is to determine the contribution of intragenic and assortative (entire gene) recombination to allelic and genotypic variation in diverse types of bacteria by estimating the frequency and extent of exchange for a large variety of genes.

NUCLEOTIDE SEQUENCE ANALYSIS OF POPULATION STRUCTURE

By providing full resolution of the allelic diversity of genes in multiple isolates, the technology of rapid nucleotide sequencing of DNA amplified by the polymerase chain reaction (PCR) has opened a new era for population genetics; and the availability of a rapidly increasing body of sequence data has stimulated the development of new statistical methods for data analysis and hypothesis testing in studies of the genetic structure of natural populations and the evolutionary processes that affect rates of nucleotide and amino acid substitution.

In application to bacteria, comparative sequence analysis has already provided evidence of intragenic recombination of horizontally transferred DNA for one or more genes in several diverse species. As a result, the evolutionary role of recombination in bacteria is currently a major focus of bacterial population genetics research (Milkman and Bridges, 1990, 1993; Bisercic et al., 1991; Cohan et al., 1991; Dykhuizen and Green, 1991; Maynard Smith et al., 1991, 1993; Hall and Sharp, 1992; Nelson and Selander, 1992; Souza et al., 1992; Spratt et al., 1992; Wang et al., 1992a; Maiden, 1993; Roberts and Cohan, 1993; Whittam and Ake, 1993; Boyd et al., 1994; Li et al., 1994).

A major theme emerging from the comparative sequence analysis of genes in bacterial populations is that the effective rate of recombination varies markedly among loci encoding proteins of different functional types (Selander and Smith, 1990; Nelson and Selander, 1992; Sibold et al., 1992; Vázquez et al., 1993; Achtman, 1994), as well as among species and subdivisions of species. The nature, extent, and evolutionary consequences of this variation may be illustrated by a consideration of the results of our research on allelic sequence diversity in several genes in natural populations of *S. enterica* and *E. coli*.

RECOMBINATIONAL BASIS OF SEROVAR DIVERSITY IN *SALMONELLA ENTERICA*

The primary basis for classification of strains of *S. enterica*[1] is the Kauffmann-White serological scheme, in which 2,324 serovars have been recognized on the basis of the antigenic properties of the phase 1 and phase 2 flagellar proteins (H1 and H2 antigens), which are encoded by the *fliC* and *fljB* genes, respectively (Macnab, 1987, 1992), and the O antigen domain of the cell-surface lipopolysaccharide, which is synthesized by enzymes specified by genes of the *rfb* region (Ewing, 1986; Popoff and Le Minor, 1992; Reeves, 1993). For the phase 1 flagellin, 52 antigenic factors and 61 serotypes (single factors or combinations of factors) have been distinguished.

The demonstration by MLEE that the same O antigen polysaccharide and flagellin serotypes may occur in distantly related strains suggested that horizontal transfer and recombination events involving *fliC*, *fljB*, and the *rfb* region genes are relatively frequent (Beltran et al., 1988; Selander et al., 1990a,b). And for *fliC*, this hypothesis subsequently was supported by partial sequencing of the gene in strains of Typhimurium (Smith and Selander, 1990) and several other serovars (Smith et al., 1990) and the identification of a plasmid-borne *fliC*-like gene (Smith and Selander, 1991).

---

1. There is now general agreement that all the salmonellae are members of a single species, for which the name *Salmonella enterica* has been proposed (Le Minor and Popoff, 1987; Old, 1992), although Reeves et al. (1989) have advocated the recognition of subspecies V as a distinct second species, *S. bongori*. In any event, the serovar names are now treated as vernaculars: thus, for example, *S. typhimurium* is designated as *S. enterica* [serovar]Typhimurium.

Table 1. Antigenic formulae of strains of 15 *S. enterica* serovars

Serovar[1]	ET[2]	Strain no.	Antigenic factors[3]	
			O	H1
Banana	Ba 1	S5332	4,12	m,t
Berta	Bc 1	S5321	1,9,12	f,g,t
Derby	De 13	S241	1,4,12	f,g
Enteritidis	En 1	S53	1,9,12	g,m
Newmexico[4]	Nm 1	S5323	9,12	g,z_{51}
Oranienburg	Or 1	S5331	6,7,14	m,t
Othmarschen	Ot 1	S5334	6,7,14	g,m
Pensacola	Pe 1	S5325	1,9,12	m,t
II 16:g,m,s,t:–	II 1	S5333	16	g,m,s,t
II 6,7:g,m,s,t:–	II 2	S5336	6,7	g,m,s,t
II 42:f,g,t:–	II 3	S2993	42	f,g,t
II 48:g,m,t:–	II 4	S5335	48	g,m,t
II 6,7:m,t:–	II 5	S5337	6,7	m,t
IV 45:g,z_{51}:–	IV 1	S3015	45	g,z_{51}
VII 1,40:g,z_{51}:–	VII 1	S3013	1,40	g,z_{51}

[1]Except as noted by Roman numerals, the serovars are of subspecies I.

[2]Electrophoretic type (distinctive multilocus enzyme genotype).

[3]O, cell-surface polysaccharide; H1, phase 1 flagella.

[4]Newmexico is the only serovar examined that is biphasic (alternately expresses H1 and H2 flagella); its complete antigenic formula is 9,12:g,z_{51}:1,5.

Li *et al.* (1994) recently obtained the complete 1,518-1,527 base-pair (bp) sequence of *fliC* for strains of 15 *S. enterica* serovars of subspecies I, II, IV, and VII that express combinations of six phase 1 flagellar antigenic factors of the g series (f, g, m, s, t, and z_{51}) (Ewing, 1986; Masten and Joys, 1993) (Table 1). This study clearly demonstrated that recombination is a major evolutionary mechanism generating both allelic variation in *fliC* and serovar diversity in natural populations.

Sequence Variation in Relation to Flagellin Structure

Sequence variation in the *fliC* gene and its flanking regions among the 15 strains of *S. enterica* is shown in Figure 2. In this species, as in other bacteria (Wilson and Beveridge, 1993), the N- and C-terminal regions of the flagellin molecule are strongly conserved in both amino acid sequence and length, whereas the central region is highly variable. The terminal regions are involved in secretion and polymerization, and amino acid substitutions in the central region, which forms a knob-like protrusion from the flagellar core, produce the antigenic variation assayed in serotyping (Joys, 1988; Newton *et al.*, 1991; Okazaki *et al.*, 1993; Li *et al.*, 1993).

For purposes of analysis, three regions of *fliC* are distinguished: C1 (conserved region 1) includes codons 1-181; V (variable region) is the central segment composed of codons 182-370; and C2 (conserved region 2) consists of codons 371-505. C1 and C2 are invariant in size; but in the V region, electrophoretic type (ET) II 2 has a deletion of codon 319, and ETs Nm 1, IV 1, and VII 1 share a deletion of codon 216. Six other ETs have three additional codons between codons 222 and 223. As shown in Figure 2, both nonsynonymous (replacement) and synonymous (silent) substitutions occur much more frequently in the V region than in C1 and C2.

5' flanking region   *fliC* (coding region)
<------------   ------------>

```
                 2222222111111                   111111111222222222222222222333333333333333333333333334444444444444455555555555555
                 5554200965211             1567777788990245677880011113444457990022334555666667788990122344556678001122233444444
ET  Serotype     2102241954087             6873026894749201951406140369404950173615368478012392547362736847925401706258483689

En 1   g,m       ATCCAGACTTGTC     ATCTTAATTTTCCTCCCTTCTTCTCCTTTTCTCGTCCGCCTCTATCGGAACTCTGATTTCCCTTTACCGCCTGTTTAGGATG
Be 1   f,g,t     .....CG.....      ..........I...C....ICC.CI.I.G.....C..................................
De 13  f,g       .............     ............TC.....................................G.....................
Pe 1   m,t       .............     ...................................................................
Ba 1   m,t       .............     ...................................................................
Nm 1   g,z51     TGATG.G.C.A.T     ...CGTCCA.C.TC.TTCCT.G...TC..C.C...TT.TCTCGCTATCG.CTC.G...TTTGCC......TGA.C.....
Or 1   m,t       .............     ..........TC......C.................................................
Ot 1   g,m       .............     ...................................................................
II 1   g,m,s,t   ...T..G.C...      ..T........C.......TCC.CT.T..A............GCC....G.......
II 2   g,m,s,t   ...T..G.CG...     ..T........C.......TCC.C.C..A............AGCC...G........A..C....
II 3   f,g,t     ...T..G.CG...     ..T........C....T....C...TCC.CT.T..A............GCC....A.....
II 5   m,t       ...T..G....G.     .CT.A.....CC...TT.........T.C....TA.A.............CA......T.......
II 4   g,m,t     .............     GCT.........T.....................C..T..A............GCC.........A.....
IV 1   g,z51     ...T..G.C.AA.     ...C......CT.C..TTGC.C.TG.TCCCGT.T..A...T.GCTATCGTCTC.G...TTTGCCGTTATTG.CG.TCACCT
VII 1  g,z51     .......T.....     .................TG..............T.......T.GCTATCGTCTC.G...TTTGCCGTTATTG.CG.TCACCT
                                  R  R  R                         RR                     R  R
```

```
                 5555555555555556666666666666666666666666666666    66666666677777777777777777777777777777777777777
                 5566666778888999000011122233334444555555566666    67777889900111223334444555556666677778888999
                 2614673625814802390451480369269784567890234       6      7235847790201780928914780135902451347803693294

En 1   g,m       TCACGCGTGTCCCGACGCGAATAATTCTTGTGTGCAGCACACG---------GAGATTACGTAACGGATGTCCTGACATATATCTCCTAGCAGATGG
Be 1   f,g,t     .G.......GTT..TTT.A..CGGCCAGG.ACA.CA.T.T.AACTGTTCCT...CCT..........C....................
De 13  f,g       .G..........................T....---------A..........C...............
Pe 1   m,t       .........GTT..TTT.A..CGGCCAGG..ACA.CA.T.T.AACTGTTCCT...CCT....GC..C.CA..G.ATT...G..ATACG.TGTCAAC
Ba 1   m,t       .........GTT..TTT.A..CGGCCAGG..ACA.CA.T.T.AACTGTTCCT...CCT....GC..C.C..TG.ATT...G..ATACT.GTCAAC
Nm 1   g,z51     ..........TGAT.TTACTC....TAA---.AT..TGTAC---------.GATC....CC.TACG.A..A..AGA.ATA.AA...TT.A...
Or 1   m,t       .........GTT..TTT.A..CGGCCAGG..ACA.CA.T.T.AACTGTTCCT...CCT....GC..C..TG.ATT...G..ATACT.GTCAAC
Ot 1   g,m       ............................................---------..........
II 1   g,m,s,t   ........G.T..........A...A..AT....---------.........G.T.C..A..AC.......A.C.....AA
II 2   g,m,s,t   ........G.T..........A...A..AT....---------.........G.T.C..A..AC.......A.T.....AA
II 3   f,g,t     ..................................---------.........
II 5   m,t       .........GTT..TTT.A..G.CCAGG..ACA.CA.T.T.AACAGTTCCT...C.T...G.T.C..A..G.ATT...G..ATACT.TGTCGAC
II 4   g,m,t     ...............TTT.A..G.CCAGG..AC..CA.T.T.AACAGTTCCT...G.T.C..........G.T........
IV 1   g,z51     GT.TATACT.TTG.TTTT.GT....TAA---.AT..TGTAC---------.GATCC.AACC.TACGCA..A..A.A.ATA.CA...TT.AT.AC
VII 1  g,z51     GT.TATACT.TTG.TTTT.GT....TAA---.AT..TGTAC---------.GATCC.AACC.TACGCA..A..A.A.ATA.CA...TT.AT.AC
                 R      RR RRR R       RR RR R R       RR  R  R  R       RRRRR RRR R       RR
```

```
                                                                       1111111111111111111111111111
                 77778888888888888888888888888888888999999999999999999999999999999999000000000000000000011111111
                 99990001111234444566677778899022223333444444445555555566666677778889900011223444556789000112362878
                 5678147035695403492459040126784901478012601234568012345670145691235478692580143884787987921473

En 1   g,m       TGCCTATTGAAACTCGATTAACTGACAGTCCGATGCGCTGCCACTGCCGACGGATTAGCTTCCTACAATATAGACTTACCGACTGCGCATATGATCA
Be 1   f,g,t     .....G..............CA........A.C..I.G.A..G...............................G
De 13  f,g       .....G..............A........A.G..I.G.A..G................A...C.....
Pe 1   m,t       AT.TC.AATTT..T.T..G....C...A.AATAG..AG.GAT.AG.AA.CAG.A.GAGATTGGCT.TAGT..C................G
Ba 1   m,t       AT.TC.CAATTT..T.T..G....C...A.AATAG.G.G.GAT.A.A..CA.AGCGAGATT.G.TCTAGT..C................G
Nm 1   g,z51     C..T....A...T.TTTCCG.AC.C.CTC7G.G..T...A.....T.G....C....G....T...A..C.....ATAT.CGCCGATG
Or 1   m,t       ..T.TC.CAATTT..T.T..G....C...A.AATAG.G.G.GAT.A.A..CA.AGCGAGATT.G.TCTAG...C................G
Ot 1   g,m       ....................................A.......................
II 1   g,m,s,t   ..T..G..A.G.....AT......A..TC..T.......A...A...G..A...............T......TG....
II 2   g,m,s,t   ..T..G..A.G.....AT......A.TC..T.......G.GG...TGT.GT---..A.........T......TG....
II 3   f,g,t     ..T..G..A.C..........A..A..........A..G..I.G.A.....A.......T.....TG....
II 5   m,t       AT.T....AACTT.C.AT...G....TC......AATAG...A..GAT.A..A..CA..GCGAGATTGGCTCTAG.....
II 4   g,m,t     .....A.....AT..........A.....................................T......TG....
IV 1   g,z51     C..T....A...T.TTT.CG.ACAC.CA..T.G....CA...G.T..G.A..C....GA....T..........C.G.AGTAATAT.CGCT.ATG
VII 1  g,z51     C..T....A...T.TTT.CG.ACAC.CA..T.G....CA...G.T..G.A..C....GA....T..........C.G.AGTAATAT.CGCT.ATG
                 RR      RR      R    RR RR RR R    R RR RR RR R R RR R   RRR RRR  R              R        R
```

3' flanking region

```
                 1111111111111111111111111111111111111111111111111111111111111111    1111111111
                 1111111111111112222222222222233333333333333333333334444444444444444    5555555566
                 344556668889001233334556999001122234445556667777890112233345566779    3334488901
                 909021425891032170395171036564703651480364582347091039251234581439    6 1344502580

En 1   g,m       CGGGACTCTAAGTATCCCACACCGTCATTCATACTATCCTCTCCTAACCGGTAGTTTTAAATTTTCGTA    TCAGGATTCC
Be 1   f,g,t     ...G...........T....T.........................................G.     ..........
De 13  f,g       ..G.A.C...I....IT.............................................       .........
Pe 1   m,t       ..G...........T....T.........................................       ...A.......
Ba 1   m,t       ..G...................................................       ..........
Nm 1   g,z51     AAAGTAT.GC...TC.TTTTGAT..TTAA.CCGTCTCTTG.CATCTCTTATCGC.CC.CGC.C.....G  G....CCCTT
Or 1   m,t       ..G...........................................C...C...G.       ..........
Ot 1   g,m       ......................................................       ..........
II 1   g,m,s,t   .....A.C....T....TT..T.G..G...G................................CT...  .........
II 2   g,m,s,t   .....A.C....T....TT..T.G..G....G...........................C.CT...    .........
II 3   f,g,t     .....A.C....T....TT..T.G..G...G...............................       .........
II 5   m,t       ...G..........T....TT.T.G.................................C.....     ...C......
II 4   g,m,t     .....A.C....T....TT..T.G..G....G..............................C.     .........
IV 1   g,z51     A.A.TAT.G.ACTCTTT.TG.TC.TTAATT.G..T..TGT.AT...CT.A..GCCCC.CGC.AG.TA..  ..C.A..CT.
VII 1  g,z51     A.A.TAT.G..CTCTTT.TG.TC.TTAATT.G..T..TGT.AT...CT.A..GCCCC.CGC.AG.TA..  .T......CT.
                 R  R     R            R               R       R   RR    R         RRR  R
```

Figure 2. Nucleotide sequence variation in the *fliC* gene and flanking regions among strains of 15 serovars of *S. enterica*. Only the polymorphic sites are shown. Dots indicate identity to the sequence of the strain of En 1. R indicates the occurrence of a substitution that results in an amino acid replacement. Arrows indicate points of division of the C1, V, and C2 regions. Nucleotides of the distinctive m,t sequence are in boldface type.

23

Evidence of Recombination

It is instructive to compare evolutionary trees based on (a) MLEE analysis, indexing the overall genetic relatedness of the strains (Figure 3A), (b) the nucleotide sequence of the combined C1 and C2 regions of *fliC* (Figure 3B), and (c) the sequence of the central region of the gene (Figure 3C). If the evolution of *fliC* has involved little or no recombination, trees for the V region and the C1 + C2 regions should be topologically similar (although rates of substitution may differ) and both gene-region trees would be expected to resemble the MLEE tree. In contrast, recombination of DNA acquired through horizontal exchange among strains would be indicated by a clustering of *fliC* sequences specifying the same flagellin serotype, regardless of the overall genetic relatedness of the strains in which they occur. [See relevant discussions of tree comparisons in Dykhuizen and Green (1991) and Valdés and Piñero (1992).]

Serotype g,m. Strains of serovars Enteritidis (En 1) and Othmarschen (Ot 1) express serotype g,m (Table 1). They are divergent in chromosomal genetic character (Figure 3A), yet their C1 + C2 *fliC* sequences are identical (Figures 2 and 3B), and their V regions differ by only a single nonsynonymous substitution (Figures 2 and 3C).

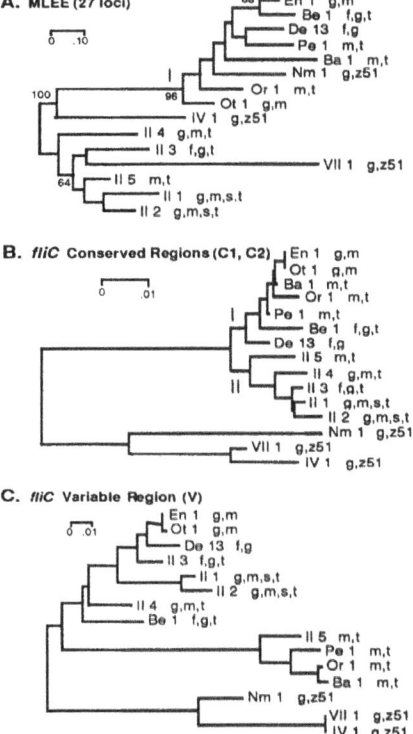

Figure 3. Unrooted evolutionary trees for 15 strains of *S. enterica* constructed by the neighbor-joining method (Saitou and Nei, 1987; Kim *et al.*, 1993) from matrices of pairwise genetic distance. A. MLEE tree (based on 27 enzyme loci). The distance measure is Nei's standard genetic distance. For nodes, bootstrap values greater than 50% are indicated. B. Sequence tree for the conserved (C1 + C2) regions of the *fliC* gene, based on both synonymous and nonsynonymous sites. The distance measure is that of Jukes and Cantor (1969). Clusters of strains of subspecies I and II are labeled. Bootstrap values for all nodes are greater than 50%. C. Sequence tree for the central variable (V) region of the *fliC* gene. Note that the scale is approximately half that of the conserved-sequence tree (B).

Serotype g,z$_{51}$. Although strains Nm 1, IV 1, and VII 1 (serotype g,z$_{51}$) represent three subspecies (Figure 3A), they cluster together in the C1 + C2 sequence tree (Figure 3B). Note in Figure 2 that the 5' half of C1 in VII 1 is similar to that of the subspecies I ET De 13 and distinctly different from that of IV 1 or Nm 1. In sequence of the V region, ETs IV 1 and VII 1 are identical and Nm 1 is more similar to those two strains than it is in the C1+C2 tree.

Serotype m,t. ETs Pe 1, Ba 1, Or 1, and II 5 are of serotype m,t. The three ETs of subspecies I are not closely allied, and all three are distantly related to II 5 (Figure 3A). On the basis of the C1 + C2 sequence, II 5 is still relatively distant; but in the V region tree (Figure 3C), the four ETs form a tight cluster that is strongly divergent from other clusters and lineages. It is apparent that the V region and part of C2 have been exchanged among the four ETs and that part of this distinctive sequence has also been transferred to ETs Be 1 and II 4.

Other Serotypes. Complex mosaic structures are shown by the *fliC* genes of several strains, most notably those of Be 1 (f,g,t) and II 4 (g,m,t). In Be 1, the C1 region is a mixture of small segments of subspecies I and II; the 5' part of region V is the distinctive m,t serotype sequence until just beyond the position of the three extra codons, where it becomes subspecies I-like, with a short segment of subspecies II in the middle; and the 3' part of V is subspecies II sequence. ET II 4 also has the m,t sequence in the 5' part of V, and the middle part is somewhat like the m,t sequence; but the 3' part of the region is subspecies I-like.

Evolutionary Sources of Diversity

The occurrence of each of the three flagellin serotypes g,m; m,t; and g,z$_{51}$ in distantly related strains is clearly attributable to horizontal exchange rather than to convergence in amino acid sequence or retention of ancestral sequences. Recombination of segments of the central region has been especially extensive among ETs of subspecies II and has generated great diversity in flagellin and polysaccharide serotype combinations, as well as new flagellin serotypes. But notwithstanding the considerable sequence diversity among *fliC* alleles of strains expressing g complex antigenic factors, all of them are more similar to one another than to the serotype a, c, d, i, and r alleles of Typhimurium, Typhi, and other serovars (Li, unpublished data). The occurrence of markedly divergent families of alleles in *S. enterica* suggests that sequences are occasionally recruited from the flagellin genes of other species or, even, from other types of genes. It is perhaps noteworthy that the central region of the flagellin of serovar Paratyphi A shows some similarity in amino acid sequence and structure to comparable regions of certain serine proteases (Grewal and Salunke, 1993).

Minimal Basis for Serotype Variation in Phase 1 Flagellin

In other research on the molecular genetic basis of flagellar antigenic variation in *S. enterica*, the *fliC* gene was sequenced in 23 strains of the serovars Antarctica, Blegdam, Dublin, Enteritidis, Moscow, Naestved, and Rostock, which exhibit nine combinations of eight flagellar antigenic factors of the g complex (Li *et al.*, in preparation). Because the *fliC* alleles of these strains exhibit more than 99% nucleotide sequence homology, the molecular genetic basis of some of the flagellin epitopes can be determined by direct comparison of the inferred amino acid sequences. The results demonstrate that single amino acid substitutions and small-scale duplications are sufficient to modify the antigenic profile of g complex flagellins (see Newton *et al.*, 1991). Two examples are shown in Table 2. The molecular genetic basis for the u epitope may be determined by comparison of the flagellin sequences of serovars Rostock (g,p,u) and Dublin (g,p), for which the only difference is a

Table 2. Variation in amino acid sequence of the phase 1 flagellin among three serovars of *S. enterica*

Serovar	ET	Serotype	Strain	Amino acid at indicated polymorphic position	
				254	314
Dublin	Du 1	g,p	S4699	Thr	Ala
Rostock	Ro 1	g,p,u	S5319	Thr	**Gly**
Naestved	Na 1	g,p,s	S5324	**Ala**	Ala

Gly/Ala-314 replacement in Rostock. Similarly, the only difference between the flagellins of Naestved (g,p,s) and Dublin (g,p) is a Ala/Thr substitution at codon 254.

Adaptive Basis of Flagellar Polymorphism

The prevailing view is that the extensive flagellar antigenic polymorphism in *S. enterica* and other bacteria is adaptive in permitting the reinfection of hosts (Brunham *et al.*, 1993), and Reeves (1992) has suggested that antigenic variation in both flagellin and the cell-surface polysaccharide is subject to "niche-specific selection." But Joys (1988) advocated a model in which amino acid substitutions accumulate by mutation in the central region of the flagellin molecule in the absence of functional constraint and resultant counter-selection, with antigenic diversity having no adaptive function per se. A corollary of this neutral mutation hypothesis (Kimura, 1983) is that effective (realized) rate of recombination may also be relatively high.

The problem is to determine to what extent flagellar antigenic diversity is due to neutral amino acid changes and to what degree it is promoted by frequency dependent or other types of diversifying selection of mutants and recombinants. Unfortunately, however, there is little available information on the role of antigenic variation among *S. enterica* serovars in immune system evasion (Macnab, 1987; Joys, 1988; Wilson and Beveridge, 1993). The observation that sensitivity to flagellotropic phage may be serotype dependent (Iino, 1977) suggests another possible adaptive basis for flagellin polymorphism.

Population Genetic Rationale for the Kauffmann-White Serological Scheme

For more than 60 years, serotyping of the salmonellae has provided a valuable and convenient marker system for distinguishing strains that have distinct host ranges and pathogenicities, including those of closely related populations such as Dublin (g,p) and Enteritidis (g,m) (Selander *et al.*, 1992). Population genetics research has now demonstrated that serotyping of *S. enterica* owes its success to the basic clonal structure of the species. In general, serovars are families of closely related clonal lineages possessing distinctive combinations of genes that determine host range, disease syndrome, and virulence; and for many of the medically important serovars, such as Typhi, Paratyphi B, Gallinarum, Typhimurium, and Dublin, only a single clone is globally predominant at any one time (Beltran *et al.*, 1988; Selander *et al.*, 1990 *a,b*; Selander *et al.*, 1992; Li *et al.*, 1993).

Table 3. Housekeeping genes sequenced in strains of *S. enterica* and *E. coli*

Gene and region	No. of strains *S. enterica*	No. of strains *E. coli*	Base pairs sequenced
Malate dehydrogenase (*mdh*) coding region	24	20	864[1]
Proline permease (*putP*)			
Coding region	16	12	1,467
Control region	16	12	416-422
Glyceraldehyde-3-phosphate dehydrogenase (*gapA*) coding region	16	13	924
Isocitrate dehydrogenase phosphatase/kinase (*aceK*)			
Coding region	2	16	1,734[2]
aceA-aceK intergenic region	2	77	48-280
aceK-iclR intergenic region	2	16	16-204
6-Phosphogluconate dehydrogenase (*gnd*) coding region	36	33	1,335

[1]849 bp sequenced in strains of *S. enterica*.
[2]Polymorphic in size; 1,722 bp in some strains.

SEQUENCE ANALYSIS OF GENES ENCODING ENZYMES AND OTHER HOUSEKEEPING PROTEINS

In the past few years, we have been generating a database of nucleotide sequences of multiple genes in diverse strains selected from the *Escherichia coli* Reference Collection (ECOR) (Ochman and Selander, 1984b; Herzer *et al.*, 1990) and the *Salmonella enterica* Reference Collections (Beltran *et al.*, 1991; Boyd *et al.*, 1993) to represent the ranges of

Table 4. Mean sequence divergence in five genes between pairs of strains of *S. enterica* and *E. coli*

Gene	*S. enterica* No. strains	*S. enterica* Mean % difference	*E. coli* No. strains	*E. coli* Mean % difference	Between species Mean % difference	CAI[1]
Nucleotides						
mdh	24	4.5	20	1.1	14.6	0.58
putP	16	4.6	12	2.4	16.3	0.33
gapA	16	3.8	13	0.2	15.6	0.83
aceK	—	—	16	2.9	—	0.34
gnd	36	3.9	33	7.2	15.6	0.50
			30[2]	5.2		
Amino acids						
mdh	24	1.3	20	0.3	4.3	
putP	16	1.3	12	0.3	5.5	
gapA	16	1.1	13	0.1	3.9	
aceK	—	—	16	0.8	—	
gnd	36	0.7	33	1.8	3.9	
			30[2]	1.0		

[1]Codon adaptation index.
[2]Strain E83085? and ECOR strains EC16 and EC71 omitted.

genomic diversity in these species previously revealed by DNA-DNA hybridization and MLEE (Table 3). The core sample consists of two strains of each of the eight subspecies (I, II, IIIa, IIIb, IV, V, VI, and VII) of *S. enterica* and 13 strains of *E. coli*. DNA templates generated by the polymerase chain reaction (PCR) are sequenced in both orientations, since, in our experience, data of sufficient accuracy for population genetics analysis cannot be obtained by sequencing only one strand (Nelson and Selander, 1994).

One objective of our comparative studies is to examine the hypothesis that recombination has played a variable role in the evolution of different genes, depending to large extent on the functional type of gene product. Evidence of the exchange of entire genes or major segments of them may be obtained by comparison of the branching order of strains in trees for different genes. And as a guide in identifying segments of genes that have been horizontally exchanged among strains (intragenic recombination), we have employed the statistical tests of nonrandom distribution of polymorphic nucleotide sites developed by Stephens (1985), Sawyer (1989), and Maynard Smith (1992).

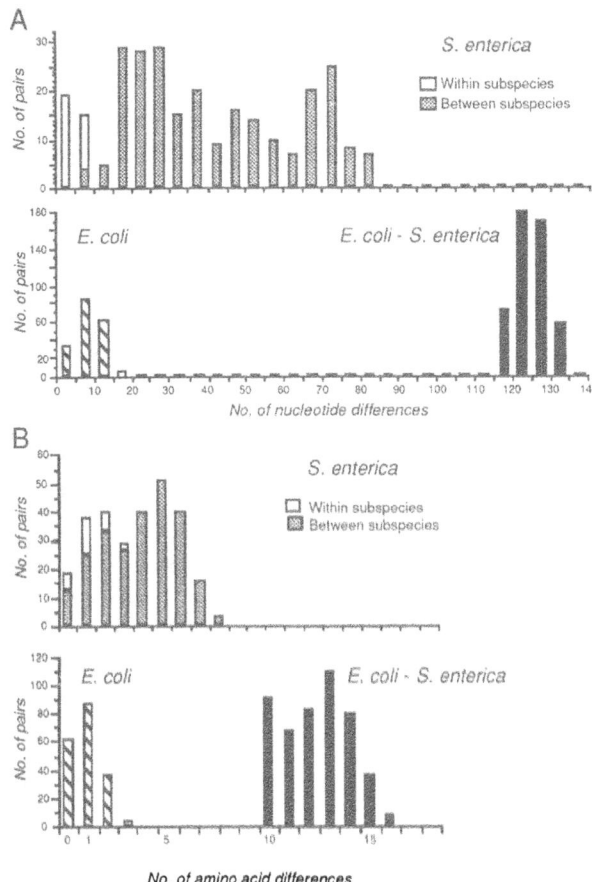

Figure 4. Pairwise mismatch distributions of (A) *mdh* nucleotides and (B) inferred malate dehydrogenase amino acids for 24 strains of the eight subspecies of *S. enterica* and 20 strains of *E. coli*. Plotted are the numbers of pairs of strains showing the indicated number of differences at nucleotide sites or amino acid positions, based on the sequence of an 849-bp segment (codons 12-294) of the *mdh* coding region.

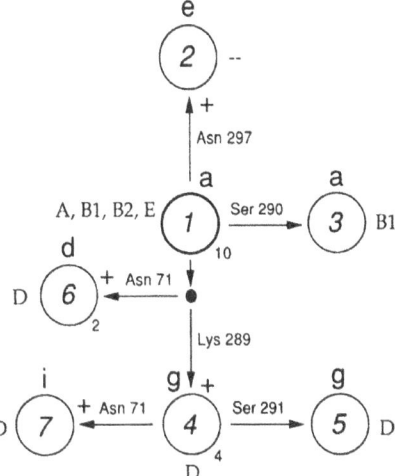

Figure 5. Hypothetical evolutionary scheme for the malate dehydrogenase allozymes (lower case letters above circles) and distinctive amino acid sequences (circled numbers) of strains of *E. coli*. Upper case letters indicate ECOR group assignments of strains having the various sequences. For sequences represented by more than one strain, the number of strains is indicated to the lower right of the circle. The dot is a postulated intermediate ancestral form.

Here we will consider sequence variation and the evidence for recombination in four genes (*mdh, putP, gapA,* and *gnd*) in both *S. enterica* and *E. coli* and a fifth gene (*aceK*) for which our analysis has thus far been limited largely to the latter species (Table 3).

Malate Dehydrogenase (*mdh*)

Nucleotide sequences of the *mdh* gene encoding the metabolic enzyme malate dehydrogenase were determined for 24 strains of *S. enterica* and 20 strains of *E. coli* (Boyd *et al.*, 1994). There were 217 polymorphic nucleotide sites among all 44 sequences, and strains of the two species differed at an average of 124 sites (14.6%) (Table 4). In all, 30 amino acid positions were polymorphic, with an average species difference of 12.4 (4.3%), which includes nine positions at which there were fixed differences between the species. Nucleotide sequence diversity was four times greater in *S. enterica* than in *E. coli* (Figure 4), and in both species the rate of amino acid substitution was lower in the NAD$^+$-binding domain than in the catalytic domain of the enzyme.

By matching the sequences with the electromorphs of the strains, we determined that enzyme electrophoresis of malate dehydrogenase detected 57% of the 21 distinctive inferred amino acid sequences and that, on average, the generation of a new allozyme has involved 2.6 amino acid substitutions (Boyd *et al.*, 1994). For *E. coli*, a hypothetical evolutionary scheme of the relationships between allozymes and amino acid sequences is shown in Figure 5. Of the seven distinctive amino acid sequences, sequence 1 is regarded as ancestral because it occurs in 10 strains representing all ECOR groups except D. The scenario postulates six amino acid replacements at five codon positions; each of four of these substitutions involved a charge change and generated a new allozyme. One parallel change is indicated: from the common ancestor of ECOR group D strains, sequence 6 (allozyme d)

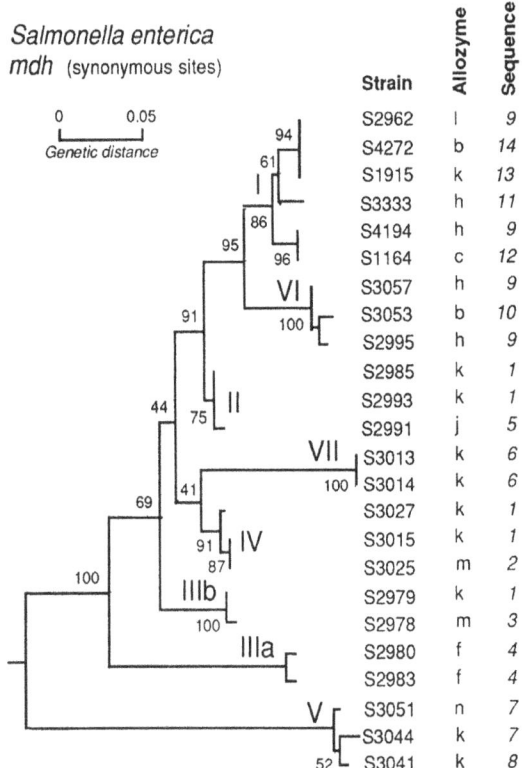

Figure 6. Evolutionary tree for the *mdh* gene in 24 strains of *S. enterica*, constructed by the neighbor-joining method from a matrix of pairwise distances based on synonymous nucleotide sites. Subspecies clusters are indicated by Roman numerals. Strain numbers, allozyme designations (letters), and amino acid sequence numbers are indicated.

was derived by an Asn-71 —> Asp replacement before the occurrence of a Gln-289 —> Lys substitution that produced sequence 4 (allozyme g); subsequently, a second independent Asp-71 —> Asn replacement created sequence 7 (allozyme i).

For both *S. enterica* and *E. coli*, the evolutionary relationships among strains indicated by synonymous sites of the *mdh* sequences are generally similar to those estimated by MLEE and sequence analysis of other genes (see discussion beyond). In the *mdh* tree for *S. enterica* (Figure 6), the genes cluster by subspecies, of which V is the most divergent.

Proline Permease (*putP*)

Virtually complete sequences of the proline permease gene (*putP*) and complete sequences of the control region of the proline utilization operon were determined for 16 strains of *S. enterica* and 12 strains of *E. coli* (Nelson and Selander, 1992). Proline permease is an integral membrane-spanning protein that mediates the transport of proline into the cell. Activity of the *put* operon is required only when proline is the sole nitrogen source, and expression of *putP* is tightly regulated in relation to the concentration of exogenous proline.

Mean levels of interstrain divergence in the sequence of *putP* and the *put* control region within and between species are shown in Table 4, and a tree based on these sequences is presented in Figure 7.

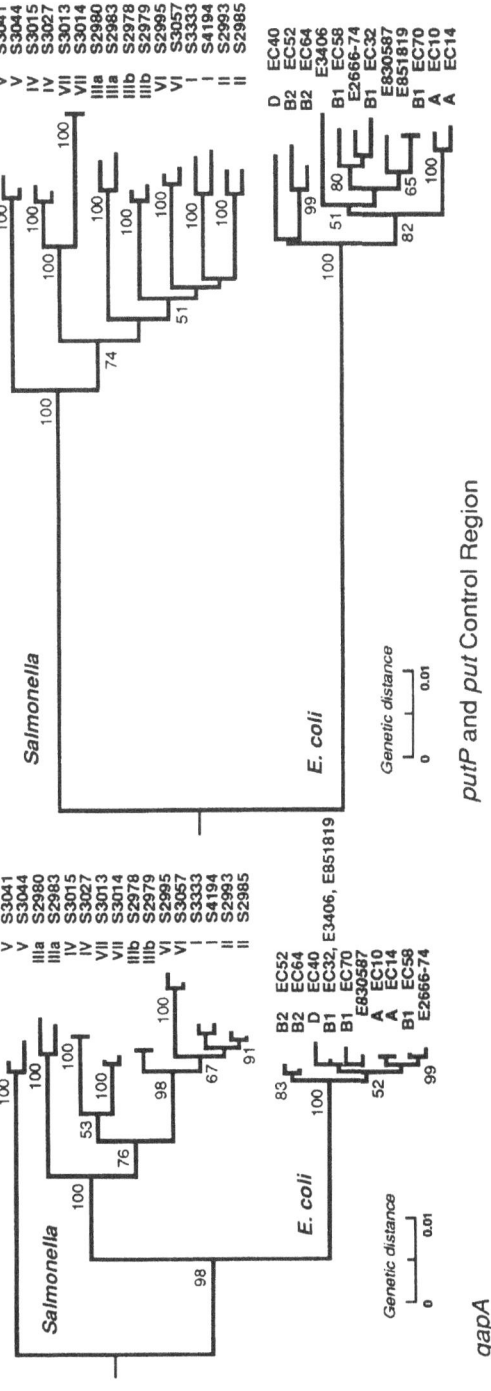

Figure 7. Neighbor-joining trees for *gapA* and the *put* operon, based on nucleotide sequences (all sites) of 16 strains of *S. enterica* and 12 strains of *E. coli*. A number adjacent to a node indicates the percentage of bootstrap trees that contained that node.

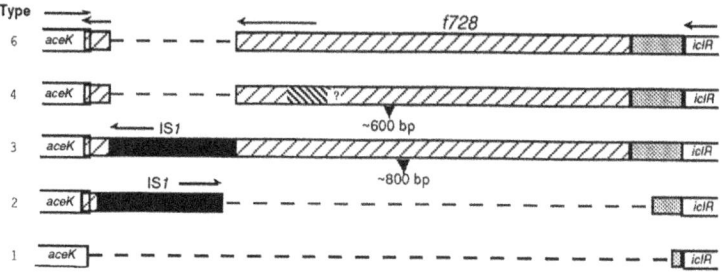

Figure 8. Diagram of the *aceK-iclR* intergenic region, showing five of the eight types, defined on the basis of size and composition, that were observed in a sample of 77 strains of *E. coli*. See text for explanation.

Glyceraldehyde-3-phosphate Dehydrogenase (*gapA*)

Sequences of the *gapA* gene, which encodes the key glycolytic enzyme glyceraldehyde-3-phosphate dehydrogenase, were obtained for the core sample of strains of *S. enterica* and *E. coli* (Nelson et al., 1991b). Consistent with the expectation that this enzyme is under especially strong selective constraint for amino acid replacement and the observation that the *gapA* gene has a high codon adaptation index (Sharp and Li, 1987), the degree of within-species sequence divergence is relatively small, particularly in the case of *E. coli* (Table 7).

Isocitrate Dehydrogenase Kinase/Phosphatase (*aceK*)

The *aceK* gene encodes a bifunctional enzyme (IDH kinase/phosphatase) that regulates the activity of isocitrate dehydrogenase (IDH). It is part of the acetate (*ace*) operon,

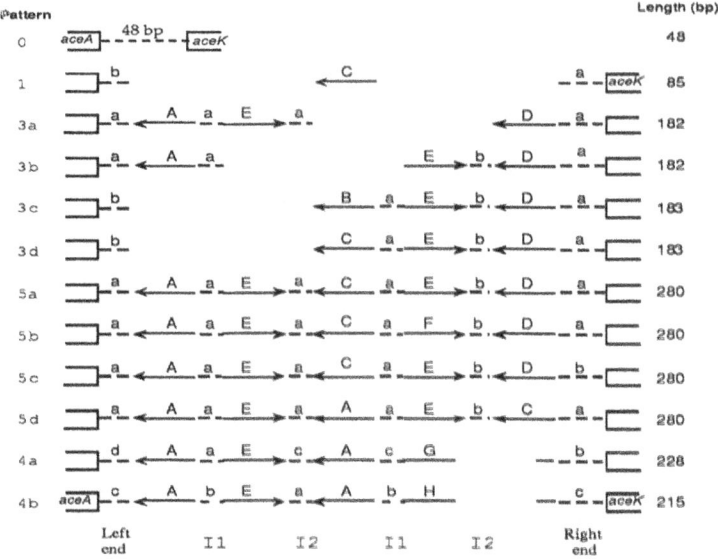

Figure 9. The 12 length and sequence patterns identified by sequence analysis of the *aceA-aceK* intergenic region in 77 strains of *E. coli*. Variant REP sequences (arrows) are identified by capital letters, and variant internal sequence elements I1 and I2 and end elements are designated by lower case letters.

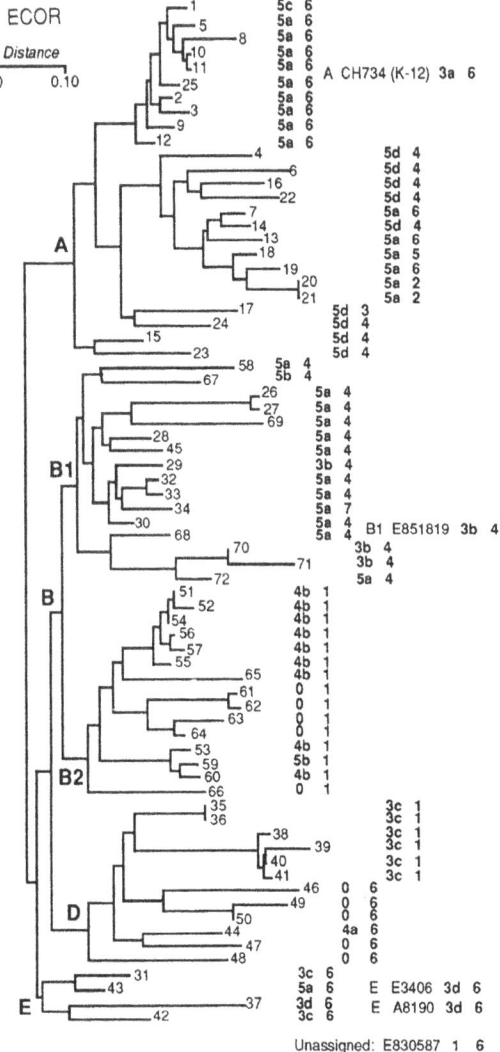

Figure 10. Distribution of size and sequence variants of the *aceA-aceK* and *aceK-iclR* intergenic regions among the 72 strains of ECOR and five additional strains. The neighbor-joining tree is based on MLEE (38 loci), and the numbers at the ends of the branches designate the ECOR strains. Shown in boldface type for each strain is the length-sequence pattern of the *aceA-aceK* intergenic region (see Figure 9) and the type of *aceK-iclR* intergenic region, based primarily on length. See text for explanation.

which also includes the *aceB* and *aceA* genes encoding, respectively, the glyoxylate bypass enzymes malate synthase and isocitrate lyase, and is expressed only during the growth of cells on acetate (review by LaPorte, 1993). Expression of the *ace* operon is controlled by both culture conditions and the product of *iclR*, the repressor gene of the *ace* operon, which is located downstream from *aceK*.

In *E. coli*, the *aceK* gene is polymorphic in size and both adjacent intergenic regions are highly variable in length (Figures 8 and 9) (Nelson *et al.*, in preparation). The coding sequence was 1,734 bp long in 13 of the 16 strains sequenced, but in three strains it was 12 bp shorter (1,722 bp) and the stop codon was TAA rather than TGA. Strains with the shorter *aceK* gene also lacked an open reading frame (*f728*) downstream between *aceK* and the *iclR* gene that was present in the other 13 strains (Figure 8). PCR amplification of the *aceK-iclR* intergenic region in the 72 ECOR strains revealed that the combination of a truncated *aceK* gene and an absence of *f728* (type 1 in Figure 8) is the condition in all 15 strains of the B2 group and in 6 of the 12 strains of the D group (Figure 10). Other variant conditions in this region include the presence of IS*1* elements in three strains and large deletions in two strains (Figure 8).

The region between *aceA* and *aceK* varied in length from 48 to 280 bp, depending largely on the number of repetitive extragenic palindromic (REP) sequences present, which ranged from 0 to 5. Analysis of the sequences of the *aceA-aceK* intergenic region in 77 strains identified 12 patterns based on the number of REPs present and variation in their sequences and those of the internal sequence elements and end segments (Figure 9). These patterns showed a high degree of phylogenetic association with the ECOR lineages (Figure 10), which permitted partial reconstruction of their evolutionary history.

Evidence of Recombination

For *S. enterica*, the eight subspecies that have been distinguished on the basis of biochemical characteristics (Le Minor, 1984; Le Minor *et al.*, 1986), DNA-DNA hybridization (Crosa *et al.*, 1973; Le Minor *et al.*, 1986), and MLEE (Reeves *et al.*, 1989; Selander *et al.*, 1991; Boyd *et al.*, in preparation) (Figure 11) are similarly identified by the nucleotide sequences of *mdh*, *putP*, *gapA*, and *aceK*. Moreover, the results of sequence analyses of these genes fully confirm other lines of evidence indicating that subspecies V is the most divergent lineage (Reeves *et al.*, 1989) and that subspecies IV and VII are relatively closely allied groups.

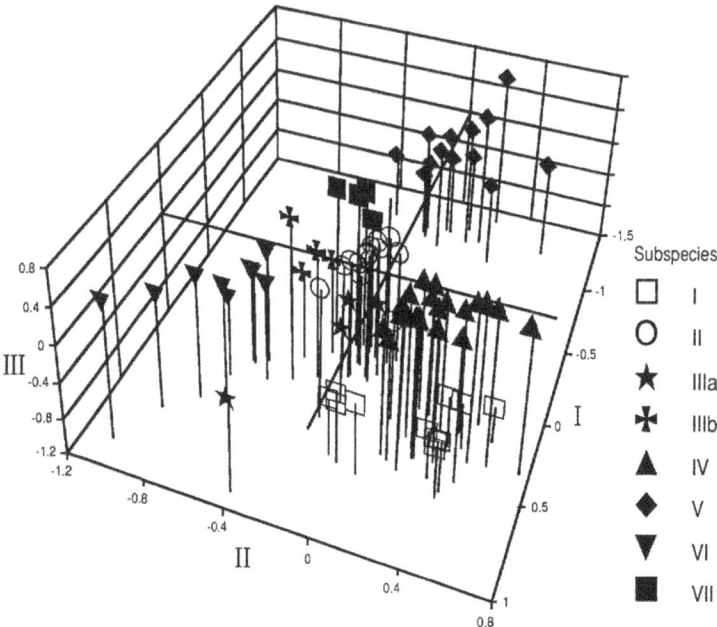

Figure 11. Genetic diversity within and among the eight subspecies of *S. enterica*. For each of 96 strains, scores for the first three axes derived from a principal coordinates analysis of MLEE allele profiles (24 loci) are plotted.

Table 5. Intragenic recombination events detected in five housekeeping genes

Gene and species	Evidence of recombination event
mdh	
S. enterica	(1) 336-bp segment (12 unique sites) from unknown donor in strains of subspecies IIIa.
	(2) 243-bp segment (5 unique sites) in strains of subspecies VI.
	(3) 6-bp segment (2 unique sites) in strains of subspecies V and VII.
E. coli	354-bp segment (5 unique sites) from unknown donor in strain EC35.
putP and *put* control region	
S. enterica	(1) 1,073-bp segment (25 unique polymorphic sites) from unknown donor in strains of subspecies VII.
	(2) 387-bp segment virtually identical in strains of subspecies I and V.
E. coli	(1) 21-bp segment (7 unique sites) shared by strains EC40 and EC64.
	(2) 866-bp segment (14 unique sites) from unknown donor in control region of strain EC40.
	(3) 227-bp segment (7 unique sites) in strain EC58.
gapA	
S. enterica	50-bp segment (15 unique sites) from *Klebsiella pneumoniae* in strains of subspecies V.
E. coli	None, but gene only weakly polymorphic.
aceK	
E. coli	(1) 57-bp (10 unique sites) acquisition from unknown donor by strain EC52.
	(2) 33-bp segment (4 unique sites) shared by group A strains EC14 and EC17 and three group B1 strains.
	(3) EC10 and CH734 differ from EC14 and EC17 (all group A strains) in first 2/3 of gene but not in the last 1/3.
	(4) Strain E830587 is a complex mosaic of segments apparently derived from group E and group A strains.
gnd	
S. enterica	Extensive recombination (see text).
E. coli	Extensive recombination, including acquisition of large parts or all of the gene from *Klebsiella* and *Citrobacter* (see text).

In Table 5, we have listed the apparent intragenic recombination events that were identified by comparative analyses of housekeeping gene sequences conducted with the
guidance of the statistical tests of nonrandom clustering of polymorphic sites mentioned
above. The lengths of segments involved in intragenic recombination ranged from 6 to
1,073 bp, but a number of the segments are in the 200-400 bp range, which, incidentally is
the size reported for recombined segments in the *phoA* alkaline phosphatase gene of *E. coli*
(DuBose *et al.*, 1988).

For the core series of strains of *S. enterica* and *E. coli*, close comparison of the trees
for *mdh* (Figure 6), *putP*, *gapA* (Figure 7), and *aceK* (not shown) reveal several cases in
which the order of branching of strains is discordant -- the topologies are different in one or
more details. Some of these are attributable to intragenic recombination events listed in
Table 5, but others apparently reflect the exchange of entire genes.

putP **and** *gapA*. Trees for these genes (Figure 7) show several differences in topology.
In the *putP* tree, *S. enterica* subspecies V clusters with the other seven subspecies; but in
the *gapA* tree, it forms a branch apart from both the other salmonellae and *E. coli*, which is

inconsistent with all other lines of evidence relating to the evolutionary genetic relationships of these organisms. Our suggestion (Nelson *et al.*, 1991*b*) that the unusual degree of divergence of *gapA* in subspecies V is a consequence of the acquisition of a segment of the gene from a source outside of both *S. enterica* and *E. coli* has since been strongly supported by the discovery of a region of almost identical sequence in *Klebsiella pneumoniae* (Nelson and Selander, in preparation).

Apart from the position of subspecies V, the topologies of the *putP* and *gapA* trees for *S. enterica* are generally similar, with subspecies I, II, IIIb, and VI showing the same relationships. But the positions of the branch leading to subspecies IV and VII and that of subspecies IIIa are reversed in the two trees; in *gapA*, subspecies IIIa is separated from subspecies I, II, IIIb, VI, IV and VII at the second node of the *S. enterica* cluster, whereas the subspecies IV and VII branch occupies a comparable position in the *putP* tree. This difference in branching order is attributable to the occurrence of a cluster of 25 unique polymorphic sites that defines a 1,073-bp segment in the central part of the *putP* sequence in strains of subspecies VII (Table 5). (The association of subspecies IV with subspecies VII remains because the *putP* sequences of strains of these two subspecies are otherwise closely similar.) This part of *putP* of subspecies VII apparently was acquired by horizontal transfer, but we have yet to identify the source. It is clear, however, that the donor was a fairly close relative of the known types of *S. enterica*, because interspersed among the 25 unique polymorphic sites in the sequence of subspecies VII are 17 other polymorphic sites that are shared with one or more of the *S. enterica* subspecies.

The total extent of diversity in DNA sequence is much less among strains of *E. coli* than among those of *S. enterica* and, consequently, relationships are harder to define; but analyses of both *put* and *gapA* have substantiated the distinctiveness of the A and B2 groups of ECOR, as originally defined by MLEE.

mdh. The only distinctive feature of the topology of the *S. enterica mdh* tree (Figure 6) is that subspecies I is more similar to subspecies VI than to subspecies II, whereas subspecies I and II cluster together in the trees for *putP* and *gapA* (Figure 7). This difference suggests a recombination event, which was detected by the nonrandom clustering of polymorphic synonymous sites in the 5' region of the gene.

Relationships among strains of *E. coli* indicated by synonymous sites of the *mdh* sequences are similar to those previously determined for *putP* and *gapA*. ECOR group A strains are closely related in all three data sets, as are group B2 strains; and group D strains occur in a single cluster with the B2 strains. The nucleotide sequences of all three genes have indicated only a weak association among strains of ECOR group B1. The only evidence of recombination in *mdh* was the presence in strain EC35 of a long segment acquired from an unidentified donor (Table 5).

aceK. In application to the distribution of the 146 polymorphic synonymous sites detected among the 16 *E. coli* sequences of *aceK*, Stephens' test identified several probable recombination events (Table 5). Near the center of the gene, there is a distinctive segment of 33 bp containing seven polymorphic sites (four of which are unique) that is present in all five group B1 strains and in group A strains EC14 and EC17 but not in group A strains EC10 and CH734. Additionally, EC14 and EC17 differ from EC10 and CH734 in the 5' two-thirds of the gene but are identical in sequence in the 3' third. Consequently, the *aceK* locus is unusual in that strains of group A do not form a single cluster. The sequence of strain E830587 is a mosaic of segments apparently derived by recombination from several sources. In the 5' third of the gene, there is a distinctive 261-bp segment that bears some resemblance to the sequence of the group E strains. Just downstream from this region, another segment is shared with group A strains EC14 and EC17; and farther along, a third sequence element is shared with strains of group E.

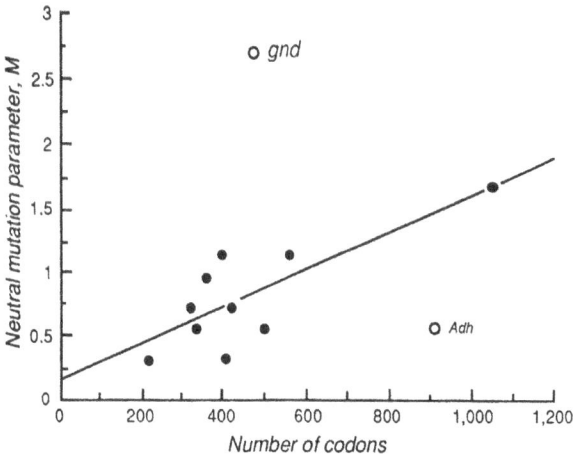

Figure 12. Relationship between the magnitude of the neutral mutation parameter (which indexes allelic diversity) and protein size (number of codons) for 12 enzyme-encoding genes in *E. coli.* (From Whittam and Ake, 1993)

In sum, our comparative studies of nucleotide sequence variation in the *mdh, putP, gapA, aceK* have identified several cases for which the most plausible explanation for the nonrandom distribution of polymorphic sites and non-concordance of gene tree topologies is horizontal gene transfer and recombination. Of these four loci, *aceK* is the most variable in sequence (Table 4), and it is our impression that recombination has been somewhat more extensive in *aceK* than in the other loci. This is consistent with evidence that *aceK* is a non-essential gene that is rarely expressed (LaPorte, 1993).

In evaluating the evolutionary significance of this evidence of recombination, two observations bearing on the problem of the frequency with which recombination events occur should be noted. First, the mere fact that segments of substantial length can be clearly identified as recombinant elements in contemporary clonal lineages indicates that episodes of intragenic recombination in the *mdh, putP, gapA,* and *aceK* have been relatively rare. Second, the simplest explanation for the fact that, for all of these genes, strains of the same subspecies of *S. enterica* invariably shared the same recombined segment indicates that these recombination events are old, antedating the time of divergence of the clonal lines within the various subspecific phylogenetic lineages.

6-Phosphogluconate Dehydrogenase (*gnd*)

This is an enzyme of the pentose phosphate pathway, which is one of two central and constitutive routes of intermediary carbohydrate metabolism in enteric bacteria, glycolysis being the other. Because the enzyme functions in an essential metabolic pathway and the *gnd* locus exhibits a moderate degree of codon bias (Table 4), the expectation from evolutionary theory is that both nucleotide and amino acid sequences should be relatively conserved. However, in both *S. enterica* and *E. coli, gnd* is an atypical enzyme locus: for its size, the enzyme is three times more variable, as assessed by enzyme electrophoresis, than expected (Figure 12).

We have sequenced the 1,335-bp *gnd* gene in 88 strains of enteric bacteria, including 36 strains of *S. enterica* and 33 strains of *E. coli* (Nelson and Selander, in preparation). For strains of *S. enterica*, evolutionary relationships deduced from *gnd* sequences are for the most part similar to those indicated by other genes and by MLEE analysis (Figure 13). Thus, subspecies V is the most divergent, followed by IIIa. However, assortative

6-Phosphogluconate dehydrogenase *(gnd)*

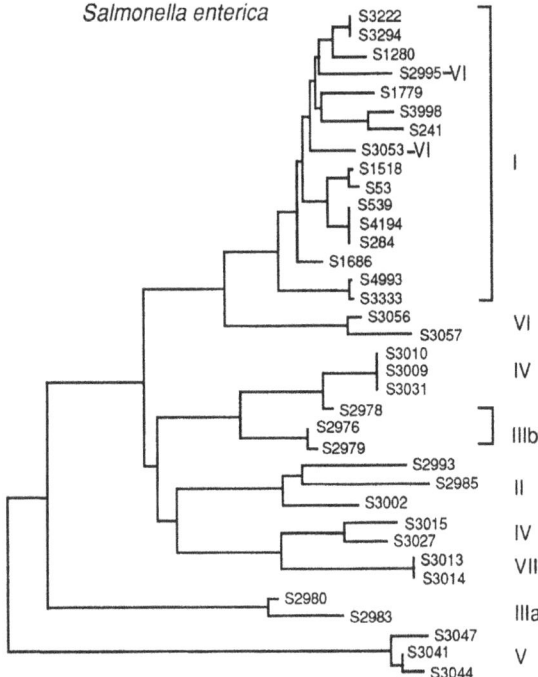

Figure 13. Evolutionary tree for the *gnd* gene in strains of *S. enterica*, based on synonymous sites. Roman numerals are subspecies designations. Note the occurrence of two subspecies VI strains within the large cluster of subspecies I strains and the presence of subspecies IV strains in two separate clusters.

recombination has occurred between strains of several subspecies (notably, I and VI), and intragenic recombination has also produced positional changes for many strains. For example, the sequences of *gnd* indicate a close relationship between the serovars Paratyphi A (S4993) and Typhi (S3333), which is inconsistent with the evidence from MLEE (Boyd *et al.*, 1993) and the sequences of the other genes we have studied.

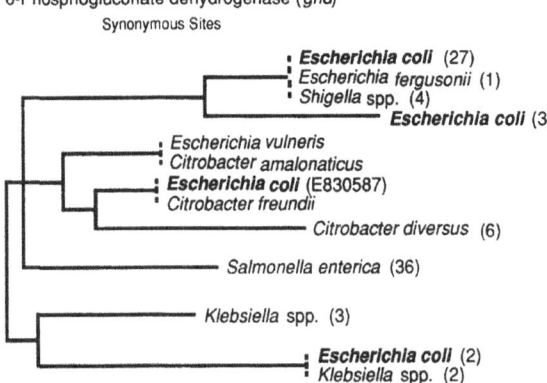

Figure 14. Neighbor-joining tree for *gnd*, based on synonymous sites. Numbers of strains sequenced are indicated in parentheses.

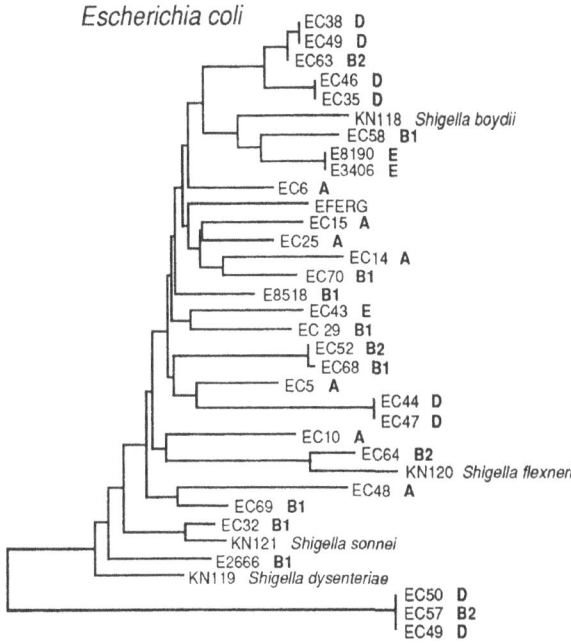

6-Phosphogluconate dehydrogenase *(gnd)*

Escherichia coli

Figure 15. Neighbor-joining tree based on synonymous nucleotide sites of *gnd* sequences of strains of *E. coli* and *Shigella* spp. ECOR group affinities of strains are indicated in boldface type.

In *E. coli*, recombination – both intragenic and assortative – has been rampant. Three strains have acquired *gnd* alleles from other species of enteric bacteria: the *gnd* allele of *E. coli* strain E830587 is similar in sequence to that of *Citrobacter freundii*, and *E. coli* strains EC16 and EC71 have *Klebsiella* spp. sequences (Figure 14). In addition, recombination in *gnd* among strains of *E. coli* has been so extensive that a tree based on this locus bears little resemblance to other gene trees or to an MLEE tree (compare Figure 15 and Figure 7). As shown in Figure 15, strains of each of the various ECOR groups are widely distributed in the *gnd* tree. Several laboratories have reported evidence of recombination in *E. coli* at the *gnd* locus (Barcak and Wolf, 1988; Dykhuizen and Green, 1991; Bisercic *et al.*, 1991).

DISCUSSION AND CONCLUSIONS

Sequence analysis of the *mdh, putP, gapA*, and *aceK* genes and several additional enzyme loci in *E. coli* (Hall and Sharp, 1992) strongly suggest that the effective rate of recombination is low for genes encoding most metabolic enzymes and other types of housekeeping proteins. Hall and Sharp (1992) could find no evidence of recombination for three enzyme genes sequenced in ECOR strains of *E. coli*, but recombination of small segments of DNA at a low rate in *phoA* (DuBose *et al.*, 1988) and the *trp* region (Milkman and Bridges, 1993) of this species has been reported. Thus, the findings for housekeeping loci in general are consistent with the results of MLEE analysis and other lines of evidence indicating that the population structure of *S. enterica* and *E. coli* is basically clonal.

In contrast, for the highly polymorphic flagellin *fliC* locus, we have shown that horizontal transfer and recombination is a major source of allelic and serovar diversity. Entire *fliC* genes and parts or all of the epitope-determining central region have been frequently exchanged within and between subspecies, and it is likely that all *fliC* alleles in the salmo-

Table 6. Proteins and polysaccharides for which there is evidence that the encoding or mediating genes are subject to horizontal transfer and intragenic or assortative recombination for which an adaptive basis has been suggested

Protein or structure (gene)	Species	References
Class 1 OMP (*porA*)	*Neisseria meningitidis*	Feavers *et al.* (1992)
M1 protein (*imm*)	*Streptococcus pyogenes*	Harbaugh *et al.* (1993)
Pilin (*pilE* and *pilS*)	*Neisseria gonorrhoeae*	Seifert *et al.* (1988); Seifert and So (1988)
Pili (*pap* and *prs* clusters)	*Escherichia coli*	Plos *et al.* (1989); Arthur *et al.* (1990); Marklund *et al.* (1992)
Capsular polysaccharide (*cap*)	*Haemophilus influenzae*	Kroll and Moxon (1990)
IgA protease (*iga*)	*Neisseria gonorrhoeae* *Neisseria meningitidis* *Haemophilus influenzae*	Halter *et al.* (1989) Lomholt *et al.* (1992); Morelli *et al.* (1994) Poulsen *et al.* (1992)
Penicillin-binding proteins	*Neisseria* spp. *Streptococcus* spp.	Spratt *et al.* (1992); Bowler *et al.* (1994) Dowson *et al.* (1990); Laible *et al.* (1991); Martin *et al.* (1992); Sibold *et al.* (1992
O antigen (*rfb* cluster)	*Salmonella enterica*	Reeves (1993); Xiang *et al.* (1993)
Flagellin (*fliC*)	*Salmonella enterica*	Smith *et al.* (1990); Li *et al.* (1994)
R-M enzymes (*hsd*)[1]	*Escherichia coli*	Sharp *et al.* (1992); Murray *et al.* (1993)
Colicin immunity proteins	*Escherichia coli*	Riley (1993)
Pyrogenic exotoxins (*speA*, *speC*)	*Streptococcus pyogenes*	Nelson *et al.* (1991a); Kapur *et al.* (1992)

[1]R-M, restriction and modification system.

nellae have a mosaic structure consisting of recombined segments derived from various sources. Because flagellin is a highly antigenic cell-surface protein that interacts directly with the external environment, recombinants may have an immediate adaptive advantage (by presenting altered cell-surface structures to host defense mechanisms and phages) and be brought to high frequency in local populations by natural selection and then transferred to other lineages (Selander and Smith, 1990; Selander *et al.*, 1991; Nelson and Selander, 1992). Similar explanations in terms of environmental adaptation may apply to a number of genes in other bacteria for which evidence of horizontal transfer and recombination is available (Table 6). But for housekeeping genes that encode polypeptides for which there would seem to be no premium on diversity in amino acid sequence per se, it is unlikely that either intragenic or assortative recombination would result in a selective advantage to the recipient cell. And the probable fate of deleterious or selectively neutral recombinants is loss from the population through purifying selection and genetic drift (Kimura, 1983).

In sum, the evidence suggests that differing modes and strengths of natural selection among loci encoding or mediating the synthesis of products of different functional type affect the realized rate of recombination within and among populations. Although the 6-phosphogluconate dehydrogenase gene (*gnd*) is a conspicuous exception to the generalization that the effective rate of recombination is low for metabolic enzyme genes, it is one that may prove the rule. The *gnd* gene is located near the genes of the *rfb* region, which mediate synthesis of the highly polymorphic polysaccharide O antigen domain of the cell-surface lipopolysaccharide. There is reason to believe that these genes are subject to strong frequency-dependent selection for the production of O antigen diversity (Reeves, 1992, 1993), and it has been suggested that the proximity of *gnd* to the *rfb* genes diminishes the chance of recombinant *gnd* alleles being lost by genetic drift (Bisercic *et al.*, 1991; Dykhuizen and Green, 1991; Murray *et al.*, 1993). We have previously noted two cases in which an O antigen serotype and a *gnd* allozyme apparently have been horizontally co-transferred between strains of *E. coli* (Selander *et al.*, 1987; Whittam *et al.*, 1993); and another example, based on more direct evidence, of the co-transfer of the *gnd* and *rfb* region genes has recently come to light. Reeves (1993) reported that the sequences of the *rfb* genes are similar in *S. enterica* serovars Paratyphi A and Typhi, and, as we noted above, the *gnd* sequences of these serovars are also closely similar. Yet these serovars show no close relationship on the basis of MLEE analysis (Selander *et al.*, 1990b) or sequences of the *gapA*, *putP*, and *mdh* genes.

In a statistical analysis of DNA sequences of two highly polymorphic human class I genes of the major histocompatibility complex (MHC) of humans, Hughes *et al.* (1993) found evidence that relatively frequent inter-allelic recombination has been a feature of the long-term evolution of the HLA-B locus but not that of the HLA-A locus. For these loci, Hughes *et al.* (1993) advanced the same hypothesis we have derived from our studies of various loci in *S. enterica* and *E. coli*: differing modes and strengths of natural selection among loci encoding products of different functional type determine the extent of realized intragenic recombination within and among populations.

From the comparative sequence analysis of genes, it is possible to obtain useful estimates of rates of recombination in natural populations, provided certain simplifying assumptions are made and sample sizes are large. Two such estimates are available for *E. coli*. Using the sequence data available for 11 genes in *E. coli*, Whittam and Ake (1993) calculated Hudson's (1987) estimator of the neutral recombination parameter, C, which is based on the variance in the number of site differences between pairs of sequences in a random sample. Their results indicated that the rate of recombination at the average locus is several times greater than the mutation rate. For individual loci, the substitution rate due to recombination was estimated to be 10^{-9} to 10^{-10} per generation. Employing a different approach, Milkman and Bridges (1990) estimated that the replacement rate is 5×10^{-12} per nucleotide site per generation, which is similar to the estimate obtained by Whittam and Ake (1993), given that the length of the average gene is about 1,000 bp. The conclusion is that mutation and recombination make more or less equal contributions to the generation of new alleles (Whittam and Ake, 1993) and that, consequently, recombination does not preclude the maintenance of clonal structure in natural populations.

In conclusion, one obvious lesson to be learned from comparative sequence analyses of bacterial genes is that recombination rates for entire genomes cannot be determined from data for single loci. It is also apparent that extrapolation from one bacterial species to another or even between different subdivisions, populations, or strains of the same species may yield erroneous conclusions. For example, whereas populations of *Neisseria meningitidis* serogroups B and C show only weak linkage disequilibrium and thus may be considered only short-term or "epidemically" clonal (Zhou and Spratt, 1992; Maynard

Table 7. Factors reducing rate of genetic exchange and recombination between differentially adapting cell lineages

1.	Decreasing frequency of contact due to ecological or geographic separation.
2	Decreasing ability to conjugate or to transform.
3.	Decreasing extent of sharing of plasmids and phage species.
4.	Differentiation in restriction and modification (R-M) systems.
5.	Increasing effect of mismatch-repair systems.
6.	Increasing probability that recombinant genes decrease fitness

Smith *et al.*, 1993), those of serogroup A exhibit strong disequilibrium and are among the more highly clonal bacterial populations yet studied (Wang *et al.*, 1992*a*; Achtman, 1994). Strains of *Bacillus subtilis* from natural populations exhibit a high degree of variation in level of competence for transformation (Roberts and Cohan, 1993); and in *E. coli*, there are individual lineages (such as that represented by strain E830587) that apparently have participated (at least as recipients) in unusually frequent episodes of genetic exchange and recombination.

Our research has yielded no evidence of genetic exchange between *S. enterica* and *E. coli*, although DNA-DNA hybridization studies and gene sequence analysis indicate a close relationship between them. Yet, surprisingly, we have identified at least three episodes of recombinational exchange between these species and *Klebsiella* and one involving *Citrobacter*. Confirming the conclusion that *S. enterica* and *E. coli* are essentially closed genetic systems (Rayssiguier *et al.*, 1989; Krawiec and Riley, 1990), Bisercic and Ochman (1993) have recently demonstrated a lack of genetic communication between them from a survey of the distribution of certain insertion sequences. For example, IS*200* occurs in both species but the nucleotide sequences are very different, suggesting divergence from a common ancestor rather than horizontal transfer. And within each species, the distribution of insertion sequence elements agrees rather well with the phylogenetic relationships of strains indicated by MLEE. A particularly instructive example of closed genetic systems is the apparent failure of *Neisseria gonorrhoeae* and *N. meningitidis* to exchange genetic material, although they share ≥98% sequence similarity in housekeeping genes, both are naturally transformable, and recombination occurs relatively frequently within populations of *N. gonorrhoeae* and those of most serogroups of *N. meningitidis* (Vázquez *et al.*, 1993). Factors that may lead to a progressive decrease in the frequency of genetic exchange and recombination in the course of the evolutionary radiation of bacterial lineages are listed in Table 7.

RELEVANCE OF POPULATION GENETICS TO IDENTIFICATION, TYPING, AND CLASSIFICATION

Explicit or implicit in the application of all schemes of bacterial typing and identification, from biotyping to fingerprinting by PCR, is the notion that it is distinctive organisms or cell lineages rather than merely individual traits that are being typed or identified and that knowledge of the state of one character provides information on the probable states of other, perhaps many, traits. And for clonally structured populations, this is indeed the case; any polymorphic genetic character that is not subject to convergence may usefully mark some or all clonal lineages or clone families because all characters in clonal populations are correlated. This is the reason, as we have shown, for the success of the Kauffmann-White scheme of serological typing of the salmonellae. Yet for bacterial populations or species approaching linkage equilibrium as the result of relatively frequent recombination, typing may amount to no more than the meaningless cataloguing of individual, unassociated traits. For this reason, knowledge of the genetic structure of populations and species is a fundamental requirement for the rational application of any scheme of typing or identification.

All typing systems seek to index the information contained in the nucleotide sequences of genes, but the ultimate level of genetic resolution is obtained only when the sequences are directly determined in full. The power of sequence data is that each nucleotide is a potential unit of information and that, over any reasonable stretch of sequence, convergence can be ruled out.

Microbiologists have interests in the nucleotide sequences of genes at several levels. When the typing of strains is the objective, the basic concern is whether two isolates are distinguishable, the quest is for stable phenotypic or genetic traits that will provide the maximum degree of strain discrimination, and the result may be little more than the qualitative sorting of isolates into categories. Employing multivariate statistics and the algorithms of phylogenetic reconstruction, taxonomists seek to estimate degrees of genomic similarity among strains for the purpose of establishing hierarchical classifications, mostly at the species level and above, that reflect phylogeny. These are the basic evolutionary reference frameworks without which biology would be little more than a collection of largely unrelated facts. But microbial taxonomists must be especially concerned with evidence bearing on the nature and frequency of horizontal genetic exchange (additive, assortative, and intragenic) (Whittam and Ake, 1993), because conventional bifurcating evolutionary trees constructed for strains or taxa among which genetic material has been transferred may not be interpretable as organismal phylogenies. Evolutionary population geneticists are concerned with the processes that mediate the mode and tempo of evolutionary change in populations. The extent and organization of genetic variation within and between populations are determined as a basis for estimating rates of mutation, recombination, and migration (dispersal), assessing the relative effects of natural selection and random genetic drift on allele frequencies, and elucidating the genetic basis for the evolution of new traits, clonal lineages, and species. Research methods involve the application of mathematical theory, large-scale sampling of multiple loci in natural populations, and the experimental manipulation of populations in the laboratory (Dykhuizen, 1990; DuBose and Hartl, 1991; Mikkola and Kurland, 1992; Korona and Levin, 1993; Zambrano et al., 1993) and in the field (Marshall et al., 1990).

It is unfortunately (but, in part, understandably) true that there is limited communication between population geneticists and microbiologists who are concerned with typing, identification, and classification, even when their research involves the same organisms and they may be employing some of the same molecular genetic protocols and computer programs for data analysis in their laboratories! This is so in large part because of the disparate ways in which genetic variation within populations is perceived and the uses to which it is put. For example, allelic polymorphism is the basic subject matter of evolutionary population genetics, but for microbiologists who must deal with the practical problems of detecting, identifying, and classifying bacterial isolates at the species level, intraspecific variation may be viewed as unwanted noise, since the objective is to find distinctive species-invariant characters. In 1980, enzyme electrophoresis was all but dismissed as a useful tool for taxonomic microbiology (Williams and Shah, 1980) precisely for the reason that it had already become (and was to remain for 25 years) the primary tool of evolutionary population geneticists studying higher organisms (Lewontin, 1974, 1991; Watt, 1994): it is a superbly sensitive, efficient method of detecting polymorphic variation at individual genetic loci within populations and species.

The inclusion of bacterial population genetics in the Granada symposium is a mark of recognition of the contribution this new discipline has already made to an understanding of the evolutionary processes that generate microbial genetic diversity and the ways in which it is organized in natural populations. We also take it as an encouraging sign that at least some microbiologists are convinced that communication among all those whose research deals with the sometimes bewilderingly complexities of microbial diversity – at whatever level and for whatever purpose – should and can be improved.

Acknowledgments

We thank S. Plock for expert technical laboratory assistance and S. W. Schaeffer and T. S. Whittam for computer programs, assistance in data analysis, and helpful discussion. Research in our laboratory is supported by grant AI-22144 from the National Institute of Health.

REFERENCES

Achtman, M. (1994) Clonal spread of serogroup A meningococci: a paradigm for the analysis of microevolution in bacteria. Mol. Microbiol. 11, 15-22.

Achtman, M. and Pluschke, G. (1986) Clonal analysis of descent and virulence among selected *Escherichia coli*. Annu. Rev. Microbiol. 40, 185-210.

Achtman, M., Mercer, A., Kusecek, B., Pohl, A., Heuzenroeder, M., Aaronson, W., Sutton, A. and Silver, R. P. (1983) Six widespread bacterial clones among *Escherichia coli* K1 isolates. Infect. Immun. 39, 315-335.

Achtman, M., Kusecek, B., Morelli, G., Eickmann, K., Wang, J., Crowe, B., Wall, R.A., Hassan-King, M., Moore, P.S. and Zollinger, W. (1992) A comparison of the variable antigens expressed by clone IV-1 and subgroup III of *Neisseria meningitidis* serogroup A. J. Infect. Dis. 165, 53-68.

Arthur, M., Arbeit, R.D., Kim, C., Beltran, P., Crowe, H., Steinbach, S., Campanelli, C., Wilson, R.A., Selander, R.K. and Goldstein, R. (1990) Restriction fragment length polymorphisms among uropathogenic *Escherichia coli* isolates: *pap*-related sequences compared with *rrn* operons. Infect. Immun. 58, 471-479.

Ashton, F.E., Ryan, J.A., Borczyk, A., Caugant, D.A., Mancino, L. and Huang, D. (1991) Emergence of a virulent clone of *Neisseria meningitidis* serotype 2a that is associated with meningococcal group C disease in Canada. J. Clin. Microbiol. 29, 2489-2493.

Barcak, G.J. and Wolf, R.E. Jr. (1988) Comparative nucleotide sequence analysis of growth-rate-regulated *gnd* alleles from natural isolates of *Escherichia coli* and from *Salmonella typhimurium* LT-2. J. Bacteriol. 170, 372-379.

Beltran, P., Musser, J.M., Helmuth, R., Farmer, J.J. III, Frerichs, W.M., Wachsmuth, I.K., Ferris, K., McWhorter, A.C., Wells, J.G., Cravioto, A. and Selander, R.K. (1988) Toward a population genetic analysis of *Salmonella*: genetic diversity and relationships among strains of serotypes *S. choleraesuis*, *S. derby*, *S. dublin*, *S. enteritids*, *S. heidelberg*, *S. infantis*, *S. newport*, and *S. typhimurium*. Proc. Natl. Acad. Sci. USA 85, 7753-7757.

Beltran, P., Plock, S.A., Smith, N.H., Whittam, T.S., Old, D.C. and Selander, R.K. (1991) Reference collection of strains of the *Salmonella typhimurium* complex from natural populations. J. Gen. Microbiol. 137, 601-606.

Bisercic, M. and Ochman, H. (1993) Natural populations of *Escherichia coli* and *Salmonella typhimurium* harbor the same classes of insertion sequences. Genetics. 133, 449-454.

Bisercic, M., Feutrier, J.Y. and Reeves, P.R. (1991) Nucleotide sequences of the *gnd* genes from nine natural isolates of *Escherichia coli*: evidence of intragenic recombination as a contributing factor in the evolution of the polymorphic *gnd* locus. J. Bacteriol. 173, 3894–3900.

Bowler, L.D., Zhang, Q.-Y., Riou, J.-Y. and Spratt, B.G. (1994) Interspecies recombination between the *penA* genes of *Neisseria meningitidis* and commensal *Neisseria* species during the emergence of penicillin resistance in *N. meningitidis*: natural events and laboratory simulation. J. Bacteriol. 176, 333-337.

Boyd, E.F., Wang, F.-S., Beltran, P., Plock, S.A., Nelson, K. and Selander, R.K. (1993) *Salmonella* reference collection B (SARB): strains of 37 serovars of subspecies I. J. Gen. Microbiol. 139, 1125-1132.

Boyd, E.F., Nelson, K., Wang, F.-S., Whittam, T.S. and Selander, R.K. (1994) Molecular genetic basis of allelic polymorphism in malate dehydrogenase (*mdh*) in natural populations of *Escherichia coli* and *Salmonella enterica*. Proc. Natl. Acad. Sci. USA 91, 1280-1284.

Brunham, R.C., Plummer, F.A. and Stephens, R.S. (1993) Bacterial antigenic variation, host immune response, and pathogen-host coevolution. Infect. Immun. 61, 2273-2276.

Caugant, D.A., Levin, B.R. and Selander, R.K. (1981) Genetic diversity and temporal variation in the *E. coli* population of a human host. Genetics 98, 467-490.

Caugant, D.A., Frøholm, L.O., Bøvre, K., Holten, E., Frasch, C.E., Mocca, L.F., Zollinger, W.D. and Selander. R.K. (1986) Intercontinental spread of a genetically distinctive complex of clones of *Neisseria meningitidis* causing epidemic disease. Proc. Natl. Acad. Sci. USA 83, 4927-4931.

Caugant, D.A., Mocca, L.F., Frasch, C.E., Frøholm, L.O., Zollinger, W.D. and Selander, R.K. (1987) Genetic structure of *Neisseria meningitidis* populations in relation to serogroup, serotype, and outer membrane protein pattern. J. Bacteriol. 169, 2781-2792.

Cohan, F.M., Roberts, M.S. and King, E.C. (1991) The potential for genetic exchange by transformation within a natural population of *Bacillus subtilis*. Evolution 45, 1383-1421.

Crosa, J.H., Brenner, D.J., Ewing, W.H. and Falkow, S. (1973) Molecular relationships among the salmonelleae. J. Bacteriol. 115, 307-315.

Denamur, E., Picard, B., Decoux, G., Denis, J.-B. and Elion, J. (1993) The absence of correlation between allozyme and *rrn* RFLP analysis indicates a high gene flow rate within human clinical *Pseudomonas aeruginosa* isolates. FEMS Microbiol. Lett. 110, 275-280.

Dowson, C.G., Hutchison, A., Woodford, N., Johnson, A.P., George, R.C. and Spratt, B.G. (1990) Penicillin-resistant viridans streptococci have obtained altered penicillin-binding protein genes from penicillin-resistant strains of *Streptococcus pneumoniae*. Proc. Natl. Acad. Sci. USA 87, 5858-5862.

DuBose, R.F. and Hartl, D.L. (1991) Evolutionary and structural constraints in the alkaline phosphatase of *Escherichia coli*, in "Evolution at the Molecular Level" (Selander, R.K, Clark, A.G. and Whittam, T.S., Ed.), pp. 58-79. Sinauer Associates, Sunderland, Massachusetts.

DuBose, R.F., Dykhuizen, D.E. and Hartl, D.L. (1988) Genetic exchange among natural isolates of bacteria: recombination within the *phoA* gene of *Escherichia coli*. Proc. Natl. Acad. Sci. USA 85, 7036-7040.

Dykhuizen, D.E. (1990) Experimental studies of natural selection in bacteria. Annu. Rev. Ecol. Syst. 21, 373-398.

Dykhuizen, D.E. and Green, L. (1991) Recombination in *Escherichia coli* and the definition of biological species. J. Bacteriol. 173, 7257-7268.

Eardly, B.D., Materon, L.A., Smith, N.H., Johnson, D.A., Rumbaugh, M.D. and Selander, R.K. (1990) Genetic structure of natural populations of the nitrogen-fixing bacterium *Rhizobium meliloti*. Appl. Environ. Microbiol. 56, 187-194.

Ewing, W.H. (1986) "Edwards and Ewing's Identification of Enterobacteriaceae, 4th Edition." Elsevier, New York.

Feavers, I.M., Heath, A.B., Bygraves, J.A. and Maiden, M.C.J. (1992) Role of horizontal genetic exchange in the antigenic variation of the class 1 outer membrane protein of *Neisseria meningitidis*. Mol. Microbiol. 6, 489-495.

Grewal, N. and Salunke, D.M. (1993) The antigenic domain of flagellin from *S. paratyphi* shares a structural fold with subtilisin. FEBS Lett. 322, 111-114.

Hall, B.G. and Sharp, P.M. (1992) Molecular population genetics of *Escherichia coli*: DNA sequence diversity at the *celC*, *crr*, and *gutB* loci of natural isolates. Mol. Biol. Evol. 9, 654-665.

Halter, R., Pohlner, J. and Meyer, T.F. (1989) Mosaic-like organization of IgA protease genes in *Neisseria gonorrhoeae* generated by horizontal genetic exchange *in vivo*. EMBO J. 9, 2737-2744.

Harbaugh, M.P., Podbielski, A., Hügl, S. and Cleary, P.P. (1993) Nucleotide substitutions and small-scale insertion produce size and antigenic variation in group A streptococcal M1 protein. Mol. Microbiol. 8, 981-991.

Herzer, P.J., Inouye, S., Inouye, M. and Whittam, T.S. (1990) Phylogenetic distribution of branched RNA-linked multicopy single-stranded DNA among natural isolates of *Escherichia coli*. J. Bacteriol. 172, 6175- 6181.

Hudson, R.R. (1987) Estimating the recombination parameter of a finite population model without selection. Genet. Res. Camb. 50, 245-250.

Hughes, A.L., Hughes, M.K. and Watkins, D.I. (1993) Contrasting roles of interallelic recombination at the HLA-A and HLA-B loci. Genetics 133, 669-680.

Iino, T. (1977) Genetics of structure and function of bacterial flagella. Annu. Rev. Genet. 11,161-182.

Istock, C.A., Duncan, K.E., Ferguson, N. and Zhou, X. (1992) Sexuality in a natural population of bacteria – *Bacillus subtilis* challenges the clonal paradigm. Mol. Ecol. 1, 95-103.

Joys, T.M. (1988) The flagellar filament protein. Can. J. Microbiol. 34, 452-458.

Jukes, T.H. and Cantor, C.R. (1969) Evolution of protein molecules, in "Mammalian Protein Metabolism" (Munro, H. N., Ed.), pp. 21-132. Academic Press. New York.

Kapur, V., Nelson, K., Schlievert, P.M., Selander, R.K. and Musser, J.M. (1992) Molecular population genetic evidence of horizontal spread of two alleles of the pyrogenic exotoxin C gene (*speC*) among pathogenic clones of *Streptococcus pyogenes*. Infect. Immun. 60, 3513-3517.

Kim, J., Rohlf, F.J. and Sokal, R.R. (1993) The accuracy of phylogenetic estimation using the neighbor-joining method. Evolution 47, 471-486.

Kimura, M. (1983) "The Neutral Theory of Molecular Evolution." Cambridge University Press, London.

Koch, A.L. (1974) The pertinence of the periodic selection phenomenon to prokaryotic evolution. Genetics 77, 127-142.

Korona, R. and Levin, B.R. (1993) Phage-mediated selection and the evolution and maintenance of restriction-modification. Evolution 47, 556-575.

Krawiec, S. and Riley, M. (1990) Organization of the bacterial chromosome. Microbiol. Rev. 54, 502-539.

Kroll, J.S. and Moxon, E.R. (1990) Capsulation in distantly related strains of *Haemophilus influenzae* type b: genetic drift and gene transfer at the capsulation locus. J. Bacteriol. 172, 1374-1379.

Laible, G., Spratt, B.G. and Hakenbeck, R. (1991) Interspecies recombinational events during the evolution of altered PBP 2x genes in penicillin-resistant clinical isolates of *Streptococcus pneumoniae*. Mol. Microbiol. 5, 1993-2002.

LaPorte, D.C. (1993) The isocitrate dehydrogenase phosphorylation cycle: regulation and enzymology. J. Cell. Biochem. 51, 14-18.

Le Minor, L. (1984) Genus III. Salmonella *Lignières 1900, 389*, in "Bergey's Manual of Systematic Bacteriology" Vol. 1. (Krieg, N.R. and Holt, J.G., Ed.), pp. 427-458. Williams & Wilkins, Baltimore.

Le Minor, L. and Popoff, M.Y. (1987) Designation of *Salmonella enterica* sp. nov., nom. rev., as the type and only species of the genus *Salmonella*. Int. J. Syst. Bacteriol. 37, 465-468.

Le Minor, L., Popoff, M.Y., Laurent, B. and Hermant, D. (1986) Individualisation d'une septième sous-espèce de *Salmonella*: *S. choleraesuis* subsp. *indica* subsp. nov. Ann. Inst. Pasteur/Microbiol. 137B, 211-217.

Levin, B.R. (1981) Periodic selection, infectious gene exchange and the genetic structure of *E. coli* populations. Genetics 99, 1-23.

Lewontin, R.C. (1974) "The Genetic Basis of Evolutionary Change." Columbia University Press, New York.

Lewontin, R.C. (1991) Electrophoresis in the development of evolutionary genetics: milestone or millstone? Genetics 128, 657-662.

Li, J., Smith, N.H., Nelson, K., Crichton, P.B., Old, D.C., Whittam, T.S. and Selander, R.K. (1993) Evolutionary origin and radiation of the avian-adapted non-motile salmonellae. J. Med. Microbiol. 38, 129-139.

Li, J., Nelson, K., McWhorter-Murlin, A.C., Whittam, T.S. and Selander, R.K. (1994) Recombinational basis of serovar diversity in *Salmonella enterica*. Proc. Natl. Acad. Sci. USA 91, 2552-2556.

Lomholt, H., Poulsen, K., Caugant, D.A. and Kilian, M. (1992) Molecular polymorphism and epidemiology of *Neisseria meningitidis* immunoglobulin A1 proteases. Proc. Natl. Acad. Sci. USA 89, 2120-2124.

Macnab, R.M. (1987) Flagella, in "*Escherichia coli* and *Salmonella typhimurium*: Cellular and Molecular Biology" Vol. 1. (Neidhardt, F.C., Ingraham, J.L., Low, K.B., Magasanik, B., Schaechter, M. and Umbarger, H.E., Ed.), pp. 70-83. American Society for Microbiology, Washington, D.C.

Macnab, R.M. (1992) Genetics and biogenesis of bacterial flagella. Annu. Rev. Genet. 26,131-158.

Maiden, M.C.J. (1993) Population genetics of a transformable bacterium: the influence of horizontal genetic exchange on the biology of *Neisseria meningitidis*. FEMS Microbiol. Lett. 112, 243-250.

Marklund, B.-I., Tennent, J.M., Garcia, E., Hamers, A., Bâga, M., Lindberg, F., Gaastra, W. and Normark, S. (1992) Horizontal gene transfer of the *Escherichia coli pap* and *prs* pili operons as a mechanism for the development of tissue-specific adhesive properties. Mol. Microbiol. 6, 2225-2242.

Marshall, B., Petroski, D. and Levy, S.D. (1990) Inter- and intraspecies spread of *Escherichia coli* in a farm environment in the absence of antibiotic usage. Proc. Natl. Acad. Sci. USA 87, 6609-6613.

Martin, C., Sibold, C. and Hakenbeck, R. (1992) Relatedness of penicillin-binding protein 1a genes from different clones of penicillin-resistant *Streptococcus pneumoniae* isolated in South Africa and Spain. EMBO J. 11, 3831-3836.

Masten, B.J. and Joys, T.M. (1993) Molecular analyses of the *Salmonella g...* flagellar antigen complex. J. Bacteriol. 175, 5359-5365.

Maynard Smith, J. (1992) Analyzing the mosaic structure of genes. J. Mol. Evol. 34, 126-129.

Maynard Smith, J., Dowson, C. G. and Spratt, B. G. (1991) Localized sex in bacteria. Nature 349, 29-31.

Maynard Smith, J., Smith, N. H., O'Rourke, M. and Spratt, B. G. (1993) How clonal are bacteria? Proc. Natl. Acad. Sci. USA 90, 4384-4388.

Mikkola, R. and Kurland, C.G. (1992) Selection of laboratory wild-type phenotype from natural isolates of *Escherichia coli* in chemostats. Mol. Biol. Evol. 9, 394-402.

Milkman, R. and Bridges, M.M. (1990) Molecular evolution of the *Escherichia coli* chromosome. III. Clonal frames. Genetics 126, 505-517.

Milkman, R. and Bridges, M.M. (1993) Molecular evolution of the *Escherichia coli* chromosome. IV. Sequence comparisons. Genetics 133, 455-468.

Morelli, G., del Valle, J., Lammel, C.J., Pohlner, J., Müller, K., Blake, M., Brooks, G.F., Meyer, T.F., Koumaré, B., Brieske, N. and Achtman, M. (1994) Immunogenicity and evolutionary variability of epitopes within IgA1 protease from serogroup A *Neisseria meningitidis*. Mol. Microbiol. 11, 175-187.

Murray, N.E., Daniel, A.S., Cowan, G.M. and Sharp, P.M. (1993) Conservation of motifs within the unusually variable polypeptide sequences of type I restriction and modification enzymes. Mol. Microbiol. 9, 133-143.

Musser, J.M. and Kapur, V. (1994) Molecular population genetics of *Haemophilus influenzae*, in "Development and Clinical Uses of *Haemophilus* b Vaccines" (Ellis, R.W. and Granoff, D.M., Ed.), pp. 181-208. Marcel Dekker, New York.

Musser, J.M., Granoff, D.M., Pattison, P.E. and Selander, R.K. (1985) A population genetic framework for the study of invasive diseases caused by serotype b strains of *Haemophilus influenzae*. Proc. Natl. Acad. Sci. USA 82, 5078-5082.

Musser, J.M., Barenkamp, S.J., Granoff, D.M. and Selander, R.K. (1986) Genetic relationships of serologically nontypable and serotype b strains of *Haemophilus influenzae*. Infect. Immun. 52, 183-191.

Musser, J.M., Bemis, D.A., Ishikawa, H. and Selander, R.K. (1987) Clonal diversity and host distribution in *Bordetella bronchiseptica*. J. Bacteriol. 169, 2793-2803.

Musser, J.M., Kroll, J.S., Moxon, E.R. and Selander, R.K. (1988*a*) Clonal population structure of encapsulated *Haemophilus influenzae*. Infect. Immun. 56, 1837-1845.

Musser, J.M., Kroll, J.S., Moxon, E.R. and Selander, R.K. (1988*b*) Evolutionary genetics of the encapsulated strains of *Haemophilus influenzae*. Proc. Natl. Acad. Sci. USA 85, 7758-7762.

Musser, J.M., Kroll, J.S., Granoff, D.M., Moxon, E.R., Brodeur, B.R., Campos, J., Dabernat, H., Frederiksen, W., Hamel, J., Hammond, G., Høiby, E.A., Jonsdottir, K.E., Kabeer, M., Kallings, I., Khan, W.N., Kilian, M., Knowles, K., Koornhof, H.J., Law, B., Li, K.I., Montgomery, J., Pattison, P.E., Piffaretti, J.-C., Takala, A.K., Thong, M.L, Wall, R.A., Ward, J.I. and Selander, R.K. (1990) Global genetic structure and molecular epidemiology of encapsulated *Haemophilus influenzae*. Rev. Infect. Dis. 12, 75–111.

Nelson, K. and Selander, R.K. (1992) Evolutionary genetics of the proline permease gene (*putP*) and the control region of the proline utilization operon in populations of *Salmonella* and *Escherichia coli*. J. Bacteriol. 174, 6886-6895.

Nelson, K. and Selander, R.K. (1994) Analysis of genetic variation by polymerase chain reaction-based nucleotide sequencing. Methods in Enzymology 235, 174-183.

Nelson, K., Schlievert, P.M., Selander, R.K. and J.M. Musser. (1991a) Characterization and clonal distribution of four alleles of the *speA* gene encoding pyrogenic exotoxin A (scarlet fever toxin) in *Streptococcus pyogenes*. J. Exp. Med. 174, 1271-1274.

Nelson, K., Whittam, T.S. and Selander, R.K. (1991b) Nucleotide polymorphism and evolution in the glyceraldehyde–3–phosphate dehydrogenase gene (*gapA*) in natural populations of *Salmonella* and *Escherichia coli*. Proc. Natl. Acad. Sci. USA 88, 6667–6671.

Newton, S.M.C., Wasley, R.D., Wilson, A., Rosenberg, L.T., Miller, J.F. and Stocker, B.A.D. (1991) Segment IV of a *Salmonella* flagellin gene specifies flagellar antigen epitopes. Mol. Microbiol. 5, 419-425.

Ochman, H., and Selander, R.K. (1984a) Evidence for clonal population structure in *Escherichia coli*. Proc. Natl. Acad. Sci. USA 81, 198-201.

Ochman, H. and Selander, R.K. (1984b) Standard reference strains of *Escherichia coli* from natural populations. J. Bacteriol. 157, 690-693.

Ochman, H., Whittam, T.S., Caugant, D.A. and Selander, R.K. (1983) Enzyme polymorphism and genetic population structure in *Escherichia coli* and *Shigella*. J. Gen. Microbiol. 129, 2715-2726.

Okazaki, N., Matsuo, S., Saito, K., Tominaga, A. and Enomoto, M. (1993) Conversion of the *Salmonella* phase 1 flagellin gene *fliC* to the phase 2 gene *fljB* on the *Escherichia coli* K-12 chromosome. J. Bacteriol. 175, 758-766.

Old, D.C. (1992) Nomenclature of *Salmonella*. Med. Microbiol. 37, 261-263.

O'Rourke, M. and Stevens, E. (1993) Genetic structure of *Neisseria gonorrhoeae* populations: a non-clonal pathogen. J. Gen. Microbiol. 139, 2603-2611.

Ørskov, F., Ørskov, I., Evans, D.J., Sack, R.B. and Wadstrom, T. (1976) Special *Escherichia coli* serotypes among enterotoxigenic strains from diarrhoea in adults and children. Med. Microbiol. Immunol. 162, 73-80.

Ørskov, I., Ørskov, F., Jann, B. and Jann, K. (1977) Serology, chemistry, and genetics of O and K antigens of *Escherichia coli*. Bacteriol. Rev. 41, 667-710.

Plos, K., Hull, S.I., Hull, R.A., Levin, B.R., Ørskov, I., Ørskov, F. and Svandborg-Edén, C. (1989) Distribution of the P-associated-pilus (*pap*) region among *Escherichia coli* from natural sources: evidence for horizontal gene transfer. Infect. Immun. 57, 1604-1611.

Popoff, M.Y. and Le Minor, L. (1992) "Antigenic Formulas of the *Salmonella* Serovars," 6th ed., WHO Collaborating Centre for Reference and Research on *Salmonella*, Institut Pasteur, Paris.

Poulsen, K., Reinholdt, J. and Kilian, M. (1992) A comparative genetic study of serologically distinct *Haemophilus influenzae* type 1 immunoglobulin A1 proteases. J. Bacteriol. 174, 2913-2921

Quentin, R., Goudeau, A., Wallace, R.J. Jr., Smith, A.L., Selander, R.K. and Musser, J.M. (1990) Urogenital, maternal and neonatal isolates of *Haemophilus influenzae*: identification of unusually virulent serologically non-typable clone families and evidence for a new *Haemophilus* species. J. Gen. Microbiol. 136, 1203-1209.

Rayssiguier, C., Thaler, D.S. and Radman, M. (1989) The barrier between *Escherichia coli* and *Salmonella typhimurium* is disrupted in mismatch-repair mutants. Nature 342, 396-401.

Reeves, P.R. (1992) Variation in O-antigens, niche-specific selection and bacterial populations. FEMS Microbiol. Lett. 100, 509-516.

Reeves, P.R. (1993) Evolution of *Salmonella* O antigen variation by interspecific gene transfer on a large scale. Trends Genet. 9:17-22.

Reeves, M.W., Evins, G.M., Heiba, A.A., Plikaytis, B.D. and Farmer J.J.III. (1989) Clonal nature of *Salmonella typhi* and its genetic relatedness to other salmonellae as shown by multilocus enzyme electrophoresis, and proposal of *Salmonella bongori* comb. nov. J. Clin. Microbiol. 27, 311-320.

Riley, M.A. (1993) Positive selection for colicin diversity in bacteria. Mol. Biol. Evol. 10, 1048-1059.

Roberts, M.S. and Cohan, F.M. (1993) The effect of DNA sequence divergence on sexual isolation in *Bacillus*. Genetics 134, 401-408.

Sacchi, C.T., Pessoa, L.L., Ramos, S.R., Milagres, L.G., Camargo, M.C.C., Hidalgo, N.T.R, Melles, C.E.A., Caugant, D.A. and Frasch, C.F. (1992a) Ongoing group B *Neisseria meningitidis* epidemic in São Paulo, Brazil, due to increased prevalence of a single clone of the ET-5 complex. J. Clin. Microbiol. 30, 1734-1738.

Sacchi, C.T., Zanella, R.C., Caugant, D.A., Frasch, C.E., Hidalgo, N.T., Milagres, L.G., Pessoa, L.L., Ramos, S.R., Camargo, M.C.C. and Melles, C.E.A. (1992*b*) Emergence of a new clone of serogroup C *Neisseria meningitidis* in São Paulo, Brazil. J. Clin. Microbiol. 30, 1282-1286.

Saitou, N. and Nei, M. (1987) The neighbor-joining method: a new method for reconstructing phylogenetic trees. Mol. Biol. Evol. 4, 406-425.

Sawyer, S. (1989) Statistical tests for detecting gene conversion. Mol. Biol. Evol. 6, 526-538.

Seifert, H.S. and So, M. (1988) Genetic mechanisms of bacterial antigenic variation. Microbiol. Rev. 52, 327-336.

Seifert, H.S., Ajioka, R.S., Marchal, C., Sparling, P.F. and So, M. (1988) DNA transformation leads to pilin antigenic variation in *Neisseria gonorrhoeae*. Nature 336, 392-395.

Selander, R.K. and Levin, B.R. (1980) Genetic diversity and structure in *Escherichia coli* populations. Science 210, 545-547.

Selander, R.K. and Musser, J.M. (1990) Population genetics of bacterial pathogenesis, in "Molecular Basis of Bacterial Pathogenesis" (Iglewski, B.H. and Clark, V.L., Ed.), pp. 11-36. Academic Press, San Diego.

Selander, R.K. and Smith, N.H. (1990) Molecular population genetics of *Salmonella*. Rev. Med. Microbiol. 1, 219–228.

Selander, R.K., McKinney, R.M., Whittam, T.S., Bibb, W.F., Brenner, D.J., Nolte, F.S. and Pattison, P.E. (1985) Genetic structure of populations of *Legionella pneumophila*. J. Bacteriol. 163, 1021-1037.

Selander, R.K., Caugant, D.A., Ochman, H., Musser, J.M., Gilmour, M.N. and Whittam, T.S. (1986) Methods of multilocus enzyme electrophoresis for bacterial population genetics and systematics. Appl. Environ. Microbiol. 51, 873-884.

Selander, R.K., Caugant, D.A. and Whittam, T.S. (1987) Genetic structure and variation in natural populations of *Escherichia coli*, in "*Escherichia coli* and *Salmonella typhimurium*: Cellular and Molecular Biology" Vol. 2. (Neidhardt, F.C., Ingraham, J.L., Low, K.B., Magasanik, B., Schaechter, M. and Umbarger, H.E., Ed.), pp. 1625-1648. American Society for Microbiology, Washington, D.C.

Selander, R.K., Beltran, P., Smith, N.H., Barker, R.M., Crichton, P.B., Old, D.C., Musser, J.M. and T.S. Whittam. (1990*a*) Genetic population structure, clonal phylogeny, and pathogenicity of *Salmonella paratyphi* B. Infect. Immun. 58, 1891-1901.

Selander, R.K., Beltran, P., Smith, N.H., Helmuth, R., Rubin, F.A., Kopecko, D.J., Ferris, K., Tall, B.D., Cravioto, A. and Musser, J.M. (1990*b*) Evolutionary genetic relationships of clones of *Salmonella* serovars that cause human typhoid and other enteric fevers. Infect. Immun. 58, 2262-2275.

Selander, R.K., Beltran, P. and Smith, N.H. (1991) Evolutionary genetics of *Salmonella*, in "Evolution at the Molecular Level" (Selander, R.K., Clark, A.G. and Whittam, T.S., Ed.), pp. 25–57. Sinauer Associates, Sunderland, Massachusetts.

Selander, R.K., Smith, N.H., Li, J., Beltran, P., Ferris, K.E., Kopecko, D.J. and Rubin, F.A. (1992) Molecular evolutionary genetics of the cattle-adapted serovar *Salmonella dublin*. J. Bacteriol. 174, 3587-3592.

Sharp, P.M. and Li, W.-H. (1987) The rate of synonymous substitution in enterobacterial genes is inversely related to codon usage bias. Mol. Biol. Evol. 4, 222-230.

Sharp, P.M., Kelleher, J.E., Daniel, A.S., Cowan, G.M. and Murray, N.E. (1992) Roles of selection and recombination in the evolution of type 1 restriction-modification systems in enterobacteria. Proc. Natl. Acad. Sci. USA 89, 9836-9840.

Sibold, C., Wang, J., Henrichsen, J. and Hackenbeck, R. (1992) Genetic relationships of penicillin-susceptible and -resistant *Streptococcus pneumoniae* isolated on different continents. Infect. Immun. 60, 4119-4126.

Smith, N.H. and Selander, R.K. (1990) Sequence invariance of the antigen-coding central region of the phase 1 flagellar filament gene (*fliC*) among strains of *Salmonella typhimurium*. J. Bacteriol. 172, 603-609.

Smith, N.H. and Selander, R.K. (1991) Molecular genetic basis for complex flagellar antigen expression in a triphasic serovar of *Salmonella*. Proc. Natl. Acad. Sci. USA. 88, 956-960.

Smith, N.H., Beltran, P. and Selander, R.K. (1990) Recombination of *Salmonella* phase 1 flagellin genes generates new serovars. J. Bacteriol. 172, 2209-2216.

Souza, V., Nguyen, T.T., Hudson, R.R., Piñero, D. and Lenski, R.E. (1992) Hierarchical analysis of linkage disequilibrium in *Rhizobium* populations: evidence for sex? Proc. Natl. Acad. Sci. USA 89, 8389-8393.

Spratt, B.G., Bowler, L.D., Zhang, Q.-Y., Zhou, J. and Maynard Smith, J. (1992) Role of interspecies transfer of chromosomal genes in the evolution of penicillin resistance in pathogenic and commensal *Neisseria* species. J. Mol. Evol. 34, 115-125.

Stephens, J.C. (1985) Statistical methods of DNA sequence analysis: detection of intragenic recombination or gene conversion. Mol. Biol. Evol. 2, 539-556.

Tibayrenc, M., Kjellberg, F. and Ayala, F.J. (1990) A clonal theory of parasitic protozoa: the population structures of *Entamoeba, Giardia, Leishmania, Naegleria, Plasmodium, Trichomonas*, and *Trypanosoma* and their medical and taxonomical consequences. Proc. Natl. Acad. Sci. USA 87, 2414-2418.

Tibayrenc, M., Neubauer, K., Barnabé, C., Guerrini, F., Skarecky, D. and Ayala, F.J. (1993) Genetic characterization of six parasitic protozoa: parity between random-primer DNA typing and multilocus enzyme electrophoresis. Proc. Natl. Acad. Sci. USA 90, 1335-1339.

Valdés, A.M. and Piñero, D. (1992) Phylogenetic estimation of plasmid exchange in bacteria. Evolution 46, 641-656.

Vázquez, J.A., de la Fuente, L., Berron, S., O'Rourke, M., Smith, N.H., Zhou, J. and Spratt, B.G. (1993) Ecological separation and genetic isolation of *Neisseria gonorrhoeae* and *Neisseria meningitidis*. Current Biol. 3, 567-572.

Wang, J.-F., Caugant, D.A., Li, X., Hu, X., Poolman, J.T., Crowe, B.A. and Achtman, M. (1992a) Clonal and antigenic analysis of serogroup A *Neisseria meningitidis* with particular reference to epidemiological features of epidemic meningitis in the People's Republic of China. Infect. Immun. 60, 5267-5282.

Wang, J.-F, Caugant, D.A., Morelli, G., Koumaré, B. and Achtman, M. (1992b) Antigenic and epidemiologic properties of the ET-37 complex of *Neisseria meningitidis*. J. Infect. Dis. 167, 1320-1329.

Watt, W.B. (1994) Allozymes in evolutionary genetics: self-imposed burden or extraordinary tool? Genetics 136, 11-16.

Wells, J.G., Davis, B.R., Wachsmuth, I.K., Riley, L.W., Remis, R.S., Sokolow, R. and Morris, G.K. (1983) Laboratory investigation of hemorrhagic colitis outbreaks associated with a rare *Escherichia coli* serotype. J. Clin. Microbiol. 18, 512-520.

Whittam, T.S. (1989) Clonal dynamics of *Escherichia coli* in its natural habitat. Antonie von Leeuwenhoek 55, 23-32.

Whittam, T.S. (1992) Sex in the soil. Current Biol. 2, 676-678.

Whittam, T.S. and Ake, S.E. (1993) Genetic polymorphisms and recombination in natural populations of *Escherichia coli*, in "Mechanisms of Molecular Evolution" (Takahata, N. and Clark, A.G., Ed.), pp. 223-245. Sinauer Associates, Sunderland, Massachusetts.

Whittam, T.S. and Wilson, R.A. (1988) Genetic relationships among pathogenic *Escherichia coli* of serogroup O157. Infect. Immun. 56, 2467-2472.

Whittam, T.S., Ochman, H. and Selander, R.K. (1983a) Multilocus genetic structure in natural populations of *Escherichia coli*. Proc. Natl. Acad. Sci. USA 80, 1751-1755.

Whittam, T.S., Ochman, H. and Selander, R.K. (1983b) Geographic components of linkage disequilibrium in natural populations of *Escherichia coli*. Mol. Biol. Evol. 1, 67-83.

Whittam, T.S., Wachsmuth, I.K. and Wilson, R.A. (1988) Genetic evidence of clonal descent of *Escherichia coli* O157:H7 associated with hemorrhagic colitis and hemolytic uremic syndrome. J. Infect. Dis. 157, 1124-1133.

Whittam, T.S., Wolfe, M.L., Wachsmuth, I.K., Ørskov, F., Ørskov, I. and Wilson, R.A. (1993) Clonal relationships among *Escherichia coli* strains that cause hemorrhagic colitis and infantile diarrhea. Infect. Immun. 61, 1619-1629.

Williams, R.A.D. and Shah, H.N. (1980) Enzyme patterns in bacterial classification and identification, in "Microbiological Classification and Identification" (Goodfellow, M. and Board, R.G., Ed.), pp. 299-318. Academic Press, London.

Wilson, D.R. and Beveridge, T.J. (1993) Bacterial flagellar filaments and their component flagellins. Can. J. Microbiol. 39, 451-472.

Xiang, S.-H., Haase, A.M. and Reeves, P.R. (1993) Variation of the *rfb* gene clusters in *Salmonella enterica*. J. Bacteriol. 175, 4877-4884.

Zambrano, M.M., Siegele, D.A., Almirón, M., Tormo, A. and Kolter, R. (1993) Microbial competition: *Escherichia coli* mutants that take over stationary phase cultures. Nature 129, 1757-1760.

Zhou, J. and Spratt, B.G. (1992) Sequence diversity within the *argF*, *fbp* and *recA* genes of natural isolates of *Neisseria meningitidis*: interspecific recombination within the *argF* gene. Mol. Microbiol. 6, 2135-2146.

IDENTIFICATION AND TYPING OF BACTERIA BY PROTEIN ELECTROPHORESIS

Karel Kersters, Bruno Pot, Dirk Dewettinck, Urbain Torck,
Marc Vancanneyt, Luc Vauterin and Peter Vandamme

Laboratorium voor Microbiologie
Universiteit Gent
K.L. Ledeganckstraat 35, B-9000 Gent, Belgium

INTRODUCTION

Polyacrylamide gel electrophoresis (PAGE) of bacterial whole-cell proteins yields complex banding patterns, which can be considered as highly specific fingerprints of the strains investigated. These protein electrophoregrams are highly reproducible, provided that standardised techniques are used and that strains are cultivated in similar conditions. One-dimensional (1D) protein electrophoresis proved to be a powerful, relatively simple and inexpensive method, which has been used for over 20 years for classification, identification and typing of diverse bacterial taxa (for reviews see: Garber and Rippon, 1968; Kersters and De Ley, 1980; Jackman, 1985, 1987; Kersters, 1985; Costas, 1992; Vauterin et al., 1993; Pot et al., 1994). The complex banding patterns can be quantified and compared by computer-assisted techniques. Databases for automated identification of bacterial strains can be constructed. One of the most popular electrophoretic techniques used is 1D-electrophoresis of whole-cell proteins (or cell-envelope proteins) in the presence of the denaturing anionic detergent sodium dodecylsulphate (SDS) (Laemmli, 1970). The protein electrophoretic technique is only applicable to strains which can be isolated as viable cultures; non-culturable or dormant microbial cells can thus not be analysed by the SDS-PAGE method.

The rationale for the application of 1D-electrophoresis of cellular proteins in microbial systematics is the excellent correlation which is usually found between the groupings of strains obtained by PAGE and the intrageneric genomic relationships estimated by DNA-DNA hybridizations (Owen and Jackman, 1982; Willems et al., 1989; Vauterin et al., 1990, 1992). This is illustrated in Figure 1 where protein fingerprints of reference strains of the three species of the genus *Hydrogenophaga* are shown together with the DNA relatedness values of these strains versus the type strains of *Hydrogenophaga pseudoflava* LMG 5945 and *Hydrogenophaga palleronii* Stanier 362. The differences between the protein patterns of the strains of the three species are much more significant than between the profiles of strains of the same species. Visually the pattern of *Hydrogenophaga flava* DSM 619T displays more similarities to the patterns of *H. pseudoflava* than to those of *H. palleronii*, which is in agreement with the relatively high degree of DNA-binding (58% D) between *H. flava* DSM 619^T and *H. pseudoflava* LMG 5945^T.

The electrophoretic separation of cellular proteins is a sensitive technique, mainly providing information on the similarity of strains within the same species or subspecies. Depending on the variability of the protein patterns within a given taxon, individual strains often can be recognised by small, but reproducible differences in part of their protein fingerprints.

In this paper we shall focus on improvements in the computer-assisted analysis of protein electrophoretic patterns, and some applications of the SDS-PAGE fingerprinting technique in bacterial systematics will be discussed.

ELECTROPHORESIS OF BACTERIAL PROTEINS

The application of protein electrophoresis for classification, identification and typing of bacteria requires very standardised and reproducible experimental conditions (Kersters and De Ley, 1975; Feltham and Sneath, 1979; Jackman, 1987; Costas, 1992; Pot *et al.*, 1994). Polyacrylamide gel electrophoresis in the presence of the anionic detergent SDS separates proteins by the more conserved parameter of molecular weight, and detects broader taxonomic relationships at the (sub)species level, than those electrophoretic techniques which are based on charge parameters. SDS-PAGE combines high-resolution with good reproducibility, and thus offers advantages over the lower resolving non-denaturing gel systems and the lower reproducibility of, for example, gradient PAGE and isoelectric focusing (Jackman, 1987).

Experimental Procedures

The various experimental steps for obtaining protein patterns of bacterial strains are in chronological order: cultivation of bacteria, preparation of cell-free extracts, preparation of polyacrylamide gels, sample application, electrophoresis, staining and destaining of gels and drying of gels. The methodology will not be described here in detail; we refer to the literature (e.g. Jackman, 1987; Costas, 1992; Vauterin *et al.*, 1993; Pot *et al.*, 1994). In order to attain high reproducibility levels it is of course necessary to monitor all experimental steps in minute detail.

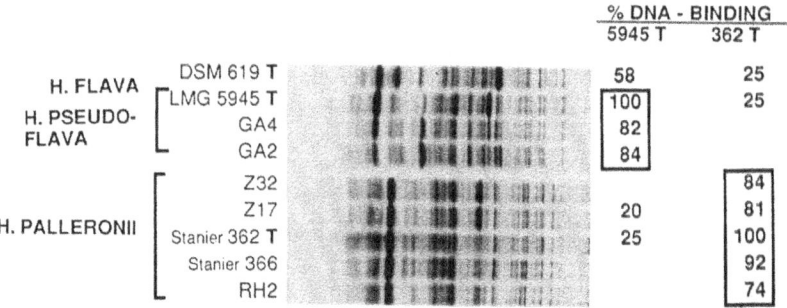

Figure 1. Correlation between SDS-PAGE protein patterns and DNA-DNA hybridizations. Percentage DNA-binding of various *Hydrogenophaga* strains versus the type strains (indicated by T) of *Hydrogenophaga pseudoflava* and *Hydrogenophaga palleronii* are indicated. Data from Willems *et al.* (1989).

Protein extraction from Gram-negative bacteria is usually done by simple chemical means such as SDS and heat, but Gram-positive bacteria can pose much greater problems and may require mechanical cell breakage (e.g. grinding with glass beads or ultrasonic treatment). The protein samples can usually be stored for further reference for at least one to two years at – 80°C. All discontinuous 1D-SDS-PAGE systems used in bacterial systematics are based on the method first described by Laemmli (1970). No two laboratories use exactly the same experimental procedures, making comparisons of results between two laboratories extremely difficult. An interlaboratory comparison of both protein patterns and analytical systems was carried out using the same set of *Campylobacter* strains (Costas *et al.*, 1990c).

Visual Comparison of Whole-cell Protein Patterns

Visual comparison is by far the most frequently used method for the interpretation of similarities or differences between bacterial protein electrophoretic fingerprints. It is always advisable to compare results of computer analysis (see further) with the original gel electrophoretic protein profiles or their photographs, even when computer-assisted analysis is thought to be the most objective method for interpretation. The human eye is a very sensitive and fast instrument to detect similarities or small differences between 2 to 5 complex banding patterns. However, the human brain cannot easily remember large numbers of complex patterns, especially when many different groups of bacteria are compared, which produce a variety of protein electrophoretic types. Visual comparison of patterns can be the method of choice to differentiate between culture contaminants or colony variants. When the patterns contain a limited number of protein bands, as for example in studies on outer membrane proteins of Gram-negative bacteria, visual assessment coupled with a calculation of the molecular weight or the mobility of the major proteins can be relevant in the grouping of strains (Dijkshoorn *et al.*, 1987; 1990). Normalized (length-corrected) photographs of gels and plots of electrophoretic traces are also useful for comparison.

Computer-assisted Analysis of Protein Electrophoretic Fingerprints

When large numbers of strains are investigated, it becomes impossible to visually compare all patterns with one another and to group them in a reliable manner. Also, estimation of quantitative resemblance is required to objectively cluster large numbers of protein patterns. Computer programs have been developed to allow data-capture, standardisation, normalisation (length correction), comparison (calculation of similarity and cluster analysis) and storage of data (Jackman *et al.*, 1983; Kersters, 1985; Costas *et al.*, 1987a; Jackman, 1987; Pot *et al.*, 1989; Vauterin and Vauterin, 1992). Some of these programs have been commercialised, e.g. GelManager (Biosystematica, Prague, Czech Republic) and GelCompar (Applied Maths, Kortrijk, Belgium). The first step towards computer-assisted processing of protein electrophoretic fingerprints is the recording of the protein concentration along the trace. This can be achieved by densitometers equipped with a conventional or laser light source or by more modern image processors. Figure 2 gives a schematic overview of the production, recording and computer-assisted comparison of protein patterns (Vauterin and Vauterin, 1992; Pot *et al.*, 1994). The major steps are: (A) cultivation of cells in standardised conditions; (B) preparation of protein extracts; (C) SDS-PAGE of proteins and staining of gels; (D) densitometric scanning of the protein patterns; (E) normalisation of traces; (F) storage of normalised patterns in databases and (G) computation of similarity between traces, and grouping and identification of patterns.

Various aspects of computer treatment of the patterns have been discussed in detail by Jackman (1987), Costas (1992), Vauterin and Vauterin (1992), Vauterin *et al.* (1993) and Pot *et al.* (1994). One of the most crucial parts of such software packages is the normalisation algorithm which compensates for discrepancies within and between gels. The reliability of the clustering of patterns and subsequent identification depends largely on the accuracy of the normalisation method. Originally two reference proteins (ovalbumin and serum albumin) were added to each extract, and the densitometric trace between these reference peaks was stretched or compressed in order to obtain an equal number of points in the densitograms (Kersters and De Ley, 1975). Nowadays, one or more lanes with reference proteins or with a reference bacterial extract are placed in the middle lane or in several different lanes of the polyacrylamide slab gel. We use the protein extract of *Psychrobacter immobilis* LMG 1125, whose profile consists of a number of well-separated sharp protein bands, which are scattered over the total length of the pattern (Figure 2D). The sharp peaks of such reference patterns can be used to delineate segments in the pattern either semi-manually (Costas *et al.*, 1987a, 1989) or automatically (Pot *et al.*, 1989). Interpolation algorithms correct the length of each segment by expansion or contraction and reduce at the same time the raw data curve of, for example 1000 digitised points to a 400 points trace. The same transformation is then carried out on all traces which are in the vicinity of the reference pattern. Vauterin and Vauterin (1992) have developed an advanced pattern recognition concept that compares the whole contours of the curves rather than peaks. This automated procedure allows the normalisation of even aberrant gels. For complex banding patterns such as protein electrophoregrams, the Pearson product moment correlation coefficient (r) has proven to be the most objective and reliable measure of similarity. Grouping of patterns is usually done by hierarchical cluster analysis or by principal component analysis.

EXAMPLES OF APPLICATIONS

In the last 20 years hundreds of publications have appeared in the literature on the use of PAGE of proteins for classification, identification and typing of bacteria. The evaluation

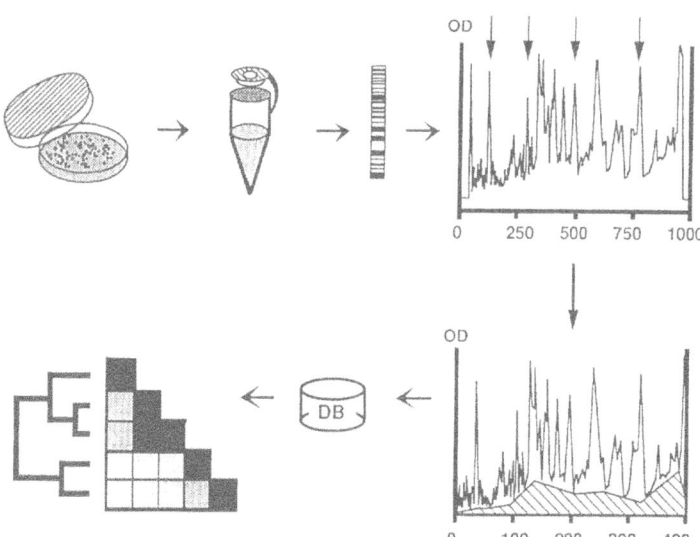

Figure 2. Major steps in the production, recording and computer-assisted comparison of protein electrophoregrams (see text for explanation).

Table 1. Some contributions of computer-aided comparisons of SDS-PAGE protein patterns to the classification, identification and typing of different bacterial groups

Bacterial genus, species or group investigated[1]	Number of strains	References
"*Achromobacter*" groups B, E, and F	32	Holmes *et al.* (1990)
Acidovorax species	36	Willems *et al.* (1990)
Acinetobacter species[2]	120	Dijkshoorn *et al.* (1990)
Actinomyces species	22	McCormick *et al.* (1985)
Aeromonas species	35	Millership and Want (1989)
Arcobacter species	83	Vandamme *et al.* (1992*c*)
Azoarcus species	12	Reinhold-Hurek *et al.* (1993)
Bacillus licheniformis and *B. polymyxa*	30	Raspoet *et al.* (1991)
Bacteroides forsythus	31	Tanner *et al.* (1986)
Bacteroides ureolyticus	42	Taylor *et al.* (1987)
Campylobacter cinaedi	17	Vandamme *et al.* (1990)
Campylobacter concisus	31	Vandamme *et al.* (1989)
Campylobacter hyointestinalis	66	On *et al.* (1993)
Campylobacter jejuni subsp. "*doylei*"	27	Owen *et al.* (1988)
Campylobacter species	29	Costas *et al.* (1987*c*;1990*c*)
Campylobacter species	>150	Vandamme *et al.* (1991*b*; 1992*a*)
Clostridium botulinum	53	Bom *et al.* (1986)
Comamonas species	43	Willems *et al.* (1991)
Corynebacterium group JK	102	Jackman and Pelczynska (1986)
Fecal streptococci	371	Niemi *et al.* (1993)
Haemophilus influenzae	34	Bruce and Jordens (1991)
Helicobacter pylori	21	Owen *et al.* (1989)
Helicobacter species	53	Costas *et al.* (1993)
Helicobacter pylori and *H. mustelae*	17	Morgan *et al.* (1991)
Hydrogenophaga species	20	Willems *et al.* (1989)
Lactobacillus acidophilus group	32	Pot *et al.* (1993)
Lactobacillus casei group	62	Hertel *et al.* (1993)
Lactobacillus species	30	Dicks *et al.* (1987)
Leuconostoc species	75	Dicks *et al.* (1990)
Moraxella canis	10	Jannes *et al.* (1993)
Moraxella lincolnii	10	Vandamme *et al.* (1993)
Morganella morganii	72	Costas *et al.* (1990*b*)
Mycobacterium tuberculosis	40	Millership and Want (1992)
Mycoplasma mycoides group	26	Costas *et al.* (1987*b*); Leach *et al.* (1989)
Propionibacterium species	62	Riedel and Britz (1992)
Proteus mirabilis	81	Holmes *et al.* (1991)
Proteus penneri and *Proteus vulgaris*	67	Costas *et al.* (1993*a*)
Providencia alcalifaciens	28	Holmes *et al.* (1988)
Providencia species	90	Costas *et al.* (1987*a*; 1990*a*)
Pseudomonas and *Xanthomonas* species and pathovars	37	Van Zyl and Steyn (1990)
Rhizobium and *Bradyrhizobium*	171	Moreira *et al.* (1993)
Rothia dentocariosa	36	Fotos *et al.* (1984)
Staphylococcus aureus	50	Costas *et al.* (1989)
Streptomyces and *Streptoverticillium* species	37	Manchester *et al.* (1990)
Xanthomonas	>400	Vauterin *et al.* (1990; 1991*a,b*; 1992*a,b,c*)

[1] Whole-cell proteins, unless indicated otherwise.
[2] Cell-envelope proteins were investigated.

of protein pattern resemblances is increasingly done by computer-assisted numerical analysis. Table 1 is an incomplete list of taxonomic studies that have applied numerical analysis of protein electrophoretic fingerprints. Instead of trying to review even part of these, we shall give below some selected examples of applications of PAGE in various areas of bacterial systematics.

Verification of Strain Authenticity

The protein electrophoretic technique is useful to verify the authenticity of bacterial cultures in one single experiment. This is illustrated in Figure 3 which displays the protein fingerprints of two strains of *Bordetella avium*. The upper part of the figure shows the protein patterns of three subcultures of the same strain Hinz 591-77, which we received in 1979 from Dr. K.-H. Hinz, who isolated it in Hannover (Germany) from a turkey suffering from rhinotracheitis. We accessed the strain as LMG 1852, and included it together with various other *Bordetella*-like strains from birds in a polyphasic taxonomic study. Apparently such isolates belonged to a new species of the genus *Bordetella*, that we named *Bordetella avium* and strain LMG 1852 was proposed as the type strain (Kersters *et al.*, 1984). The National Collection of Type Cultures (NCTC, London, Great Britain) requested the type strain as well as another *B. avium* strain (LMG 3549) in 1987, and sent back lyophilised cultures (NCTC 12033 and NCTC 12034, respectively) for confirmation of strain identity. Instead of performing numerous phenotypic tests, we compared the protein electrophoretic fingerprints of the respective subcultures of the two strains. Strain authenticity was easily confirmed because the patterns of the respective subcultures were almost identical (Figure 3), indicating that they were highly related. At the same time Figure 3 shows also that minor individual differences can be seen in the fingerprints of the two strains (indicated by arrows). Numerical analysis indicates that patterns of the same strain cluster above 0.95 *r*, whereas patterns of the two strains grouped only at 0.85 *r*,

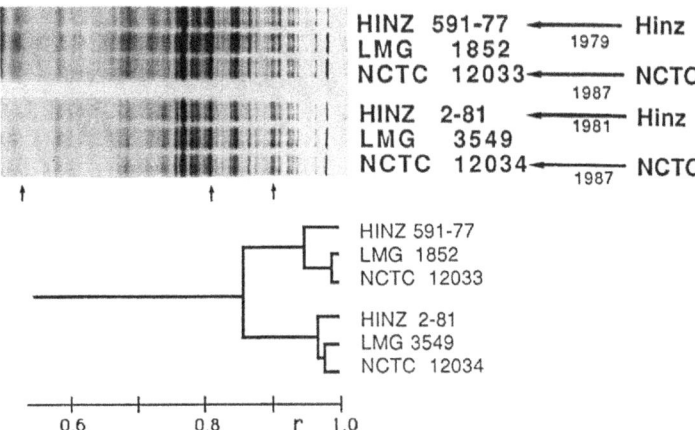

Figure 3. Protein electrophoretic patterns used for verification of the authenticity of two strains of *Bordetella avium*, which were received from K.-H. Hinz, accessed at LMG, sent from LMG to NCTC and again received from NCTC in 1987.

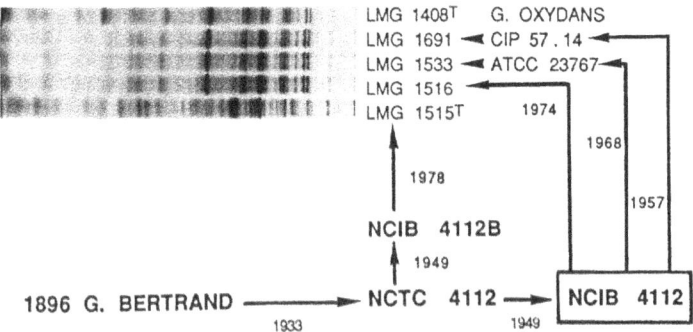

LMG 1408ᵀ   G. OXYDANS
LMG 1691 ◄ CIP 57.14 ◄
LMG 1533 ◄ ATCC 23767◄
LMG 1516 ◄
LMG 1515ᵀ   1974

                                    1968

                1978
                                         1957

            NCIB   4112B
                ↑ 1949

1896 G. BERTRAND ──────► NCTC 4112 ──► NCIB 4112
                  1933   1949

Figure 4. Rediscovery of the authentic type strain of *Acetobacter xylinum*. Protein patterns of the following strains are shown (from top to bottom): the type strain of *Gluconobacter oxydans* LMG 1408; three subcultures (NCIB 4112, CIP 57.14 and ATCC 23767) of the so-called type strain of *A. xylinum*, and the revived authentic type strain NCIB 4112B (= LMG 1515) of *A. xylinum* (data from Gillis *et al.*, 1983).

emphasizing again the small but reproducible differences between the two strains (Figure 3).

The second example concerns the authenticity of the type strain of *Acetobacter xylinum*, where PAGE of cellular proteins was extremely helpful in unravelling the perplexing finding that in 1978 none of the three culture collection strains (NCIB 4112, ATCC 23767 and CIP 57.14) was found to be a member of the genus *Acetobacter* (Gillis *et al.*, 1983). In 1896, G. Bertrand isolated in France from the juice of mountain-ash berries a bacterium which oxidized sorbitol to L-sorbose, and consequently he named this bacterium "la bactérie du sorbose" (Bertrand, 1896, 1898). Morphological, physiological and biochemical properties indicated that this strain belonged to *Acetobacter xylinum*. Among other features, this strain formed a thick characteristic leathery cellulose pellicle in liquid culture. Historically the strain is important because it led to the discovery of ketogenesis (Bertrand, 1904) and later to the well-known rule of Bertrand-Hudson concerning the stereospecific oxidation of sugar alcohols to ketoses by acetic acid bacteria. In 1933, G. Bertrand deposited this strain at the NCTC (NCTC 4112), from where it came in 1949 to the National Collection of Industrial Bacteria (Aberdeen, Scotland) as NCIB 4112. The typical properties of this strain were confirmed in 1942 by Tosic and later also by Kulka *et al.* (1949). According to the Approved Lists of Bacterial Names (Skerman *et al.*, 1980) strain NCIB 4112 is the type strain of *Acetobacter xylinum*. In the framework of a taxonomic study on the acetic acid bacteria, encompassing the genera *Acetobacter* and *Gluconobacter* (Gillis and De Ley, 1980), we accessed the NCIB strain in 1974 as LMG 1516 (Figure 4), and much to our surprise the strain did not form the characteristic cellulose pellicle, and more importantly DNA-rRNA hybridisations indicated that strain NCIB 4112 belonged to the genus *Gluconobacter*. In order to verify our results we accessed in 1976 the type strain from the Collection de l'Institut Pasteur (CIP, Paris, France) and from the ATCC, which received the strain from the NCIB in 1957 and 1968, respectively. The protein electrophoregrams of these three strains were identical (Figure 4), indicating that the strain did not change between 1957 and 1976. Figure 4 shows also that the protein patterns of strains NCIB 4112 and the parent cultures are very similar (but not identical) to the pattern of the type strain of *Gluconobacter oxydans* NCIB 9013, confirming the above-mentioned results of DNA-rRNA hybridizations (Gillis and De Ley, 1980). Thus, strains NCIB 4112,

ATCC 23767 and CIP 57.14 do not correspond to the original sorbose bacterium, which must have been interchanged either at NCTC or NCIB between 1933 and 1957 with a strain of *G. oxydans*. As a consequence of these results a Sordelli-dried culture (prepared in 1949 either at NCTC or NCIB) was revived at NCIB in 1978. The subculture (designated NCIB 4112B) was found to be pure and capable of producing a thick cellulose pellicle, and its protein pattern was completely different from that of strain NCIB 4112 (Figure 4), but similar to that of other *A. xylinum* strains (data not shown). DNA-rRNA hybridizations proved that strain NCIB 4112B definitely belonged to the genus *Acetobacter*.

The conclusion of our experiments is that strain NCIB 4112B is most probably a descendant of the culture originally deposited by Gabriel Bertrand at the NCTC in 1933. A new accession number (NCIB 11664) was given to the revived sorbose bacterium and a request for an opinion was issued to designate the latter strain as the type strain of *A. xylinum* (synonym *Acetobacter aceti* subsp. *xylinus*) in place of strain NCIB 4112 (Gillis *et al.*, 1983). This request was accepted in 1985 as opinion nr. 59 of the Judicial Commission.

Differentiation between Colony Variants and Contaminants

SDS-PAGE of whole-cell bacterial proteins appears to be a very useful and simple technique to unravel whether observed differences in colony morphology in a bacterial strain are due to colony variants or contaminants. This is illustrated for two lactic acid bacterial strains in Figure 5, where the protein patterns of three stable colony types of *Streptococcus thermophilus* strain LMG 7953 are indistinguishable from each other, indicating that they are most likely simple colony variants. However, the protein fingerprints of the two colony types observed in strain LMG 7940 of *Leuconostoc lactis* are completely different from each other. Comparison with protein patterns of other lactic acid bacterial strains showed that one of these cultures belonged to *Lactobacillus plantarum*, clearly indicating that the original culture of strain LMG 7940 was contaminated. In general, one-dimensional SDS-PAGE of cellular proteins cannot discriminate between different strains that are genotypically highly related.

Classification, Identification and Typing of Clinical Aerotolerant Campylobacters

Comparison of bacterial protein electrophoretic patterns has especially been applied in bacterial systematics for the differentiation and identification of various taxa (see Table 1). The more strains are involved in such studies, the more obvious the application of SDS-PAGE as a fast and simple screening method becomes. In this area, one of the most appreciated features is the possibility to decrease the number of strains required for genotypic analyses by preliminary grouping of similar, i. e. closely related, strains. In this section we shall briefly describe how we used SDS-PAGE of cellular proteins to unravel thetaxonomy of aerotolerant campylobacters in a polyphasic approach. Moreover, SDS-

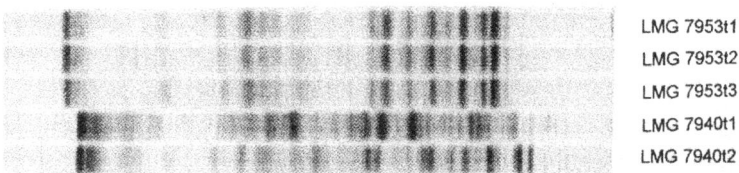

LMG 7953t1
LMG 7953t2
LMG 7953t3
LMG 7940t1
LMG 7940t2

Figure 5. A computer-generated print-out of the normalised protein patterns of different colony types of *Streptococcus thermophilus* strain LMG 7953 and *Leuconostoc lactis* LMG 7940. Colony types are indicated by the suffixes t1, t2 and t3.

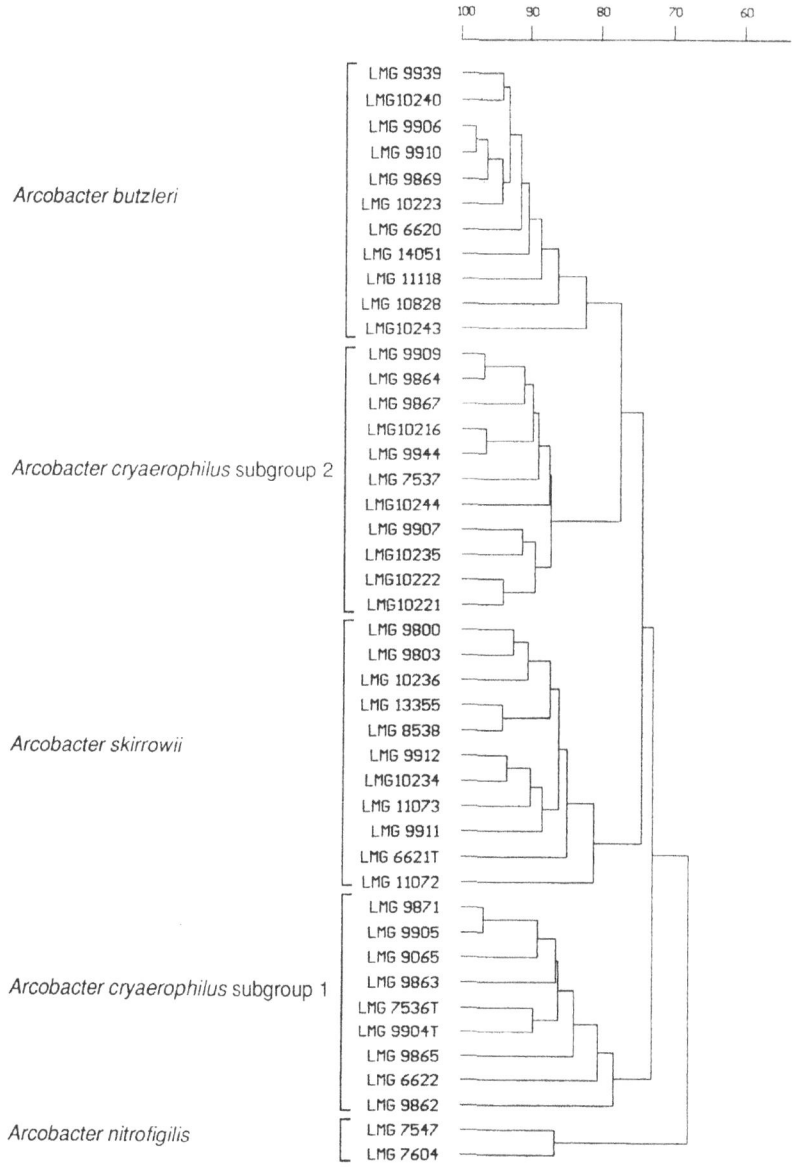

Figure 6. Dendrogram derived from the unweighted pair group linkage of similarity values (r x 100) of whole-cell protein patterns of *Arcobacter* strains.

PAGE was also useful in an epidemiological study for typing large numbers of *Arcobacter* strains.

The genus *Arcobacter* was created in 1991 for two species previously known as aerotolerant campylobacters (Vandamme *et al.*, 1991a). *Arcobacter nitrofigilis*, the type species, is an awkward member of the large group of *Campylobacter*-like organisms as it has never been isolated from clinical specimens. Until now, these bacteria have exclusively been found in or near the roots of *Spartina alterniflora*, a salt-water plant (McClung *et al.*, 1983). *Arcobacter cryaerophilus*, the second species has been isolated from aborted foetuses and diarrhoeic faeces of several livestock animals (Vandamme *et al.*, 1992c). Strains belonging to the latter species, which was originally known as *Campylobacter cryaerophila* (Neill *et al.*, 1985), were first isolated in the late 1970s. Neill *et al.* (1985) performed an extensive phenotypic analysis on a large collection of strains and reported a remarkable phenotypic heterogeneity within this species. The identification of a number of closely related strains which showed only low or non-significant DNA binding values towards the type strain (Vandamme *et al.*, 1991a) urged us to collect a large number of *C. cryaerophila* isolates. We used whole-cell protein pattern analysis as an initial screening method. The outcome of the cluster analysis revealed five major clusters, one of which comprised the two reference strains of *A. nitrofigilis* (Figure 6). The four remaining clusters contained aerotolerant strains from animal and human origin. From each cluster, representative strains were selected to perform an extensive DNA-DNA hybridisation analysis and to determine the phylogenetic position of each of these taxa. The DNA-DNA hybridisation analysis showed that two of the clusters constituted a single species containing the type strain of *A. cryaerophilus* (Vandamme *et al.*, 1992c). These two groups were referred to as *A. cryaerophilus* subgroup 1 and *A. cryaerophilus* subgroup 2. The two remaining clusters each constituted a separate species for which the names *A. skirrowii* and *A. butzleri* were proposed (Vandamme *et al.*, 1992c). DNA-rRNA hybridizations confirmed that these new species were genuine members of *Arcobacter* which constitutes together with *Campylobacter* the family *Campylobacteraceae* (Vandamme and De Ley, 1991). These five taxa within *Arcobacter* were further characterised by analyses of their fatty acid components and biochemical features (Vandamme *et al.*, 1992c). The two electrophoretic

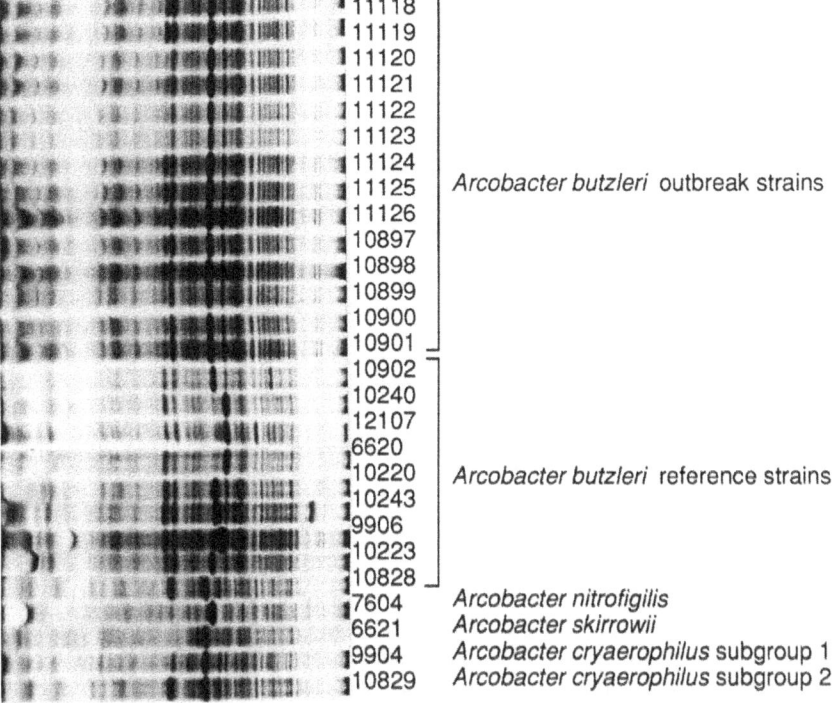

Figure 7. Protein patterns of *A. butleri* strains and various *Arcobacter* reference strains. All strain numbers are LMG numbers.

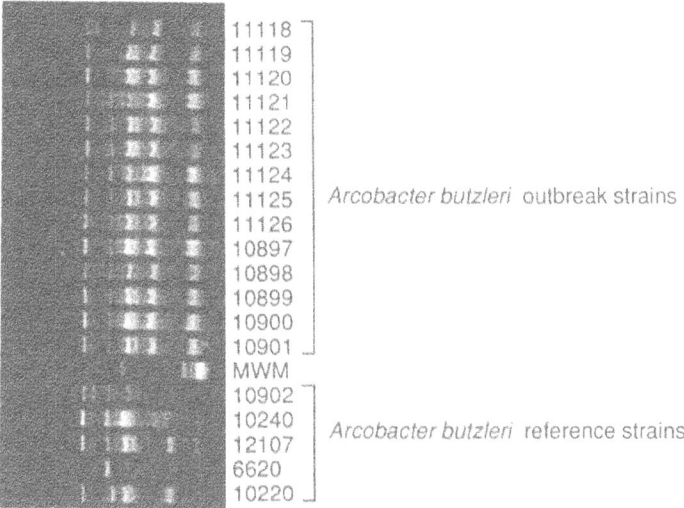

11118 ⎫
11119 │
11120 │
11121 │
11122 │
11123 │
11124 │
11125 │ *Arcobacter butzleri* outbreak strains
11126 │
10897 │
10898 │
10899 │
10900 │
10901 ⎭
MWM
10902 ⎫
10240 │ *Arcobacter butzleri* reference strains
12107 │
6620 │
10220 ⎭

Figure 8. DNA amplification fingerprints of *A. butzleri* outbreak strains and reference strains using the random sequence 1267 as primer. All strain numbers are LMG numbers.

subgroups within *A. cryaerophilus* also differed in their fatty acid composition and in several biochemical features. Surprisingly, protein analysis was the only phenotypic method that allowed separation of *A. cryaerophilus* subgroup 2 from *A. butzleri*. Our database for the identification of campylobacters and related bacteria (i. e. *Campylobacter, Arcobacter, Helicobacter* and *Wolinella*) was expanded with entries for the new *Arcobacter* taxa and soon proved to be useful in an epidemiological study.

In the early 1980s, several children of a primary and a nursery school in the Rovigo area in Italy were affected by an abnormal syndrome characterised by recurrent abdominal cramps without the presence of diarrhoea. Some children had to be hospitalised. The only

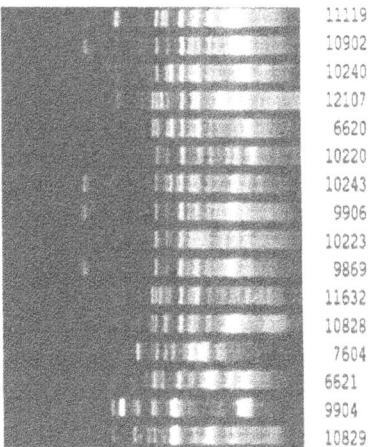

11119
10902
10240
12107
6620
10220
10243
9906
10223
9869
11632
10828
7604
6621
9904
10829

Figure 9. DNA amplification fingerprints of *A. butzleri* strains (nrs. 11119 to 10828) and reference strains of *A. nitrofigilis* (LMG 7604), *A. skirrowii* (LMG 6621), and *A. cryaerophilus* subgroup 1 (LMG 9904) and subgroup 2 (LMG 10829) using the repetitive ERIC sequence as primer. All strain numbers are LMG numbers.

abnormal organism present in their stools was an aerotolerant *Campylobacter*-like organism which was tentatively identified as *Campylobacter coli*. Fourteen outbreak related strains were maintained and deposited ten years later in our culture collection. The protein analysis of all strains revealed that they had a virtually identical protein pattern which was identified as *A. butzleri* (Vandamme *et al.*, 1992b). All fourteen strains were also identical in classical biotyping and serotyping analyses. However, the resolution of these techniques is not very high as most *A. butzleri* strains belong to only a few biotypes or serotypes. Protein analysis has been applied for epidemiological typing of several genera. The protein patterns of campylobacters and related bacteria are characterised by the presence of one or a few dense protein bands with a molecular weight that is highly variable within species (Costas, 1992). Such an interstrain variability, due to this variable dense band region, is also seen within *A. butzleri* (Figure 7). Furthermore, additional interstrain variability is also present for several minor protein bands in the high molecular weight region (Figure 7). Interestingly, each of the *A. butzleri* reference strains is characterised by a unique protein fingerprint.

In order to evaluate the resolution of this electrophoretic protein typing, we also performed a random amplified polymorphic DNA analysis on the same set of strains (Figure 8 and 9). With this method, genomic DNA is primed with short (about 10 to 20 basepairs), random or repetitive, DNA fragments in a polymerase chain reaction assay. The arrays of DNA fragments obtained after amplification are separated in an agarose gel and visualised by ethidium bromide staining (Lupski and Weinstock, 1992). This method yields reproducible and highly specific DNA fingerprints (Lupski and Weinstock, 1992). The result of this DNA fingerprinting method is entirely similar to the result of the protein analysis. Virtually no variability was found between the fourteen outbreak strains (Figure 8), whereas each of the reference strains is characterised by a unique DNA fingerprint (Figure 9). Clearly, the high variability in protein content and the high reproducibility of the method confirm the feasibility of protein analysis for epidemiological typing studies as well.

CONCLUSIONS

High-resolution one-dimensional SDS-PAGE combined with computer analysis of whole-cell protein patterns allows the identification of a number of protein electrophoretic types in various bacterial species. Taxonomically this method differentiates at the species and subspecies level.

Although conditions for the cultivation of bacterial strains need to standardized, electrophoresis of cellular proteins has a number of attractive features:
- an electrophoristic protein pattern is a stable and characteristic fingerprint of a bacterial strain, which is fairly simple to obtain in a cost-effective way
- comparison of protein electrophoretic patterns provides a reliable measure of taxonomic relatedness of the investigated strains
- the patterns can be stored in data bases as normalised densitometric records and compared by techniques of numerical analysis
- once electrophoretic groups of reliable taxonomic status have been established, libraries of reference patterns can be contructed to identify new isolates.

ACKNOWLEDGMENTS

Part of our research on protein fingerprinting was carried out in the framework of contracts BIOT-CT91-0263 and BIOT-CT91-0294 of the Commission of the European Communities. PV and LV are indebted to the National Fund for Scientific Research (Belgium) for positions as senior research assistant and research assistant, respectively. KK is grateful to the Fund for Medical Scientific Research (Belgium) and the Ministerie van de Vlaamse Gemeenschap, Bestuur Wetenschappelijk Onderzoek (Belgium) for research and personnel grants.

REFERENCES

Bertrand, G. (1896) Préparation biochimique du sorbose. C.R. Acad. Sci. 122, 900-903.

Bertrand, G. (1898) Recherches sur la production biochimique du sorbose. Ann. Inst. Pasteur (Paris) 12, 385-399.

Bertrand, G. (1904) Etude biochimique de la bactérie du sorbose. Ann. Chim. Phys. 3, 181-288.

Bom, I., Smelt, J.P.P.M., Kersters, K. and Verrips, C.T. (1986) Identification and grouping of *Clostridium botulinum* strains by numerical analysis of their electrophoretic protein patterns. J. Appl. Bacteriol. 60, 483-490.

Bruce, K.D. and Jordens, J.Z. (1991) Characterization of noncapsulate *Haemophilus influenzae* by whole-cell polypetide profiles, restriction endonuclease analysis, and rRNA gene restriction patterns. J. Clin. Microbiol. 29, 291-296.

Costas, M. (1992) Classification, identification and typing of bacteria by the analysis of their one-dimensional polyacrylamide gel electrophoretic protein patterns, in "Advances in Electrophoresis" (Chrambach, A, Dun, M.J. and Radola, B.J., Eds.), pp. 351-408. VCH Verlaggesellschaft mbH, Weinheim.

Costas, M., Holmes, B. and Sloss, L.L. (1987a) Numerical analysis of electrophoretic protein patterns of *Providencia rustigianii* strains from human diarrhoea and other sources. J. Appl. Bacteriol. 63, 319-328.

Costas, M., Leach, R.H. and Mitchelmore, D.L. (1987b) Numerical analysis of PAGE protein patterns and the taxonomic relationships within the "*Mycoplasma mycoides* cluster". J. Gen. Microbiol. 133, 3319-3329.

Costas, M., Owen, R.J. and Jackman, P.J.H. (1987c) Classification of *Campylobacter sputorum* and allied campylobacters based on numerical analysis of electrophoretic protein patterns. Syst. Appl. Microbiol. 9, 125-131.

Costas, M., Cookson, B.D., Talsania, H.G. and Owen, R.J. (1989) Numerical analysis of electrophoretic protein patterns of methycillin-resistant strains of *Staphylococcus aureus* . J. Clin. Microbiol. 27, 2574-2581.

Costas, M., Holmes, B. and Wood, A.C. (1990a) Numerical analysis of electrophoretic protein patterns of *Providencia stuartii* strains from urine, wound and other clinical sources. J. Appl. Bacteriol. 68, 505-518.

Costas, M., Holmes, B. and Wood, A.C. (1990b) Numerical analysis of electrophoretic protein patterns of *Morganella morganii* strains from faeces, wound, urine and other clinical sources. J. Appl. Bacteriol. 69, 426-438.

Costas, M., Pot, B., Vandamme, P., Kersters, K., Owen, R.J. and Hill, L.R. (1990c) Interlaboratory comparative study of the numerical analysis of one-dimensional sodium dodecyl sulphate-polyacrylamide gel electrophoretic protein patterns of *Campylobacter* strains. Electrophoresis 11, 467-474.

Costas, M., Holmes, B., Frith, K.A., Riddle, C. and Hawkey, P.M. (1993a) Identification and typing of *Proteus penneri* and *Proteus vulgaris* biogroups 2 and 3, from clinical sources, by computerized analysis of electrophoretic protein patterns. J. Appl. Bacteriol., 75, 489-498.

Costas, M., On, S.L.W., Owen, R.J., Lopez-Urquijo, B. and Lastovica, A.J. (1993b) Differentiation of *Helicobacter* species by numerical analysis of their one-dimensional electrophoretic protein patterns. Syst. Appl. Microbiol. 16, 396-404.

Dicks, L.M.T., van Vuuren H.J.J. and Dellaglio, F. (1987) Relatedness of homofermentative *Lactobacillus* species revealed by numerical analysis of total soluble cell protein patterns. Int. J. Syst. Bacteriol. 37, 437-440.

Dicks, L.M.T., van Vuuren, H.J.J. and Dellaglio, F. (1990) The taxonomy of *Leuconostoc* species, particularly *Leuconostoc oenos*, as revealed by numerical analysis of total soluble cell protein patterns, DNA base composition, and DNA:DNA hybridizations. Int. J. Syst. Bacteriol. 40: 83-91.

Dijkshoorn, L., van Vianen, W., Degener, J.E. and Michel, M.F. (1987) Typing of *Acinetobacter calcoaceticus* strains isolated from hospital patients by cell envelope protein profiles. Epidemiol. Infect. 99, 659-667.

Dijkshoorn, L., Tjernberg, I., Pot, B., Michel, M.F., Ursing, J. and K. Kersters. (1990) Numerical analysis of cell envelope protein profiles of *Acinetobacter* strains classified by DNA-DNA hybridization. Syst. Appl. Microbiol. 13, 338-344.

Feltham, R.K.A. and Sneath, P.H.A. (1979) Quantitative comparison of electrophoretic traces of bacterial proteins. Comput. Biomed. Res. 12, 247-263.

Fotos, P.G., Gerencser, M.A. and Yelton, D.B.. (1984) Strain differentiation of *Rothia dentocariosa* and related isolates by sodium dodecyl sulphate-polyacrylamide gel electrophoresis. Int. J. Syst. Bacteriol. 34, 102-106.

Garber, E.D. and Rippon, J.W. (1968) Proteins and enzymes as taxonomic tools. Adv. Appl. Microbiol. 10, 137-154.

Gillis, M. and De Ley, J. (1980) Intra- and intergeneric similarities of the ribosomal ribonucleic acid cistrons of *Acetobacter* and *Gluconobacter*. Int. J. Syst. Bacteriol. 30, 7-27.

Gillis, M., Kersters, K., Gosselé, F., Swings, J., De Ley, J., MacKenzie, A.R. and Bousfield, I.J. (1983) The rediscovery of Bertrand's sorbose bacterium (*Acetobacter aceti* subsp. *xylinum*): proposal to designate NCIB 11664 in place of NCIB 4112 (ATCC 23767) as the type strain of *Acetobacter aceti* subsp. *xylinum*. Request for an opinion. Int. J. Syst. Bacteriol. 33, 122-124.

Hertel, C., Ludwig, W., Pot, B., Kersters, K. and Schleifer, K.-H. (1993) Differentiation of lactobacilli occurring in fermented milk products by using oligonucleotide probes and electrophoretic protein profiles. Syst. Appl. Microbiol. 16, 463 - 467.

Holmes, B., Costas, M. and Sloss, L.L. (1988) Numerical analysis of electrophoretic protein patterns of *Providencia alcalifaciens* strains from human faeces and veterinary specimens. J. Appl. Bacteriol. 64, 27-35.

Holmes, B., Costas, M., Wood, A.C. and Kersters, K. (1990) Numerical analysis of electrophoretic protein patterns of *Achromobacter* group B, E and F strains from human blood. J. Appl. Bacteriol. 68, 495-504.

Holmes, B., Costas, M., and Wood, A.C. (1991) Typing of *Proteus mirabilis* from clinical sources by computerized analysis of electrophoretic protein patterns. J. Appl. Bacteriol. 71, 467-476.

Jackman, P.J.H. (1985) Bacterial taxonomy based on electrophoretic whole-cell protein patterns, in "Chemical Methods in Bacterial Systematics" (Goodfellow, M. and Minnikin, D.E., Eds.), pp. 115-128. Academic Press, London.

Jackman, P.J.H. (1987) Microbial systematics based on electrophoretic whole-cell protein patterns, in "Methods in Microbiology" (Colwell, R.R. and Grigorova, R., Eds.), Vol. 19, pp. 209-225. Academic Press, London.

Jackman, P.J.H. and Pelczynska, S. (1986). Characterization of *Corynebacterium* group JK by whole-cell protein patterns. J. Gen. Microbiol. 132, 1911-1915.

Jackman, P.J.H., Feltham, R.K.A. and Sneath, P.H.A. (1983) A program in Basic for numerical taxonomy of micro-organisms based on electrophoretic protein patterns. Microbios Lett. 23, 87-98.

Jannes, G., Vaneechoutte, M., Lannoo, M., Gillis, M., Vancanneyt, M., Vandamme, P., Verschraegen, G., Van Heuverswyn, H. and Rossau, R. (1993) Polyphasic taxonomy leading to the proposal of *Moraxella canis* sp. nov. for *Moraxella catarrhalis*-like strains. Int. J. Syst. Bacteriol. 43, 438-449.

Kersters, K. (1985) Numerical methods in the classification of bacteria by protein electrophoresis, in "Computer-assisted Bacterial Systematics" (Goodfellow, M., Jones, D. and Priest, F.G., Eds.), pp. 337-368. Academic Press, London.

Kersters, K. and De Ley, J. (1975) Identification and grouping of bacteria by numerical analysis of their protein patterns. J. Gen. Microbiol. 87, 333-342.

Kersters, K. and De Ley, J. (1980) Classification and identification of bacteria by electrophoresis of their proteins, in "Microbiological Classification and Identification" (Goodfellow, M. and Board, R.G., Eds.), pp. 273-297. Academic Press, London.

Kersters, K., Hinz, K.-H., Hertle, A., Segers, P., Lievens, A., Siegmann, O. and De Ley, J. (1984) *Bordetella avium* sp. nov., isolated from the respiratory tracts of turkeys and other birds. Int. J. Syst. Bacteriol. 34, 56-70.

Kulka, D., Preston, J.M. and Walker, T.K. (1949) Giant colonies of *Acetobacter* species as an aid to identification. J. Inst. Brew. London 55, 141-146.

Laemmli, U.K. (1970) Cleavage of structural proteins during the assembly of the head of bacteriophage T4. Nature (London) 227, 680-685.

Leach, R.H., Costas, M. and Mitchelmore, D.L. (1989) Relationship between *Mycoplasma mycoides* subsp. *mycoides* ('large-colony strains) and *M. mycoides* subsp. *capri*, as indicated by numerical analysis of one-dimensional SDS-PAGE protein patterns. J. Gen. Microbiol. 135, 2993-3000.

Lupski, J. R. and Weinstock, G.E. (1992) Short, interspersed repetitive DNA sequences in prokaryotic genomes. J. Bacteriol. 174, 4525-4529.

Manchester, L., Pot, B., Kersters, K., and Goodfellow, M. (1990) Classification of *Streptomyces* and *Streptoverticillium* species by numerical analysis of electrophoretic protein patterns. Syst. Appl. Microbiol. 13, 333-337.

McClung, C.R., Patriquin, D.G. and Davis, R.E. (1983) *Campylobacter nitrofigilis* sp. nov., a nitrogen-fixing bacterium associated with roots of *Spartina alterniflora* Loisel. Int. J. Syst. Bacteriol. 33, 605-612.

McCormick, S., Mengoli, H.F. and Gerencser, M.A. (1985) Polyacrylamide gel electrophoresis of whole-cell preparations of *Actinomyces* spp. Int. J. Syst. Bacteriol. 35, 429-433.

Millership, S.E. and Want, S.V. (1989) Typing of *Aeromonas* species by protein fingerprinting: comparison of radiolabelling and silver staining for visualising proteins. J. Med. Microbiol. 29, 29-32.

Millership, S.E. and Want, S.V. (1992) Whole-cell protein electrophoresis for typing *Mycobacterium tuberculosis*. J. Clin. Microbiol. 30, 2784-2787.

Moreira, F., Gillis, M., Pot, B., Kersters, K. and Franco, A. (1993) Characterization of rhizobia isolated from different divergence groups of tropical *Leguminosae* by comparative polyacrylamide gel electrophoresis of their total proteins. Syst. Appl. Microbiol. 16, 135-146.

Morgan, D.R., Fox, F.G. and Leunk, R.D. (1991). Comparison of isolates of *Helicobacter pylori* and *Helicobacter mustelae*. J. Clin. Microbiol. 29, 395-397.

Neill, S.D., Campbell, J.N., O'Brien, J.J., Weatherup, S.T.C. and Ellis, W.A. (1985) Taxonomic position of *Campylobacter cryaerophila* sp. nov. Int. J. Syst. Bacteriol. 35, 342-356.

Niemi, R.M., Niemelä, S.I., Bamford, D.H., Hantula, J., Hyvärinen, T., Forsten, T. and Raateland, A. (1993) Presumptive fecal streptococci in environmental samples characterized by one-dimensional sodium dodecyl sulfate-polyacrylamide gel electrophoresis. Appl. Environm. Microbiol. 59, 2190-2196.

On, S.L.W., Costas, M. and Holmes, B. (1993) Identification and intra-specific heterogeneity of *Campylobacter hyointestinalis* based on numerical analysis of electrophoretic protein profiles. Syst. Appl. Microbiol. 16, 37-46.

Owen, R.J. and Jackman, P.J.H. (1982) The similarities between *Pseudomonas paucimobilis* and allied bacteria derived from analysis of deoxyribonucleic acids and electrophoretic protein patterns. J. Gen. Microbiol. 128, 2945-2954.

Owen, R.J., Costas, M. and Sloss, L.L. (1988). Electrophoretic protein typing of *Campylobacter jejuni* subsp. *"doylei"* (nitrate-negative campylobacter-like organism) from human faeces and gastric mucosa. Eur. J. Epidemiol. 4, 277-283.

Owen, R.J., Costas, M., Morgan, D.D., On, S.L.W., Hill, L.R., Pearson, A.D. and Morgan, D.R. (1989) Strain variation in *Campylobacter pylori* detected by numerical analysis of one-dimensional electrophoretic protein patterns. Antonie van Leeuwenhoek 55, 253-267.

Pot, B., Gillis, M., Hoste, B., Van de Velde, A., Bekaert, F., Kersters, K. and De Ley, J. (1989). Intra- and intergeneric relationships of the genus *Oceanospirillum*. Int. J. Syst. Bacteriol. 39, 23-34.

Pot, B., Hertel, C., Ludwig, W., Descheemaeker, P., Kersters, K. and Schleifer, K.-H. (1993) Identification and classification of *Lactobacillus acidophilus*, *L. gasseri* and *L. johnsonii* strains by SDS-PAGE and rRNA-targeted oligonucleotide probe hybridization. J. Gen. Microbiol. 139, 513 - 517.

Pot, B., Vandamme, P. and Kersters, K. (1994). Analysis of electrophoretic whole-organism fingerprints, in "Chemical Methods in Bacterial Systematics" (Goodfellow, M. and O'Donnell, A.G., Eds.). J. Whiley & Sons, Chichester (in press).

Raspoet, D., Pot, B., De Deyn, D., De Vos, P., Kersters, K. and De Ley, J. (1991) Differentiation between 2,3-butanediol producing *Bacillus licheniformis* and *B. polymyxa* strains by fermentation product profiles and whole-cell protein electrophoretic patterns. Syst. Appl. Microbiol. 14, 1-7.

Reinhold-Hurek, B., Hurek, T., Gillis, M., Hoste, B., Vancanneyt, M., Kersters, K. and De Ley, J. (1993) *Azoarcus* gen. nov., nitrogen-fixing Proteobacteria associated with roots of Kallar grass (*Leptochloa fusca* (L.) Kunth), and description of two species, *Azoarcus indigens* sp. nov., and *Azoarcus communis* sp. nov. Int. J. Syst. Bacteriol. 43, 574-584.

Riedel, K.-H.J. and Britz, T.J. (1992) Differentiation of "classical" *Propionibacterium* species by numerical analysis of electrophoretic protein profiles. Syst. Appl. Microbiol. 15, 567-572.

Skerman, V.B.D., McGowan, V. and Sneath, P.H.A. (1980) Approved lists of bacterial names. Int. J. Syst. Bacteriol. 30, 225-420.

Tanner, A.C.R., Listgarten, M.A., Ebersole, J.L. and Strzempko, M.N. (1986) *Bacteroides forsythus* sp. nov., a slow-growing, fusiform *Bacteroides* sp. from the human oral cavity. Int. J. Syst. Bacteriol. 36, 213-221.

Taylor, A.J., Costas, M. and Owen, R.J. (1987) Numerical analysis of electrophoretic protein patterns of *Bacteroides ureolyticus* clinical isolates. J. Clin. Microbiol. 25, 660-666.

Vandamme, P. and De Ley, J. (1991) Proposal for a new family, *Campylobacteraceae*. Int. J. Syst. Bacteriol. 41, 451-455.

Vandamme, P., Falsen, E., Pot, B., Hoste, B., Kersters, K. and De Ley, J. (1989) Identification of EF Group 22 campylobacters from gastroenteritis cases as *Campylobacter concisus*. J. Clin. Microbiol. 27, 1775-1781.

Vandamme, P., Falsen, E., Pot, B., Kersters, K. and De Ley, J. (1990) Identification of *Campylobacter cinaedi* isolated from blood and feces of children and adult females. J. Clin. Microbiol., 28, 1016-1020.

Vandamme, P., Falsen, E., Rossau, R., Hoste, B., Segers, P., R. Tytgat and De Ley, J. (1991*a*) Revision of *Campylobacter*, *Helicobacter*, and *Wolinella* taxonomy: emendation of generic descriptions and proposal of *Arcobacter* gen. nov. Int. J. Syst. Bacteriol. 41, 88-103.

Vandamme, P., Pot, B. and Kersters, K. (1991*b*) Differentiation of campylobacters and *Campylobacter*-like organisms by numerical analysis of one-dimensional electrophoretic protein patterns. Syst. Appl. Microbiol. 14, 57-66.

Vandamme, P., Dewettick, D. and Kersters, K. (1992*a*) Application of numerical analysis of electrophoretic protein profiles for the identification of thermophilic campylobacters. Syst. Appl. Microbiol. 15, 402-408.

Vandamme, P., Pugina, P., Benzi, G., Van Etterijck, R., L. Vlaes, Kersters, K., Butzler, J.-P., Lior, H. and Lauwers, S. (1992*b*) Outbreak of recurrent abdominal cramps associated with *"Arcobacter butzleri"* in an Italian school. J. Clin. Microbiol. 30, 2335-2337.

Vandamme, P., Vancanneyt, M., Pot, B., Mels, L., Hoste, B., Dewettinck, D., Vlaes, L., Van Den Borre, C., Higgins, R., Hommez, J., Kersters, K., Butzler, J.-P. and Goossens, H. (1992c) Polyphasic taxonomic study of the emended genus *Arcobacter* with *Arcobacter butzleri* comb. nov. and *Arcobacter skirrowii* sp. nov., an aerotolerant bacterium isolated from veterinary specimens. Int. J. Syst. Bacteriol. 42, 344-356.

Vandamme, P., Gillis, M., Vancanneyt, M., Hoste, B., Kersters, K. and Falsen, E. (1993) *Moraxella lincolnii* sp. nov., isolated from the human respiratory tract, and reevaluation of the taxonomic position of *Moraxella osloensis*. Int. J. Syst. Bacteriol. 43, 474 - 481.

Van Zyl, E. and Steyn, P. (1990) Differentiation of phytopathogenic *Pseudomonas* and *Xanthomonas* species and pathovars by numerical taxonomy and protein gel electrophoregrams. Syst. Appl. Microbiol. 13, 60-71.

Vauterin, L. and Vauterin, P. (1992) Computer-aided objective comparison of electrophoresis patterns for grouping and identification of microorganisms. European Microbiology 1, 37-41.

Vauterin, L., Vantomme, R., Pot, B., Hoste, B., Swings, J. and Kersters, K. (1990) Taxonomic analysis of *Xanthomonas campestris* pv. *begoniae* and *X. campestris* pv. *pelargonii* by means of phytopathological, phenotypic, protein electrophoretic and DNA hybridization methods. Syst. Appl. Bacteriol. (In press).

Vauterin, L., Swings, J. and Kersters, K. (1991a) Grouping of *Xanthomonas campestris* pathovars by SDS-PAGE of proteins. J. Gen. Microbiol. 137, 1677-1687.

Vauterin, L., Yang, P., Hoste, B., Vancanneyt, M., Civerolo, E.L., Swings, J. and Kersters, K. (1991b) Differentiation of *Xanthomonas campestris* pv. citri by sodium dodecyl sulfate-polyacrylamide gel electrophoresis of proteins, fatty acid analysis, and DNA-DNA hybridization. Int. J. Syst. Bacteriol. 41, 535-542.

Vauterin, L., Yang, P., Hoste, B., Pot, B., Swings, J. and Kersters, K. (1992) Taxonomy of xanthomonads from cereals and grasses based on SDS-PAGE of proteins, fatty acids analysis and DNA hybridization. J. Gen. Microbiol. 138, 1467-1477.

Vauterin L., Swings, J. and Kersters, K. (1993) Protein electrophoresis and classification, in "Handbook of New Bacterial Systematics" (Goodfellow, M. and O'Donnell, A.G., Eds.), pp. 251-280. Academic Press, London.

Willems, A., Busse, J., Goor, M., Pot, B., Falsen, E., Jantzen, E., Hoste, B., Gillis, M., Kersters, K. and De Ley, J. (1989) *Hydrogenophaga*, a new genus of hydrogen-oxidizing bacteria including *Hydrogenophaga flava* comb. nov. (formerly *Pseudomonas flava*), *Hydrogenophaga palleronii* (formerly *Pseudomonas palleronii*), *Hydrogenophaga pseudoflava* (formerly *Pseudomonas pseudoflava* and "*Pseudomonas carboxydoflava*") and *Hydrogenophaga taeniospiralis* (formerly *Pseudomonas taeniospiralis*). Int. J. Syst. Bacteriol. 39, 319-333.

Willems, A, Falsen, E., Pot, B., Jantzen, E., Hoste, B., Vandamme, P., Gillis, M., Kersters, K. and De Ley, J. (1990) *Acidovorax*, a new genus for *Pseudomonas facilis*, *Pseudomonas delafieldii*, E. Falsen (EF) group 13, EF group 16, and several clinical isolates, with the species *Acidovorax facilis* comb. nov., *Acidovorax delafieldii* comb. nov., and *Acidovorax temperans* sp. nov. Int. J. Syst. Bacteriol. 40, 384-398.

Willems, A., Pot, B., Falsen, E., Vandamme, P., Gillis, M., Kersters, K. and De Ley, J. (1991) Polyphasic taxonomic study of the emended genus *Comamonas*: relationships to *Aquaspirillum aquaticum*, E. Falsen group 10, and other clinical isolates. Int. J. Syst. Bacteriol. 41, 427-444.

CHARACTERIZATION AND IDENTIFICATION OF MICRO-ORGANISMS BY FT-IR SPECTROSCOPY AND FT-IR MICROSCOPY

Dieter Naumann, Dieter Helm, and Christian Schultz

Abteilung für Cytologie
Robert Koch-Institut des Bundesgesundheitsamtes
Nordufer 20, D-13353 Berlin, Germany

INTRODUCTION

A large series of physical techniques has been tested for the characterization of microbial cells. These techniques include gas chromatographic (GC) and high-performance liquid chromatographic (HPLC) techniques, gel electrophoresis, pyrolysis mass spectroscopy, fluorescence and chemiluminescence techniques, flow cytometry, impedance measurements, microcalorimetry, circular intensity differential scattering, nuclear magnetic resonance, and infrared as well as Raman spectroscopy (for review see Nelson, 1991). All these efforts have provided additional biochemical and structural information, which might - in some areas - be helpful for taxonomic characterizations.

Spectroscopic techniques are extremely specific and provide a wealth of qualitative and quantitative information about a given sample. The infrared spectrum, in particular, of any compound is known to express an unique 'fingerprint' and it is this characteristic that enables IR-spectroscopy to be used in the identification of unknown samples using spectral data libraries (for example, in forensic medicine and the pharmaceutical industry). The advent of modern interferometric and Fourier-transform techniques has augmented the specifications of IR-spectroscopy and, thus, paved the way for novel applications of IR in many fields of microbiological research.

The FT-IR technique recently developed (Naumann et al., 1991 a,b; Helm et al., 1991 a,b) essentially takes advantage of the spectral fingerprints of intact bacteria. These patterns comprise the vibrational features of all cell constituents, that is, DNA/RNA, proteins, membrane and cell wall components. Consequently, FT-IR probes the total composition of a given organism in a single experiment. As all cell compounds depend on the expression of smaller or larger parts of the genome, the spectra display in a specific way a phenotypic and a genetic fingerprint of the micro-organisms under study. This is why the selectivity of the new technique is extremely high, allowing differentiation down to the strain and/or serogroup/serotype level (Helm *et al.*, 1991 *a*).

EXPERIMENTAL PROCEDURES

Owing to the multitude of cellular components, only broad and superimposed spectral bands are observed within the mid-infrared range ($\approx$ 4,000-600 reciprocal wavelengths = wavenumbers [cm^{-1}]). Some spectral sub-ranges are dominated by particular cell components, for example, fatty acids, polysaccharides or proteins. In order to extract specific information from the broad spectral contour, it is advantageous to analyse microbial FT-IR spectra by considering carefully selected single sub-ranges or defined combinations of sub-ranges and to apply sophisticated mathematical filter techniques (*i.e.* calculation of the first or second derivatives, band-pass filtering or Fourier-selfdecon-volution).

Since intact cells are being tested, complicated and time-consuming procedures for decomposing cells into single chemical or structural components are avoided. Consequently, simple and uniform procedures are available that are applicable to all bacteria that can be grown in culture. In short, small amounts of cells ($\approx$ 10-60 μg) grown under standardized conditions (concerning growth medium, streak pattern, and temperature and time of growth) are transferred from the culture to an IR-sample holder. After a short drying procedure, the sample holder is exposed to the IR-beam, the spectrum is recorded and automatically stored in a data file. The characterization of micro-organisms is accomplished with the aid of comprehensive, computer-stored reference databases using search algorithms, cluster analysis or other multivariate statistics (Naumann *et al.*, 1988 *a,b*; Helm *et al.*, 1991 *a,b*). Data evaluation is performed either by a library search, or by a comparison of all spectra collected from identical experimental set-ups. Results are presented by hit lists, dendrograms or by 'factorial maps' displaying graphically the spectral similarities and classifications of the strains under study.

FT-IR SPECTROSCOPIC CLASSIFICATION OF STAPHYLOCOCCI

Staphylococci are Gram-positive bacteria which inhabit the skin and mucous membranes of mammals (including man) and birds. Most of the staphylococci are known as harmless saprophytes, however some strains can cause severe infections in man and animals (Taufer and Zangger, 1981). The well circumscribed genus (Mordarski *et al.*, 1991), currently comprising 30 species (Kloos *et al.*, 1992; Hájek *et al.*, 1992; Chesneau *et al.*, 1993), has recently gained increased taxonomic interest due to the discovery of new species and a wide host range (*e.g.* Freney *et al.*, 1988; Varaldo *et al.*, 1988; Igimi *et al.*, 1989; Hájek *et al.*, 1992). In addition, with the development of new tools for the determination of phenotypic and/or genetic relationships, the systematics of the staphylococci has undergone numerous revisions in the past decades (for review see Kloos, 1980; Zakrzewska-Czerwinska *et al.*, 1991; Mordarski *et al.*, 1991).

Such a new tool is the FT-IR spectroscopy of intact bacteria grown on agar plates. This chapter is about the classification of some species of staphylococci using spectroscopic information and cluster analysis. Clustering of staphylococcal species on the

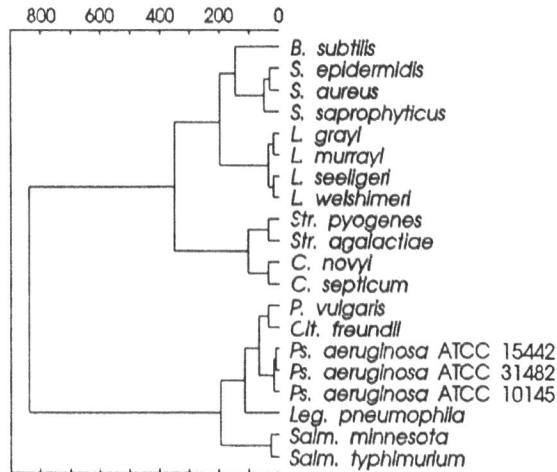

Heterogeneity

Figure 1. FT-IR spectroscopic classification of three staphylococcal species in respect to some other Gram-positive and Gram-negative species. Cluster analysis was performed using the similarities between the first derivatives of the spectra considering the five spectral sub-ranges previously defined (Helm *et al.*, 1991 *b*). The fatty acid range, *i.e.* 3000-2800 cm^{-1}, was given double weight in order to yield clear separation between Gram-positive and Gram-negative species. Ward's algorithm was used for cluster analysis (Ward, 1983). Abbreviations: *B.*, *Bacillus*; *C.*; *Clostridium*; *Cit.*, *Citrobacter*; *L.*, *Listeria*; *Leg.*, *Legionella*; *P.*, *Proteus*; *Ps.*, *Pseudomonas*; *S.*, *Staphylococcus*; *Salm.*, *Salmonella*; *Str.*, *Streptococcus*.

basis of their spectra has yielded a new classification which parallels to some extent the biochemical properties of the strains under study.

Intrageneric Relationships of Staphylococci

Figure 1 depicts the result of an FT-IR spectroscopic classification of three staphylococcal species in respect to closely related bacteria and to unrelated bacteria, respectively. In order to reveal the "overall" spectroscopic similarity between the strains investigated, all five spectral sub-ranges defined by Helm *et al.* (1991 *b*) were considered. This encompasses information about fatty acid components of the membrane, proteins and peptides, phosphate carrying compounds, polysaccharides of the cell wall and, in a less dominant way, since their specific absorbance bands are comparatively weak, nucleic acids. Thus, the result is an "overall" chemotaxonomic image of the intergeneric relationships of the staphylococci. Two major clusters were formed. The Gram-positive cluster, which was less dense than the Gram-negative one, consisted of specimens of the genera *Bacillus*, *Listeria*, *Staphylococcus*, *Streptococcus* and *Clostridium*. This is in agreement with the affiliation of the genus *Staphylococcus* to the *Bacillus-Lactobacillus-Streptococcus-(Clostridium)* branch of the eubacteria (Goodfellow, 1987; Ludwig *et al.*, 1981). From an FT-IR spectroscopic point of view, the staphylococcal species seem to be more closely related to *Listeria* and *Bacillus* than to *Streptococcus* and *Clostridium*. It can be assumed that FT-IR primarily reveals the biochemical similarity between strains and - in a more indirect way - the phenotypic and genetic relationships, too. Although FT-IR cannot clarify the phylogenetic relationships of bacteria, FT-IR spectroscopic classification is phenetically sound and, moreover, is useful at different levels of taxonomic discrimination (Hedrick *et al.*, 1991; Helm *et al.*, 1991 *a*). All the strains of *Staphylococcus, Streptococcus, Clostridium* and *Listeria* investigated here formed compact genus clusters.

The Gram-negative cluster was composed of *Proteus, Citrobacter, Pseudomonas, Legionella* and *Salmonella* strains. Grouping of *Salmonella* strains distinct from other *Enterobacteriaceae* has previously been established using FT-IR spectroscopic analysis (Helm *et al.*, 1991 *a*). Within the Gram-negative cluster, the sub-cluster comprising three different *Pseudomonas aeruginosa* strains shows the resolution of the method for different strains of the same species.

Intrageneric Relationships of Staphylococci

The intrageneric relationships of staphylococci, as seen by FT-IR spectroscopy, are depicted in Figure 2. As in the preceding study, all five spectral ranges were considered

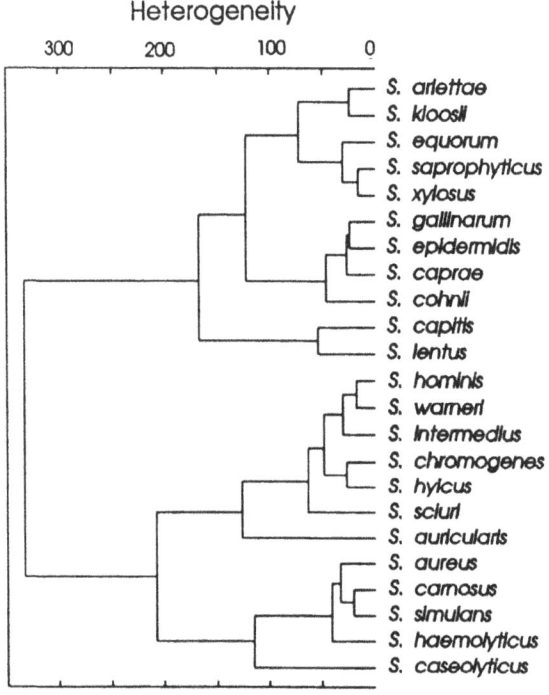

Figure 2.   An "overall" FT-IR spectroscopic classification of the type strains from 23 staphylococcal species. Cluster analysis was performed using the similarities between the first derivatives of the spectra considering the five spectral sub-ranges[*] previously defined (Helm *et al.*, 1991 *b*), *i.e.* 3000-2800 cm^{-1}, 1800-1500 cm^{-1}, 1500-1200 cm^{-1}, 1200-900 cm^{-1} and 900-700 cm^{-1}. All ranges have been equally weighted. Ward's algorithm was used for cluster analysis (Ward, 1963).

---

[*] Please note that the magnitude of heterogeneity depends on (i) the number of the spectra in the analysis, (ii) the similarity between them, and (iii) the number and kind of the spectral sub-ranges selected.

Heterogeneity

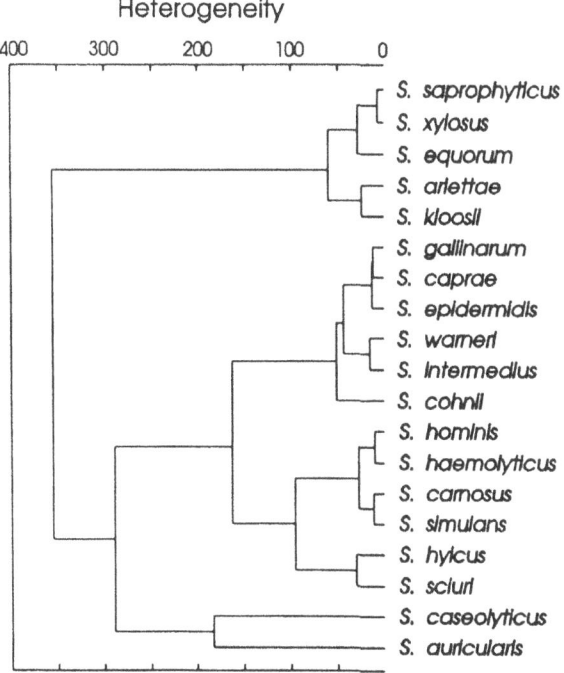

Figure 3. FT-IR spectroscopic classification of the type strains from 19 staphylococcal species . Cluster analysis was performed using the similarities between the first derivatives of the spectra considering the spectral range 1300-800 cm^{-1}. Ward's algorithm was used for cluster analysis (Ward, 1963).

in the analysis. At a heterogeneity of about 100, five species clusters of staphylococci could be detected. In most cases, these groups differed from those six species groups which have been recognized on the basis of DNA relationships and extensive phenotypic character analysis (Kloos *et al.*, 1992). The first species cluster, as defined by FT-IR, consisted of *S. saprophyticus, S. xylosus, S. arlettae, S. kloosii* and *S. equorum*. The corresponding species group determined by genomic and phenotypic analysis additionally included the species *S. cohnii* and *S. gallinarum*. The second FT-IR species cluster was composed of *S. epidermidis, S. gallinarum* and *S. caprae. S. cohnii* was at the margin of this cluster. The third group was composed of *S. capitis* and *S. lentus*. The fourth FT-IR species group contained *S. hominis, S. warneri, S. intermedius, S. hyicus* and *S. chromogenes. S. sciuri* was more peripherally grouped. The close spectroscopic resemblance between *S. hyicus* and *S. chromogenes* is in accordance with the genetic and phenotypic similarity between these species (Kloos *et al.*, 1992; Hájek *et al.*, 1986). The fifth species cluster consisted of *S. aureus, S. carnosus, S. simulans* and *S. haemolyticus*.

A clear separation of *S. aureus* from other species, as indicated by genetic and phenotypic analysis (Kloos *et al.*, 1992), was not supported by FT-IR spectroscopy. However, as described by Kloos *et al.* (1992), *S. auricularis* and *S. caseolyticus* were not included in any of the species groups.

Grouping of Staphylococci According to the Type of Cell Wall Teichoic Acid

A specific grouping of some staphylococcal type strains could be achieved by the selection of that part of the FT-IR spectrum which represents information about the phosphate and sugar components of the cell, *i.e.* the spectral range of 1300-800 cm^{-1}. The result was a classification in good agreement with the composition of the cell wall teichoic acids (Schleifer and Kroppenstedt, 1990; Kloos *et al.*, 1992). At a heterogeneity of about 100, five clusters and two unclustered species could be detected (Figure 3). The first cluster contained species which synthesize both, poly(glycerolphosphate) and poly(ribitolphosphate). The second cluster was composed of strains containing poly(glycerolphosphate) with glucose and N-acetylglucosamine as substituents.

The third cluster was composed of strains which produce poly(glycerolphosphate) with either N-acetylglucosamine or N-acetylgalactosamine. Quite distinct from these, the type strains of *S. hyicus* and *S. sciuri* were grouped together. Both species produce poly(glycerolphosphateglycosylphosphate). *Staphylococcus caseolyticus* and *S. auricularis*, whose cell walls contain the rather atypical teichoic acid poly(N-acetylglucosaminylphosphate), were both grouped peripherally, but failed to form a dense cluster. It is evident from Figure 3 that the sugar and/or amino sugar substituents of the cell wall teichoic acids, despite being the bulk material, play only a minor role in the classification observed. This is predictable because, for example, N-acetylglucosamine and N-acetylgalactosamine, are merely steric isomers.

FT-IR SPECTROSCOPIC DIFFERENTIATION OF LPS-MUTANT STRAINS OF *SALMONELLA MINNESOTA*

Salmonellae, like other Gram-negative bacteria, possess a second membrane layer additional to the cytoplasmic membrane, the outer membrane (OM). The OM plays an important role in the defence against hydrophilic high-molecular solutes and a variety of hydrophobic drugs (Rietschel *et al.*, 1984; Nikaido and Vaara, 1985; Sukupolvi and Vaara, 1989). The asymmetrically designed outer membrane is assembled from phospholipids (inner leaflet), lipopolysaccharides (LPS) (outer leaflet), and unspecific, as well as specific, pore proteins. The LPS are essential for the function of the OM as an effective permeation barrier. It was shown that in mutant strains with defective LPS-structures the barrier function of the OM against hydrophobic drugs was dramatically reduced (Vaara and Vaara, 1983 *a,b*). Thus, the increased permeability of the so-called "deep rough" LPS-mutants reflects the crucial role of the LPS for an intact organization of the OM. FT-IR spectroscopic investigations of intact cells of *Salmonella minnesota* yielded new insights into the membrane organization of LPS-mutants and the biophysical differences between them (Schultz, 1993; Schultz and Naumann, 1993). Additionally, cluster analysis on the basis of selected IR-spectral sub-ranges yielded meaningful classifications which may be helpful in the characterization of LPS-mutants of *Salmonella* and other bacteria.

The Effect of Different LPS-Structures in the Sugar and Phosphate Region

Figure 4A shows the clusters obtained by FT-IR spectroscopic analysis of the wild type and different LPS-mutant strains of *S. minnesota* when using that spectral region

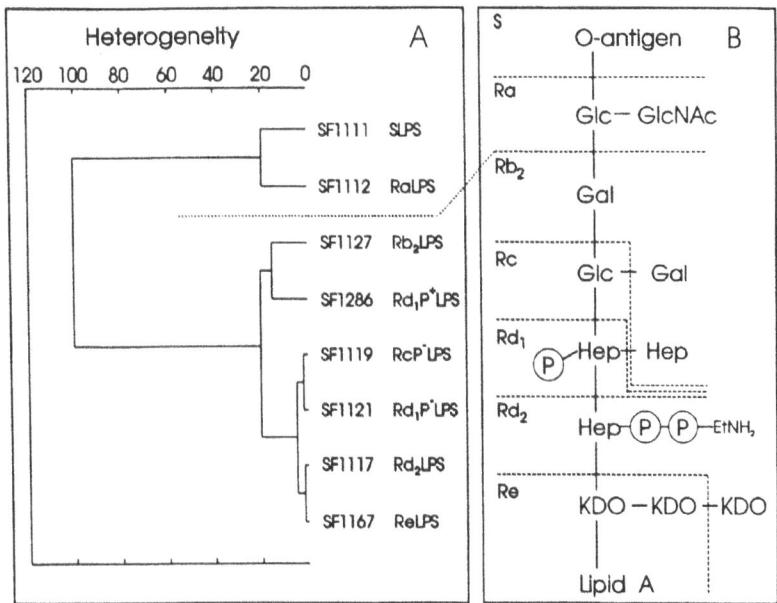

Figure 4.   A) FT-IR spectroscopic classification of wild type and different LPS-mutant strains of *Salmonella minnesota* using the spectral region of sugar and phosphate vibrations (1200-900 cm^{-1}). Cluster analysis was performed using Ward's algorithm (Ward, 1963).
B) For clarity and better legibility the structure of intact and defective LPS is given together with the commonly used nomenclature for defective LPS-structures.

which is dominated by sugars and phosphate carrying compounds (*i.e.* 1200-900 cm^{-1}). For explanation, Figure 4B describes the principles of LPS-structure together with the commonly used designations for smooth and rough strains (S, Ra, Rb, Rc, Rd, and Re). The dendrogram resolved two well separated main clusters which contained strains with intact (S) or less defective LPS-structures (Ra LPS) and those with more defective LPS-structures, respectively (see Figure 4A). The classification was strongly correlated with the number of sugar substituents in the O-specific chain (S), the outer core (Ra, Rb) and the inner core region (Rc, Rd, Re). Sub-clusters were formed by the Rb$_2$- and Rd$_1$P$^+$-mutants and the strains with unphosphorylated LPS, respectively. The spectral differences between the Rd$_1$P$^-$, RcP$^-$, Rd$_2$ and ReLPS-mutants were not sufficient to discriminate between these strains. However, from an FT-IR spectroscopic point of view, the sugar characteristics of the modified LPS-structures determined the classification of the LPS-mutants.

The Effect of Different LPS-Structures in the Fatty Acid and Protein Region

The influence of different LPS-structures on the grouping of the strains under study when using two other spectral sub-ranges, namely the "fatty acyl chain" region (3000-2800 cm^{-1}) and the "protein" region (1700-1620 cm^{-1}), is depicted in the Figures 5A and

B, respectively. The dendrogram shown in Figure 5A divided the strains into two major clusters in a similar way as in Figure 4. Two groups were obtained which comprised the strains with intact or less defective LPS and those with more defective LPS, respectively. The classification into two groups seemed to be influenced by the presence or absence of a single phosphate group located in the inner core region (heptoses) of LPS (see Figure 4B). This observation agrees well with the presence or absence of an intact barrier function of the outer membrane against hydrophobic drugs (Vaara and Vaara, 1983 a,b) and correlates directly with the described observation of the appearance of phospholipids in the outer leaflet of these mutants. Also documented in Figure 5A is the effect of different growth media (peptone or endo agar) on the classification. The differences caused by the different media are sufficient to divide each main cluster into two sub-clusters.

In the "protein" region (see Figure 5B) the cluster analysis yielded a dendrogram which also revealed two main groups. These two clusters correlated with the results of Nikaido and Vaara (1985), who found reduced content of outer membrane proteins (porins) in the more defective LPS-mutants (Rc through Re), whereas the strains with lesser defects possessed greater amounts of porins. For the "protein" region only small spectral differences between the strains were observed but they were, however, highly reproducible and reflected the unique variations of membrane protein content caused by defective LPS-structures.

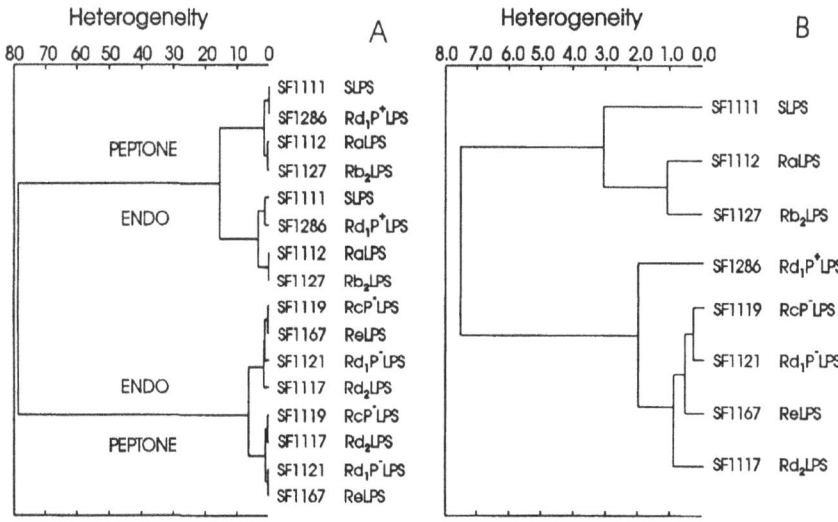

Figure 5. Classification of wild type and LPS-mutant strains of *Salmonella minnesota* using two different FT-IR spectroscopic sub-ranges (*i.e.* 3000-2800 cm^{-1} (A) and 1700-1620 cm^{-1} (B)). Spectra were baseline-corrected and normalized in respect to the intensity of the band near 2959 cm^{-1} (-CH$_3$ asymmetric stretching band) (A), or Fourier-selfdeconvoluted with a Lorentzian line shape function of 18 cm^{-1} half-width and a resolution enhancement factor k of 2.4 cm^{-1} (B). Cluster analysis was performed using Ward's algorithm (Ward, 1963).

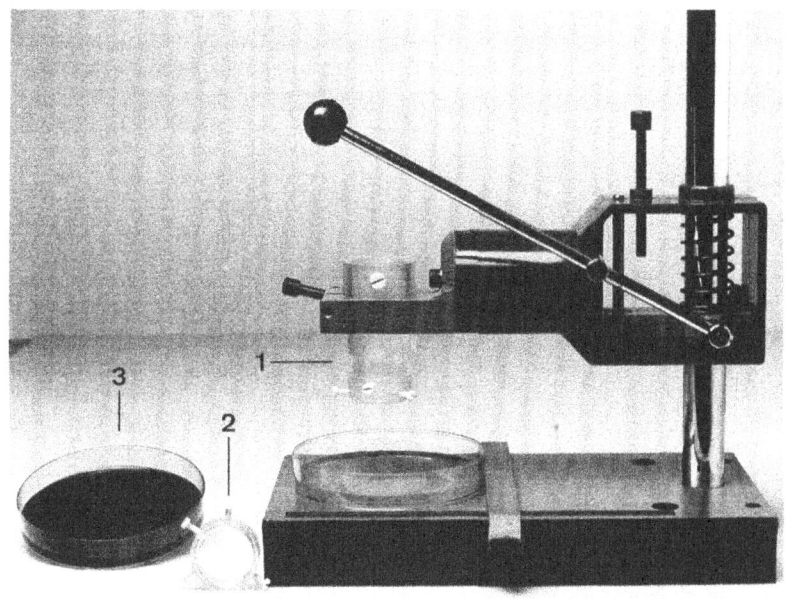

Figure 6. The home-made apparatus for obtaining prints from single microcolonies.
(1) The stamp holder, (2) The stamp with a BaF_2-window, (3) Agar plate.

DETECTION, ENUMERATION AND DIFFERENTIATION OF MICRO-ORGANISMS BY MEANS OF FT-IR MICROSCOPY

Having established that the Fourier-transform infrared spectroscopic patterns of intact bacterial cells can be used to differentiate rapidly a variety of bacterial strains and taxa, we were encouraged by the recent advent of high-quality FT-IR microscopes to develop a technique which combines the advantages of light microscopy with the sensitivity and selectivity of FT-IR. We aimed to be able to detect and differentiate/identify bacterial microcolonies in the order of 20 to 100 μm in diameter (approx. 10^2 to 10^3 bacterial cells) and to enumerate microcolony-forming cells.

A Typical Protocol for Obtaining FT-IR Spectra of Bacterial Microcolonies

In order to measure bacterial microcolonies, the following procedure has been elaborated: An aliquot of a sufficiently diluted bacterial suspension (*e.g.* urine) is plated on an appropriate solid growth medium (agar plate) and is incubated for 6 to 8 hours at 37 °C. A round, IR-transparent BaF_2-plate (typically 25 mm in diameter) is then pressed gently onto the agar surface using a special, home-made device (see Figure 6). Thus, small amounts (replica) of the bacterial microcolonies (approx. 2 to 3 bacterial layers per spot) are transferred to the IR-plate. These are dried to transparent films within seconds at ambient conditions and are subjected to FT-IR microscopy. Spectra are recorded applying round apertures of 40, 60, 80 and 100 μm in diameter, respectively. The optical properties (size etc.) of replica and the number of spots to be estimated are controlled by the operator or by imaging techniques. Figure 7 gives a typical micrograph obtained from microcolonies by the apparatus shown in Figure 6.

Differentiation Test

Figure 8 shows the dendrogram of a cluster analysis and the factorial map of a correspondence analysis performed on 40 FT-IR spectra of some selected bacterial spots from an artificial mixed culture containing three different kind of bacteria, namely *Escherichia coli* (isolated from an urine specimen) and two different strains of *Staphylococcus aureus* (from our strain collection). Identification of *E. coli* was attained by conventional microbiological techniques. Data elaboration and cluster analysis were carried out as described previously (Naumann *et al.*, 1988 *a,b*, 1991 *a*; Helm *et al.*, 1991 *a*).

From the results depicted in Figure 8 we conclude that (i) bacterial microcolonies (40 to 100 μm in diameter) can be transferred from solid culture plates onto IR-transparent plates by a special replica (printing) technique for subsequent FT-IR measurements, (ii) colony spots can be enumerated, and (iii) the data accessible from the light-microscopic images (numbers, size and different shapes etc.) and from the FT-IR spectra of bacterial colonies (cell composition, structural data etc.) can be successfully used to differentiate and classify bacteria and to characterize colony growth.

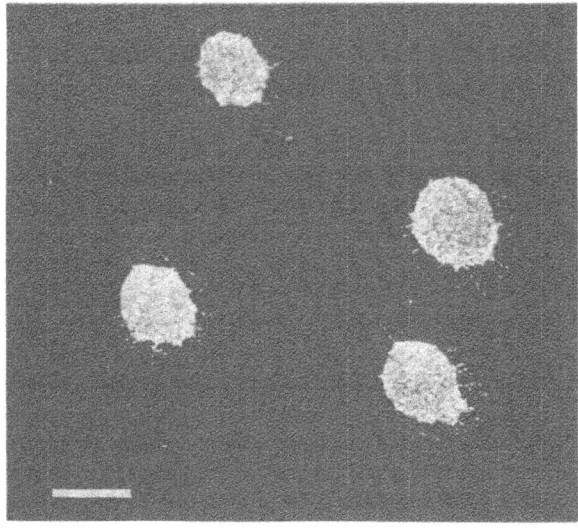

Figure 7.   Micrograph of four colony spots of *Staphylococcus aureus* Pelzer obtained by the stamping technique. Time of growth: 8 hrs; medium: peptone agar; techniques and physical parameters for obtaining the FT-IR spectra from microcolonies are given by Naumann *et al.* (1991 *a*). Bar: 100 μm.

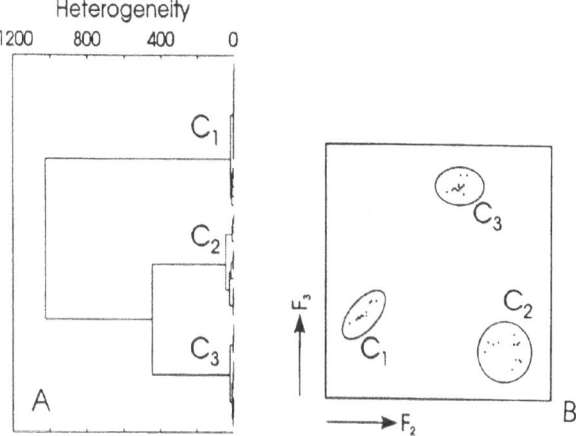

Figure 8.   Classification of FT-IR microscopic spectra obtained from microcolonies of a mixed culture. A) Dendrogram of the classification of 40 spectra. Spectra were recorded applying round apertures of 40, 60, 80 and 100 μm in diameter, respectively. Cluster analysis was performed with band-pass filtered spectra considering the two equally weighted spectral ranges 1200 -1500 cm^{-1} and 900 - 1200 cm^{-1}. Ward's algorithm was applied (Ward, 1963).
B) Factorial map for the same 40 spectra. Each spectrum is depicted by a point. Correspondence analysis was performed using peak normalization, modulation metric and 12 *eigen*-vectors (van Heel and Keegstra, 1981; van Heel, 1984). Spectral sub-ranges and weighting factors are given in Figure 8A.
C_1 = *Escherichia coli*, C_2 = *Staphylococcus aureus* Pelzer, C_3 = *S. aureus* SG511.

Furthermore, the application of multivariate statistical image analysis to both, the 2D-images of colony prints (replica) and 1D-"images" of bacterial FT-IR spectra, may strongly enhance the capacity and versatility of this technique.

The technique can therefore be used to combine the three fundamental steps of microbiological analysis of bacteria, namely detection, enumeration and differentiation.

MONITORING OF ANTIBIOTIC-INDUCED ALTERATIONS BY FT-IR SPECTROSCOPY

Chloramphenicol (CAP), originally isolated from the actinomycete *Streptomyces venezuelae*, is a bacteriostatic compound which inhibits the peptidyl transferase activity of the 50S ribosomal subunits (Nierhaus and Wittmann, 1980). Since CAP inhibits protein biosynthesis, but does not primarily interfere with the incorporation of peptides and amino sugars into the peptidoglycan sacculus, staphylococci respond to application of CAP with up to tenfold thickened cell walls in comparison to untreated cells (Giesbrecht and Ruska, 1968). Additionally, the content of O-acetyl groups in the cell wall was found to be considerably increased after treatment with CAP (Johannsen *et al.*, 1983). FT-IR has already been proved to be useful for the monitoring of antibiotic treatment, since it reveals subtle changes in the cell composition. (Naumann *et al.*, 1991 a). We have therefore tested the value of selected FT-IR marker bands for improved analysis of alterations of some particular cellular components of *S. aureus* under the influence of CAP.

Detection of Subtle Changes in Cell Composition after Treatment with CAP

Staphylococcus aureus SG511 was grown in bactopeptone broth pH 7.2 containing 2.5% bactopeptone (Difco) and 0.5% NaCl. CAP was added at zero time to give a final concentration of 20 μg/ml. Cells were harvested, centrifugated and washed twice with physiological sodium chloride solution and distilled water, respectively. After appropriate dilution of the bacterial suspension, spectra were obtained with a maximum absorbance in the range of 0.4-0.6 absorbance units. Therefore, the standard error of repeated measurements of the same suspension was > 4%. Prior to the calculation of the second derivatives, spectra were base-line corrected and normalized to unity area. Six suitable peaks or peak systems were selected to serve as marker bands for particular cell components (Table 1). Peak areas were calculated as indicated in Figure 9 and compared with those determined for the control culture (Figure 10A-F). Due to the complex contour of bacterial spectra (mainly caused by superimposed peaks) it is more sensible to determine the peak areas rather than the mere peak heights.

All the peaks analysed showed only slowly increasing areas when compared with the control, indicating a generally reduced metabolism which could previously be demonstrated by flow microcalorimetry (Krüger and Giesbrecht, 1989). Moreover, the marker band for fatty acids of the membrane was slightly decreased immediately after application of CAP. This is an indication of changes in the order of the membrane caused by a greater number of CH_2-groups of long-chain fatty acids in the *cis*-position. The markers for both, tyrosine and carboxylate remained almost constant in CAP-treated cells. Given that tyrosine occurs exclusively in proteins and not in the peptide component of the peptidoglycan, this result corresponds to the sudden cessation of protein biosynthesis.

Because of the nearly identical course of both markers, it can be concluded that the carboxylate peak was mainly or exclusively caused by (free carboxyl groups of) aspartic acid and glutamic acid in peptide chains rather than by dissolved amino acids. In contrast to peptides and fatty acids, the marker bands for phosphates, esters, and polysaccharides of the cell wall were less drastically affected. This suggests an almost undisturbed

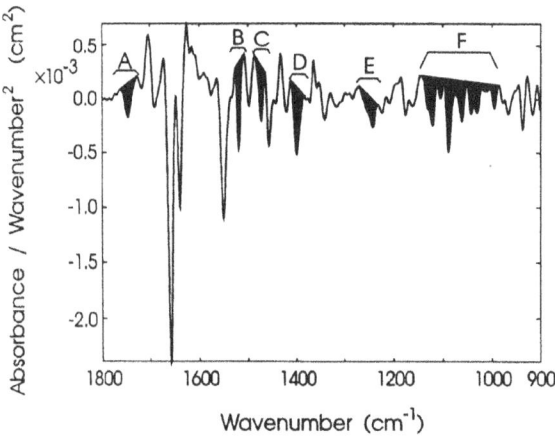

Figure 9.   Selected peaks and peak systems which can be used for the detection of subtle changes in cell composition and metabolism under the influence of antibiotics. A second derivative spectrum from *Staphylococcus aureus* SG511 is shown. The peak designations A through F correspond to those used in Table 1 and Figure 10. Note that in second derivative spectra the peaks are directed downwards.

Table 1. Selected FT-IR spectroscopic marker bands for the determination of alterations of particular cell components in cells of *Staphylococcus aureus*

Peak	Position Wavenumber [cm^{-1}]	Functional group(s) / compound	Usable as marker for:
A	≈1740	>C=O stretching vibrations of esters	Lipoteichoic acids, Phospholipids, O-acetyl groups
B	≈1516	Tyrosine	Proteins and peptides
C	≈1467	C—H scissoring vibrations of >CH$_2$ methylene in fatty acids	State of order of the membrane, fatty acid composition of the membrane
D	≈1400	—COO$^-$ carboxylate	Aspartic acid and glutamic acid in peptides, free amino acids, free fatty acids
E	≈1245	P=O stretching vibrations of >PO$_2^-$ in phosphodiesters	Phospholips, Nucleic acids
F	≈1085	C—O—C, C—O, dominated by ring vibrations of polysaccharides	Polysaccharides of the cell wall

incorporation of teichoic acids and peptidoglycan into the cell envelope during the first 30-60 min after application of the antibiotic.

The curve obtained for the marker band for esters brought no evidence for an increase of O-acetyl groups. However, this is understandable since the >C=O groups in both, teichoic acid and O-acetylated muramic acid contribute to this peak. Additional analysis of the amide I peak, which is derived from both, the proteins and O-acetyl groups, will probably lead to a more detailed picture. The most important result of this analysis, however, is the detection of alterations in cell composition and metabolism within the first 30 min after application of the antibiotic. Even at this early stage of CAP action, FT-IR spectroscopy is capable of detecting significant changes in cell composition. Therefore, we consider that FT-IR is a very useful method for the screening and testing of potentially antibiotic drugs.

DETECTION AND MEASUREMENT OF INTRACELLULAR POLY-β-HYDROXYBUTYRATE IN *LEGIONELLA PNEUMOPHILA*

Poly-β-hydroxyfatty acids are energy and carbon reserves in prokaryotes. Poly-β-hydroxybutyrate (PHB), a homopolymer of D(-)-3-hydroxybutyrate is a common storage compound in many bacteria (*e.g. Azotobacter, Bacillus, Chromatium, Pseudomonas*) (Dawes and Senior, 1973). Production of PHB in some *Legionella* species was recently detected by FT-IR spectroscopy (Helm *et al.*, 1991 *a*). In most cases, PHB is accumulated under nutrient limitation but the supply of energy and carbon is in excess. Under conditions of starvation the compound may be used as an energy and/or carbon source (Merrick, 1978; Dawes, 1986). Intracellular reserve PHB is located within light-

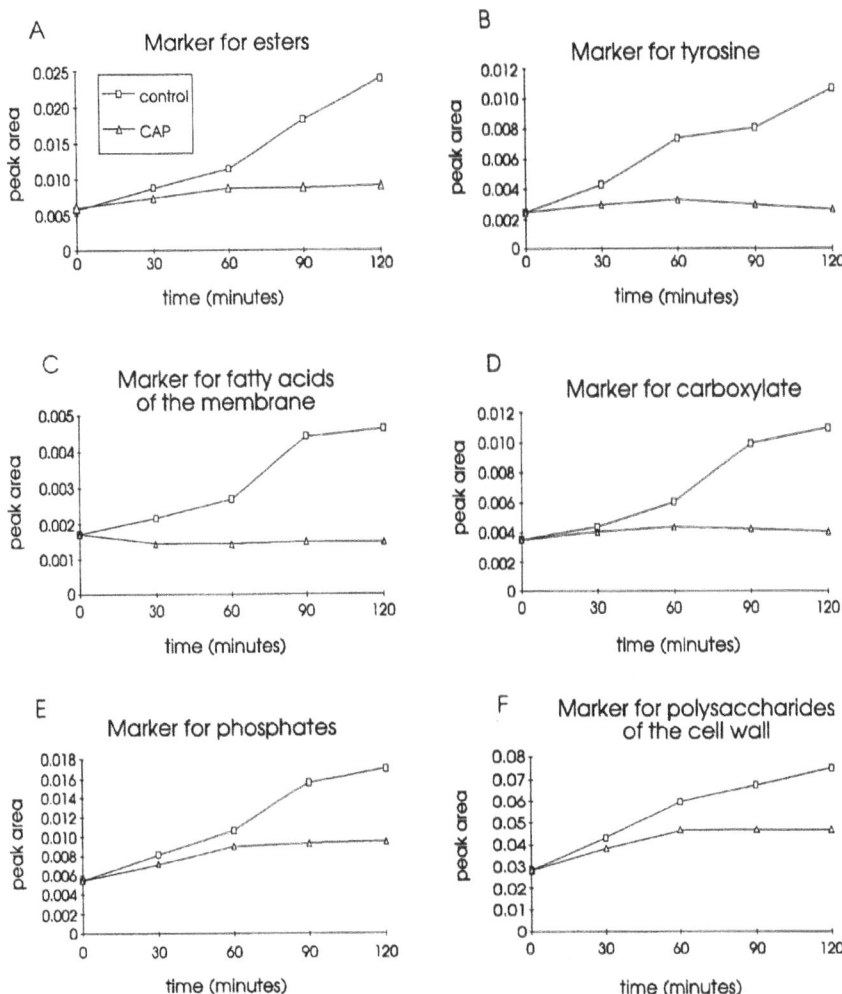

Figure 10. Changes in the amount of some cellular components of *Staphylococcus aureus* SG511 growing under the influence of 20 μg/ml chloramphenicol (CAP) as determined by FT-IR. The antibiotic was added at minute zero. For selection of the marker bands used refer to Figure 9 and Table 1.

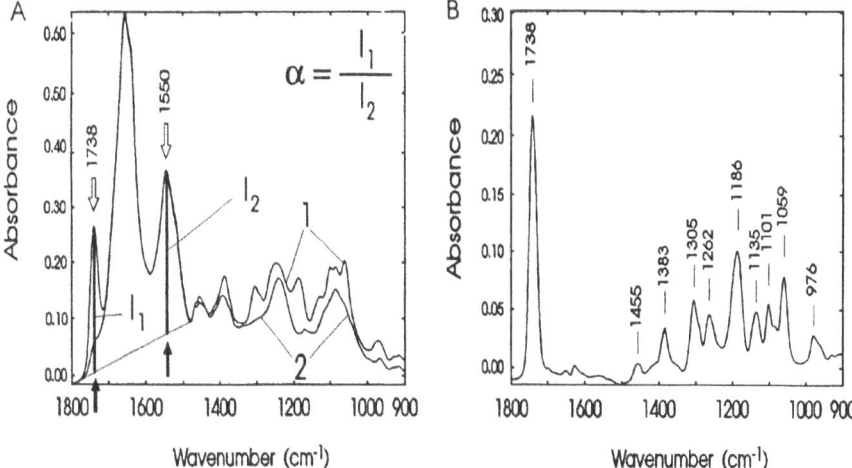

Figure 11. (A) FT-IR spectra of *Legionella pneumophila*, strain II8, grown on CYE-agar plates for 48 h (1) and for 120 h (2). The two peaks used for determination of α are marked by arrows. Base line correction was performed as indicated (black arrows).
(B) Difference spectrum (1 minus 2). Ten typical peaks are labelled. For assignments see Table 2.

microscopically visible granules which are surrounded by a membrane (Lundgren and Pfister, 1964). The enzyme PHB synthase was found to be granule-bound (Dawes and Senior, 1973). The task of this study was to determine production of PHB by FT-IR spectroscopy. PHB was identified and measured quantitatively in *Legionella pneumophila* strain II8 *in situ*.

Difference Spectroscopy

Figure 11A shows the FT-IR spectra of *L. pneumophila* strain II8 grown on CYE-agar (charcoal/yeast extract agar) for 48 h (PHB-rich cells) and 72 h (cells almost devoid of PHB granules), respectively. Examination by eye revealed some remarkable differences in the carbonyl ester peak and, less prominent, in the region between 1500-900 cm^{-1}. By the aid of digital subtraction techniques a difference spectrum was calculated which should represent a spectrum of PHB-granules rather than that of the pure compound. However, the resulting difference spectrum (Figure 11B) resembled FT-IR spectra recorded for isolated and synthetic PHB (Szewcyk and Mikucki, 1989). Ten peaks could be identified and assigned to typical polyester absorbance features (Table 2). Interestingly, no peaks could be found within the difference spectrum which could be assigned to the various enzymes associated with native PHB-granules. Perhaps the enzymes involved in PHB metabolism of *Legionella pneumophila* are produced constitutively, as it was found in *Alcaligenes eutrophus* (Haywood *et al.*, 1989).

Quantitative Analysis of the Relative PHB-Content

Quantitative determination of PHB was assessed by calculation of the ratio $\alpha = I_1/I_2$, where I_1 is the intensity of the ester carbonyl peak at 1738 cm^{-1} (used as measure of PHB-content), and I_2 is the intensity of the amide II peak at 1550 cm^{-1} (used as measure of cell mass). Because the amide II peak remains constant, α gives the amount of PHB in relation to the cell mass. Since other esters, like, *e.g.* the phosphlipids present in most bacterial membranes, contribute to the ester carbonyl peak, too, α will always be greater than zero. In practise, spectra of PHB-free *Legionella pneumophila* cells yielded an α of $\approx 0.14 \pm 0.03$.

Accumulation of PHB in *L. pneumophila* started during the early stages of growth and was possibly not caused by nutrient limitations. Maximum PHB-content was attained after 48 hr (incubation at 37 °C, CYE-medium). PHB was subsequently degraded during the later stages of the culture. When PHB-rich cells were transferred to fresh medium, the polymer was degraded during the *lag*-phase of the new culture. Smaller inocula and decreased growth temperature led to delayed accumulation of the polymer. Degradation of the polymer was not detectable when PHB-rich cells were held in pure water.

Table 2. Identification and assignment of some peaks essential for PHB-containing *Legionella pneumophila*

Peak #	Peak Position Wavenumber [cm^{-1}]	Assignment
1	$\approx$1738	C=O stretching vibrations in esters
2	$\approx$1455	C-H bending (symmetrical) of >CH$_2$ methylene and C-H bending (asymmetrical) of -CH$_3$ methyl groups
3	$\approx$1383	C-H bending (asymmetrical) of -CH$_3$ methyl groups
4	$\approx$1305	?
5	$\approx$1262	?
6	$\approx$1186	C-O stretching vibrations in esters
7	$\approx$1135	
8	$\approx$1101	C-O-C stretching vibrations (asymmetrical)
9	$\approx$1059	
10	$\approx$976	?

CONCLUSIONS

FT-IR spectroscopy is an extremely specific technique which can provide a wealth of qualitative and quantitative information about micro-organisms. Therefore, FT-IR is useful in many fields of microbiological research. Cluster analysis of bacteria on the basis of their spectra has yielded new classifications which parallel to some extent the biochemical and structural properties of the strains under study. Although FT-IR cannot reveal the phylogenetic relationships of bacteria, FT-IR spectroscopic classification is phenetically sound and, moreover, is useful at different taxonomic levels *(i.e.* genus, species, strain).

Specific grouping of bacteria can be obtained by the selection of those parts of the spectrum which represent information about particular cell components (*e.g.* fatty acids, polysaccharides or proteins). An FT-IR spectroscopic grouping of staphylococci was in good agreement with the composition of cell wall teichoic acids. Investigations of smooth and rough mutants of *Salmonella minnesota* yielded meaningful classifications which may be helpful in the characterization of lipopolysaccharide (LPS) mutants.

An FT-IR microscope was used to combine the fundamental steps of microbiological analysis of bacteria, namely detection, enumeration and differentiation. We have shown that FT-IR microscopy enables the detection and differentiation of bacterial microcolonies in the order of 20-100 μm in diameter (approx. 10^2-10^3 bacterial cells) and the enumeration of microcolony-forming units.

Selected FT-IR peaks can be used for the analysis of changes in cell composition and metabolism under the influence of antibiotics. We have tested the value of this technique after application of chloramphenicol to a culture of *Staphylococcus aureus*. The analysis revealed significant alterations within the first 30 min after application. Therefore, we consider that FT-IR is a very useful method for the screening and testing of potentially antibiotic drugs.

Another task of our study was to determine production of poly-β-hydroxybutyrate (PHB) by FT-IR spectroscopy. PHB, an intracellular storage material, was identified and measured quantitatively in a *Legionella pneumophila* strain *in situ*.

Acknowledgements

Parts of this work were supported by a grant of the Deutsche Forschungsgemeinschaft (Na 226/1-1). We thank Andrea Stäuble for technical assistance, Angelika Brauer for preparing the typescript, Frank Remke for his contribution to the chloramphenicol study and Joachim Löffler for the measurement of poly-β-hydroxybutyrate.

REFERENCES

Chesneau, O., Morvan, A., Grimont, F., Labischinski, H. and El Solh, N. (1993) *Staphylococcus pasteuri* sp. nov., isolated from human, animal, and food specimens. Int. J. Syst. Bacteriol. 43, 237-244.

Dawes, E.A. (1986) "Microbial Energetics", p. 159. Blackie & Son Ltd., Glasgow.

Dawes, E.A. and Senior, P.J. (1973) The role and regulation of energy reserve polymers in micro-organisms. Adv. Microbial Physiol. 10, 135-266.

Freney, J., Brun, Y., Bes, M., Meugnier, H., Grimont, F., Grimont, P.A.D., Nervi, C. and Fleurette, J. (1988) *Staphyloccus lugdunensis* sp. nov. and *Staphylococcus schleiferi* sp. nov., two new species from human clinical specimens. Int. J. Syst. Bacteriol. 38, 168-172.

Giesbrecht, P. and Ruska, H. (1968) Über Veränderungen der Feinstruktur von Bakterien unter der Einwirkung von Chloramphenicol. Klin. Wochenschrift 46, 575-582.

Goodfellow, M. (1987) Taxonomy of coagulase-negative staphylococci. Problems and perspectives. Zbl. Bakt. Suppl. 16, 1-14.

Hajék, V., Devriese, L.A., Mordarski, M., Goodfellow, M., Pulverer, G. and Varaldo, P.E. (1986) Elevation of *Staphylococcus hyicus* subsp. *chromogenes* (Devriese *et al.*, 1978) to species status: *Staphylococcus chromogenes* (Devriese *et al.*, 1978) comb. nov. Syst. Appl. Microbiol. 8, 169-173.

Hájek, V., Ludwig, W., Schleifer, K.H., Springer, N., Zitzelberger, W., Kroppenstedt, R.M. and Kocur, M. (1992) *Staphylococcus muscae*, a new species isolated from flies. Int. J. Syst. Bacteriol. 42, 97-101.

Haywood, G.W., Anderson, A.A., and Dawes, E.A. (1989) The importance of PHB-synthase substrate specifity in polyhydroxyalkanoate synthesis by *Alcaligenes eutrophus*. FEMS Microbiol. Letts. 57, 1-6.

Hedrick, D.B., Nivens, D.E., Stafford, C. and White, D.C. (1991) Rapid differentiation of archaebacteria from eubacteria by diffuse reflectance Fourier-transform IR spectroscopic analysis of lipid preparations. J. Microbiol. Methods 13, 67-73.

Helm, D. (1992) Ph.D. thesis. Freie Universität Berlin.

Helm, D., Labischinski, H., Schallehn, G. and Naumann, D. (1991 *a*) Classification and identification of bacteria by Fourier-transform infrared spectroscopy. J. Gen. Microbiol. 137, 69-79.

Helm, D., Labischinski, H. and Naumann, D. (1991 *b*) Elaboration of a procedure for identification of bacteria using Fourier-transform IR spectral libraries: a stepwise correlation approach. J. Microbiol. Methods 14, 127-142.

Igimi, S., Kawamura, S., Takahashi and Mitsuoka, T. (1989) *Staphylococcus felis*, a new species from clinical specimens from cats. Int. J. Syst. Bacteriol. 39, 373-377.

Johannsen, L., Labischinski, H., Reinicke, B. and Giesbrecht, P. (1983) Changes in the chemical structure of walls of *Staphylococcus aureus* grown in the presence of chloramphenicol. FEMS Microbiol. Letts. 16, 313-316.

Kloos, W.E. (1980) Natural populations of the genus *Staphylococcus*. Ann. Rev. Microbiol. 34, 559-592.

Kloos, W.E., Schleifer, K.-H.and Götz, F. (1992) The Genus *Staphylococcus*, in "The Prokaryotes. A Handbook on the Biology of Bacteria: Ecophysiology, Isolation, Identification, Applications", Vol. II, 2nd Edition, (Balows, A., Trüper, H.G., Dworkin, M., Harder, W. and Schleifer, K.-H., Eds.), pp. 1369-1420. Springer, New York.

Krüger, D. and Giesbrecht, P. (1989) Flow microcalorimetry as a tool for an improved analysis of antibiotic activity: The different stages of chloramphenicol action. Experientia 45, 322-325.

Ludwig, W., Schleifer, K.-H., Fox, G.E., Seewaldt, E., and Stackebrandt, E. (1981) A phylogenetic analysis of staphylococci, *Peptococcus saccharolyticus* and *Micrococcus mucilaginosus*. J. Gen. Microbiol. 125, 357-366.

Lundgren, D.G. and Pfister, R.M. (1964) Structure of poly-β-hydroxybutyric acid granules. J. Gen. Microbiol. 34, 441-446.

Merrick, J. M. (1978) Metabolism of reserve materials, in "Photosynthetic Bacteria", pp. 199-219, (Clayton, R. and Sistrom W.R., Eds), Plenum Press, New York.

Mordarski, M., Goodfellow, M. and Pulverer, G. (1991) Staphylococcal systematics: order out of chaos. Zbl. Bakt. Suppl. 21, 13-20.

Naumann, D., Fijala, V. and Labischinski, H. (1988 *a*) The differentiation and identification of pathogenic bacteria using FT-IR and multivariate statistical analysis. Mikrochim. Acta (Wien) I, 373-377.

Naumann, D., Fijala, V., Labischinski, H. and Giesbrecht, P. (1988 *b*) The rapid differentiation and identification of pathogenic bacteria using Fourier transform infrared spectroscopic and multivariate statistical analysis. J. Mol. Structure 174, 165-170.

Naumann, D., Helm, D., Labischinski, H. and Giesbrecht, P. (1991 *a*) The characterization of microorganisms by Fourier-transform infrared spectroscopy (FT-IR), in "Modern Techniques for Rapid Microbiological Analysis" (Nelson, W.H., Ed.), pp. 43-96. VCH Publishers, New York.

Naumann, D., Helm, D. and Labischinski, H. (1991 *b*) Microbiological characterizations by FT-IR spectroscopy. Nature 351, 81-82.

Nelson, W.H. (1991) "Modern Techniques for Rapid Microbiological Analysis", VCH Publishers, New York.

Nierhaus, K.H. and Wittmann, H.G. (1980) Ribosomal function and its inhibition by antibiotics in prokaryotes. Naturwissenschaften 67, 234-250.

Nikaido, H. and Vaara, M. (1985) Molecular basis of bacterial outer membrane permeability. Microbiol. Rev. 49, 1-32.

Rietschel, E.Th., Wollenweber, H.-W., Brade, H., Zähringer, U., Lindner, B., Seydel, U., Bradaczek, H., Barnickel, G., Labischinski, H. and Giesbrecht, P. (1984) Structure and conformation of the lipid A component of lipopolysaccharides, in "Handbook of Endotoxins", Vol. 1, "Chemistry of Endotoxins" (Rietschel, E.Th., Ed.), pp. 187-220. Elsevier, Amsterdam.

Schleifer, K.H. and Kroppenstedt, R.M. (1990) Chemical and molecular classification of staphylococci. J. Appl. Bacteriol. Symp. Suppl.1990, 9S-24S.

Schultz, Chr. (1993) Ph.D. thesis. Freie Universität Berlin.

Schultz, Chr. and Naumann, D. (1993) The state of order in the outer membrane of live cells of different LPS-mutant strains of *Salmonella minnesota* and *Salmonella typhimurium*: An FT-IR study. Direct biophysical evidence for an improved model of outer membrane permeability. J. Bact. (in the Press).

Sukupolvi, S. and Vaara, M. (1989) *Salmonella typhimurium* and *Escherichia coli* mutants with increased outer membrane permeability to hydrophobic compounds. Biochim. Biophys. Acta 988, 377-387.

Szewcyk, E. and Mikucki, J. (1989) Poly-β-hydroxybutyric acid in staphylococci. FEMS Microbiol. Letts. 61, 279-284.

Taufer, M. and Zangger, J. (1981) Staphylococci-absolutely not 'saprophytes' of the skin. Zbl. Bakt. Suppl. 10, 961-965.

Vaara, M. and Vaara, T. (1983 *a*) Polycations sensitize enteric bacteria to antibiotics. Antimicrob. Agents Chemother. 24, 107-113.

Vaara, M. and Vaara, T. (1983 *b*) Polycations as outer membrane-disorganizing agents. Antimicrob. Agents Chemother. 24, 114-122.

van Heel, M. (1984) Multivariate statistical classification of noisy images (randomly oriented biological macromolecules). Ultramicroscopy 13, 165-184.

van Heel, M. and Keegstra, W. (1981) IMAGIC: a fast, flexible and friendly image analysis software system. Ultramicroscopy 7, 113-130.

Varaldo, P.E., Kilpper-Bälz, R., Biavasco, F., Satta, G. and Schleifer, K.-H. (1988) *Staphylococcus delphini* sp. nov., a new coagulase-positive staphylococcal species isolated from dolphins. Int. J. Syst. Bacteriol. 38, 436-439.

Ward, J.H. (1963) Hierarchical grouping to optimize an objective function. J. Amer. Stat. Ass. 58, 236-244.

Zakrzewska-Czerwinska, J., Mordarski, M., Goodfellow, M. and Pulverer, G. (1991) Relatedness amongst staphylococcal species based on DNA homology data. Zbl. Bakt. Suppl. 21, 83-84.

CURIE POINT PYROLYSIS MASS SPECTROMETRY
AND ITS APPLICATION TO BACTERIAL SYSTEMATICS

Michael Goodfellow,[1] Jongsik Chun,[1] Ekrem Atalan [1]
and Jean-Jacques Sanglier [2]

[1]Department of Microbiology, The Medical School, Framlington Place
 Newcastle upon Tyne, U.K. and
[2]Preclinical Research, Sandoz Pharma Ltd.
 CH-4002 Basle, Switzerland

INTRODUCTION

The need to classify, identify and type microorganisms is an ever present theme in microbiology, notably in clinical and industrial microbiology. Here, identification is critical for distinguishing between potential pathogens or spoilage organisms, and commensals or contaminants of no significance. Similarly, the choice of microorganisms for industrial screening programmes, especially those with a low throughput, is primarily a problem of distinguishing between known organisms and recognising new ones. Further, effective typing procedures are essential in epidemiological tracing and in eliminating sources of microbial contamination.

Conventional techniques used for purposes such as those outlined above are often taxon specific, require a varied assortment of methods, media and reagents, and are generally slow given the time constraints within which decisions have to be made. These well known limitations fuel the requirement for the development of rapid, reproducible and cost effective methods for classifying, identifying and typing microorganisms. This review is concerned with the application and potential of Curie point pyrolysis mass spectrometry in the ever changing, controversial and exciting fields of pure and applied bacterial systematics.

Pyrolysis is the thermal degradation of complex organic material in an inert atmosphere or vacuum. Thermal intramolecular vibrations cause molecules to cleave at their weakest point to produce a mixture of low molecular weight, predominantly volatile organic compounds, the pyrolysate (Irwin, 1982). The components of the pyrolysate are then separated by mass spectrometry on the basis of their mass-to-charge ratio (m/z), to give a pyrolysis mass spectrum, which can be used as a 'chemical fingerprint' of the original material. The resultant data are complex and need to be analysed using suitable statistical routines (Magee, 1993, 1994). One of the major advantages of pyrolysis mass spectrometry (PyMS) over other taxonomic methods, such as conventional chemotaxonomic procedures

(Suzuki *et al.*, 1993) and nucleic acid probes (Schleifer *et al.*, 1993), is that it is rapid with respect to single and multiple samples. Typical sample time is less than 2 minutes and up to 300 samples can be analysed per batch. Pyrolysis techniques, notably Curie point pyrolysis mass spectrometry, are being introduced into some diagnostic and industrial screening laboratories (Sanglier *et al.*, 1992; Sisson *et al.*, 1992a).

Initial studies on Curie point PyMS of microorganisms were hampered given the expensive, cumbersome hardware and inadequate software for the analysis of pyrolysis data (Gutteridge, 1987). Manual loading of samples together with prolonged processing time meant than PyMS was labour intensive with a low sample throughput per day. Attempts to compare new PyMS data with results held in data libraries were bedevilled by inherent machine instability leading to a lack of reproducibility over time. In co-operative interlaboratory studies PyMS failed to give consistent results and the technique fell into disrepute. It was only with the development of low-cost, fully automatic pyrolysis mass spectrometry (Aries *et al.*, 1986) and improved data handling facilities that the technique has been recognised as a force for the classification, identification and typing of microorganisms (Magee, 1993, 1994).

CURIE POINT PYROLYSIS MASS SPECTROMETRY

The steps involved in Curie point pyrolysis mass spectrometric analysis have been considered in detail by Magee (1993, 1994). The procedure involves four stages: smearing a small sample (*ca.* 10-100 µg dry weight) onto a metal carrier; pyrolysis, in which the carrier and sample are heated rapidly under vacuum to a fixed temperature (Curie point) in the range of 358 to 1000 °C resulting in the thermal degradation of the sample to a mixture of volatile compounds; mass spectrometry, in which the volatile compounds are separated by molecular weight and quantified; and computation, in which the spectra are compared using multivariate statistical methods. A schematic representation of an automatic Curie point pyrolysis mass spectrometer is shown in Figure 1.

The organisms to be examined should be grown on the same batch of medium under highly standardised conditions. Samples can be taken directly from culture plates using a disposable loop and smeared onto each of three pyrolysis foils. A broad sweep of colonies should be taken, especially for actinomycetes showing colonial variation, e.g. *Rhodococcus* and *Streptomyces* species. The coated foils are dried at 80 °C to dehydrate the sample, and to destroy enzyme activity that might otherwise alter sample composition. In Curie point pyrolysis, the foil plus sample are heated by passing a high-frequency, high amperage oscillating current through the pyrolysis coil for a set time, usually 3 seconds. This produces an intensive magnetic field that penetrates the foil causing induction heating. When the temperature of the foil reaches the Curie point of the iron-nickel alloy, the latter ceases to be ferromagnetic, and the field no longer induces heating. The alloy then cools slightly, ferromagnetism is restored, and heating is resumed.

The thermostatic mechanism outlined above maintains the carrier temperature close to the Curie point. In microbiological work, an alloy with a Curie point of 530 °C is usually used, but foils with Curie points of 358 and 770 °C can be purchased. Pyrolysis at low temperatures gives pyrolysates with a large proportion of high-boiling point tarry products. High temperature pyrolysis gives greater pyrolysate yields, but low molecular weight products predominate, and these tend to be highly unreproducible. The temperature rise time needs to be carefully standardised to minimise the formation of non-volatile products. Heating rates, which depend on the field strength, the thermal capacity of the alloy and heat dissipation are rapid, e.g. 0.6 seconds to 530 °C for the Horizon Instruments PYMS 200X pyrolysis system (Ottley and Maddock, 1986).

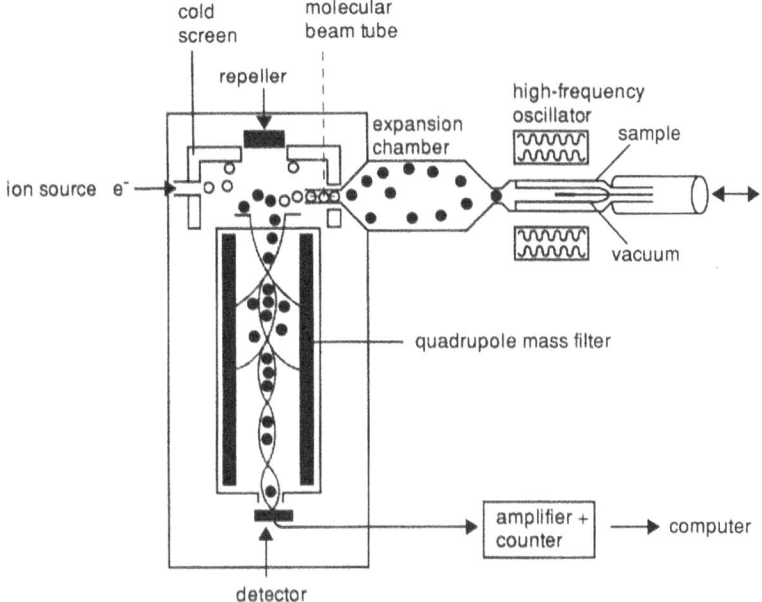

Figure 1. Schematic representation of an automated Curie point pyrolysis mass spectrometer based on a quadrupole mass analyser.

After pyrolysis, the pyrolysate enters the expansion chamber, which is designed to optimise flow mixing, so that early, low molecular weight fragments and late, high molecular weight products of the pyrolysis mix evenly. As the products leave through the molecular beam tube, they meet an interacting ray of electrons (Figure 1). The resultant collisions yield ions. The low energy of the electrons (30 eV) maximises the yield of molecular ions, that is, the loss of one electron per molecule to yield M^+ ions. The sample tube, expansion chamber and molecular beam tube are heated at 100 to 140 °C to prevent condensation of the products (Windig *et al.*, 1979); the sides of the expansion chamber are gold-plated to avoid reaction with the pyrolysate. Unionised molecules are condensed on a trap cooled with liquid nitrogen to prevent contamination of the mass spectrometer. The ions are accelerated out of the electron beam by a positively charged repeller plate through an aperture into an electrostatic lens which focuses the molecular ion beam.

The focused beam passes through a quadrupole mass filter (Meuzelaar and Kistemaker, 1973) which separates the ions according to their mass:charge ratio and delivers them to an electron multiplier. Here each ion collides with a target, releasing a shower of electrons which undergo further acceleration and collision steps thereby amplifying the signal, which is registered on an electronic counter. The mass filter controls and counter outputs are monitored by a computer, which scans the mass filter repeatedly, registering the ion counts at each mass interval. Ion counts for each interval are summed over multiple scans and presented as a pyrolysis mass spectrum, resolved at intervals of 1 amu. In the minute following pyrolysis, an instrument can repeatedly scan from masses 11 to 200 (PYMS 200X) or 11 to 400 (RAPyD-400, Horizon Instruments) accumulating ion counts at 0.1 mass intervals. The data from PyMS are usually displayed as quantitative mass spectra (Figure 2) in which the abscissa represent the m/z ratio and the ordinates contain information on the ion count for any particular m/z value ranging from 51 to 200. The data are automatically recorded to a hard disc during the run.

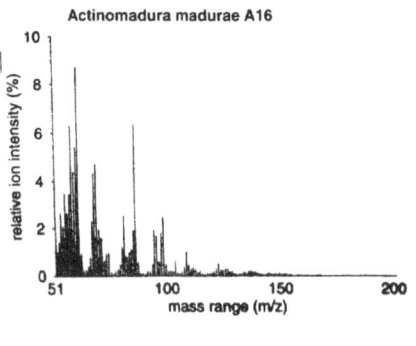

Actinomadura madurae A16

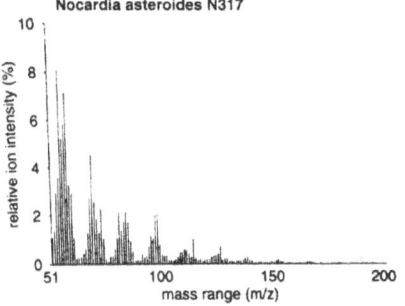

Nocardia asteroides N317

Figure 2. Representative pyrolysis mass spectra.

DATA ANALYSIS

Standard Statistical Procedures

Pyrolysis mass spectra contain quantitative information on the pyrolysis products of the test strains. Most bacteria generate essentially the same peaks but quantities of the materials in the peaks, represented by the peak heights, vary significantly and reproducibly from one organism to another (Figure 2). Comparison of pyrograms between organisms thereby provides the basis for determining the relationships among a group of isolates. It is, however, necessary to compare the data in the pyrograms quantitatively using multivariate statistical procedures (see Magee, 1994 for details). The necessary data handling routines can be performed rapidly on personal computers using commercially available statistical packages such as GENSTAT (Nelder, 1979) and SIMCA (Wold and Söderström, 1987). In essence, the overall aim of data analysis is the selection of the most reproducible and discriminating peaks for the separation of samples. The major steps involved are discussed below and outlined in Figure 3.

Ion counts from all samples are normalised to compensate for variation in sample size. Data reduction steps applied prior to discriminant analysis include the selection of the most "characteristic" masses (Gutteridge, 1987), that is, the selection of masses which have a high ratio of discrimination (between group variation) to reproducibility (within group variation), and selection is based on reproducibility (Magee, 1993, 1994). In practice, these different approaches often yield similar results. Principle component (PC) and canonical variate (CV) analyses are then used to maximise discrimination between the groups; the results of PCCV analyses are usually displayed on ordination diagrams (Figure 8). Eigenvectors are then used

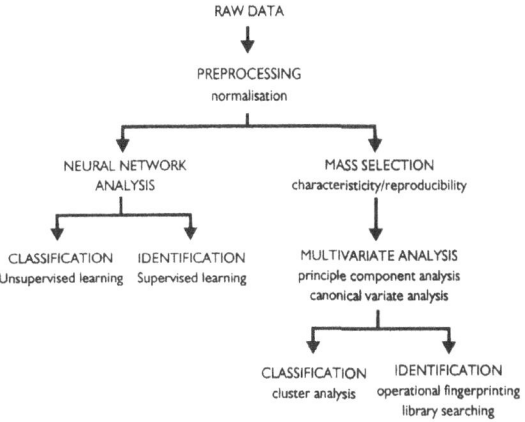

Figure 3. Major steps in handling pyrolysis mass spectrometric data.

to generate a distance matrix which contains information on Mahalanobis distances. The latter can be transformed to a percentage similarity matrix using Gower's coefficient S_G (Gower, 1971) and used to generate a dendrogram (Figure 9) with the unweighted pair group method with arithmetic averages (UPGMA) algorithm.

Both ordination diagrams and corresponding dendrograms are examined to determine the relatedness of the strains pyrolysed, which may be inferred from the relative proximity of the PyMS-derived datapoints. Data for any isolate shown to be distinct on PyMS analysis can be deleted from the dataset, which can then be re-analysed to further determine the relationships between the remaining isolates.

Neural Networks

Artificial neural networks (ANNs) are a well known means of uncovering complex, non-linear relationships in multivariate data (Simpson, 1990; Hertz *et al.*, Zurada, 1992). They have been widely applied in knowledge representation studies designed to emulate the pattern recognition abilities of the human brain. ANNs have an analogous architecture to the human brain. The building blocks of ANNs are processing elements (PEs) or nodes which correspond to neurons (Figure 4). Individual PEs are connected to form the neural network. Many inputs, equivalent to dendrites, enter each PE where they are combined mathematically to give the internal activation level of the PE. This summed input is then transferred by means of a simple transfer function, which is usually sigmoid (Figure 5), to give the output of the PE, which is connected to the input paths of other PEs. These input paths are weighted (equivalent to synaptic efficiency) hence values of input signals to a PE are modified prior to summation in the PE. A typical neural network has three layers of PEs, the input layer, a hidden layer and an output layer (Figure 6).

In contrast to conventional computer systems, neural networks do not follow explicit instructions in the form of a program but "learn" by being shown examples. Learning is achieved by exposing the network to these examples. This causes modification of the connection weights between PEs to give the desired outputs. In such supervised learning the training set, a data set in which the inputs and desired outputs are known, is presented to the network and the weights repeatedly adjusted using a wide spectrum of data. The effectiveness of this training process is defined in terms of the root mean squared (RMS) error between the actual output averaged over the training set and the desired outputs. Training is effected using the backpropagation algorithm (Figure 7).

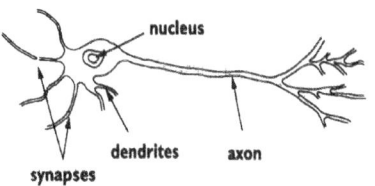

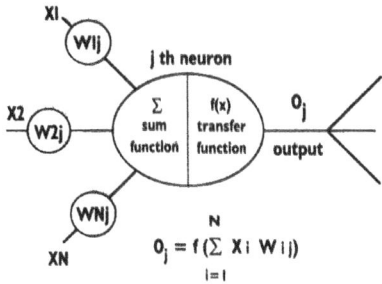

Figure 4. The basic building block of artificial neural networks, consisting of inputs (i), a processing element (PE) and outputs(o). Summation (S_j) of the weighted inputs occurs within the PE, $S_j = \Sigma W_{ji} x_i$. A transfer function is applied to the summation, $y_j = f(S_j)$.

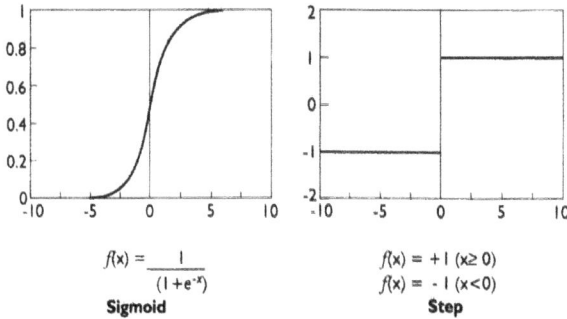

$$f(x) = \frac{1}{(1+e^{-x})}$$

Sigmoid

$$f(x) = +1 \ (x \geq 0)$$
$$f(x) = -1 \ (x < 0)$$

Step

Figure 5. Two typical transfer functions.

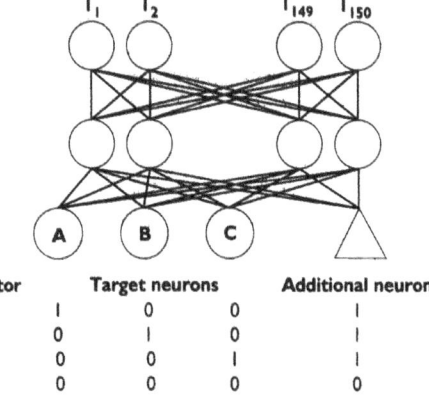

Target vector		Target neurons		Additional neuron
Group A	1	0	0	1
Group B	0	1	0	1
Group C	0	0	1	1
Unknown	0	0	0	0

Figure 6. A neural network consisting of 150 inputs and 4 outputs connected to each other by one hidden layer. In the architecture shown, adjacent layers of the network are fully interconnected although other architectures are possible.

Step 1	Initialise weights with small random values.
Step 2	Set δ (learning rate) and β (momentum term).
Step 3	Present a single set of inputs, and propagate data forward to obtain the predicted output (o).
	h=f(W1i+bias1) o=f(W2h+bias2)
Step 4	Calculate output layer error vector.
	d=o(1-o)(o-t)
Step 5	Calculate hidden layer error vector.
	e=h(1-h)W2d
Step 6	Adjust weights and bias between hidden and output layers.
	$\Delta W2_t = \delta hd + \beta \Delta W2_{t-1}$, bias2=$\delta$d
	W2=W2+ΔW2
Step 7	Adjust weights and bias between input and hidden layers.
	$\Delta W1_t = \delta ie + \beta \Delta W1_{t-1}$, bias1=$\delta$e
	W1=W1+ΔW1
Step 8	Repeat steps 3 to 7 until the error between output and target vectors is within the desired tolerance.

• The weight adjustments can be made after the entire set of input vectors has been propagated (Epoch-based backpropagation).

Figure 7. Backpropagation algorithm (Rumelhart *et al.*, 1986). Symbols: i, input, h, hidden, o, output layer, t, target vectors; d, output , e, hidden layer error vectors; bias1 and bias2, bias vectors; W1 and W2, weight matrices.

Once the specified RMS error is reached the trained ANN is challenged with new known inputs, but ones to which it has been never been exposed (challenge set). The ANN will then immediately process these new inputs and output the optimal best fit to the previously specified outputs. If the outputs from the challenge set are accurate, that is, in agreement with the known answers, the ANN is said to have generalised. A generalised network may then be exposed to unknown inputs to determine recognition of the presence or absence of the character specified in the desired outputs. ANNs are particularly attractive for the analysis of PyMS data as it has been shown mathematically that a neural network consisting of only one hidden layer with an arbitrary large number of nodes can learn any arbitrary mapping to an arbitrary degree of accuracy (White, 1990).

Commercial software is available for ANN analyses (see Morris and Boddy, 1993) though "in house" dedicated software is often preferable. Programming ANNs is fairly easy as only simple vector and matrix manipulations are needed. Most programming languages can be used without difficulty, for example, the C, C++ language (Chun *et al.*, 1993a,b) and PASCAL (Rataj and Schindler, 1991). Several good programming guidelines are available (Freeman and Skapura, 1991; Blum, 1992).

PYROLYSIS MASS SPECTROMETRY AND BACTERIAL SYSTEMATICS

Early applications of PyMS to bacterial classification, identification, and epidemiological typing have been reviewed (Gutteridge and Norris, 1979; Gutteridge *et al.*, 1985; Shute *et al.*, 1985) and more recent work has been considered by Magee (1993). Most applications in classification have been designed to evaluate the integrity of taxa circumscribed using conventional criteria. In general, good agreement has been found between classifications derived using PyMS and more standard procedures, for example, in comparative studies on *Bacteroides* (Duerden *et al.*, 1989), *Corynebacterium* (Hindmarch *et al.*, 1990), *Fusobacterium* (Adriaans and Shah, 1988; Magee *et al.*, 1989a) and *Streptococcus* (Winstanley *et al.*, 1992). It is important in such studies to ensure that underlying correlations are not obscured by intraspecific effects and differences in the order of clusters between conventional and pyroclassifications. Strategies to confirm the homogeneity of taxa highlighted in such comparative studies have been discussed by Magee

(1993). It has also been shown that PyMS and DNA pairing procedures give similar profiles of relatedness (Goodacre *et al.*, 1991; Sanglier *et al.*, 1992; Chun and Goodfellow, unpublished data).

In some instances, as foreseen by Gutteridge and Norris (1979), pyroclassifications have differed from traditional taxonomies thereby casting a new perspective on the relationships between the test strains. Such situations are not surprising especially with groups such as *Thermus* (Donnison *et al.*, 1986) and *Fusobacterium* (Magee *et al.*, 1989a) where conventional methods are inadequate. Studies on a few well chosen taxa are needed to validate pyroclassification strategies.

The ability to compare many isolates within a single batch quickly and accurately makes PyMS an attractive technique for strain identification under standardised conditions. Pyrolysis studies on mycobacteria have yielded particularly promising results. The technique has been used to distinguish strains within the *M. tuberculosis* complex (Sisson *et al.*, 1991a) and to separate *M. xenopi* strains from those belonging to the *M. avium-M. intracellulare* complex (Sisson *et al.*, 1992b). Pyrolysis mass spectrometry of primary isolates of slow-growing mycobacteria may enable rapid identification provided that appropriate reference strains with equivalent growth characteristics are included for comparative purposes. Pyrolysis mass spectrometry and a genetic fingerprinting technique have been successfully used in the identification of *Bradyrhizobium japonicum* strains isolated from Italian soils (Kay *et al.*, 1994). However, the value of this approach to identification is currently restricted given the need to compare unknown organisms with reference strains in a procedure known as 'operational fingerprinting' (Meuzelaar *et al.*, 1982).

The speed and reproducibility of PyMS and its applicability to a wide range of bacteria make it an attractive method for epidemiological studies. Pyrolysis mass spectrometry is not a typing method *per se* as a permanent type designation is not assigned to the examined organisms, but it has proved to be a quick and effective method of interstrain comparison of bacteria that commonly cause outbreaks. This conclusion is based on studies of many medically important organisms, recent examples include *Bacteroides ureolyticus* (Duerden *et al.*, 1989), *Corynebacterium jeikeium* (Hindmarch *et al.*, 1990), *Legionella pneumoniae* (Sisson *et al.*, 1991b), *Listeria monocytogenes* (Freeman *et al.*, 1991a; Low *et al.*, 1992), *Pseudomonas aeruginosa* (Sisson *et al.*, 1991c), *Salmonella enteritidis* (Freeman *et al.*, 1990), *Streptococcus pneumoniae* (Freeman *et al.*, 1991b), *Staphylococcus aureus* (Gould *et al.*, 1991), *Staphylococcus pyogenes* (Magee *et al.*, 1989b, 1991; Freeman *et al.*, 1990) and *Xanthomonas maltophilia* (Orr *et al.*, 1991).

It is evident that PyMS can be used to discriminate between strains as accurately as routine typing systems (Freeman *et al.*, 1991a; Sisson *et al.*, 1991c). Indeed, in some cases it has been used to separate isolates beyond the resolution of such systems (Freeman *et al.*, 1991b; Gould *et al.*, 1991). The results of PyMS analyses have also been shown to correspond with those from molecular based techniques, such as restriction length fragment polymorphism (Sisson *et al.*, 1991c; Low *et al.*, 1992). Similarly, PyMS studies on isolates from outbreaks of infection due to untypable isolates have been found to agree with epidemiological data that have subsequently become available (Orr *et al.*, 1991; Cartmill *et al.*, 1992).

The discriminating capacity of PyMS has also been explored by examination of very closely related strains of bacteria. The technique has been used to detect small genotypic changes in *Escherichia coli* (Goodacre and Berkeley, 1990). It has also proved possible to distinguish between representatives of staphylococcal species by PyMS of extracted DNA (Mathers *et al.*, 1994). These results challenge previous assumptions that PyMS is restricted to detecting phenotypic differences, although the basis of the differentiation of the DNA extracts has still to be determined.

94

Table 1. Some applications of artificial neural networks in chemistry and biology

Application	Reference
Analysis of nucleic acid sequences	Arrigo et al. (1991)
	Delmeler and Zhou (1991)
	Snyder and Stormo (1993)
Analysis of protein sequences	Colloch et al. (1993)
	Qian and Sejnowski (1988)
	Vieth and Kolinski (1991)
Bacterial systematics	Chun et al. (1993 a,b)
	Freeman et al. (1994a)
	Goodacre et al. (1994)
	Kennedy and Thakur (1993)
	Millership (1993)
	Rataj and Schindler (1991)
Fermentation	Cleran et al. (1992)
	Willis et al. (1991)
Flow cytometry	Balfoort et al. (1992)
	Boddy and Morris (1993)
	Morris et al. (1992)
Infrared spectroscopy	Fessenden and Gyorgyi (1991)
	Munk et al. (1991)
	Rob and Munk (1990)
	Tanabe et al. (1992)
	Weigel and Herges (1992)
Mass spectrometry	Sellers et al. (1990)
Nuclear magnetic resonance spectroscopy	Anker and Jurs (1992)
	Kjaer and Poulsen (1991)
	Meyer et al. (1991)
Pyrolysis mass spectrometry	Goodacre et al. (1992)
	Goodacre and Kell (1993)
	Goodacre et al. (1993b)
Raman spectroscopy	Schulze et al. (1994)

NEURAL NETWORK ANALYSIS OF PYMS DATA

Chemical, and to some extent biological, data have been successfully analysed using neural networks (Table 1). Spectral identification of chemical compounds has been achieved using ANNs for the analysis of chemical data obtained from infrared, nuclear magnetic resonance, Raman spectroscopy, flow cytometry and mass spectrometry. The combination of PyMS and ANNs has recently been applied to the analysis of biological samples (Goodacre et al., 1992, 1994; Chun et al., 1993a,b). In particular, the use of ANNs allows PyMS data to be used to detect a single character, other differences being ignored. This approach has been applied to determine the purity of olive oil (Goodacre et al., 1992, 1993a), to assess indole production from *Escherichia coli* (Goodacre and Kell, 1993) and to analyse mixtures of casamino acids in glycogen (Goodacre et al., 1993b). These findings indicate that PyMS-ANN analyses provide a new and powerful way of determining the concentrations of key substrates, metabolites and products in bioprocesses. Neural networks can also be used to analyse biochemical (Rataj and Schindler, 1991; Kennedy and Thakur, 1993) and whole-organism protein electrophoretic patterns (Millership, 1993) for diagnostic purposes and to identify microorganisms by flow cytometric patterns (Morris et al., 1992; Boddy and Morris, 1993).

Initial studies indicate that use of PyMS and various types of neural networks provide an objective, rapid and accurate way of classifying and identifying bacteria. Neural nets have been used to identify members of putatively novel *Streptomyces* spp. (Chun *et al.*, 1993a,b) and *Propionibacterium acnes* isolates from dogs (Goodacre *et al.*, 1994). The classification of the canine isolates by Kohonen's self-organising feature map (Kohonen, 1989) was found to give similar results to multivariate statistical techniques (Goodacre *et al.*, 1994). This is an interesting finding as Kohonen's self-organising feature map, an unsupervised learning algorithm, may provide an alternative way of grouping and identifying pyrolysis mass spectra. Similarly, the successful separation of *Mycobacterium bovis* and *Mycobacterium tuberculosis* (Freeman *et al.*, 1994a) exemplifies the power of PyMS-ANN analyses in diagnostic bacteriology. This approach may prove to be an important, powerful and rapid way of detecting clinically significant products, such as exotoxins, in otherwise relatively homogeneous bacterial populations (Freeman *et al.*, 1994b).

INDUSTRIAL SCREENING PROGRAMMES

The choice of microorganisms, notably actinomycetes, for microbial products discovering programmes is essentially a problem of distinguishing between known organisms and recognising members of novel, that is, previously undiscovered taxa (Okami and Hotta, 1988; Goodfellow and O'Donnell, 1989). To date, the search for novel actinomycetes owes more to custom and practice than to any rational design. Large numbers of organisms have been isolated and screened in anticipation that commercially successful products will be found. This empirical approach has been successful but, increasingly, the same actinomycetes and bioactive compounds are reisolated and rediscovered. The effectiveness of industrial screening can therefore be improved by the elimination of representatives of well studied taxa early in the programme, and by the recognition of target and novel organisms on primary isolation plates.

Pyrolysis mass spectrometry has been used to classify and identify industrially significant actinomycetes (Saddler *et al.*, 1988; Sanglier *et al.*, 1992). In this latter study, members of representative actinomycete genera were pyrolysed in order to determine the effects of medium design, incubation time and sample preparation on experimental data; it was concluded that reproducible results could be obtained given rigorous standardisation of growth and pyrolysis conditions. Sanglier and his colleagues also showed that PyMS data could be used to objectively select strains for pharmacological screens, as unknown or putatively novel actinomycetes appeared as outliers on ordination diagrams. They were also able to distinguish between actinomycete strains at and below the species level. In particular, representatives of three closely related *Streptomyces* species, namely *S. albidoflavus*, *S. anulatus* and *S. halstedii*, were distinguished. The separation of these numerically circumscribed streptomycete species indicated that PyMS can provide a rapid way of establishing the taxonomic integrity of established or putatively novel actinomycete species (Sanglier *et al.*, 1992).

In an extension of these pilot studies, carefully chosen representatives of three putatively novel streptomycete species, labelled *Streptomyces* groups A, B and C, were examined to determine their taxonomic status using the PyMS procedure. The test strains represented actinomycetes which had been repeatedly isolated from soil taken from the Palace Leas site at Cockle Park Experimental Farm, Northumberland, U.K. using a standard dilution plate technique and two computer-formulated selective isolation media. *Streptomyces* group A and B strains were isolated on raffinose-histidine agar plates supplemented with cycloheximide (50 μg/ml) and nystatin (50 μg/ml) and incubated at 25 °C for 14 days (Vickers *et al.*, 1984), and the *Streptomyces* group C strains from similarly

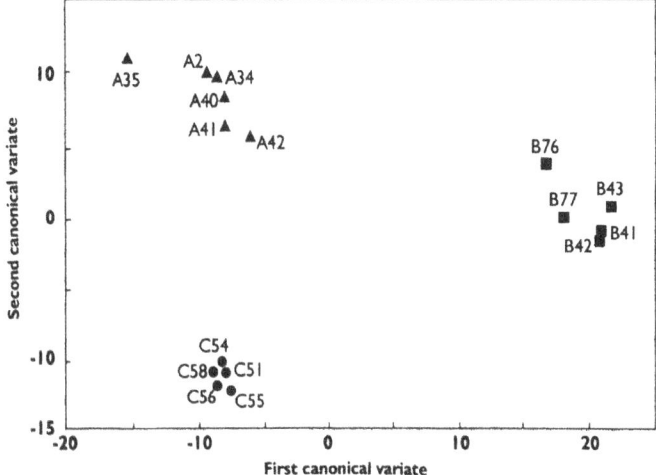

Figure 8. Ordination plot along the first two canonical variate axes showing the mean position of the duplicated *Streptomyces* group A, B and C strains isolated from soil at Cockle Park Experimental Farm, Northumberland, U.K. The first two axes accounted for 66% of the variation between the strains. Key : *Streptomyces* group A; *Streptomyces* group B and *Streptomyces* group C.

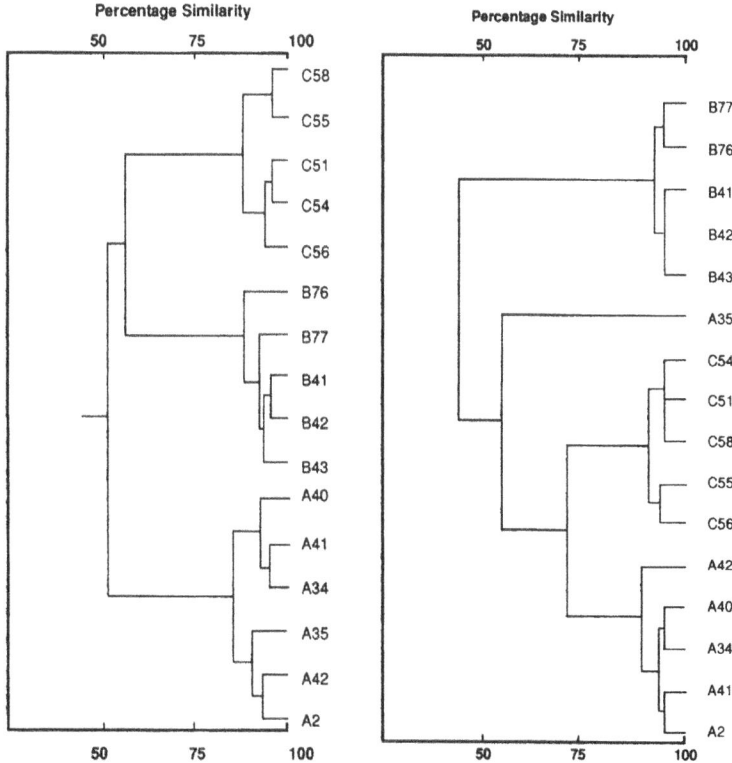

Figure 9. Dendrograms, a and b, showing the relationships between the representatives of *Streptomyces* groups A, B and C based on pyrolysis mass data from a replicated study. Data were analysed using the GENSTAT software with clustering achieved by applying the unweighted-pair group method with arithmetic averages.

97

Table 2. Identification of streptomycetes by artificial neural network analysis of pyrolysis mass spectral data using the backpropagation algorithm

Taxa	Number of strains	*Streptomyces* Group A	*Streptomyces* Group B	*Streptomyces* Group C	Unknown strains
1. Strains identified with the simple ANN :					
Streptomyces A	6(36)	6(36)	0(0)	0(0)	0(0)
Streptomyces B	5(30)	0(0)	5(30)	0(0)	0(0)
Streptomyces C	5(30)	0(0)	0(0)	5(30)	0(0)
Actinomadura	8(24)	7(19)	0(0)	0(1)	1(4)
Mycobacterium	34(102)	32(94)	0(0)	2(5)	0(3)
Nocardia	40(120)	23(66)	0(0)	15(38)	2(16)
Nocardiopsis	9(27)	0(0)	1(3)	6(19)	2(5)
Saccharomonospora	28(84)	19(57)	0(2)	0(0)	9(25)
Streptosporangium	19(57)	0(6)	19(50)	0(0)	0(1)
2. Strains identified by the ANN with an additional neuron :					
Streptomyces A	6(36)	6(36)	0(0)	0(0)	0(0)
Streptomyces B	5(30)	0(0)	5(30)	0(0)	0(0)
Streptomyces C	5(30)	0(0)	0(0)	5(30)	0(0)
Actinomadura	8(24)	0(0)	0(0)	0(0)	8(24)
Mycobacterium	34(102)	0(0)	0(0)	0(0)	34(102)
Nocardia	40(120)	0(0)	0(0)	0(0)	40(120)
Nocardiopsis	9(27)	0(0)	0(0)	0(0)	9(27)
Saccharomonospora	28(84)	0(0)	0(0)	0(0)	28(84)
Streptosporangium	19(57)	0(0)	0(0)	0(0)	19(57)

Figures in parenthesis indicate the number of spectra studied. When two or more spectra were identified to one target group the test strain was assigned to that group, otherwise it was not identified (Modified from Chun *et al.*, 1993b).

treated starch casein plates supplemented with novobiocin (25 µg/ml) and the two antifungal antibiotics. Extensive numerical phenetic, molecular fingerprinting and DNA relatedness studies showed that the *Streptomyces* group A, B and C strains had a profile of properties that separated them from one another and from representatives of established streptomycete species (Atalan, 1993; Manfio *et al.*, unpublished data). The recovery of the representative isolates in three distinct clusters corresponding to *Streptomyces* groups A, B and C (Figure 8) underlines the potential of PyMS as a rapid and reliable technique for underpinning the taxonomic status of putatively novel actinomycete species. Similar results were obtained when data were expressed in a dendrogram based on Mahalanobis distances (Figure 9a) though in a replicate experiment one of the duplicated strains, *Streptomyces* group A strain A35, formed a single membered cluster (Figure 9b).

It is essential in industrial screening programmes that large numbers of strains are handled quickly. To this end, representatives of the three putatively novel *Streptomyces* species mentioned above were examined by PyMS and the first data set used for standard multivariate statistical analysis and as a training set for an ANN (Chun *et al.*, 1993a). A second set of data, the challenge set, was then prepared for 'operational fingerprinting' (Meuzelaar *et al.*, 1982) and for testing the ANN. All of the test strains were correctly identified using the ANN but one of the sixteen strains was misidentified using the conventional operational fingerprinting procedure. It was concluded from this preliminary study that PyMS-ANN analysis offers considerable advantages over current practices used

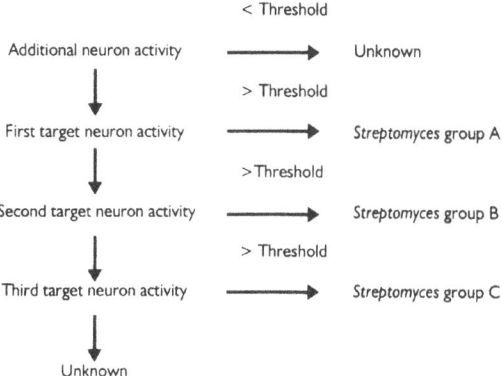

Figure 10. The procedure used for the identification of *Streptomyces* group A, B and C strains. When two or more target neurons showed activities over the threshold value of 0.8, the organism was considered to be unidentified.

to examine PyMS data as it is not necessary to study duplicated of triplicated samples or to include reference and unknown strains in a single batch. Further, once an ANN is trained, identifications can be made across PyMS runs using single samples thereby making it possible to handle many strains in a short range of time.

Once neural networks have been trained they can be used to discriminate between target organisms (Chun *et al.*, 1993a; Freeman *et al.*, 1994a). A serious limitation of these existing ANNs is that the networks are not trained to recognise unknown patterns which consequently may be misidentified. This proved to be the case when the ANN described above was challenged with spectral data from the original streptomycetes, the training set, from additional members of the putatively novel taxa, and from over a hundred strains representing six other actinomycete genera. All of the streptomycetes were correctly identified but many of the other actinomycetes were misidentified (Table 2). These results were not unexpected as it is well known that supervised learning generates neural networks which behave poorly when challenged with unknown patterns outside the range of the training data (Freeman and Skapura, 1991).

A network topology was developed to overcome the problem of the misidentification of the strains that did not belong to the three target groups. The network had the same architecture as the simple ANN but included an additional, fourth neuron. In addition, the training set was extended to include mass spectra derived from representatives of some of the additional genera. Identification was achieved using the procedure outlined in Figure 10.

It is very encouraging that the use of the ANN with the additional neuron led to the successful identification of the target strains. This ANN also, without exception, recognised the differences between the target mass spectral patterns and those derived from actinomycetes other than streptomycetes, including representatives of the genera *Mycobacterium*, *Saccharomonospora* and *Streptosporangium* which were not part of the training set. These important, albeit preliminary, findings raise the exciting prospect that a nest of neural networks might be generated and used for the sequential identification of diverse taxa and for the identification of species in genera, such as *Mycobacterium* and *Streptomyces*, that encompass many centres of variation. Indeed, it is possible that the long awaited goal of an objective choice of actinomycetes for industrial screening programmes may be possible through the judicious application of PyMS-ANN analyses.

CONCLUSIONS

The main advantages of Curie point PyMS as a tool for the characterisation, identification and typing of bacteria is that it is rapid (<2 minutes per sample), sensitive, reproducible and widely applicable. Sample preparation is straightforward and in many cases biomass can be taken directly from culture plates. Sample sizes are often sufficient to allow the analysis of single colonies. The routine operation of PyMS procedures on low cost, automated machines such as the PyMS 200X and RAPyD-400 require minimal operator experience, although an understanding of mass spectrometry is necessary to maintain the instrument and to effect regular servicing.

The speed, comparatively low running costs and versatility of PyMS make it ideal for the initial screening of suspected outbreaks of infection thereby allowing expensive and time-consuming methods to be used more cost-effectively for formal typing. These features also make the technique attractive for evaluating established taxonomies and for selecting strains for metabolite screening programmes. In addition, the PyMS procedure has led to considerable progress towards the ultimate goal of diagnosticians, the identification of freshly isolated bacteria within minutes of their arrival in the laboratory.

The most serious challenge to the more widespread use of PyMS is the lack of understanding over what pyrolysis mass spectra represent and hence why they can be used to differentiate microorganisms. Detailed chemical analyses of microbial pyrolysis products are needed to characterise the cell constituents that contribute to the spectra and hence to the differentiation of organisms.

The use of neural network systems for identification and quantitative determination using PyMS data is a method at an early stage of development but with long-term potential. The combination of PyMS and ANNs provides a novel, powerful and practical means for identifying large numbers of isolates for ecological and industrial purposes and for analysis of metabolites and fermentation products. It is also likely to have important benefits for the diagnosis and treatment of bacterial diseases. The prospects for neural computing are enormous though there are drawbacks, notably the slow learning time, which is exacerbated with the increased size and complexity of data, and the lack of dedicated neural network hardware.

In conclusion, it is fair to say that the ability to handle small amounts of bacterial biomass with minimal sample preparation to obtain, in minutes, chemical fingerprints that can be used for identification and typing is unparalleled by corresponding methods, including the use of molecular techniques.

ACKNOWLEDGEMENTS

The authors are indebted to Mr. C.S. Hetherington and Dr. A.C. Ward for much helpful discussion and to Mr. G.P. Manfio for critically reading the manuscript. In addition, one of us (J.C.) is grateful to the British Council (Seoul, Republic of Korea) for financial support.

REFERENCES

Adriaans, B. and Shah, H. (1988) *Fusobacterium ulcerans* sp. nov. from tropical ulcers. Int. J. Syst. Bacteriol. 38, 447-448.

Anker, L.S. and Jurs, P.C. (1992) Prediction of carbon-13 nuclear magnetic resonance chemical shifts by artificial neural networks. Anal. Chem. 62, 1157-1164.

Aries, R.E., Gutteridge, G.S. and Ottley, T.W. (1986) Evaluation of a low cost, automated pyrolysis mass spectrometry. J. Anal. Appl. Pyrol. 9, 81-98.

Arrigo, P., Giuliano, F. ,Scalia, F., Rapallo, A. and Damiani, G. (1991) Identification of a new motif on nucleic acid sequence data using Kohonen's self-organizing map. Comp. Appl. Biol. Sci. 7, 353-357.

Atalan, E. (1993) Selective Isolation, Characterisation and Identification of Some *Streptomyces* Species. Ph. D. Thesis, University of Newcastle upon Tyne.

Balfoort, H.W., Snoek, J., Smits, J.R.M., Breedveld, L.W., Hofstraat, J.W. and Ringelberg, J. (1992) Automatic identification of algae: Neural network analysis of flow cytometric data. J. Plank. Res. 14, 575-589.

Blum, A. (1992) Neural Networks in C++. John Wiley and Sons, New York.

Boddy, L. and Morris, C.W. (1993) Neural network analysis of flow cytometry data, in "Flow Cytometry in Microbiology" (Lloyd, D., ed.), pp. 159-169. Springer-Verlag, London.

Cartmill, T.D.I., Orr, K., Freeman, R. and Sisson, P.R. (1992) Nosocomial infection with *Clostridium difficile* investigated by pyrolysis mass spectrometry. J. Med. Microbiol. 37, 352-356.

Chun, J., Atalan, E., Ward, A.C. and Goodfellow, M. (1993a) Artificial neural network analysis of pyrolysis mass spectrometric data in the identification of *Streptomyces* strains. FEMS Microbiol. Lett. 107, 321-326.

Chun., J., Atalan, E., Kim, S-B., Kim, H-J., Hamid, M.E., Trujillo, M.E., Magee, J.G., Manfio, G.P., Ward, A.C. and Goodfellow, M. (1993b) Rapid identification of streptomycetes by artificial neural network analysis of pyrolysis mass spectra. FEMS Microbiol. Lett. 114, 115-120.

Cleran, Y.,Thibault, J., Cheruy, A. and Corrieu, G. (1991) Comparison of prediction performances between models obtained by the group method of data handling and neural networks for the alcoholic fermentation rate in enology. J. Ferment. Biotech. 71, 356-362.

Colloch, N., Etchebest, C., Thoreau, E., Henrissat, B. and Mornon, J.P. (1993) Comparison of 3 algorithms for the assignment of secondary structure in proteins - the advantages of a consensus assignment. Protein Eng. 6, 377-382.

Demeler, B. and Zhou, G. (1991) Neural network optimization for *E. coli* promoter prediction. Nucleic Acids Res. 19, 1593-1599.

Donnison, A.M., Gutteridge, C.S., Norris, J.R., Morgan, H.W. and Daniel, R. (1986) A preliminary grouping of New Zealand *Thermus* strains by pyrolysis mass spectrometry. J. Anal. Appl. Pyrol. 9, 281-285.

Duerden, B.I., Eley, A., Goodwin, L., Magee, J.T., Hindmarch, J.M. and Bennett, K.W. (1989) A comparison of *Bacteroides ureolyticus* isolates from different clinical sources. J. Med. Microbiol. 29, 63-73.

Fessenden, R.J. and Gyorgyi, L. (1991) Identifying functional groups in IR spectra using an artificial neural network. J. Chem. Soc. Perkin Trans. 2, 1755-1762.

Freeman, J.A. and Skapura, D.M. (1991) Neural Networks: Algorithms, Applications, and Programming Techniques. Addison-Wesley Publishing Company, Reading, MA.

Freeman, R., Goodfellow, M., Gould, F.K., Hudson, S.J. and Lightfoot, N.F. (1990) Pyrolysis mass spectrometry for the rapid epidemiological typing of clinically significant bacterial pathogens. J. Med. Microbiol. 32, 283-286.

Freeman, R., Sisson, P.R., Lightfoot, N.F. and McLauchlin, J. (1991a) Analysis of epidemic and sporadic strains of *Listeria monocytogenes* by pyrolysis mass spectrometry. Lett. Appl. Microbiol. 12, 133-136.

Freeman, R., Gould, F.K., Sisson, P.R. and Lightfoot, N.F. (1991b) Strain differentiation of capsule type 23 penicillin-resistant *Streptococcus pneumoniae* from nosocomial infections by pyrolysis mass spectrometry. Lett. Appl. Microbiol. 13, 28-31.

Freeman, R., Goodacre, R., Sisson, P.R., Magee, J.G., Ward, A.C. and Lightfoot, N.F. (1994a) Rapid identification of species within the *Mycobacterium tuberculosis* complex by artificial neural network analysis of pyrolysis mass spectra. J. Med. Microbiol. (in press).

Freeman, R., Law, D., Sisson, P.R., Ward, A.C. and Lightfoot, N.F. (1994b) Rapid detection of vero cytotoxin production in *Escherichia coli* by artificial neural network analysis of pyrolysis mass spectra. J. Med. Microbiol. (in press).

Goodacre, R. and Berkeley, R.C.W. (1990) Detection of small genotypic changes in *Escherichia coli* by pyrolysis mass spectrometry. FEMS Microbiol. Lett. 71, 133-138.

Goodacre, R. and D.B. Kell (1993) Rapid and quantitative analysis of bioprocesses using pyrolysis mass spectrometry and neural networks: application to indole production. Anal. Chim. Acta 279, 17-26.

Goodacre, R., Hartmann, A., Beringer, J.E. and Berkeley, R.C.W. (1991) The use of pyrolysis mass spectrometry in the characterization of *Rhizobium meliloti*. Lett. Appl. Microbiol. 13, 157-160.

Goodacre, R., Kell, D.B. and Bianchi, G. (1992) Neural networks and olive oil. Nature 359, 594.

Goodacre, R., Kell, D.B. and Bianchi, G. (1993a) Rapid assessment of virgin olive oils and other seed oils using pyrolysis mass spectrometry and artificial neural networks. J. Sci. Food Agric. 63, 297-307.

Goodacre, R., Edmonds, A.N. and Kell, D.B. (1993b) Quantitative analysis of the pyrolysis-mass spectra of complex mixtures using artificial neural networks: application to amino acids in glycogen. J. Anal. Appl. Pyrol. 26, 93-114.

Goodacre, R., Neal, M.J., Kell, D.B., Greenham, L.W., Noble, W.C. and Harvey, R.G. (1994) Rapid identification using pyrolysis mass spectrometry and artificial neural networks of *Propionibacterium acnes* isolated from dogs. J. Appl. Bacteriol. 76, 124-134.

Goodfellow, M. and O'Donnell, A.G. (1989) Search and discovery of industrially significant actinomycetes, in "Microbial Products: New Approaches" (Baumberg, S., Hunter, I.S. and Rhodes, P.M., eds.), pp. 343-383, Cambridge University Press, Cambridge.

Gould, F.K., Freeman, R., Sisson, P.R., Cookson, B.D. and Lightfoot, N.F. (1991) Inter-strain comparison by pyrolysis mass spectrometry in the investigation of *Staphylococcus aureus* nosocomial infection. J. Hosp. Infect. 19, 41-48.

Gower, J.C. (1971) A general coefficient of similarity and some of its properties. Biometrics 27, 857-874.

Gutteridge, C.S. (1987) Characterization of microorganisms by pyrolysis mass spectrometry. Meth. Microbiol. 19, 227-272.

Gutteridge, C.S. and Norris, J.R. (1979) The application of pyrolysis techniques to the identification of microorganisms. J. Appl. Bacteriol. 47, 5-43.

Gutteridge, C.S., Vallis, L. and MacFie, H.J.H. (1985) Numerical methods in the classification of micro-organisms by pyrolysis mass spectrometry, in "Computer-Assisted Bacterial Systematics" (Goodfellow, M., Jones, D. and Priest, F.G., eds.), pp. 369-401. Academic Press, London.

Hertz, J., Krogh, A. and Palmer, R.G. (1991) Introduction to the Theory of Neural Computation. Addison-Wesley, Redwood City, CA.

Hindmarch, J.M., Magee, J.T., Hadfield, M.A. and Duerden, B.I. (1990) A pyrolysis-mass spectrometry study of *Corynebacterium* spp.. J. Med. Microbiol. 31, 137-149.

Irwin, W.J. (1982) Analytical Pyrolysis: A Comprehensive Guide. Marcel Dekker, New York.

Kay, H.E., Coutinho, H.L.C., Fattori, M., Manfio, G.P., Goodacre, R., Nuti, M.P., Basaglia, M. and Beringer, J.E. (1994) The identification of *Bradyrhizobium japonicum* strains isolated from Italian soils. Microbiology (in press).

Kennedy, M.J. and Thakur, M.S. (1993) The use of neural networks to aid in microorganism identification: A case study of *Haemophilus* species identification. Antonie van Leeuwenhoek 63, 35-38.

Kjaer, M. and Poulsen, F.M. (1991) Identification of 2D H-1-NMR antiphase cross peaks using a neural network. J. Magn. Reson. 94, 659-663.

Kohonen, T. (1989) Self-Organization and Associative Memory. Springer-Verlag, Berlin.

Low, J.C., Chalmers, R.M., Donachie, W., Freeman, R., McLauchlin, J. and Sisson, P.R. (1992) Pyrolysis mass spectrometry of *Listeria monocytogenes* isolates from sheep. Res. Vet. Sci. 53, 64-67.

Magee, J.T. (1993) Whole-organism fingerprinting, in "Handbook of New Bacterial Systematics" (Goodfellow, M. and O'Donnell, A.G., eds.), pp. 383-427. Academic Press, London.

Magee, J.T. (1994) Analytical fingerprinting methods, in "Chemical Methods in Prokaryotic Systematics" (Goodfellow, M. and O'Donnell, A.G., eds.), pp. 523-553. John Wiley and Sons Ltd., Chichester.

Magee, J.T., Hindmarch, J.M., Bennett, K.W., Duerden, B.I. and Aries, R.E. (1989a) A pyrolysis mass spectrometry study of fusobacteria. J. Med. Microbiol. 28, 227-236.

Magee, J.T., Hindmarch, J.M., Burnett, I.A. and Pease, A. (1989b) Epidemiological typing of *Streptococcus pyogenes* by pyrolysis mass spectrometry. J. Med. Microbiol. 30, 273-278.

Magee, J.T., Hindmarch, J.M. and Nicol, C.D. (1991) Typing of *Staphylococcus pyogenes* by pyrolysis mass spectrometry. J. Med. Microbiol. 35, 304-306.

Mathers, K., Freeman, R., Sisson, P.R. and Lightfoot, N.F. (1994) Differentiation between bacterial species and sub-species by pyrolysis mass spectrometry of extracted DNA. Zbl. Bakt. (in press).

Meuzelaar, H.L.C. and Kistemaker, P.G. (1973) A technique for fast and reproducible fingerprinting of bacteria by pyrolysis mass spectrometry. Anal. Chem. 45, 587-590.

Meuzelaar, H.L.C., Haverkamp, J. and Hileman, F.D. (1982) Pyrolysis Mass Spectrometry of Recent and Fossil Biomaterials. Compendium and Atlas. Elsevier, Amsterdam.

Meyer, B.T., Hansen, T., Nute, D., Albersheim, P., Darvill, A., York, W. and Sellers, J. (1991) Identification of the 1H-NMR spectra of complex oligosaccharides with artificial neural networks. Science 251, 542-544.

Millership, S. (1993) Use of a neural network for analysis of bacterial whole cell protein fingerprints. Binary 5, 126-131.

Morris, C.W. and Boddy, L. (1993) Some neural network packages: A comparative review. Binary 5, 7-9.

Morris, C.W., Boddy, L. and Allman, R. (1992) Identification of basidiomycete spores by neural network analysis of flow cytometry data. Mycol. Res. 96, 697-701.

Munk, M.E., Madison, M.S. and Robb, E.W. (1991) Neural network models for infrared spectrum interpretation. Mikrochim. Acta (Wien) 2, 505-514.

Nelder, J.C. (1979) Genstat Reference Manual. Scientific and Social Service Program Library, University of Edinburgh.

Okami, Y. and Hotta, K. (1988) Search and discovery of new antibiotics, in "Actinomycetes in Biotechnology" (Goodfellow, M., Williams, S.T. and Mordarski, M., eds.), pp. 33-67. Academic Press, London.

Orr, K., Gould, F.K., Sisson, P.R., Lightfoot, N.F., Freeman, R. and Burdess, D. (1991) Rapid interstrain comparison by pyrolysis mass spectrometry in nosocomial infection with *Xanthomonas maltophilia*. J. Hosp. Infect. 17, 187-195.

Ottley, T.W. and Maddock, J. (1986) The use of pyrolysis mass spectrometry. Lab. Proc., October, 53-55.

Qian, N. and Sejnowski, T.S. (1988) Predicting the secondary structure of globular proteins using neural network models. J. Mol. Biol. 202, 865-884.

Rataj, T. and J. Schindler (1991) Identification of bacteria by a multilayer neural network. Binary 3, 159-164.

Robb, E.W. and Munk, M.E. (1990) A neural network approach to infrared spectrum interpretation. Mikrochim. Acta 1, 131-155.

Rumelhart, D.E., McClelland, J.L and the PDP Research Group (1986) Parallel Distributed Processing. MIT Press, Cambridge.

Saddler, G.S., Falconer, C. and Sanglier, J.J. (1989) Preliminary experiments for the selection and identification of actinomycetes by pyrolysis mass spectrometry. Actinomycetologia 2, S3-S4.

Sanglier, J-J., Whitehead, D., Saddler, G.S., Ferguson, E.V. and Goodfellow, M. (1992) Pyrolysis mass spectrometry as a method for the classification, identification and selection of actinomycetes. Gene 115, 235-242.

Schleifer, K.H., Ludwig, W. and Amann, R. (1993) Nucleic acid probes, in "Handbook of New Bacterial Systematics" (Goodfellow, M. and O'Donnell, A.G., eds.), pp. 463-524. Academic Press, London.

Schulze, H.G, Blades, M.W., Bree, A.V., Gorzalka, B.B., Greek, L.S.and Turner, R.F.B. (1994) Characteristics of backpropagation neural networks employed in the identification of neurotransmitter Raman spectra. Appl. Spectr. 48, 50-57.

Sellers, J., York, W., Albersheim, P., Darvill, A. and Meyer, B. (1990) Identification of the mass spectra of partially methylated alditol acetates by artificial neural networks. Carbohydr. Res. 207, c1-c5.

Shute, L.A., Berkeley, R.C.W., Norris, J.R. and Gutteridge, C.S. (1985) Pyrolysis mass spectrometry in bacterial systematics, in "Chemical Methods in Bacterial Systematics" (Goodfellow, M. and Minnikin, D.E., eds.), pp. 95-114. Academic Press, London.

Simpson, P.K. (1990) Artificial Neural Systems. Pergamon Press, Oxford.

Sisson, P.R., Freeman, R., Magee, J.G. and Lightfoot, N.F. (1991a) Differentiation between mycobacteria of the *Mycobacterium tuberculosis* complex by pyrolysis mass spectrometry. Tubercle 72, 206-209.

Sisson, P.R., Freeman, R., Lightfoot, N.F. and Richardson, I.R. (1991b) Incrimination of an environmental source of a case of Legionnaires' disease by pyrolysis mass spectrometry. Epidermiol. Infect. 107, 127-132.

Sisson, P.R., Freeman, R., Gould, F.K. and Lightfoot, N.F. (1991c) Strain differentiation of nosocomial isolates of *Pseudomonas aeruginosa* by pyrolysis mass spectrometry. J. Hosp. Infect. 19, 137-140.

Sisson, P.R., Lightfoot, N.F. and Freeman, R. (1992a) Pyrolysis mass spectrometry of micro-organisms. PHLS Microbiology Digest 9, 65-68.

Sisson, P.R., Freeman, R., Magee, J.G. and Lightfoot, N.F. (1992b) Rapid differentiation of *Mycobacterium xenopi* from mycobacteria of the *Mycobacterium avium-intracellulare* complex by pyrolysis mass spectrometry. J. Clin. Pathol. 45, 355-357.

Snyder, E.E. and Stormo, G.D. (1993) Identification of coding regions in genomic DNA sequences: an application of dynamic programming and neural networks. Nucleic Acids Res. 21, 607-613.

Suzuki, K., Goodfellow, M. and O'Donnell, A.G. (1993) Cell envelopes and classification, in "Handbook of New Bacterial Systematics" (Goodfellow, M. and O'Donnell, A.G., eds.), pp. 195-250. Academic Press, London.

Tanabe, K., Tamura, T. and Uesaka, H. (1992) Neural network system for the identification of infrared spectra. Appl. Spectr. 46, 807-810.

Vickers, J.C., Williams, S.T. and Ross, G.W. (1984) A taxonomic approach to selective isolation of streptomycetes from soil, in "Biological, Biochemical and Biomedical Aspects of Actinomycetes" (Ortiz-Ortiz, L., Bojalil, L.F. and Yakoleff, V., eds.), pp. 553-561. Academic Press, Orlando.

Vieth, M. and A. Kolinski (1991) Prediction of protein secondary structure by an enhanced neural network. Acta Biochim. Polon. 38:335-351.

Weigel, U.M. and Herges, R. (1992) Automatic interpretation of infrared spectra - recognition of aromatic substitution patterns using neural networks. J. Chem. Inform. Comp. Sci. 32, 723-731.

White, H. (1990) Connectional nonparametric regression: Multilayer feedforward networks can learn arbitrary mappings. Neural Networks 3, 535-549.

Willis, M.J., Di Massimo, C., Montague, G.A., Tham, M.T. and Morris, A.J. (1991) Artificial neural networks in process engineering. IEE Proc.-D 138, 256-266.

Windig, W., Kistemaker, P.G., Haverkamp, J. and Meuzelaar, H.L.C. (1979) The effects of sample preparation, pyrolysis and pyrolysate transfer conditions on pyrolysis mass spectra. J. Anal. Appl. Pyrol. 1, 39-52.

Winstanley, T.G., Magee, J.T., Limb, D.I., Hindmarch, J.M., Spencer, R.C., Whiley, R.A., Beighton, D. and Hardie, J.M. (1992) A numerical taxonomic study of the *'Streptococcus milleri'* group based upon conventional phenotypic tests and pyrolysis mass spectrometry. J. Med. Microbiol. 36, 149-155.

Wold, S. and Söderström, M.J. (1977) SIMCA: A method for analysing chemical data in terms of similarity and analogy, in "Chemometrics: Theory and Application" (Kowalski, B., ed.), Series no. 52, pp. 243-282. ACS Symposium Washington, DC: American Chemical Society.

Zurada, J.M. (1992) Introduction to Artificial Neural Systems. West Publishing Company, St. Paul.

NEW METHODS FOR DIAGNOSIS AND EPIDEMIOLOGICAL STUDIES OF TUBERCULOSIS BASED ON PCR AND RFLP

Carlos Martín[1], Sofía Samper[1], Isabel Otal[1], Pilar Asensio[1],
Rafael Goméz-Lus[1], Gabriela Torrea[2], Brigitte Gicquel[2]

[1]Unidad Microbiología
Facultad de Medicina
Universidad Zaragoza

[2]Unité Génétique Mycobactérienne
Institut Pasteur, Paris

INTRODUCTION

Tuberculosis is a major cause of morbidity and mortality in the world today, particularly in developing and tropical countries. More than one billion individuals are infected with *Mycobacterium tuberculosis,* and more than 3.5 million people die anually. After years of declining case rates, tuberculosis has returned as a major public health problem in the United States and Europe. The epidemiology of tuberculosis is changing dramatically due to socio-economical problems, the HIV pandemic and the emergence of drug-resistant strains. New methods, based on polymerase chain reaction (PCR) and restriction fragment length polymorphism (RFLP), for the diagnosis and epidemiological studies of tuberculosis are of critical importance for the development of effective control strategies for the disease.

The definitive diagnosis of tuberculosis depends on the isolation of pathogenic bacteria and their subsequent identification by biochemical tests which requires between 4 and 8 weeks. It has been shown that the presence of *M. tuberculosis* in clinical samples may be detected by PCR assays, which are more rapid and more sensitive than classical bacteriological techniques.

Taxonomy relies on biochemical and phenotypic characters (numerical taxonomy) and to genomic DNA/DNA relatedness. Results obtained from the sequencing of 16s RNA confirm the previous data. The basis for phylogenetically valid taxonomy of the genus *Mycobacterium* has been established by Rogall *et al.* (1990). By sequencing the 16S rRNA of 19 species, a high level of 16S rRNA sequence similarity within the genus (greater than 94.3%) has been shown. Tuberculosis is caused by species belonging to *M. tuberculosis* complex, i.e. *M. tuberculosis*, *M. africanum*, *M. bovis*, and *M. microti,* the levels of DNA homology shown, by total DNA hybridization, ranged from 90 to 100% (Imaeda, 1985).

Different mycobacterial DNA targets have been used for PCR detection such as the 65 kDa antigen (gene *groEL*) (Hance *et al.*, 1989) or 16S rRNA (Böddinghaus *et al.*, 1990) among others. Both can detect the presence of strains belonging to the genus *Mycobacterium* and a subsequent identification of the amplified DNA fragment is often necessary for species differentiation. The insertion sequence IS6110 has been shown to be present only in pathogenic mycobacterial species belonging to *M. tuberculosis* complex (Thierry *et al.*, 1990). Amplification of IS6110 has shown to be highly specific and sensitive for the diagnosis of tuberculosis.

Due to the long incubation period time of tuberculosis, the tracing of individuals who might have been in contact with infected people, is a major strategy for limiting the dissemination of *M. tuberculosis*. The typing of strains by RFLP from infected individuals could play an important role in tracing the sources of infection. RFLP based on IS6110 polymorphism is shown here to be useful in epidemiological studies.

METHODS FOR RAPID DIAGNOSIS BASED ON PCR

Detection and Identification of Mycobacteria by Amplification of the 65 kD Gene

Polymerase chain reaction based on the amplification of a DNA sequence present in the gene coding for the 65 kDa (GroEL) mycobacterial antigen, can detect the presence of strains belonging to the genus *Mycobacterium* (Hance *et al.*, 1989). The DNAs from four species, *M. tuberculosis*, *M. bovis*, *M. avium*, *M. paratuberculosis* and *M. fortuitum* were compared by nucleotide sequencing. Although the gene segments from these species showed considerable similarity, oligonucleotide probes which could distinguish *M. tuberculosis*/*M. bovis*, *M. avium*/*M. paratuberculosis* and *M. fortuitum* could be identified. Subsequent identification of the amplified product using oligonucleotide probes that hybridize selectively to DNA from the different species allowed species differentiation.

This method based on the amplification and hybridization of the 65 kD gene (*hsp*65) for the rapid detection of *M. tuberculosis* was used to test clinical specimens, including peripheral blood samples of patients with AIDS. The amplification procedure has important advantages over conventional methods for the diagnosis of mycobacterial infections; it is more sensitive than direct examination and more rapid than culture. The sensitivity of the technique is especially useful for samples that contain very few mycobacteria, such as cerebrospinal fluid specimens of patients with suspected tuberculous meningitis (Brisson-Noël *et al*., 1989).

Recently a method for the differentiation of mycobacterial species has been described by restriction enzyme digestion of the PCR products of *hsp*65 common amplified fragments, thus avoiding the hybridization step (Plikaytis *et al.*, 1992, Telenti *et al.*, 1993). This method involves restriction enzyme analysis with *Bst*NI and *Xho*I or *Bst*EII and *Hae*III, of PCR products obtained with primers common to the 65 kDa antigen genes of all mycobacteria. Using these two restriction enzymes combinations, mycobacterial isolates were differentiated to the species or subspecies level by PCR-restriction enzyme pattern analysis (PRA). The procedure does not involve hybridization steps, multiple probes, or the use of radioactivity and the authors claim that it can be completed within one working day. A comparison of the restriction fragment pattern of an unknown with the database of known patterns should allow identification of the unknown organism to the species level.

Detection and Identification of Mycobacteria by Amplification of rRNA

Another technique for the detection and identification of mycobacteria has been developed based on the amplification of nucleic acid sequences from 16S rRNA. Following amplification of mycobacterial nucleic acids after optional synthesis of cDNA. The subsequent identification of the amplified DNA fragment is carried out by hybridization to oligonucleotides, the specificity of which is defined at the genus or species level (Böddinghaus *et al.*, 1990).

Less-conserved regions of the 16S rRNA genes of potential use as target sites for genus- or species-specific oligonucleotide probes were identified by comparison of sequence variability within the collection of mycobacterial 16S rRNA sequences. The results obtained showed that there are small but consistent sequence differences in certain regions of the rRNA molecule between mycobacterial species which allow the definition of highly specific oligonucleotides at genus, group, or species level. Specific DNA probes have been experimentally demonstrated for *M. tuberculosis, M. avium*, and *M. intracellulare*.

A method based on the combination of enzymatic amplification and restriction analysis (amplified rDNA restriction analysis, ARDRA) for identification of *Mycobacterium* species (Vaneechoutte *et al.*, 1993) has recently been shown. The 16S rRNA genes (rDNA) of 18 different species of the genus *Mycobacterium* were PCR amplified. ADRAs obtained after restriction with *Cfo*I, *Mbo*I and *Rsa*I allowed the differentation within species.

Use of Species-Specific Repetitive DNA Sequences for Detection of *M. tuberculosis*

Repetitive segments of DNA are found in virtually all prokaryotic and eukaryotic organisms and it is therefore not surprising that these elements have been identified in mycobacteria. It has been shown that some repetitive DNA sequences display species specificity. The insertion sequence IS*6110*, which possesses similarities to insertion sequences of the IS*3* family, is detected only in species belonging to the *M. tuberculosis* complex. Because IS*6110* is found only in *M. tuberculosis* complex strains, this element is a useful target for *in vitro* amplification by PCR for the rapid and specific detection of such bacteria in clinical samples, without the need for culturing these slowly-growing microorganisms (Thierry *et al.*, 1990).

Oligonucleotides derived from this sequence can be used to detect *M. tuberculosis* in clinical specimens following *in vitro* DNA amplification. The advantages of this approach are increased sensitivity due to the presence of 1-20 copies of IS*6110* on the chromosome of *M. tuberculosis* complex strains and high specificity for the causative agent of tuberculosis since it is only present in strains of the *M. tuberculosis* complex. Figure 1 shows the specificity of the IS*6110*-derived and 65 kDa antigen-derived primers and probes in PCR experiments with 19 reference mycobacterial species. Amplification corresponding to the 65-kDa antigen gene (383-bp fragment) occurred in all the mycobacterial species tested. Amplification of the 325-bp fragment corresponding to IS*6110*, was observed for all the species corresponding to the *M. tuberculosis* complex. These results were confirmed by hybridization using the 325-bp fragment of IS*6110* as a probe.

At the present time amplification based on IS*6110* is the most universally used PCR diagnostic test for *M. tuberculosis*, due to the high sensitivity and specificity of the approach (Brisson-Nöel *et al*., 1991, Hermans *et al*., 1990, Clarridge III *et al.*, 1993). Recently, after thousands strains of *M. tuberculosis* had been studied, one strain which lacked IS*6110* was described (van Soolingen *et al.*, 1993). If this is the case for more strains the use of this element as a target for PCR should be reconsidered.

Although PCR studies are encouraging, there are still, however, a number of problems which need to be addressed before this procedure can be used routinely in laboratories, e.g., price, contamination and presence of inhibitors. Contamination derived from its extreme sensitivity, can be largely overcome by careful laboratory procedure (Kwok and Higuchi, 1989). Two methods of PCR sterilization have been described by the use of UV irradiation (sarkar and Sommer, 1990) or uracil glycosidase (Longo *et al.*, 1990). The detection of the presence of inhibitors in each specimen is critical if PCR results are to be correctly interpreted. This can be achieved by the presence in each sample of an appropiate internal control (de Wit *el al.*, 1993). The high cost of the PCR limits applications of this technique in developing countries.

Novel Methods for Diagnosis

A method for diagnosis is under development that is based on infection of viable *Mycobacterium* (*M. smegmatis* and *M. tuberculosis*) by phage carrying the luciferase gene under the control of a mycobacterial promoter. This results in luciferase expression

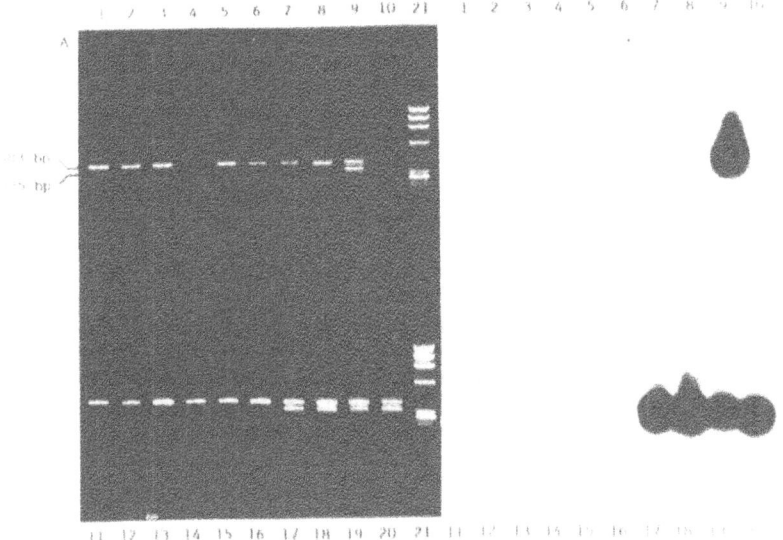

Figure 1. PCR amplification of mycobacterial DNA with IS*6110* primers (325bp product, amplifies species corresponding to the *M. tuberculosis* complex) and 65 kDa primers (383bp, amplifies all species belonging the genus *Mycobacterium*) simultaneously. (A) Ethidium bromide-stained agarose gel containing amplification reactions. (B) Southern blot hybridized with IS*6110* probe. Lanes: 1.- *M. asiaticum*, 2.- *M. avium*, 3.- *M. chelonae*, 4.- *M. flavescens*, 5.- *M. gordonae*, 6.- *M. kansasii*, 7.- *M. malmoense*, 8.- *M. marinum*, 9.- *M. tuberculosis*, 10.- control without DNA, 11.- *M. paratuberculosis*, 12.- *M. scrofulaceum*, 13.- *M. simiae*, 14.- *M. szulgai*, 15.- *M. terrae*, 16.- *M. xenoi*, 17.- *M. bovis*, 18.- *M. bovis* BCG, 19.- *M. africanum*, 20.- *M. microtii*. 21.- φX174 DNA HaeIII.

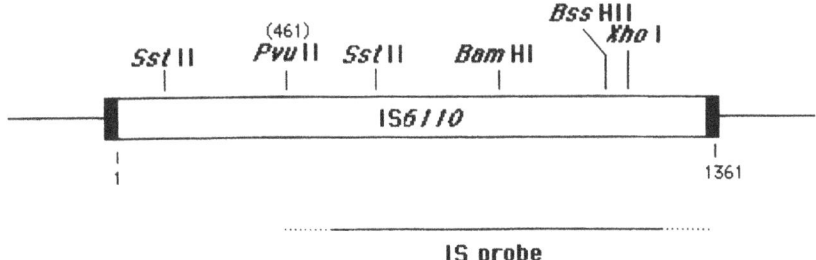

IS probe

Figure 2. Physical map of the *M. tuberculosis* insertion sequence IS*6110* . The restriction enzyme *Pvu*II cleaves at base pair 461. The closed bars represent the 28-bp inverted repeats bordering IS*6110* DNA. The lines to the left and right denote chromosomal DNA

minutes after the phage DNA gains entry into the bacterium (Jacobs *et al.*, 1993). Consequently, the infected mycobacteria emit light in the presence of ATP and luciferin. Because the detection of these light quanta is extremely sensitive even using standard instruments, only a few viable bacteria have to be infected by the phage for a positive result to be recorded. For the direct detection of pathogenic mycobacteria in clinical specimens the host range of the phage should be limited to the *M. tuberculosis* complex. Advantages of this method over PCR detection are its simplicity and absence of false positive results caused by cross-contamination with previously amplified DNA. Furthermore, the engineered phages can be used for rapid testing of drug sensitivity by monitoring the effect of mycobacterial growth on exposure to antimycobacterial drugs.testing the luciferase expression on mycobacteria. However it's use may be limited due to the variable infectivity of phage stocks.

MOLECULAR GENETIC TOOLS FOR EPIDEMIOLOGY

Typing of *M. tuberculosis* Strains by IS*6110*/RFLP (DNA "Fingerprinting")

Epidemiological studies of tuberculosis can be greatly facilitated by the application of strain-specific markers. The recently discovered transposable elements in *M. tuberculosis* have been shown to be of great potential use in strain differentiation by RFLP. Duplications of 3-4 nucleotides were found at the extremities of the IS*6110* copies, sugesting that IS*6110* RFLP is due to transposition of the IS element (Mendiola *et al.*, 1992). The basis of this technique is the detection of IS*6110* which have the potential to relocate to different positions in the chromosome but this frequency of transposition is low enough to allow the recognition of the same strain.

IS*6110* RFLP analysis of strains isolated from patients who developed tuberculosis showed identical patterns over a 2- to 3-year period. In contrast, a high degree of polymorphism was observed between strains of the *M. tuberculosis* complex isolated from different patients (Otal *et al.*, 1991).

IS*6110*/RFLPs have been shown to be very useful in clinical epidemiological studies. RFLP studies have also demonstrated the reinfection of HIV patients from a single multidrug-resistant *M. tuberculosis* strain (Small *et al.*, 1993). In order to compare the results obtained from different laboratories there are three critical parameters relevant to a standardized IS*6110*-based DNA fingerprinting system: the choice of the restriction enzyme to be used, the DNA probe, and the use of appropriate molecular mass standards.

A standardised method has been published by nine different laboratories using a nonradiactive hybridization method (Van Embdem *et al.*, 1993). The recommendations are as follows: use of *Pvu*II to generate RFLP of the total mycobacterial DNA, a DNA probe to the right of the *Pvu*II site on IS*6110* (Figure 2) and to compare the RFLPs it is recommended that a combination of external and internal standard molecular size markers

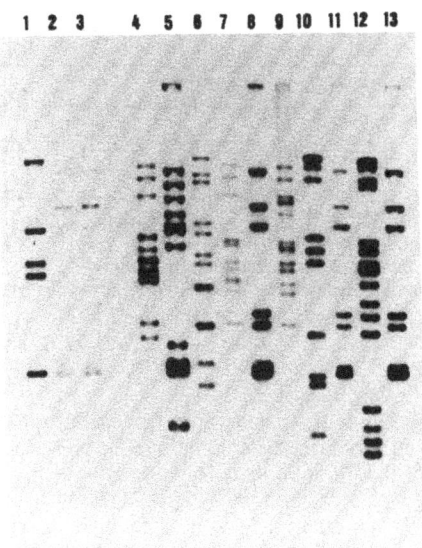

Figure 3. DNA fingerprintings of various *M. tuberculosis* isolates from Hospital Clínico Universitario de Zaragoza tuberculosis patients. Lanes 2 and 3, isolated from the same patient from different origin (sputum, urine). Lanes 4 to 13, *M. tuberculosis* isolates from different patients.

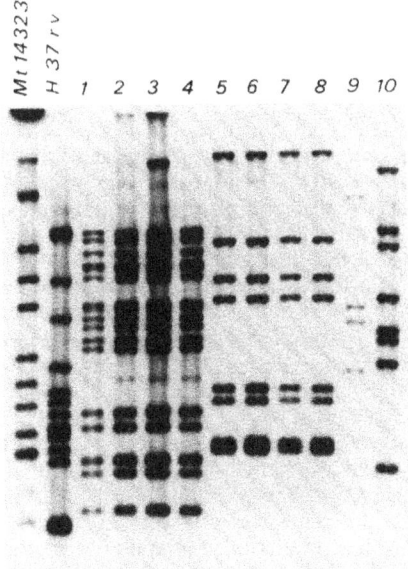

Figure 4. DNA fingerprintings of various *M. tuberculosis* isolates. Lanes 1 to 4 isolated from patients of ages between 65 and 87 years and living in the same area, show identical patterns. Lanes 5 to 10 show *M. tuberculosis* isolates from patients living in the same area, four of which are identical suggesting an outbreak between young people from 16 to 19 years old.

are used. If the above recommendations are used it would permit the comparison of DNA fingerprints of *M. tuberculosis* from different origins obtained in different laboratories. In Figures 3 and 4 some examples of the standardized RFLP technique are shown for different *M. tuberculosis* clinical isolates.

RFLP is a powerful and well-established method of "DNA fingerprinting", it is very useful in tracing the transmission of particular strains of *M. tuberculosis* during investigations of outbreaks and might provide insight into the extent of disease due to reactivation versus reinfection. The possible association between fingerprint types and properties of the strains such as pathogenicity, multiple drug resistance, HIV infection, extrapulmonar tuberculosis and geographical origin will provide information of critical importance for development of effective control strategies of tuberculosis.

Other Genetic Markers

Although most of *M. tuberculosis* complex strains carry multiple IS*6110* copies, isolates which contain only a few copies have occasionally been encountered. Analysis of isolates from epidemiologically unrelated cases of tuberculosis were found to carry only a single IS copy(van Soolingen *et al.*, 1993). These strains could not be differentiated by IS*6110* DNA fingerprinting but can be differentiated by other genetic markers as DRs and PGRS (van Soolingen *et al.*, 1993).

Numerous repeated DNA other than IS*6110* has been described in *M. tuberculosis* complex strains: the insertion sequence IS*1081* (Collins and Stephens, 1991), the major polymorphic tandem repeat (MPTR) (Hermans *et al.*, 1992), the direct repeats (DR) of 36 bp (Hermans *et al.*, 1991) or the polymorphic GC-rich repetitive sequence (PGRS) highly repetitive DNA present in pTBN12 (Ross *et al* ., 1992).

The IS*1081* element is found to be less useful for strain differentiation becuse most of the five to seven copies of IS*1081* are present at identical positions in the *M. tuberculosis* chromosome. Similarly, the MPTR sequence allows only limited strain differentiation.

PGRS and DR show polymorphism between different *M. tuberculosis* strains (Ross *et al.*, 1991). These elements are useful for distinguishing strains with low copy numbers of IS*6110* on the chromosome using *Alu*I (van Soolingen *et al.*, 1993).

NEW METHODS FOR EPIDEMILOGICAL STUDIES OF TUBERCULOSIS

The development of new PCR typing methods attempts to decrease the time for strain identification of *M. tuberculosis*. RFLP analysis of *M. tuberculosis* isolates by arbitrarly primed (AP) PCR has been described by Palittapongarnpim *et al* . (1993a). Moreover, new RFLP methods based on IS*6110* amplification by PCR have been described recently (Ross and Dwyner, 1993). Polymorphism in the patterns is obtained between different *M. tuberculosis* strains studied using oligonucleotide primers at the end of IS*6110*. Another method uses one primer specific for IS*6110* and a second primer complementary to a linker ligated to the restricted genomic DNA. This method has already allowed the differentiation of multidrug-resistant isolates from outbreaks (Haas *et al.*, 1993).

A novel approach to differentiate *M. tuberculosis* strains by combining the variability of the IS*6110* insertion sites with the conserved locations of the MPTR, named IS*6110*-ampliprinting, has been described recently by Plikaytis *et al.* (1993). Primers corresponding to portions of these sequences were used to amplify a set of DNA fragments that can differentiate *M. tuberculosis* strains into groups that correlated well with those obtained by IS*6110*-RFLP typing. This method should be useful as a rapid screening method for fingerprinting *M. tuberculosis*. However, because of the limited number and sizes of amplicons produced, and the resulting limited information, ampliprinting may provide somewhat less information about strain relatedness than the IS*6110*-RFLP fingerprinting method.

A ligation-mediated polymerase chain reaction (LMPCR) procedure to amplify the flanking sequences on both sides of the IS6110 (Palittapongarnpim *et al.*, 1993b) has recently been designed. This method uses a nonphosphorylated *BamHI*-compatible linker which was designed to contain a sequence identical to that found in IS6110 itself. The LMPCR method has been shown to be able to differentiate H37Rv and H37Ra strains and 26 clinical isolates of *M. tuberculosis*. The results were confirmed by RFLP detected by an IS6110 probe. The LMPCR method is technically easy as purification of DNA is unnecessary after restriction digestion or ligation. It also requires less DNA and tolerates greater shearing than Southern blotting and hybridization.

Based on the nature of the DNA polymorpism in the DR cluster a novel method of strain differenciation has been developed (Groenen *et al.*, 1993). Direct variable repeat polymer chain reaction (DVR-PCR) enables typing of individual *M. tuberculosis* strains in a single PCR.

Because of their rapidity, these techniques represent a significant advance in the molecular typing of *M. tuberculosis* isolates. The PCR typing methods could be a rapid method for recognising strains with important characteristics such as antibiotic resistances. New studies should be carried out to type *M. tuberculosis* from pathological samples.

ACKNOWLEDGEMENTS

We thank Julie-ann Gavigan for the critical reading of the manuscript. Our labs are supported by the European Economic Community CE-BIO/0520-92 and Spanish Ministerio de Educación y Ciencia CAICYT BIO/ 91-0925, BIO/93-0742 and Fondo de Investigaciones Sanitarias de la Seguridad Social FIS 94/0051-01 grants.

REFERENCES

Böddinghaus, B., Rogall, T., Flohr, T., Blöcker, H. and Böttger, C. (1990) Detection and identification of mycobacteria by amplification rRNA. J. Clin. Microbiol. 28, 1751-1759.

Brisson-Noël, A., Gicquel, B, Lecossier, D., Nassif, X., Lévy-Frébault, V. and Hance, A. J. (1989) Rapid diagnosis of tuberculosis by amplification of mycobacterial DNA in clinical samples. Lancet. ii , 1069-1071.

Brisson-Noël, A., Aznar, C., Chureau, C., Nguyen, S., Pierre, C., Bartoli, M., Bonete, R., Pialoux, G., Gicquel, B., and Garrigue, G. (1991) Diagnosis of tuberculosis by DNA amplification in clinical practice evaluation. Lancet. 338, 364-366.

Clarridge III, J.E., Shawar, R. M., Shinnick, T. M. and Plikaytis B. B. (1993) Large-scale use of polymerase chain reaction for detection of *Mycobacterium tuberculosis* in a routine mycobacteriology laboratory. J. Clin. Microbiol. 31, 2049-2056.

Collins, D. M., and Stephens D. M. (1991) Identification of insertion sequence, IS1081, in *Mycobacterium bovis*. FEMS Lett. 83, 11-16.

de Wit, Wootton, M. Allan, B., and Steyn, L. (1993) Simple method for production of internal control DNA for *Mycobacterium tuberculosis* polymerase chain reaction assays. J. Clin. Microbiol. 31, 2204-2207.

Groenen, P. M. A., Bunschoten, A. E., van Soolingen D. and van Embden J. (1993). Nature of DNA polymorphism in the direct repeat cluster of *Mycobacterium tuberculosis*; application for strain differenciation by a novel typing method. Molec Microbiol. 10, 1057-1065.

Hance, A.J., Grandchamps, B., Lévy-Frébault, Lecossier, D., Rauzier, J., Bocart, D., and Gicquel, B. (1989) Detection and identification of mycobacteria by amplification of mycobacterial DNA. Molec. Microbiol. 3, 843-849.

Haas, W.H., Butler, W. R., Wooley, C. L. and Crawford J. T. (1993) Mixed-linker polymerase chain reaction: a new method for rapid fingerprinting of isolates of the *Mycobacterium tuberculosis* complex. J. Clin. Microbiol. 31, 1293-1298.

Hermans, P. W. M., Van Soolingen, D., Dale, J. W., Schuitema, R. J., McAdam, R. A. Catty, D.,and Van Embdem J. D. (1990) Insertion element IS986 from *Mycobacterium tuberculosis*: a useful tool for diagnosis and epidemiology of tuberculosis. J. Clin. Microbiol. 28, 2051-2085.

Hermans, P. W. M., Van Soolingen, E. M. Bik, P.E.W. de Haas, J. W. Dale,and Van Embdem J. D. (1991) Insertion element IS987 from *Mycobacterium bovis* BCG is locatedin a hot-spot integration region for insertion elements in *Mycobacterium tuberculosis* complex strains. Infection and Immunity. 59, 2695-2705.

Hermans, P. W. M., Van Soolingen, and Van Embdem J. D. (1992) Characterization of a Major Polymorphic Tandem Repeat in *Mycobacterium tuberculosis* and its potential use in the epidemiology of *Mycobacterium kansasii* and *Mycobacterium gordonae*. J. Bacteriol. 174, 4157-4165.

Imaeda, T. (1985) Deoxyribonucleic acid relatedness among selected strains of *Mycobacterium uberculosis*, *Mycobacterium bovis*, *Mycobacterium bovis* BCG, *Mycobacterium microti* and *Mycobacterium sfricanum*.. J. Bacteriol. 35,147-150.

Jacobs, W. R., Barletta, R.G., Udani, R., Chan, J., Kalkut. G., Sosne, G., Kieser, T., Sarkis, G.J., Hatfull, G. F., and Bloom, B.R. (1993) Rapid assment of drug susceptibilities of *Mycobacterium tuberculosis* by means of luciferase repoter phages. Science. 260, 819-822.

Kwok, S. and Higuchi. (1989) Avoiding false positives with PCR. Nature (London) 339,237-238.

Longo, M. L., Berninger, M. S., and Harley J.L. (1990) The use of uracil glycosylase to control carry-over contamination in polymerase chain reactions. Gene. 93, 125-128.

Mendiola, M. V., Martín, C., Otal, I. and Gicquel, B. (1992) Analysis of the regions for IS*6110* RFLP in a single *Mycobacterium tuberculosis* strain. Res. Microbiol. 143, 767-772.

Otal, I., Martín, C., Vincent-Lévy-Frébault, V., Thierry, D., and Gicquel, B. (1991) Restriction fragment length polymorphism (RFLP) using IS*6110* as an epidemiological marker in tuberculosis. J. Clin. Microbiol. 29, 1252-1254.

Palittapongarnpim, P., Chomyc, S., Fanning, A., and Kunimoto , D. (1993a) DNA fragment length polymorphism analysis of *Mycobacterium tuberculosis* isolates by arbitrarily primed polymerase chain reaction. J. Infec. Diseases. 167, 975-978.

Palittapongarnpim, P., Chomyc, S., Fanning, A., and Kunimoto , D. (1993b) DNA fingerprinting of *Mycobacterium tuberculosis* isolates by ligation-mediated polymerase chain reaction. Nucleic Acid Research.21,761-762.

Plikaytis, B.B., Plikaytis, B.D., Yakrus, M. A., Butler, W. R., Woodley, C. L. , Silcox, V.A., and Shinnick, T. M. (1992) Differentiation of slowly growing *Mycobacterium* species, including *Mycobacterium tuberculosis*, by gene amplification and restriction fragment legth polymorphism analysis.J. Clin. Microbiol. 30, 1815-1822.

Plikaytis, B.B., Crawford, J.T.,Woodley, C. L., Butler, W. R., Eisenach K.D.,Cave, D.M. and Shinnick. (1993) Rapid, amplification-based fingerprinting of *Mycobacterium tuberculosis*. J. Gen. Microbiol. 139, 1537-1542.

Rogall, T., Wolters, J. Flohr, T. and Bottger E. (1990) Towards a phylogeny and definition of species at molecular level within the genus Mycobacterium. I. J. Systematic Bacteriol. 40,323-330.

Ross, B. Raios, K., Jackson, K., Sievers, A. and Dwyer, B. (1991) Differentation of *Mycobacterium tuberculosis* strains by use of a nonradiactive southern blot hybridization method. J. Infec. Diseases. 163, 904-907.

Ross, B. Raios, K., Jackson, K., Sievers, A. and Dwyer, B. (1992) Molecular cloning of a higly repeated DNA element from *Mycobacterium tuberculosis* and its use as epidemiological tool. J. Clin. Microbiol. 30, 942-946.

Ross, B. and Dwyer, B. (1993) Rapid, simple method for typing isolates of *Mycobacterium tuberculosis* by using the polymerase chain reaction. J. Clin. Microbiol. 31, 329-324.

Sarkar, G. and Sommer, S.S. (1990) Shedding light on PCR contamination. Nature (London). 343, 27.

Small P.,Shafer R.W., Shingh, S.P., Murphy M.J. Desmond M.D. and Sierra M. F. (1993) Exogenous reinfection with multidrug-resistant *Mycobacterium tuberculosis* in patiens with advanced HIV infection. New England J. of Medicine 328, 1137-1144.

Telenti, A., Marchesi, F., Balz, M., F. Bally, Böttger, e. C. and Bodmer, T. (1993) Rapid identification of mycobacteria to species level by polymerase chain reaction and restriction enzyme analysis. J. Clin. Microbiol. 31, 175-178.

Thierry, D., Brisson-Noel, A., Vincent-Lévy-Frébault, V., Nguyen, S., Guesdon, J-L., and Gicquel, B. (1990) Characterization of a *Mycobacterium tuberculosis* Insertion Sequence, IS*6110*, and its application in diagnosis. J. Clin. Microbiol. 28, 2668-2673.

Van Embden J. D.A. ,Cave M. D., CrawfordJ. T., DaleJ. W., Eisenach K. D., Gicquel, B. HermansP., Martín C., McAdam R., Shinnick T. and Small P. (1993) Strain identification of *Mycobacterium tuberculosis* by DNA fingerprinting: Recommendations for a standardized methodology. J. Clin. Microbiol. 31, 406-409.

Van Soolingen, D. de Haas, P.E.W., Hermans, P. W.M., Groenen P.M. A., and Van Embdem J. D. (1993) Comparison of various repewtitive DNA elements as genetic markers for strain differentiation and epidemiology of *Mycobacterium tuberculosis*. J. Clin. Microbiol. 31,1987-1995.

Vaneechoutte, M., de Beenhouwer, H., Claeys, G., Verschraegen, G.de Rouck ann, Paepe, N. Elaichouni, A. and Portaels, F. (1993) Identification of *Mycobacterium* species by using amplified ribosomal DNA restriction analysis. J. Clin. Microbiol. 31,2061-2065.

TYPING *IN SITU* WITH PROBES

Rudolf Amann and Wolfgang Ludwig
Lehrstuhl für Mikrobiologie
Technische Universität München
80290 München, Germany

INTRODUCTION

After nearly one hundred years of fruitless searches for methods for the establishment of prokaryotic systematics based on genealogical relationships of the organisms, Zuckerkandl and Pauling (1965) introduced the idea of "Molecules as documents of evolutionary history". Thereafter the "molecular revolution" (Woese, 1991) was launched by Woese and coworkers who introduced comparative sequence analysis of ribosomal RNA for the elucidation of the phylogeny of microorganisms (Fox *et al.*, 1977; Woese and Fox, 1977). Nowadays, the determination of rRNA primary structures has become a routine method for taxonomic investigations. The number of small subunit rRNA sequences which are currently available in public databases (Larsen *et al.*, 1993; Neefs *et al.*, 1993) is about 2000, that of large subunit counterparts about 200; both are rapidly growing. The larger data set allowed not only definition of groups of phylogenetically related microorganisms but also recognition of molecular signatures (Woese, 1987) for these groups. Signatures are rRNA structure elements which are unique to a particular group. Primary structure elements, such as single nucleotides or sequence stretches, insertions (Roller *et al.*, 1992) and deletions (Ludwig *et al.*, in press), as well as presence and shape of higher order structure elements (Roller *et al.*, 1992; Ludwig *et al.*, 1992) may represent signatures for higher or lower phylogenetic entities respectively. Primary structure signatures can be used as target sites for diagnostic hybridization probes, thereby allowing identification of microorganisms on the basis of their phylogenetic relationships. Since the first experimental approaches (Göbel and Stanbridge, 1984; Festl *et al.*, 1986), a reasonable number of rRNA or rDNA (DNA encoding rRNA) targeted, specific hybridization probes has been developed (reviewed by Schleifer *et al.*, 1993). Further progress in the approach was achieved by the introduction of *in situ* techniques for microbial identification at the cellular level (DeLong *et al.*, 1989; Amann *et al.*, 1990*b*). The application of such probes for the identification of culturable microorganisms in pure and mixed cultures using a variety of hybridization assays has emphasized the power of the probe approach (Schleifer *et al.*, 1993). Methods for probe design and application have 'evolved' as rapidly as the methods for rRNA based phylogenetic investigations. Nowadays, probe design following PCR (polymerase chain reaction) based partial rDNA sequence analysis is a rapid routine method. Computer programs for automated design of probes as well as of experimental parameters are under development as is a public database of probe sequences and application methods (Ludwig *et al.*, in prep.). However, following the establishment of the sequence based methods for phylogenetic analyses and identification of microorganisms, another milestone in the history of microbiology is the development and introduction of probe based techniques for *in situ* identification of microorganisms without prior cultivation (Amann *et al.*, 1991). By combined application of comparative sequence analysis of *in vitro* amplified, cloned rDNA retrieved from community DNA and whole cell hybridization with specific probes, it is at last possible to elucidate the phylogenetic position of so far nonculturable microorganisms and to analyse their distribution over space and time in complex communities. The power of this method cannot be overestimated keeping in mind that in some cases less than 1% of the organisms visualized by microscopy are accessible for investigation by cultivation (Colwell *et al.*, 1985; Brock *et al.*, 1987; Torsvik *et al.*, 1990). Not surprisingly initial results obtained by direct retrieval of rRNA sequences from environmental samples indicated that most of the bacterial diversity is yet unknown (Giovannoni *et al.*, 1990; Ward *et al.*, 1990; Schmidt *et al.*, 1991)

MOLECULAR BACKGROUND

The primary structures of rRNA molecules contain an alternating succession of regions which have changed at different rates during the course of evolution. Sequence positions wich are identical in rRNA molecules of all organisms have preserved the character states of the progenote molecule. Highly conserved positions carry information on early times in the history of life and more variable positions document on more recent evolutionary events. This is illustrated in figures 1 and 2 for part of bacterial 23S rRNA. Figure 1 shows a universal phylogenetic tree based on 23S rRNA data. Different phylogenetic levels such as the domain, phylum, group and genus levels are indicated by different patterns. Figure 2 is a potential secondary structure model of part of domain III (Höpfl *et al.*, 1989; Larsen, 1992; Ludwig *et al.*, 1992) of bacterial 23S rRNA homologous to bases 1265 - 1645 of *Escherichia coli* 23S rRNA (Brosius *et al.*, 1981). Those regions which are highly conserved below definite phylogenetic levels and therefore may be diagnostic for microorganisms of the corresponding phylogenetic entity are marked by the same patterns as in Figure 1. The 23S rRNA region shown in Figure 2 contains useful target sites for specific probes.

rRNA-targeted hybridization probes are usually synthetic oligonucleotides of 15 to 30 bases in length which are complementary to diagnostic regions of rRNA and carry a detectable label (Schleifer *et al.*, 1993). The primary structures of ideal diagnostic regions have to be identical in the rRNAs of all members of the particular specificity group but different in all other organisms. Applying suitable experimental conditions, only perfectly basepaired hybrids remain stable (Stahl and Amann, 1991) and therefore detectable. The stability of imperfectly basepaired hybrids is influenced by the length of the probe, its base composition, its sequence and the number, nature and positions of mismatches (Stahl and Amann, 1991).

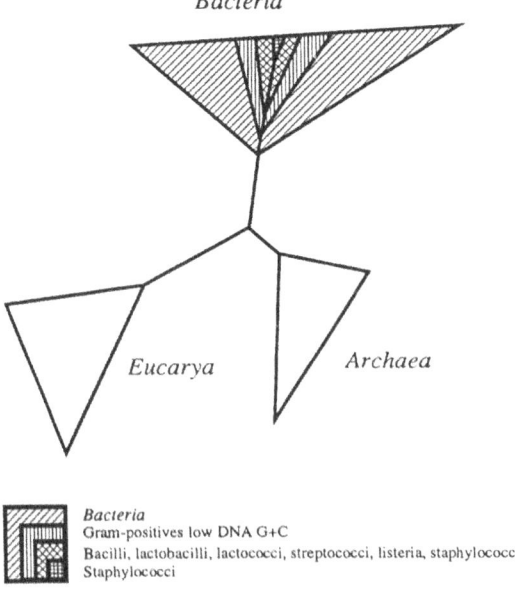

Bacteria

Bacteria
Gram-positives low DNA G+C
Bacilli, lactobacilli, lactococci, streptococci, listeria, staphylococci
Staphylococci

Eucarya   *Archaea*

Figure 1. Schematic universal phylogenetic tree based on large subunit rRNA data. Phylogenetic entities are indicated by triangles, phylogenetic levels by different patterns.

For the design of specific hybridization probes, an optimal alignment of rRNA primary structures (Ludwig and Schleifer, 1993) is a prerequisite.   First, all available sequences from a specificity group have to be compared for sequence identities, then the homologous sequences of the closest relatives are inspected for suitable differences. A further criterion for the selection of potential target sites is avoidance of self complementarities within the deduced probe sequences. Intramolecular base pairing during the hybridization experiment may hinder probes binding to their destined target sites. The comparison of potential target sites with the homologous sites of all available rRNA sequences is the next step in the evaluation procedure. To prevent nonspecific binding to other rRNA or genomic sites, a database search (EMBL sequence data base, Heidelberg, Germany) for accidental sequence similarities is recommended. Figure 3 shows a section of aligned bacterial 16S rRNA sequences highlighting base differences at a probe target site.

In general, probes can be designed complementary to either of the two strands of rRNA genes, but usually it is advantageous to select the version which is complementary to the rRNA primary structure. The sensitivity of most hybridization assays benefits from the naturally amplified target since there are up to 10^4 times more rRNA molecules than rRNA genes per microbial cell.

Both larger rRNAs (16S and 23S rRNA) have been successfully used for probe design, whereas the small 5S rRNA contains too few potential target sites. In comparison with 16S rRNAs, 23S rRNA molecules are not only twice the length, but also contain more and longer sequence stretches exhibiting a given degree of sequence conservation. Therefore, the probability of finding suitable target sites for specific probes is higher using 23S rRNA

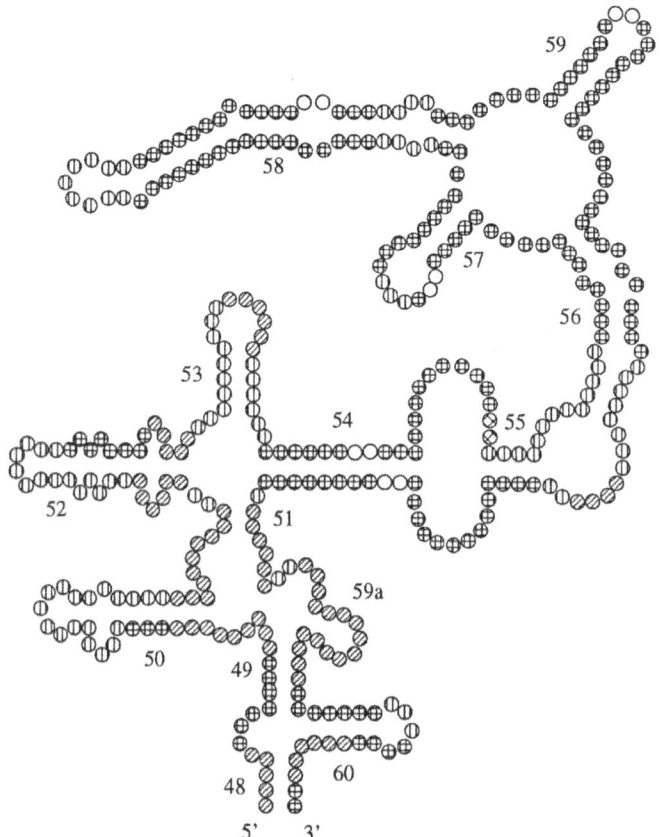

Figure 2. Schematic secondary struture model of domain III of 23S rRNA based on the consensus sequence from Gram-positive bacteria with a low DNA G+C content (Ludwig *et al.*, 1992). The patterns indicate regions which are at least 95% similar in all available primary structures from the phylogenetic entities defined in Fig. 1.

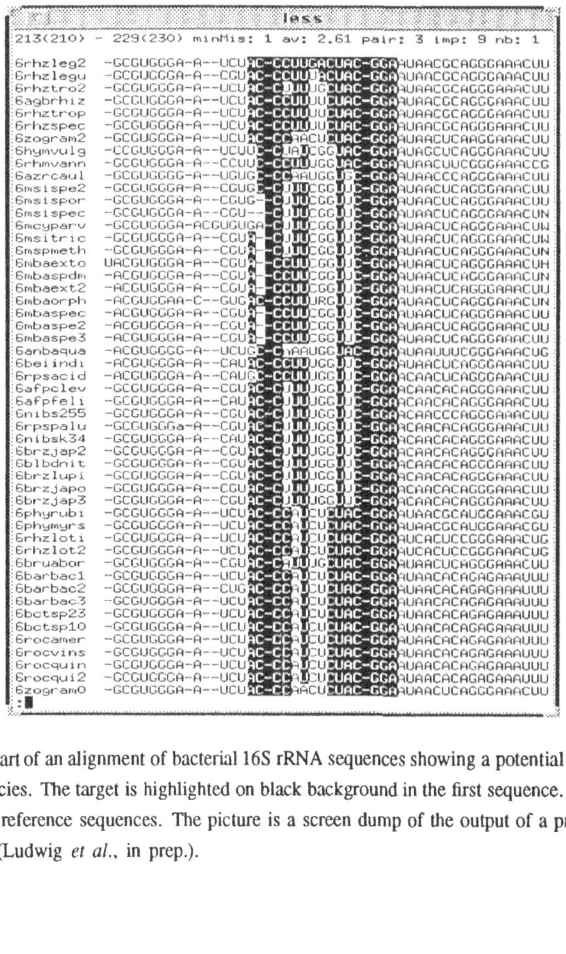

Figure 3. Part of an alignment of bacterial 16S rRNA sequences showing a potential probe target site for a *Rhizobium* species. The target is highlighted on black background in the first sequence. Identical residues are marked in the reference sequences. The picture is a screen dump of the output of a program for automated probe design (Ludwig *et al.*, in prep.).

Table 1. Sequences and specificities of rRNA targeted specific probes for *in situ* cell hybridizations

Sequence	Target	Specificity	Reference
5'-ACGGGCGGTGTGTRC-3'	16S	universal	Amann *et al.*, 1990*b*
5'-GWATTACCGCGGCKGCTG-3'	16S	universal	Giovannoni *et al.*, 1988
5'-GTGCTCCCCCGCCAATTCCT-3'	16S	*Archaea*$_{do}$	Stahl and Amann 1991
5'-TCCGGCRGGATCAACCGGAA-5'	16S	*Archaea*$_{do}$	Giovannoni *et al.*, 1988
5'-GCTGCCTCCCGTAGGAGT-3'	16S	*Bacteria*$_{do}$	Amann *et al.*, 1990*a*
5'-ACCGCTTGTGCGGGCCC-3'	16S	*Bacteria*$_{do}$	Giovannoni *et al.*, 1988
5'-TGAGCCAGGATCAAACTCT3'	16S	*Bacteria*$_{do}$	Hicks *et al.*, 1992
5'-ACCAGACTTGCCCTCC-3'	16S	*Eucarya*$_{do}$	Amann *et al.*, 1990*a*
5'-GGGCATCACAGACCTG-3'	16S	*Eucarya*$_{do}$	Giovannoni *et al.*, 1988
5'-TAGAAAGGGCAGGGA-3'	16S	*Eucarya*$_{do}$	Hicks *et al.*, 1992
5'-CGTTCGYTCTGAGCCAG-3'	16S	α-subclass$_{gr}$	Manz *et al.*, 1992
5'-GCCTTCCCACTTCGTTT-3'	23S	β-subclass$_{gr}$	Manz *et al.*, 1992
5'-GCCTTCCCACATCGTTT-3'	23S	γ-subclass$_{gr}$	Manz *et al.*, 1992
5'-CGGCGTCGCTGCGTCAGG-3'	16S	sulfate reducing bacteria$_{gr}$	Amann *et al.*, 1990*a*
5'-CTTCCGATCCGGTCG-3'	23S	fluorescent pseudomonads$_{gr}$	Schleifer *et al.*, 1992*a*
5'-GGTCCGAAGATCCCCCGCTT-3'	16S	Methylotrophic bacteria$_{gr}$	Tsien *et al.*, 1990
5'-CCCTGAGTTATTCCGAAC-3'	16S	Methylotrophic bacteria$_{gr}$	Tsien *et al.*, 1990
5'-TTCCACTTTCCTCTACCG-3'	16S	*Holospora*$_{ge}$	Amann *et al.*, 1991
5'-ACTACCCTCTCCGTGATT-3'	16S	*Magnetospirillum*$_{ge}$	Schleifer *et al.*, 1991
5'-TMCGCARACTCATCCCCAAA-3'	16S	*Desulfobacter*$_{ge}$	Amann *et al.*, 1990*a*
5'-ACCATCCTCTACCGAACT-3'	16S	magnetotactic coccus	Spring *et al.*, 1992
5'-ACCACCCTCTGCCAAACT-3'	16S	magnetotactic coccus	Spring *et al.*, 1992
5'-ACCACCCTCTGCCGGACT-3'	16S	magnetotactic coccus	Spring *et al.*, 1992
5'- GGACTCACCCTTAAACGG-3'	16S	magnetotactic bacterium	DeLong *et al.*, 1993
5'-TGGACTTGAGTTCGGAGA-3'	16S	sulfate reducing bacterium	Kane *et al.*, 1993
5'-GCCATCCCCTCGCTTACT-3'	16S	"*Magnetobacterium bavaricum*"	Spring *et al.*, 1993
5'-GCTGTACTCAAGTTACCCAGTT-CTAA-3'	16S	bacterial symbiont	Distel *et al.*, 1991
5'-AACCCGTACAGATCAAAGG-3'	16S	archaeal symbiont	Embley *et al.*, 1992*b*
5'-GACCATTCCAGGAATCTCTA-3'	16S	archaeal symbiont	Embley *et al.*, 1992*b*
5'-CAGGATACATCCACTATG3'	16S	archaeal symbiont	Embley *et al.*, 1992*a*
5'-CTGCATCGACAGGCACT-3'	16S	archaeal symbiont	Embley *et al.*, 1992*a*
5'-CTGCATCGACAGGCACT-3'	16S	archaeal symbiont	Finlay *et al.*, 1993
5'-AACACCCTACTTCTTTCG-3'	23S	*Holospora obtusa*$_{sp}$	Amann *et al.*, 1991
5'-CTTCCAATCCGGTCG-3'	23S	*Pseudomonas putida*$_{sp}$	Schleifer *et al.*, 1992*a*
5'-GTGGAAGCTTGACGGTATATCG-CAAACTCCTA-3'	16S	*Porphyromonas gingivalis*$_{sp}$	Gersdorf *et al.*, 1993*b*
5'-TAACAATCTACGCTACCC-3'	23S	*Pseudomonas cepacia*$_{sp}$	Schleifer *et al.*, 1992*a*

120

5'-ACTCTCTCCATACACCTT-3'	23S	*Pseudomonas diminuta*sp	Schleifer *et al.*, 1992*a*
5'-GCCGTCTCCCCCTACCTT-3'	23S	*Comamonas testosteroni*sp	Schleifer *et al.*, 1992*a*
5'-ACTACCCTCTCCCATACT-3'	16S	*Sarcobium lyticum*sp	Springer *et al.*, 1992
5'-TGCCCCTGAACTATCCAAGA-3'	16S	*Fibrobacter succinogenes*sp	Amann *et al.*, 1990*b*
5'-GTAGAGCTTACACTATATCGCA-AACTCCTA-3'	16S	*Bacteroides forsythus*sp	Gersdorf *et al.*, 1993*a*
5'-CCGCATCGATGAATCTTTCGT-3'	16S	*Fibrobacter intestinalis*sp	Amann *et al.*, 1990*b*
5'-GCCCCGCTGCCCATTGTACCGC-CC-3'	16S	*Fibrobacter intestinalis*sp	Amann *et al.*, 1990*b*
5'-TAAAGGACCCACCATCTCT-3'	16S	*Frankia* spec.sp	Hahn *et al.*, 1993
5'-GCAGGACCCTTACGGATCCC-3'	16S	*Frankia* spec.sp	Hahn *et al.*, 1993
5'-ATACCCGATACATCCTTTTTCCT GCATGGA-3'	16S	*Bacillus polymyxa*sp	Jurtschuk *et al.*, 1992
5'-TACAACGGGAAGCGAAGTAGT-GATATGGAG-3'	16S	*Bacillus macerans*sp	Jurtschuk *et al.*, 1992
5'-CCATACCGATAAATCTCTAGT-3'	16S	*Fibrobacter succinogenes* subsp. *succinogenes*su	Amann *et al.*, 1990*b*

do domain; gr group; ge genus; sp species; su subspecies.

sequences than using 16S rRNA sequences. Furthermore, since homologous regions of rRNA genes may have been changed at different rates in phylogenetically equivalent lineages, the most promising primary structure regions have to be defined for the particular species or groups and cannot be generally predicted.

If rRNA targeted probes are designed for application in whole cell hybridizations, the accessibility of target sites within intact or partially denatured ribosomes is an additional criterion for probe design. Sites which are excellent targets for performing probe hybridizations to purified RNAs, may be worthless for whole cell hybridizations. No general rules for this behaviour have yet been determined. Therefore, the applicability of the *in situ* approach has to be tested for the particular probe. A compilation of sequences and specificities of probes which have successfully been used for whole cell hybridizations is given in Table 1.

WHOLE CELL HYBRIDIZATION

Principles of Whole Cell Hybridization

The major task when applying the whole cell hybridization technique is to keep the cells morphologically intact and to make them at the same time permeable for the labeled probes. Several protocols have been published for the permeabilization of microbial cells (DeLong *et al.*, 1989; Amann *et al.*, 1990*a*; Amann *et al.*, 1990*b*; Spring *et al.*, 1992; Hahn *et al.*, 1993). In general, the bacteria are killed by the fixation procedures and therefore, in contrast to immunological methods based on fluorescently labeled antibodies specific for surface antigens, the *in situ* probe assay can only be applied to nonviable cells. The cells are usually immobilized on glass slides and covered with small volumes of hybridization solution containing the labeled probes. The slides are incubated in a equilibrated moisture chamber at appropriate hybridization temperature (Stahl and Amann, 1991) for one to several hours. Not or unspecifically bound probe molecules are subsequently removed by incubating the slides in washing buffer at appropriate temperatures (Stahl and Amann, 1991). After rinsing away salts with water, the slides are air dried and subsequently mounted in anti-bleaching buffer (eg. Citifluor, Citifluor Ltd, Canterbury, UK). Oligonucleotide probes are typically endlabeled with a single molecule of a fluorchrome (DeLong *et al.*, 1989; Amann *et al.*, 1990*b*). Successfull binding of the probes can be evaluated viewing the slides in an epiflourescence microscope. Using suitable filter sets individual bacterial cells can be identified as members of the defined specicifity groups and their morphology, distribution and number can be determined.

Specificity of Probes

Oligonucleotides targeted to 16S or 23S rRNA molecules can be designed to be specific for various phylogenetic levels. Domain specific (DeLong *et al.*, 1989), subclass specific (Manz *et al.*, 1992), species and subspecies specific probes (Amann *et al.*, 1990*b*; Amann *et al.*, 1991; Distel *et al.*, 1991; Schleifer *et al.*, 1992*a*, 1992*b*; Springer *et al.*, 1992; Spring *et al.*, 1992; DeLong *et al.*, 1993; Kane *et al.*, 1993; Spring *et al.*, 1993) have been successfully applied for the *in situ* approach. Although the quality of mispairings in probe target hybrids substantially influences their stabilities (Stahl and Amann, 1991), target sites can be specifically differentiated from nontarget sites differing by only one residue (Amann *et al.*, 1990*b*; Manz *et al.*, 1992).

However, proper controls have to be performed to avoid false negative or positive results. Failures to detect cells containing target sequences (false negatives) may originate from low cellular rRNA content or from inaccessibility to the target sequence. The accessibility of target molecules can be tested by hybridizing the cells with universal or domain specific probes (DeLong *et al.*, 1989) which detect rRNA from all organisms or from all representatives of a domain (Woese *et al.*, 1990; Winker and Woese, 1991; formerly called primary kingdom; Woese and Fox, 1977), respectively. False positive results can be obtained due to nonspecific binding of the probe and/or autofluorescence within the sample. Nonspecific binding of the probes may occur to highly similar sequences on RNA molecules or due to interaction of the probe label with other cellular or sample components. Also, the samples have to be checked for autofluorescence prior to hybridization. Sequence independent, nonspecific binding of the probes to cellular components can be evaluated by using 'anti' probes which are designed complementary to the specific probe sequences and therefore cannot bind to the particular rRNA target sites. Sequence dependent, nonspecific binding can be decreased by more stringent hybridization conditions. Also the use of competitor or helper probes, which are designed for sites of high similarity with the target sites of specific probes, was shown to be helpful in reducing nonspecific binding by masking these sites. The combined use of specific and helper probes results in higher specificity (Manz *et al.*, 1992)

In comparison with conventional hybridization assays, the *in situ* approach provides the advantage that probes of different specificities can be used in one experiment. Currently, fluorescein, rhodamine and aminomethylcoumarine can be used to apply three probes of different specificities in one exeriment with detection of the hybrids by green, red and blue fluorescence respectively (Trebesius *et al.*, in prep.). Different groups of one phylogenetic level (e.g different subclasses of *Proteobacteria*) can be identified in a single hybridization experiment applying a set of specific probes. Figure 4 shows an artificial mixture of bacteria hybridized to fluorescein and rhodamine-labeled probes which are specific for the β- and γ-subclasses of *Proteobacteria*. Alternatively, a set of probes with "nested" specificities could be used to analyse populations at higher and lower phylogenetic levels simultanuously. Additional dyes which are excited or emit at alternative wave lenghts and

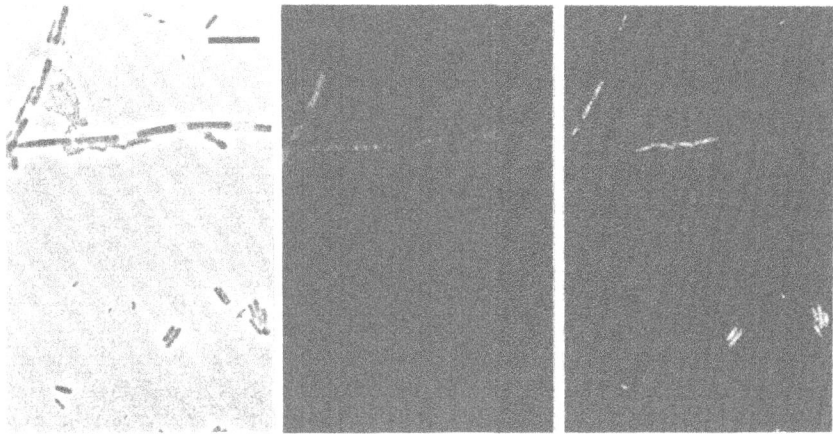

Figure 4. An artificial mixture of paraformaldehyde-fixed cells of *Sphaerotilus natans* (β-subclass of *Proteobacteria*), *Escherichia coli* (γ-subclass of *Proteobacteria*) and *Pseudomonas diminuta* (α-subclass of *Proteobacteria*) was simultaneously hybridized with a 23S rRNA-targeted, fluorescein-labeled probe specific for the β-subclass and a rhodamine-labeled, 23S rRNA-targeted probe specific for the γ-subclass. One microscopic field was viewed sequentially in phase contrast (left panel) and epifluorescence using fluorescein- (middle panel) and rhodamine-specific filter sets (right panel). As expected, the β-subclass probe bound specifically to *Sphaerotilus natans* and the γ-subclass probe to *Escherichia coli*. Bar = 10 μm.

the availability of the corresponding filters may increase the number of probes which can be used and detected in a single experiment in the near future. Computer assisted image analysis helps to resolve overlying signals resulting from more than one specific hybrid per target molecule and cell.

Sensitivity of the Approach

A major drawback of the *in situ* methods is that hybridization signals can only be visualized if sufficient (approximately 10^3) target molecules are accessible for the probe within a particular cell. There are usually enough ribosomes in exponetially growing and/or physiologically active cells to achieve this, but in many ecosystems such as soil, a major fraction of the microbial cells may be in a dormant stage and therefore remain undetectable (Hahn *et al.*, 1992).

The sensitivity can be increased by directing multiple fluorescent labels to one target molecule either by the simultanuous use of several monolabeled probes of identical specificities with alternative target sites or by the use of a probe with multiple detector groups located on the same probe molecule. The combined application of alternative probes can yield an additive increase in signal strength (Amann *et al.*, 1990a), but in other cases the effect is lower than expected (Wagner *et al.*, 1993). Since the number of useful target sites for probes of a given specificity on 16S and/or 23S molecules is limited, it is in many cases difficult or even impossible to design multiple probes of identical specificities.

Multiple flourochromes can be attached to an oligonucletide either within the hybridizing part of the molecule or on a nonhybridizing tail. The number of labels which can be attached to a particular molecule is limited for steric reasons. The increase in sensitivity is less than additive due to potential quenching effects of closely neighboured labels. Furthermore, the specificity of the assay may be negatively affected by multiple labels (Wallner *et al.*, 1993).

Autofluorescence of cellular or sample components may interfere with hybridization signals and thus hamper their detection. These effects are of major importance when investigating plant material (Hahn *et al.*, 1993), communities in soil (Hahn *et al.*, 1992) or activated sludge (Wagner *et al.*, 1993). Computer assisted image analysis again helps to restore the specificity of the approach.

Accessibility of the Target

Successful hybridization with specific probes depends on the accessibility of the specific target sequence on the rRNA molecules. The first barrier for target accessibility is the cell periphery. Cells with Gram-negative-type walls can usually be easily hybridized while those with Gram-positive cell walls may require special treatments. These methods encompass simple heat fixation (Jurtschuk *et al.*, 1992), treatment with paraformaldehyde and/or ethanol or methanol (DeLong *et al.*, 1989; Spring *et al.*, 1992), sometimes followed by incubation with cell wall lytic enzymes or solvents and/or detergents (Beimfohr *et al.*, in press; Hahn *et al.*, 1993). Problems may arise when investigating mixed samples of cells with different cell wall types. Some of the bacteria may be completely lysed and therefore no longer detectable in the microscope, while others remain impermeable to the probe.

Besides permeability of the cell envelope the accessibility of rRNA target sites within the cell is a limiting factor for specific probe hybridization. Higher order structures of the rRNA molecules, as well as of the ribosomes, may not be completely denatured by the fixation or during the hybridization experiment and thus stable hybrid formation may be prevented. Therefore, regions of the rRNA molecules which are excellent target sites for specific probes using crude or purified nucleic acids are not necessarily good targets for whole cell hybridization. The accesibility of new target sites has to be tested experimentally and the results may be different when analysing organisms from phylogenetically diverse groups.

Nonfluorescent Labels

The use of alternative labeling and detection systems may increase the sensitivity of *in situ* probing and also help to circumvent detection problems resulting from autofluorescence. Significant signal amplification was achieved using digoxigenin (DIG) labeled oligonucleotides and anti-DIG F_{ab} fragments for hybrid detection (Zarda *et al.*, 1991; Hahn *et al.*, 1993). Fluorescently labeled hybrid associated antibodies were visualized by epifluorescence microscopy. As alternative detection systems avoiding fluorescence, antibody fragments labeled either with alkaline phosphatase or horseradish peroxidase were used to detect hybrids of DIG-labeled probes and rRNA targets. The cell associated antibody complexes were detected by coloured precipitates of enzymic reactions using bright field microscopy. Besides the signal amplification resulting from enzymic reactions these approaches are not affected by autoflourescence. A disadvantage of this indirect detection system is the need to permeabilize the cell periphery for the antibodies that are considerably larger than oligonucleotides. Consequently, detection requires lysozyme (Zarda *et al.*, 1991) and/or detergent (Hahn *et al.*, 1993) treatment of the cells prior to hybridization. The applicability of antibody based approaches has been shown for bacteria with Gram-negative cell walls and a limited number of bacteria with Gram-positive cell walls and methanogens. The experimental procedures for permeabilizing the cells to the probes and antibody complexes have to be optimized for individual bacterial strains (Hahn *et al.*, 1993). The problems associated with the use of antibodies partly can be circumvented by applying probes directly attached to enzymes like horseradish peroxidase (Amann *et al.*, 1992*b*). However, special treatment is again needed to make the cells permeable for the large enzyme-oligonucleotide complex.

Flow Cytometry

The combined use of fluorescently labeled, rRNA-targeted probes and flow cytometry allows the automation of the whole cell hybridization approach. In comparison with the microscopic methods, flow cytometry offers the advantage of easier quantification of the hybridization signals of individual cells and rapid automated qualitative and quantitative analysis of many cells. The method was successfully applied to specifically separate and enumerate bacterial and yeast cells in mixed samples after simultanuous *in situ* hybridization with combinations of differently (fluorescein and rhodamin) labeled rRNA targeted probes. (Amann *et al.*, 1990*a*; Wallner *et al.*, 1993). The cells were separated according to size and red or green fluorescence. The correlation of growth rate and target rRNA content of bacterial cells could be monitored by quantification of cellular fluorescence (Wallner *et al.*, 1993). The applicability of the method for the analysis of microbial communities is limited in that only suspended cells can be analysed. Immobilized microbial communities require microscopy which might be supported and automated by computer assisted image analysis.

APPLICATIONS OF WHOLE CELL HYBRIDIZATION

The availability of rRNA sequence data from the particular strain and from closely and distantly related reference organisms is the basis for rational design of specific hybridization probes. Modern rapid sequencing methods allow easy access to new sequence data from culturable bacteria and the public databases provide a wide spectrum of reference data (Larsen *et al.*, 1993; Neefs *et al.*, 1993). In the case of culturable bacteria, problems concerning permeability and ribosome content can be minimized by applying optimal culture conditions. Culturable bacteria have been easily identified in complex mixtures (Amann *et al.*, 1992*a*; Schleifer *et al.*, 1992*a*, 1992*b*). After addition of nutrients and short incubation at appropriate conditions, culturable bacteria could be identified in soil samples (Hahn *et al.*, 1992). Therefore, analysis of culturable bacteria by whole cell hybridization is presently only limited by the availability of suitable methods for cell permeabilization.

Identification *in situ* and phylogeny of uncultured microorganisms

The introduction of whole cell hybridization methods allowed, for the first time in the history of microbiology, the *in situ* identification of single cells on the basis of phylogenetic relationships. Even more importantly, the combined application of modern molecular genetics and probe technologies allows the phylogenetic analysis of hitherto nonculturable bacteria. Keeping in mind that less than 1% of bacteria in ecosystems such as soil are culturable, the impact of the approach cannot be overestimated.

The principles of the approach (Amann *et al.*, 1991) are schematically shown in Figure 5. Ribosomal RNA genes are amplified *in vitro* applying the polymerase chain reaction technique (Saiki *et al.*, 1988) in combination with rRNA gene specific universal or conserved oligonucleotide primers (Ludwig *et al.*, 1992). This can be done starting from either purified DNA or from enriched cells or even directly from complex samples. The individual gene fragments in the resulting mixture of community rRNA genes have to be separated by cloning. By comparative sequence analysis, the individual sequences can be assigned to known phylogenetic groups or recognized as new. Specific probes can be designed and used to identify and enumerate the corresponding cells in the original sample. Only by performing the complete sequence of experimental steps can one undoubtely prove that the retrieved sequences originate from organisms present in the population and not from free (probably contaminating) DNA.

Taxonomic analyses based on genealogical relationships among the members of complex ecosystems are now possible. This approach does not allow direct analysis of the impact of a particular species or group of bacteria on a given ecosystem, but induced or spontanous changes in the community structure can now be quantitatively monitored. Population development of a definite taxon in response to changing physicochemical conditions indirectly reflects its physiological predispositions. Moreover, a quantification of cellular ribosome content allows *in situ* estimation of growth rates (Poulsen *et al.*, 1993).

Bacterial Endosymbionts. Bacterial endosymbionts in eucaryal microorganisms have been known for more than a century (Preer and Preer, 1984). In most cases these organisms could only be described from microscopic investigations of morphological characteristics. Since most of these bacteria cannot yet be grown in pure cultures, it has not been possible to perform further genetic and epigenetic analyses. However, using the rRNA approach phylogenetic analysis and identification *in situ* are now possible. Since endosymbiotic populations usually consist of a limited number of different bacteria, the host endosymbiont system represents an ecosystem of low complexity. Physically enriched samples of endosymbiotic bacteria or the complete eucaryal cells have successfully been used to amplify selectively bacterial rRNA gene fragments for phylogenetic analyses and probe design (Amann *et al.*, 1991; Springer *et al.*, 1992, 1993). Using domain specific primers the amplification of bacterial rRNA genes does not interfere with that of eucaryal rRNA genes and vice versa. However, non endosymbiotic bacteria in the sample, or bacteria ingested by the host, contribute to the complexity of the system.

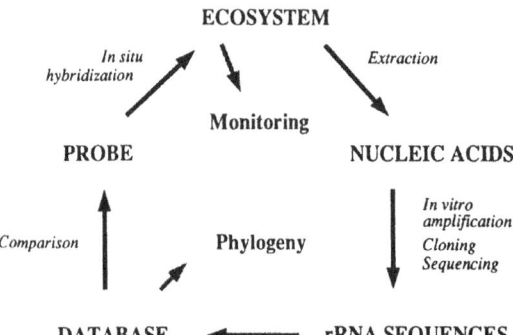

Figure 5. General approach to characterize uncultured microorganisms phylogenetically and to monitor their number and distribution in complex communities.

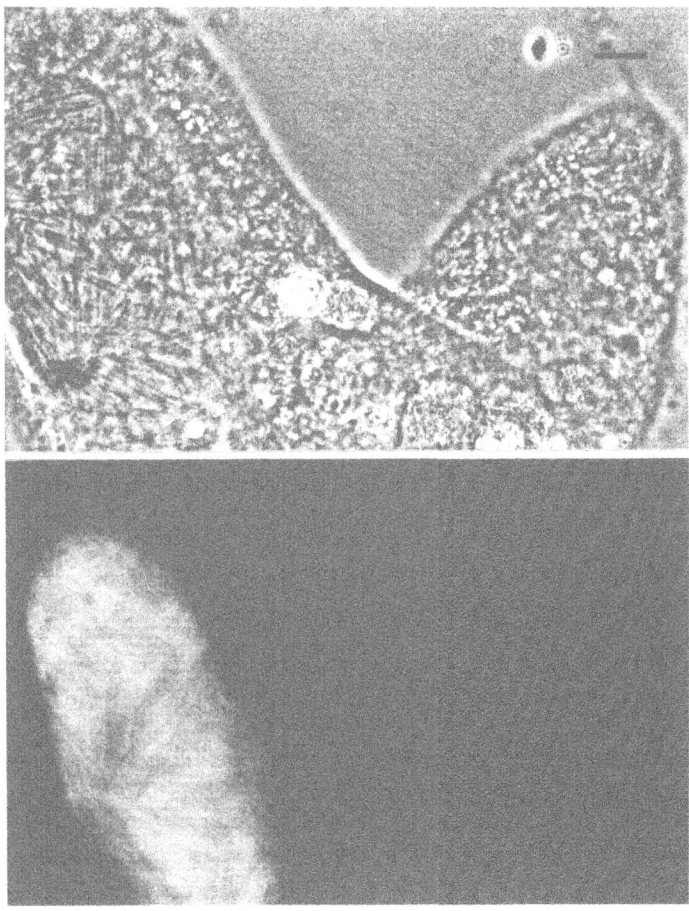

Figure 6. *In situ* identification of *Holospora obtusa* using a rRNA-targeted oligonucleotide probe. After hybridization with a tetramethylrhodamine-labeled probe specific for *Holospora obtusa* elongated rods were visualized in the macronucleus of *Paramecium caudatum* by epifluorescence using a rhodamine-selective filter set (lower panel). Bar = 10 μm.

Species of the genus *Holospora* encompass endosymbionts living in the nuclei of *Paramecium caudatum* (Preer and Preer, 1984). *Holospora obtusa* selectively infects the macronucleus whereas *Holospora elegans* only occurs in the micronucleus (Ossipov *et al.*, 1975). The two species never coexist within the same protozoal cell. Comparative sequence analysis of 16S rRNA genes which were amplified from bacterial cells enriched by centrifugation revealed the two species to be closely related to one another and moderately to the obligately cell parasitic rickettsia which are representatives of the α-subclass of *Proteobacteria* (Amann *et al.*, 1991). A genus specific 16S rRNA targeted probe, as well as species specific 23S rRNA targeted probes, were used to identify *Holospora obtusa* and *Holospora elegans* within macro- and micronuclei of fixed *Paramecium caudatum* cells, respectively (Amann *et al.*, 1991). Figure 6 shows part of a *Paramecium caudatum* containing a macronucleus infected with *Holospora obtusa* hybridized to a specific probe.

A peculiarity of *Caedibacter* species, which are also nonculturable bacterial endosymbionts of *Paramecium caudatum*, is their ability to confer killer traits upon their hosts which become toxic for susceptible paramecia (Quackenbusch *et al.*, 1986). *Caedibacter caryophila* is frequently found in the macronucleus (Schmidt *et al.*, 1987), whereas other *Caedibacter* species occur within the cytoplasm of paramecia (Quackenbusch *et al.*, 1986; Preer *et al.*, 1974). The polymerase chain reaction assisted procedure of rRNA sequence analysis was applied to crude nucleic acid preparations from enriched endosymbiotic cells. *Caedibacter caryophila* was shown to be a relative of the *Holospora* species which share a common root with the rickettsia among the proteobacteria of the α-subclass (Springer *et al.*, 1993). The *Caedibacter* cells could be clearly identified within the macronuclei of *Paramecium caudatum* and differentiated from other bacteria which were found within food vacuoles (Springer *et al.*, 1993).

Sarcobium lyticum was described as an obligate intracellular bacterial parasite of free-living amoebae (Drożański *et al.*, 1991). It was shown that *Sarcobium lyticum* is a member of the genus *Legionella* closely related to an isolate (LLAP-3) recovered from the sputum of a patient with pneumonia which had been enriched by propagation within amoebae (Fry *et al.*, 1991). A probe was designed with which *Sarcobium lyticum* cells could be monitored within *Acanthamoeba castellanii* during different stages of infection and reproduction (Springer *et al.*, 1992).

Magnetotactic Bacteria. The outstanding feature of magnetotactic bacteria is the presence of precisely formed intracellular chains of single-domain magnetic particles (magnetosomes) which allow the cells to orient themselves along the lines of magnetic fields. Since many of these organisms are typical gradient organisms, it has been postulated that magnetotactic behaviour helps the cells to find their niches in the gradients. Only a few isolates are available as axenic cultures. Three of them (*Magnetospirillum magnetotacticum*, *Magnetospirillum gryphiswaldense* and *Magnetospirillum* sp. AMB-1) have been phylogenetically analysed by comparative 16S rRNA sequencing (Eden *et al.*, 1991; Schleifer *et al.*, 1991; Kawaguchi *et al.*, 1992). They are members of the α-subclass of *Proteobacteria* and closely related to the phototrophic species *Rhodospirillum molischianum* and *Rhodospirillum fulvum*. Magnetotactic bacteria were physically enriched from sediment samples by inducing them to move along an artificial magnetic field. Several diverse morphotypes could be visualized in sediments from a large freshwater lake (Chiemsee; Bavaria, Germany). The samples of enriched bacteria appeared as pure collections of magnetotactic bacteria by microscopic inspection for the presence of magnetosomes. However, an unexpectedly diverse collection of rRNA sequences was obtained after applying the PCR based comparative sequencing technique. Only a minor part of these sequences originated from magnetotactic bacteria as was shown by *in situ* hybridizations. This illustrates problems in population studies resulting from experimental biases. There seems to exist a reasonable amount of free nucleic acids from diverse organisms within sediments. On the other hand, nucleic acids from only part of morphologically intact bacteria are accessible to *in vitro* amplification under the experimental conditions applied. This may be due to failure of cell lysis or due to cellular compounds inhibiting the polymerase chain reaction. Further discrimination may occur during the cloning procedure which is necessary to separate amplified rRNA genes. Another interesting finding was a great deal of microheterogeneity

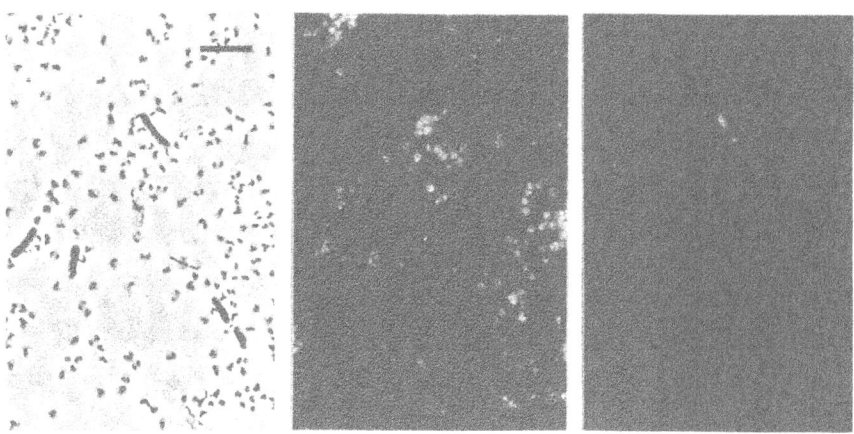

Figure 7. *In situ* identification of magnetotactic bacteria. A magnetic enrichment from the microaerobic zone of a freshwater sediment (Chiemsee, Germany) was hybridized with specific oligonucleotide probes. The fluorescein-labeled probe detected magnetotactic cocci (middle panel), whereas the rhodamine-labeled probe bound selectively to a large magnetotactic rod, tentatively named *"Magnetobacterium bavaricum"* (right panel). Bar = 10 μm.

among highly similar sequences retrieved from sediment samples. Collections of nearly identical 16S rRNA sequences were found, differing in only one or two residues. Part of these heterogeneities may result from artefacts introduced during the PCR procedure and selected by cloning, but some of the slightly different primary structures could be assigned unambiguousely to different cells by probe hybridizations.

The phylogenetic positions of some magnetotactic cocci were elucidated recently (Spring *et al.*, 1992; DeLong *et al.*, 1993). The majority of them represents a separate, deeply rooting line among the radiation of members of the α-subclass of *Proteobacteria* not closely related to *Magnetospirillum*. Some of them are phylogenetically affiliated to the δ-subclass of *Proteobacteria*. Interestingly, differences in the migration behaviour of magnetococci could be assigned to phylogenetically defined groups. Slow and fast migrating cocci could be differentiated applying specific probes (Spring *et al.*, 1992). A most remarkable giant magnetotactic bacterium was enriched from Chiemsee sediment and analysed phylogenetically. *"Magnetobacterium bavaricum"* is a rod shaped bacterium, 10 μm long and 2 μm thick. It contains up to 1000 magnetosomes. In a study analysing contiguous depth fractions of the sediment applying the specific probe assay in combination with measurements of oxygen and sulfide concentrations by microelectrodes the dominant role of this giant within the microaerophilic zone was demonstrated (Spring *et al.*, 1993). *In situ* identification of *"Magnetobacterium bavaricum"* and magnetotactic cocci enriched from freshwater sediment is shown in Figure 7.

Complex Microbial Communities. In general, the *in situ* cell hybridization approach is an excellent tool for the analysis of communities of higher organismal complexity such as soil or activated sludge. Currently, it can be used to detect and identify the physiologicallyactive fraction of the population directly (Wagner *et al.*, 1993) or after supplement of nutrients and short incubation to allow cells to switch from the dormant stage to an active one (Hahn *et al.*, 1992). However, to comprehensively analyse composition

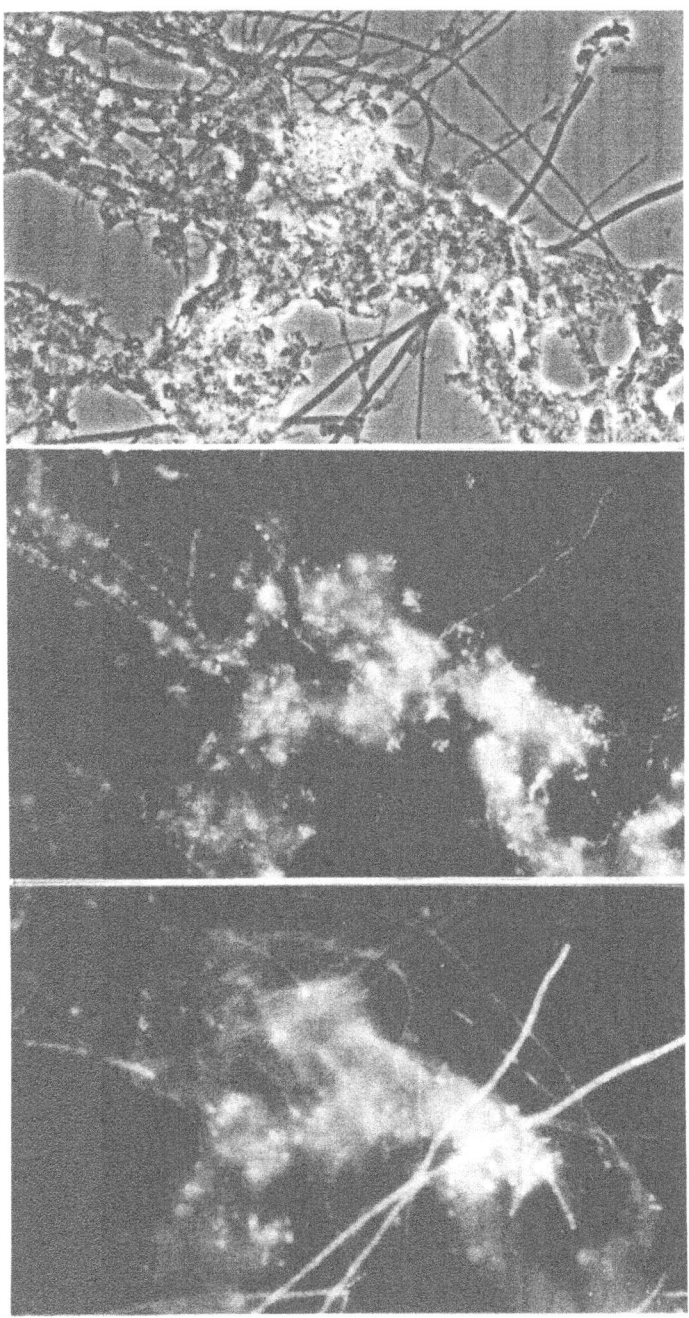

Figure 8. Activated sludge hybridized with rhodamine- and fluorescein-labeled probes specific for β- (middle panel) and γ-subclasses (lower panel) of *Proteobacteria* respectively. Upper panel: phase contrast. Bar = 10 μm.

and spacial distribution of the probably enormous number of different species, an equivalent number of specific probes has to be available. Currently, the number of available probes is rather limited and mainly based on sequence data from described culture collection strains. Extensive sequencing work would be necessary to create the basis for probe design for the silent majority of so far unculturable bacteria in a given ecosystem. A more practicable approach is to sequentially apply a limited set of probes with nested specificities. In a first step the numbers and distribution of the organisms could be determined with respect to their belonging to the domains of *Archaea, Bacteria* or *Eucarya*. In subsequent steps probes with a narrower specificity range such as phylum-, subclass-, group- and finally species-specific probes could be applied. Specific probes for the α-, β- and γ-subclasses of the *Proteobacteria* have been designed, evaluated (Manz *et al.*, 1992) and used to analyse activated sludge samples (Wagner *et al.*, 1993). In all samples from a municipal sewage plant studied, the majority of microscopically visible cells (about 80%) could be assigned to the phylum *Proteobacteria*. Samples from different installations of the plant were dominated by different proteobacterial subclasses. Only about 15% of the visible cells could be isolated and studied by conventional culture methods. The subclasses were represented differently before and after cultivation. This is not surprising since the addition of nutrients for cultivation exerts bias towards the enrichment of the best adapted organisms. In the study of Wagner *et al.* (1993) during cultivation, organisms of the β-subclass were discriminated against members of the γ-subclass. The micrographs in Figure 8 show the results of probing an activated sludge sample with β- and γ-subclass specific probes. The γ-subclass probe bound among other morphotypes to large filaments in the sludge.

Microbial growth at surfaces results in formation of biofilms of often complex composition. rRNA sequence retrieval and *in situ* cell hybridizations were performed to study a sulfidogenic biofilm in a laboratory scale bioreactor and to analyse diversity and spacial distribution of taxa in this community (Amann *et al.*, 1992c). The development of the biofilms and changes in their compositions in response to changed environmental conditions could be monitored with species or group specific probes. In a recent study (Manz *et al.*, 1993) biofilms in drinking water pipes were studied applying β- and γ-subclass (*Proteobacteria*) specific probes. Despite the oligotrophic environment, more than 70% of surface grown cells contained enough ribosomes to be detected *in situ* by fluorescent probes, whereas the corresponding fraction among planktonic cells was less than 40%. This indicates higher physiological activities of attached cells. Monitoring studies on developing biofilms showed that microcolonies containing fewer than 50 cells at early stages consisted of phylogenetically diverse bacteria. Apparently, coexistence and interaction in biofilms is of great importance for survival of bacteria in oligotrophic (drinking) water.

Monitoring of the Isolation of New Bacteria. The rRNA approach is particularly useful for ecological studies, but the physiological role that the organisms play in an ecosystem can only be indirectly estimated by monitoring their cell numbers and ribosome content in response to induced environmental changes. For further studies, cultivation methods have to be developed. There are two benfits of the rRNA based methods for advancing cultivation studies. First, the known phylogenetic affiliation of a new bacterium to its culturable relatives may help to recognize special requirements with respect to nutrients and conditions. Second, a specific probe allows monitoring of the enrichment of the particular organisms during the cultivation procedures (Kane *et al.*, 1993).

CONCLUSIONS

The combined use of comparative rRNA sequence analysis and *in situ* probing is a powerful method for analysing complex microbial communities and currently is the only way to analyse and monitor so far nonculturable microorganisms. However, there are several problems which have to be solved in the future:

1. The sensitivity of the method has to be improved to overcome the failure to detect cells with a low ribosome content. This could be achieved by improved labels and detection systems. Another attractive alternative is *in situ* (intracellular) target amplification using polymerase chain reaction based methods.

2. The range of probe specificities is limited by the degree of sequence diversity of rRNA molecules. To open the methodology for target nucleic acids other than rRNA, mRNA directed techniques could be improved. Furthermore, *in situ* PCR-based methods would allow targeting of any region of the genome for which sequence data are available.

3. Comprehensive databases of target and probe sequences have to be available. In the case of 16S rRNA comprehensive sets of properly aligned and phylogenetically analysed sequences are available to the public (Larsen *et al.*, 1993; Neefs *et al.*, 1993). For evaluated rRNA-targeted probes, a specialized database has been initiated at our institute.

REFERENCES

Amann, R., Binder, B.J., Olson, R.J., Chisholm, S.W., Devereux, R. and Stahl, D.A. (1990*a*) Combination of 16S rRNA-targeted oligonucleotide probes with flow cytometry for analyzing mixed microbial populations. Appl. Environ. Microbiol. 56, 1919–1925.

Amann, R., Krumholz, L., and Stahl, D.A. (1990*b*) Fluorescent-oligonucleotide probing of whole cells for determinative phylogenetic, and environmental studies in microbiology. J. Bacteriol. 172, 762–170.

Amann, R., Springer, N., Ludwig, W., Görtz, H.D., and Schleifer, K.H. (1991) Identification *in situ* and phylogeny of uncultured bacterial endosymbionts. Nature 351, 161–164.

Amann, R., Ludwig, W. and Schleifer, K.H. (1992*a*) Identification and *in situ* detection of individual bacterial cells. FEMS Microbiol. Lett. 100, 54–50 .

Amann, R., Zarda, B., Stahl, D.A. and Schleifer, K.H. (1992*b*) Identification of individual prokaryotic cells with enzyme labelled, rRNA targeted oligonucleotide probes. Appl. Environ. Microbiol. 58, 3007–3011.

Amann, R., Stromley, J., Devereux, R., Key, R. and Stahl, D.A. (1992*c*) Molecular and microscopic identification of sulfate-reducing bacteria in multispecies biofilms. Appl. Environ. Microbiol. 58, 614–623.

Beimfohr, C., Krause, A., Amann, R., Ludwig, W. and Schleifer, K.H. (1993) *In situ* identification of lactococci, enterococci and streptococci. Syst. Appl. Microbiol. in press

Brock, T.D. (1987) The study of microorganisms in situ: progress and problems, in "Symposium of the Society of General Microbiology 41" (Fletcher, M., Gray, T.R.G. and Jones, J.G., Eds.), pp. 1–17. Cambridge University Press, Cambridge.

Brosius, J., Dull, T.J., Sleeter, D.D., and Noller, H.F. (1981) Gene organization and primary structure of a ribosomal RNA operon from *Escherichia coli*. J. Mol. Biol. 148, 107-127.

Colwell, R.R., Brayton. P.R., Grimes, D.J., Roszak, D.R., Huq, S.A. and Palmer, L.M. (1985) Viable but nonculturable *Vibrio cholerae* and related pathogens in the environment: implications for the release of genetically engineered microorganisms. Biotechnology 3, 817–820.

DeLong, E.F., Wickham, G.S., and Pace, N.R. (1989) Phylogenetic stains: Ribosomal RNA based probes for the identification of single cells. Science 243, 1360–1363.

DeLong, E.F., Frankel, R.B. and Bazylinski, D.A. (1993) Multiple evolutionary origins of magnetotaxis in bacteria. Science. 259, 803–806.

Distel, D.L., DeLong, E.F. and Waterbury, J.B. (1991) Phylogenetic characterization and *in situ* localization of the bacterial symbiont of shipworms (*Teredinidae: Bivalvia*) by using 16S rRNA sequence analysis and oligodeoxynucleotide probe hybridization. Appl. Environ. Microbiol. 57, 2376–2382

Drożański, W. (1991) *Sarcobium lyticum* gen. nov. , sp. nov., an obligate intracellular bacterial parasite of small free-living amoebae. Int J. Syst. Bacteriol. 41, 82–87.

Eden, P.A., Schmidt, T.M., Blakemoore and Pace, N.R. (1991) Phylogenetic analysis of *Aquaspirillum magnetotacticum* using PCR-amplified 16S rRNA-specific DNA. Int. J. Syst. Bacteriol. 41, 324–325.

Embley, T.M., Finlay, B.J. and Brown, S. (1992a) RNA sequence analysis shows that the symbionts in the ciliate *Metopus contortus* are polymorphs of a single methanogen species. FEMS Microbiol. Lett. 97, 57–62.

Embley, T.M., Finlay, B.J., Thomas, R.H. and Dyal, P.L. (1992b) The use of rRNA sequences and fluorescent probes to investigate the the phylogenetic positions of the anaerobic ciliate *Metopus palaeformis* and its archaeobacterial endosymbiont. J. Gen. Microbiol. 138, 1479–1487.

Finlay, B.J., Embley, T.M. and Fenchel, T. (1993) A new polymorphic methanogen, closely related to *Methanocorpusculum parvum*, living in stable symbiosis within the anaerobic ciliate *Trimyema* sp. J. Gen. Microbiol. 139, 371–378.

Festl, H., Ludwig, W. and Schleifer, K.H. (1986) DNA hybridization probe for the *Pseudomonas fluorescens* group. Appl. Environ. Microbiol. 47, 49–55.

Fox, G.E., Pechman, K.J., and Woese, C.R. (1977) Comparative cataloging of 16S ribosomal ribonucleic acid: molecular approach to prokaryotic systematics. Int. J. Syst. Bacteriol. 27, 44-57.

Fry, N.K., Rowbotham, T.J., Saunders, N.A. and Embley, T.M. (1991) Direct amplification and sequencing of the 16S ribosomal DNA of an intracellular *Legionella* species recovered by amoebal enrichment from the sputum of a patient with pneumonia. FEMS Microbiol. Lett. 83, 165–168.

Gersdorf, H., Pelz, and Göbel, U.B. (1993a) Fluorescence in situ hybridization for direct visualization of Gram-negative anaerobes in subgingival plaque samples. FEMS Immunol. Med. Microbiol. 6, 109–114.

Gersdorf, H., Meissner, A., Pelz, K., Krekeler, G. and Göbel, U.B. (1993b) Identification of *Bacteroides gingivalis* in subgingival plaque from patients with advanced periodontitis. J. Clin. Microbiol. 31, 941–946.

Giovannoni, S.J., DeLong, E.F., Olsen, G.J. and Pace, N.R. (1988) Phylogenetic group-specific oligodeoxinucleotide probes for identification of single microbial cells. J. Bacteriol. 170, 720–726.

Giovannoni, S.J., Britschgi, T.B., Moyer, C.L. and Field, K.G. (1990) Genetic diversity in Sargasso Sea bacterioplankton. Nature 345, 60–63.

Göbel, U.G. and Stanbridge, E.J. (1984) Cloned *Mycoplasma* ribosomal RNA genes for the detection of mycoplasma contamination in tissue cultures. Science. 226, 1211–1213

Hahn, D., Amann, R., Ludwig, W., Akkermans, A.D.L. and Schleifer, K.H. (1992) Detection of microorganisms in soil after *in situ* hybridization with rRNA-targeted, fluorescently labelled oligonucleotides. J. Gen. Microbiol. 138, 879– 878

Hahn, D., Amann, R. and Zeyer, J. (1992) Whole-cell hybridization of *Frankia* strains with fluorescence- or digoxygenin-labeled, 16S rRNA targeted oligonucleotide probes. Appl. Environ. Microbiol. 59, 1709–1716.

Hicks, R.E., Amann, R. and Stahl, D. (1992) Dual staining of natural bacterioplankton with diamino-2–phenylindole and fluorescent oligonucleotide probes targeting kingdom-level 16S rRNA sequences. Appl. Environ. Microbiol. 58, 2158–2163.

Höpfl, P., Ludwig, W., Schleifer, K.H., Larsen, N. (1989) The 23S ribosomal RNA higher-order structure of *Pseudomonas cepacia* and other prokaryotes. Eur. J. Biochem. 185, 355-364.

Jurtschuk, R.J., Blick, M., Bresser, J., Fox, G.E. and Jurtschuk, jr., P (1992) Rapid in situ hybridization technique using 16S rRNA segments for detecting and differentiating the closely related Gram-positive organisms *Bacillus polymyxa* and *Bacillus macerans*. Appl. Environ. Microbiol. 58, 2571–2578.

Kane, M.D., Poulsen, L.K. and Stahl, D.A. (1993) Monitoring the enrichment and isolation of sulfate-reducing bacteria by using oligonucleotide hybridization probes designed from environmentally derived 16S rRNA sequences Appl. Environ. Microbiol. 59, 682–686.

Kawaguchi, R., Burgess, J.G. and Matsunaga, T. (1992) Phylogeny and 16S rRNA sequence of *Magnetospirillum* sp. AMB-1, an aerobic magnetic bacterium. Nucleic Acids Res. 20, 1140.

Larsen, N. (1992) Higher order interaction in 23S rRNA. Proc. Natl. Acad. Sci. USA 89, 5044-5048.

Larsen, N., Olsen, G.J., Maidak, B.L., McCaughey, M.J., Overbeek, R., Macke, T.J., Marsh, T.L. and Woese, C.R. (1993) The ribosomal database project. Nucleic Acids Res. 21, Supplement: 3021–3023.

Ludwig, W., Kirchhof, G., Klugbauer, N., Weizenegger, M., Betzl, D., Ehrmann, M., Hertel, C., Jilg, S., Tatzel, R., Zitzelsberger, H., Liebl, S., Hochberger, M., Lane, D., Wallnöfer, P.R. and Schleifer, K.H.

(1992) Complete 23S ribosomal RNA sequences of Gram-positive bacteria with a low DNA G+C content. Syst Appl Microbiol 15, 487-501.

Ludwig, W. and Schleifer, K.H. (1993) The rRNA approach in bacterial systematics. FEMS Microbiol. Rev., in press

Manz, W., Amann, R., Ludwig, W., Wagner, M. and Schleifer, K.H. (1992) Phylogenetic oligodeoxynucleotide probes for the major subclasses of proteobacteria: problems and solutions. Syst. Appl. Microbiol. 15, 593–600.

Manz, W., Szewzyk, U., Ericson, P., Amann, R., Schleifer, K.H. and Stenström T.A. (1993) *In situ* identification of bacteria in drinking water and adjoining biofilms by hybridization with 16S and 23S rRNA-directed fluorescent oligonucleotide probes. Appl. Environ. Microbiol. 59, 2293–2298.

Neefs, J.M., De Rijk, P., Van de Peer, Y., Chapelle, S. and De Wachter, R. (1993) Compilation of small ribosomal subunit RNA structures. Nucleic Acids Res. 21, 3025–3049.

Ossipov, D.V., Skobolo, I.I. and Rautian, M.S. (1975). Jota-particles, macronuclear symbiontic bacteria of the ciliate *Paramecium caudatum* clone M115. Acta Protozool. 14, 263–280.

Poulsen, L.K., Ballard, G. and Stahl, D.A. (1993) Use of rRNA flourescence in situ hybridization for measuring the activity of single cells in young and established biofilms. Appl. Environ. Microbiol. 59, 1354–1360.

Preer, J.R. and Preer, L.B. (1984) Endosymbionts of protozoa, in "Bergey's Manual of Systematic Bacteriology" (Krieg, N.R., Ed.) Vol. 1, pp. 795–942. Williams & Wilkins, Baltimore.

Preer, J.R., Preer, L.B. and Jurand, A. (1974) Kappa and other endosymbionts in *Paramecium aurelia*. Bacteriol. Rev. 38, 113–163.

Quackenbusch, R.L., Cox,, B.J. and Kanabrocki, J.A. (1986) Extra chromosomal elements of extra chromosomal elements of *Paramecium* and their extra chromosomal elements, in "Extrachromosomal Elements in Lower Eukaryotes" (Wickner, R.B., Hinnebusch, A., Lambowitz, A.M., Consalus, I.C. and Hollaender, A., Eds.), pp. 265–278. Plenum Publishing Corp., New York .

Roller, C. Ludwig, W. and Schleifer, K.H. (1992) Gram-positive bacteria with a high DNA G+C content are characterized by a common insertion within their 23S rRNA genes. J. Gen. Microbiol. 138, 1167-1175.

Saiki, R.K., Gelfand, D.H., Stoffel, S., Scharf, S.J., Higuchi, R., Horn, G.T., Mullis, K.B., Ehrlich, H.A. (1988) Primer directed enzymatic amplification of DNA with a thermostable DNA polymerase. Science 293, 487-491.

Schleifer, K.H. , Schüler, D., Spring, S., Weizenegger, M., Amann, R., Ludwig, W. and Köhler, M. (1991) The genus *Magnetospirillum* gen. nov. Description of *Magnetospirillum gryphiswaldense* sp. nov. and transfer of *Aquaspirillum magnetotacticum* to *Magnetospirillum magnetotacticum* comb. nov. Syst. Appl. Microbiol. 14, 379–385.

Schleifer, K.H., Amann, R., Ludwig, W., Rothemund, C., Springer, N. and Dorn, S. (1992a) Nucleic acid probes for the identification and in *situ detection* of pseudomonads, in "Pseudomonas: Molecular biology and biotechnology" (Galli, E., Silver, S. and Witholt, B., Eds.), pp. 127-134. ASM, Washington, D.C.

Schleifer, K.H., Ludwig, W., Amann, R., Hertel, C., Ehrmann, M., Köhler, G., and Krause, A. (1992b) Phylogenetic relationships of lactic acid bacteria and their identification with nucleic acid probes, in "Les bactéries lactiques. Actes du colloque LACTIC 91" (Novel, G., and Le Querler, J.F. Eds.), pp. 23 – 32. Centre du publications de l'université de Caen.

Schleifer, K.H., Ludwig, W. and Amann, R. (1993) Nucleic acid probes, In: Handbook of New Bacterial Systematics, (Goodfellow, and O. McDonnell Eds.), pp. 464–499 Academic Press, London - New York.

Schmidt, H.J., Görtz, H.D. and Quackenbusch, R.L. (1987) *Caedibacter caryophila* sp. nov., a killer symbiont inhabiting the macronucleus of *Paramecium caudatum*. Int. J. Syst. Bacteriol. 37, 459–462.

Schmidt, T.M., DeLong, E.F. and Pace, N.R. (1991) Analysis of marine picoplankton community by 16S rRNA genecloning and sequencing. J. Bacteriol. 173, 4371–4378.

Spring, S., Amann, R., Ludwig, W., Schleifer, K.H. and Petersen, N. (1992) Phylogenetic diversity and identification of nonculturable magnetotactic bacteria. Syst. Appl. Microbiol. 15, 116-122.

Spring, S., Amann, R., Ludwig, W., Schleifer, K.H., van Gemerden, H. and Petersen, N. (1993) Dominating role of an unusual magnetotactic bacterium in the microaerophilic zone of a freshwater sediment. Appl. Environ. Microbiol. 59, 2397-2403.

Springer, N., Ludwig, W., Drożánski, V., Amann, R. and Schleifer, K.H. (1992) The phylogenetic status of *Sarcobium lyticum*, an obligate intracellular parasite of small amoebae. FEMS Microbiol. Lett. 96, 199-202.

Springer, N., Ludwig, W., Amann, R., Schmidt, H.J, Görtz, H.D. and Schleifer, K.H. (1993) Occurrence of fragmented 16S rRNA in an obligate bacterial endosymbiont of *Paramecium caudatum*. Proc. Natl. Acad. Sci. USA. in press.

Stahl, D.A. and Amann, R. (1991) Development and application of nucleic acid probes, in "Nucleic Acid Techniques in Bacterial Systematics" (Stackebrandt, E. and Goodfellow, M. (Eds.), pp. 205–242, Wiley, Chichester

Torsvik, V., Goksoyr, J. and Daae, F.L. (1990) High diversity in DNA of soil bacteria. Appl. Environ. Microbiol. 56, 782–787.

Tsien, H.C., Bratina, B.J., Tsuji, K. and Hanson, R.S. (1990) Use of oligonucleotide signature probes for the identification of physiological groups of methylotrophic bacteria. Appl. Environ. Microbiol. 56, 2858–2865.

Wagner, M., Amann, R., Lemmer, H. and Schleifer, K.H. (1993) Probing activated sludge with oligonucleotides specific for proteobacteria: inadequacy of culture-dependent methods for describing microbial community structure. Appl. Environ. Microbiol. 59, 1520–1525.

Wallner, G., Amann, R. and Beisker, W. (1993) Optimizing fluorescent in situ hybridization with rRNA-targeted oligonucleotide probes for flow cytometric identification of microorganisms. Cytometry 14, 136–143.

Ward, D.M., Weller, R. and Bateson, M.M. (1990) 16S rRNA sequences reveal numerous uncultured microorganisms in a natural community. Nature 345, 63–65.

Winker, S., and Woese, C.R. (1991) A definition of the domains *Archaea, Bacteria* and *Eukarya* in terms of small subunit ribosomal RNA characteristics.

Woese, C.R., and Fox, G.E., (1977) Phylogenetic structure of the prokaryotic domain: the primary kingdoms. Proc. Natl. Acad. Sci. USA 74, 5088-5090

Woese, C.R. (1987) Bacterial evolution. Micribiol. Rev. 51, 221-271.

Woese, C.R. (1991) Prokaryote systematics: the evolution of a science, in: "The Prokaryotes" (Balows, A. Trüper, H.G. Dworkin, M. Harder, W. and Schleifer, K.H., Eds.) pp. 3-18, vol. I. Springer, New York.

Woese, C.R., Kandler, O., and Wheelis, M.L. (1990) Towards a natural system of organisms: proposal for the domains Archaea, Bacteria, and Eukarya. Proc. Natl. Acad. Sci. USA 87, 4576-4579.

Zuckerkandl, E., and Pauling, L. (1965) Molecules as documents of evolutionary history. J. Theor. Biol. 8, 357-366.

Zarda, B., Amann, R., Wallner, G. and Schleifer, K.H. (1991) Identification of single bacterial cells using digoxigenin-labelled, rRNA-targeted oligonucleotides. J. Gen. Microbiol. 137, 2823–2830.

THE USE OF MOLECULAR MARKERS FOR THE DETECTION AND TYPING OF BACTERIA IN SOIL

Elizabeth M. H. Wellington, Annaliesa S. Huddleston, and Peter Marsh

Department of Biological Sciences
University of Warwick
Coventry CV4 7AL, UK

INTRODUCTION

The detection of bacteria in natural environments such as soil and water requires sensitive methods which avoid the need for culturing but allow identification, preferably with some means of quantification. The *in situ* detection of specific gene sequences in soil has provided methods giving a new dimension to our understanding of the bacterial population structure. Environments such as soil contain a vast surface area for colonization, and the bacteria may be unevenly distributed within the microhabitats. In addition a range of studies have indicated that we are, as yet, unaware of the true extent of bacterial diversity (Wayne *et al.*, 1987). A large non-culturable bacterial fraction in soil was detected by the soil fractionation studies of Fægri *et al.* (1977). Further substantial evidence for previously uncultured groups was provided by the molecular ecological studies of Liesack and Stackebrandt (1992) who defined novel groups of bacteria from analysis of 16S rRNA amplified directly from soil. Previously culturable bacteria may also enter a non-culturable state when grown in soil (van Overbeek *et al.*, 1990; Turpin *et al.*, 1993a) and this makes detection using standard isolation procedures impossible unless some method can be found to resuscitate the cells. It is not clear if such cells can be detected phenotypically in soil by methods such as specific enzyme activities or the expression of marker genes such as *lux* for light production. Studies using luminescence as a marker have indicated that inactive cells in soil can be activated and will produce light if incubated with a substrate for two hours (Meikle *et al.*, 1992). These cells are likely to be starved or at least have insufficient reducing equivalents to emit light. The use of marker genes in this way illustrates the power of selective marking to facilitate studies of bacterial activity in natural habitats. The problems of culturing bacteria and obtaining a true assessment of diversity in habitats such as soil are due to a lack of understanding about bacterial activity in nature and the physiological state of cells.

Molecular markers have been used extensively in studies of gene transfer and the fate of recombinant bacteria in soil (for review see Cresswell and Wellington, 1992). Such markers must be unique to the soil environment and so genes from marine organisms, for example *lux*, have proved to be versatile in this application allowing a unique genotype and phenotype. The tracking of genes led to the development of methods for DNA and RNA extraction from soil which have now been used extensively in a range of studies and different methods developed depending on the application (Holben, 1993). The detection of indigenous genes in DNA extracted from soil has many applications enabling detection of novel, uncultured bacteria (Liesack and Stackebrandt; 1992), determination of genetic diversity (Muyzer *et al.*, 1993), detection of potential for activities such as nutrient cycling (Picard *et al.*, 1992), biodegradation (Herrick *et al.*, 1993), antibiotic production (Wellington *et al.*, 1993). and plant pathogenicity (Picard *et al.*, 1992).

The detection and identification of whole cells in soil is desirable in combination with the direct molecular detection of the genes and this is particularly important where uncultured bacterial have been revealed by extraction, amplification and sequencing of genes such as 16S rRNA. *In situ* hybridizations of probes to specific sequences on the rRNA have allowed detection and identification of bacterial endosymbionts where large numbers of cells are grouped in defined locations and are metabolically active (Amann *et al.*, 1990a;1991). There are still many unsolved problems with detection in soil such as the very large surface area, heterogeneous nature of the substrate, low metabolic activity of the cells and strong background fluorescence which interferes with the use of oligofluors used for hybridizations. Other types of *in situ* probes targeting DNA have yet to be used in ecological studies due to the lack of methods for detection at the single cell level. Some methods are available for detection of cells *in situ* without isolation and culture such as the use of *lux*-marked bacteriophage which infects only a specific group of cells in a soil suspension and results in light production from the transcription of the phage genome (Turpin *et al.*, 1993b).

The recovery of genetically marked bacteria from soil is facilitated by the use of markers with a selectable phenotype, for example antibiotic resistance or enzyme activity for recognition on isolation plates such as the *xylE* (Morgan *et al.*, 1989), *lacZY* (Drahos *et al.*, 1986) or *luxAB* (Meikle *et al.*, 1992) genes. Indigenous bacteria detected using molecular markers may have an unknown phenotype which makes detection difficult and few studies have been able to overcome this problem other than by doing rather laborious isolation, culturing and screening with probes. Flow cytometry enabled the detection, typing and enumeration of cells with rRNA labelled by oligofluors (Amann *et al.*, 1990b) but the cells would be killed by the fixation process and substrates such as soil still present many problems for cell sorting devices.

The following review of the use of molecular markers for identification and typing bacteria is confined to applications for specific genes enabling detection *in situ* either in introduced or indigenous bacterial populations in soil. Consideration is given to the problems of dealing with soils and the ways in which molecular techniques can facilitate detection.

EXTRACTION OF NUCLEIC ACIDS FROM SOIL

Nucleic acids may be present within viable or dead cells and due to their polyanionic properties free nucleic acids can become bound to soil components such as clay. Methods have been developed to investigate the source of the gene or sequence detected, but these are often organism dependant as they rely on the method of cell lysis prior to extraction. Methods for the extraction of both DNA and RNA have been applied to soil samples and currently there are many variations in the precise method adopted. Each of the studies listed in Table 1 used a different technique and this variation probably relates to the extreme heterogeneity of soil. Generally the higher the organic carbon and clay content of the soil, the more difficult the task to extract clean nucleic acids. Humic acids are copurified with DNA and inhibit various enzymes whereas DNA binds to clay and can not be extracted (Holben, 1993). A polymer, polyvinylpolypyrrolidone (PVPP), is recommended for the removal of humic contaminants. One strategy adopted has been the extraction of the total bacterial community by fractionation (Holben *et al.*, 1988), while the other involved lysis of cells *in situ* followed by extraction of nucleic acids. These methods are summarized for both DNA and RNA in Figures 1, 2 and 3. Most studies have used DNA extraction to detect rRNA genes although in theory the rRNA should provide at least a thousand-fold greater detection limit. The problems with RNA extractions are discussed below.

Extraction of DNA from Soil

The fate of DNA in the environment has been investigated in a wide range of studies and has included environments such as sea water, sediment and soil (Lorenz and Wackernagel, 1987; Ogram *et al.*, 1988; Paul *et al.*, 1990). DNA is released into the environment continually as part of the natural process of decomposition from lysing dead cells . Live cells may also release DNA during their life cycle probably as part of the process of gene transfer by transformation.

DNA turnover rates in natural environments can be quite high; Paul *et al.* (1987) used (^{3}H)thymidine to label heterotrophic bacterioplankton and found that they produced dissolved DNA by both lysis and excretion. Production was ˜4% of the ambient concentrations per day. There was on average 10-13 μg/l dissolved DNA with a turnover time around 6.5 h. Free DNA in soil has been less well studied but some approximate figures indicate a pool of ˜5μg /g soil, with a turnover time similar to that observed in aquatic environments but no comprehensive studies have been done in a range of soils. The amount of free DNA can vary markedly between soil samples and is dependant also on the extraction efficiency when methods are applied to soils with differing physicochemical properties. On average, values for total DNA in soil (cellular, prokaryotic and eukaryotic) range from 10-200 μg/g soil. Most studies have implicated DNases of cellular origin as the major mechanism for degradation. DNA can be hydrolyzed by both cell-associated and extracellular nucleases but can be protected from enzymatic attack by adsorption to sand (Lorenz and Wackernagel, 1987), clay (Greaves and Wilson, 1969) and sediments (Aardema *et al.*, 1983). Many studies have implicated divalent cations as having a crucial role in the adsorption of DNA and maintaining it in a biologically active state. Clays, and in particular montmorillonite, were capable of adsorbing large amounts of DNA but adsorption was highest at low pH (around 5) and negligible at pH 10. As DNA has a pKa of ˜5 then it will be in the ionic form at pH 10 and prevented from moving into the interlamellar space in the clay where it is adsorbed.

The extraction of DNA from environmental samples has been optimized over recent years for bacteria (for review see Sayler *et al.*, 1992) although fewer studies have been done with eukaryotes. Total DNA samples can be obtained by two methods (Figure 1) either by separation of cells first from the sample prior to lysis or lysis within the sample and direct extraction of DNA. The former method is more time consuming and may be less sensitive as mycelial bacteria and mucoidal cells will stick to soil, clay or humus in the sample and not be recovered even following homogenization. Most of the methods now developed require some further clean-up procedure to allow amplification by PCR (polymerase chain reaction). This may involve the use of various treatments as indicated by Smalla *et al.* (1993) who compared several methods and recommended caesium chloride centrifugation and potassium acetate precipitation of proteins followed by precipitation of DNA in the supernatant using glass milk or spermine-HCl.

One of the problems encountered with DNA extractions is the efficiency of extraction and the extent of cell lysis *in situ* if the direct method is used. The former is dependant on the soil type whilst the latter on the bacterial community and method of lysis used. Bead-beating may lyse more bacteria than chemical methods particularly when done in the presence of soil particles. Studies by Cresswell *et al.* (1991) showed that streptomycete spores were resistant to SDS-heat lysis methods but were lysed by bead-beating, this enabled a differential extraction of DNA from two cell types which facilitated ecological studies of the life cycle in soil. Picard *et al.* (1992) recommended a method of treatment with sonication, microwave heating and thermal shocks for lysis of spores and other resistant propagules.

Improvements in the efficiency of extraction may be achieved by pretreating the soil to disrupt the clay lattice structure and remove the electric double layer of cations surrounding them. Soil dispersion can be achieved by extracting the soil with cation-exchange resins (Macdonald, 1986). A method for extracting spores from soil was developed by Herron and Wellington (1990) using Chelex-100 (Bio-Rad) which exchanges sodium ions for the calcium ions on clay lattices. This pretreatment was incorporated into a DNA extraction protocol developed by Jacobson and Rasmussen (1992) which avoided the lengthy cell fractionation procedures of the indirect method.

Once the DNA has been purified, further analysis has been achieved by a wide range of techniques with the emphasis on detection of genes rather than expression. For the former blotting techniques and PCR have been used and the latter achieved either by transformation assays or cloning and use of suitable host cell for expression. Limits for detection are difficult to specify as these will depend on cell lysis, extraction efficiency, copy number of the gene in the cell and methods used for detection. A rough guide for a single copy gene on the chromosome of a unicell is 10^3 cells/g soil to allow detection by probing and with heavily labelled probes ˜1.0 pg can be detected in a slot blot. The detection limits for various methods used in studies with soil are summarised in Table 1.

Extraction of RNA from Soil

RNA is less stable than DNA and more readily degraded in environmental samples. mRNA has been extracted from sediments and soils (Tsai *et al.*, 1991; Sayler *et al.*, 1992) and rRNA extracted and used for studying the distribution of *Frankia* spp. in soil (Hahn *et al.*, 1990). However, methods using mRNA have only been used to

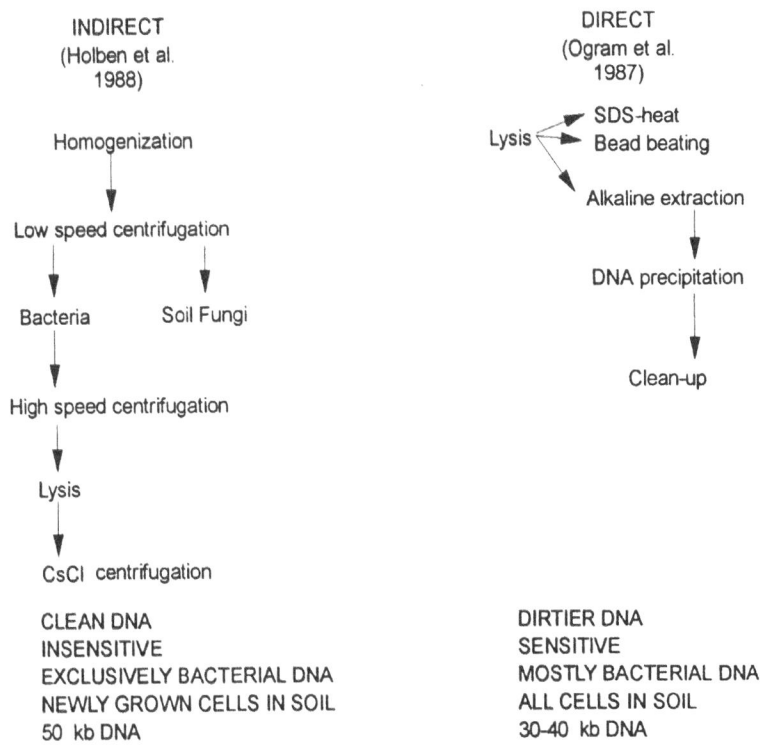

INDIRECT
(Holben et al.
1988)

Homogenization

Low speed centrifugation

Bacteria   Soil Fungi

High speed centrifugation

Lysis

CsCl centrifugation

CLEAN DNA
INSENSITIVE
EXCLUSIVELY BACTERIAL DNA
NEWLY GROWN CELLS IN SOIL
50 kb DNA

DIRECT
(Ogram et al.
1987)

Lysis   SDS-heat
Bead beating

Alkaline extraction

DNA precipitation

Clean-up

DIRTIER DNA
SENSITIVE
MOSTLY BACTERIAL DNA
ALL CELLS IN SOIL
30-40 kb DNA

Figure 1. Summary of two approaches to the extraction of DNA from soil.

detect gene expression *in situ* using soils heavily seeded with the target organism. The extraction procedure developed by Hahn *et al.* (1990) was used to detect *Frankia* species in soils and the study involved a comparison of two methods; the first was a direct extraction technique with cell lysis *in situ* by sonication in the presence of 7.5 M guanidine hydrochloride (Figure 2). Bramwell (1992) modified this method for use in the detection and identification of streptomycetes in soil where bead-beating was incorporated to lyse spores. The method is summarised in Figure 2 and can be used with detergent to facilitate cell lysis. The second method involved an indirect extraction using the cell fractionation technique used for indirect DNA isolation (Holben *et al.*, 1988) with rRNA extraction being applied to the bacterial fraction recovered. RNA yields from this indirect method were significantly lower than for the direct method (Hahn *et al.*, 1990).

A further extraction procedure was used by Bramwell (1992) (Figure 3) and Baker (unpublished data) and was based on the use of a homogenization buffer containing lithium salts, EDTA and detergents termed GOS buffer based on the method of Hughes and Galau (1988). This was developed to extract RNA from plant tissue containing phenolic compounds. The method was also applied to the detection of actinomycetes in soil and found to be useful for certain types of soil, particularly those of high humic content (Bramwell, 1992; Baker, P. pers. comm.).

One of the problems encountered with RNA extractions is the need to protect against RNase contamination and so all solutions and plasticware must be treated with DEPC, a potent inhibitor of RNase, loss of RNA during extraction may result in reduced sensitivity. Various studies (see Table 1) have indicated that the sensitivity for

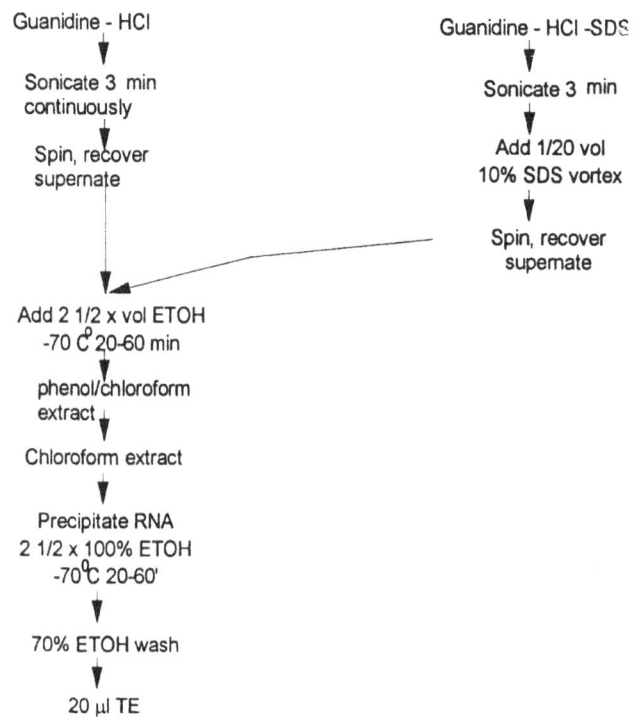

Guanidine - HCl
↓
Sonicate 3 min
continuously
↓
Spin, recover
supernate
↓

Guanidine - HCl -SDS
↓
Sonicate 3 min
↓
Add 1/20 vol
10% SDS vortex
↓
Spin, recover
supernate

Add 2 1/2 x vol ETOH
-70 C° 20-60 min
↓
phenol/chloroform
extract
↓
Chloroform extract
↓
Precipitate RNA
2 1/2 x 100% ETOH
-70°C 20-60'
↓
70% ETOH wash
↓
20 µl TE

Figure 2. Method for rRNA extraction from soil based on the method of Hahn *et al.* (1990) and modified by Bramwell (1992).

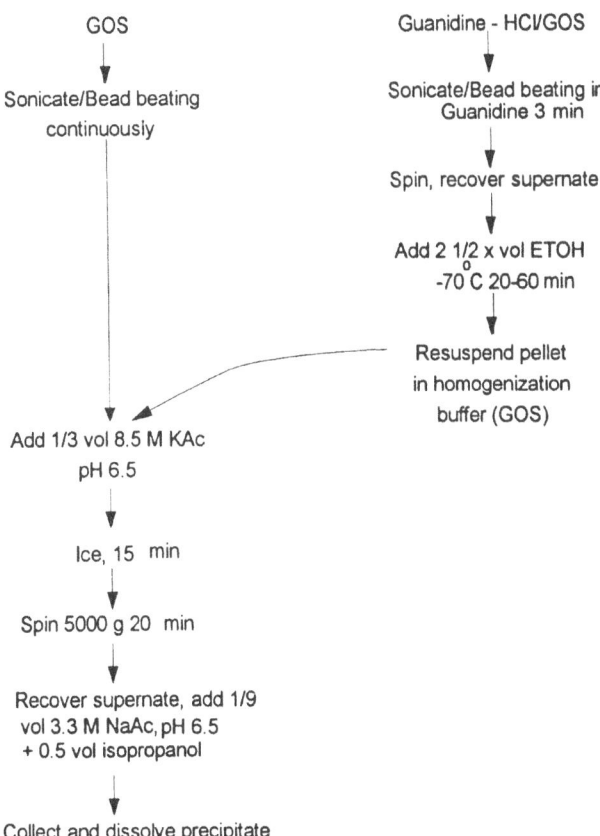

GOS   Guanidine - HCl/GOS

Sonicate/Bead beating continuously

Sonicate/Bead beating in Guanidine 3 min

Spin, recover supernate

Add 2 1/2 x vol ETOH -70°C 20-60 min

Resuspend pellet in homogenization buffer (GOS)

Add 1/3 vol 8.5 M KAc pH 6.5

Ice, 15 min

Spin 5000 g 20 min

Recover supernate, add 1/9 vol 3.3 M NaAc, pH 6.5 + 0.5 vol isopropanol

Collect and dissolve precipitate

Figure 3. RNA extraction from soil based on the method of Hughes and Galau (1988) and modified by Bramwell (1992).

in situ identification is improved compared to probing the DNA. Subsequent amplification of the target rRNA genes from DNA is usually the method of choice for analysing indigenous populations, although the detection limit even after PCR is similar to that for probing RNA directly. Detection limits for these methods are given in Table 1, where they have proved useful for the detection of actinomycetes in soil using specific 16S rRNA probes .

If rRNA genes are being targeted then the analysis of RNA does have some advantages compared to PCR of rDNA given that limits of detection for both methods are similar. The number of ribosomes in a cell does correlate with activity (Delong *et al.*, 1989) and so in certain nutrient-enriched habitats probing of an rRNA extract may indicate active members of the population. However, studies using *in situ* hybridization to fixed ribosomes in single cells in soil have indicated that, in the bulk soil at least, ribosome numbers are too low to allow significant hybridization signals (Hahn *et al.*, 1992).

DETECTION OF SPECIFIC GENES IN SOIL

Comparative studies of bacterial diversity in soil using DNA analysis and traditional isolation procedures have shown a marked discrepancy where soil DNA analysis indicated considerably greater diversity (Torsvik *et al.*, 1990). These

143

observations are supported by subsequent analysis of 16S rRNA sequences derived from PCR of rDNA (Liesack and Stackebrandt, 1992). This discrepancy is likely to be due to problems in culturing and isolating all the bacterial community. An alternative strategy is therefore needed for direct and rapid detection and identification of bacteria in the environment. A range of studies using direct analysis of DNA from soil for detection are summarised in Table 1. The genes used in such studies fall into three categories; firstly rRNA genes which allow detection and identification to genus or species, secondly the use of marker genes allowing definitive identification of an introduced bacterial strain and finally the detection of genes for specific activities such as virulence or antibiosis. The use of marker genes has been essential for the development of methodology and in tracking bacteria in natural environments. It has also provided further information on the gene pool in selected soil habitats as markers must be unique to the introduced strain.

Detection of rRNA Genes

The majority of studies with soil have exploited the 16S rRNA gene for detection *in situ,* this is a reflection of the more extensive sequence database available for this gene compared with the 23S rRNA gene. However, probes have been used for detection of pseudomonads in soil exploiting variable regions of the 23S rRNA gene (Sayler *et al.*, 1992). 16S rRNA probes have been designed which are specific for selected taxons and used mainly for detection (Table 1), only the study of Picard *et al.* (1992) provided data on enumeration. This illustrates one of the problems in direct analysis of nucleic acids, but quantification of bacteria in soil is also problematic using traditional techniques. If the number of copies of the gene are known, then reference can be made to a dilution series of the target gene and the amount of target in the soil DNA quantified by densitometric analysis of autoradiographs (Cresswell *et al.*, 1991) or by phosphorimaging the blot if ^{32}P-radiolabelled probes are used (Karagouni *et al.*, 1993). Obviously these techniques are only applicable to unamplified target DNA or RNA, once PCR has been used then quantification is more difficult. Picard *et al.* (1992) devised a method for enumeration combined with the use of PCR which estimated the number of target DNA sequences corresponding to the number of bacterial cells according to the most-probable number technique (MPN). Alternatives to this approach involve the use of a competitive template for the PCR primers which allows estimation of the precise number of targets by comparison of the two products (see review by Steffan and Atlas, 1991).

The symbiotic actinomycete genus *Frankia* is difficult to isolate from soil and detection is usually achieved using a biotest by determining nodule formation on host plants. This lengthy and selective procedure has been avoided by the use of targeted 16S rRNA probes (Hahn *et al.*, 1990; Picard *et al.*, 1992). Other actinomycetes have been identified *in situ* using 16S rRNA probes (Table 1) where recognition of the target species would be difficult and time consuming. The streptomycetes comprise an important component of the soil microbial community and are thought to be readily isolated and cultured. However, once recovered on isolation plates their identification is tedious using phenotypic analysis and chemotaxonomic data are required to assign generic identity. It is not known if certain species are more readily cultured than others, so direct analysis of soil DNA using specific probes provides an alternative strategy.

Previously unknown groups can be detected by PCR of 16S rRNA sequences using group-specific primers (Liesack and Stackebrandt, 1992). This has allowed the development of specific probes to search for novel groups so their distribution and

Table 1 Genes used for direct detection of bacteria in soil

Gene sequence	Method	Target	Detection limit /g soil	Reference
16S rRNA	Soil RNA, dot blot	*Frankia*	10^4	Hahn *et al.*,1990
		Streptomyces	10^5	Bramwell, 1992
		Actinomycetes	10^4	Baker, P. (unpub)
		Bacteria	—	Moran *et al.*, 1993
	Soil DNA MPN-PCR, gel	*Frankia*	10^4	Picard *et al.*, 1992
	Soil DNA PCR, Southern	*Escherichia coli*	5×10^2	Tsai & Olson, 1992
	Soil DNA PCR, sequence	Selected bacterial groups	—	Liesack & Stackebrandt, 1992
npt II[1]	Soil DNA, dot blot	*Bradyrhizobium japonicum*	4.3×10^3	Holben *et al.*,1988
xylE[2]	Soil DNA PCR, dot blot	*Streptomyces* spores mycelium	10^3 10	Wipat *et al.*, 1991
pat[3]	Soil DNA PCR, Southern	*Pseudomanas fluorescens*	10^3	Smalla *et al.*,1993
vir	Soil DNA, MPN-PCR, gel	*Agrobacterium tumefaciens*	10^4	Picard et al., 1992
phoP, Hin H-Li	Soil DNA, PCR, gel	*Salmonella*	—	Way *et al.*, 1993
aph D	Soil DNA PCR, Southern	*Streptomyces* sp.	10^3	Wellington *et al.*,1993

[1] Marker gene on the chromosome.
[2,3] Marker gene on a multicopy plasmid.

abundance in soil can then be investigated. Attempts have been made to analyse the genetic diversity of a biofilm by providing a profile based on the separation of PCR-generated fragments of 16S rRNA genes in soil DNA (Muyzer *et al.*, 1993). The fragments were separated by denaturing gradient gel electrophoresis.

Use of Exotic Marker Genes for Detection in Soil

Markers genes have been used extensively for the tracking of recombinant bacteria in the environment (reviewed by Saunders and Saunders, 1993) and for the detection of gene transfer in soil (Wellington *et al.*, 1992). A selection of these genes is given in Table 1 and, in addition to *lux*, they probably represent the most useful markers for soil studies. Certain soils do contain *xylE* indigenous genes which codes for the catechol 2,3-dioxygenase enzyme (Wipat *et al.*, 1991). This enzymes also provides marked bacteria with a phenotype which can be demonstrated by spraying isolation plates with catechol which results in a yellow coloration. Another selectable gene, *nptII*, appears to occur rarely but can be detected in sewage (K. Smalla, pers. comm.) and possibly where sewage sludge has been applied to land. The gene codes for resistance to neomycin and is found in clinically important bacteria usually within Tn5. Other genetic markers used include the patatin gene, *pat*, from potato which will not be present in other bacteria but is likely to be present in the soil DNA.

Genes found in marine bacteria but, as yet, not found in soil DNA are the *luxAB* genes for bacterial luminescence. These confer both a unique phenotype and

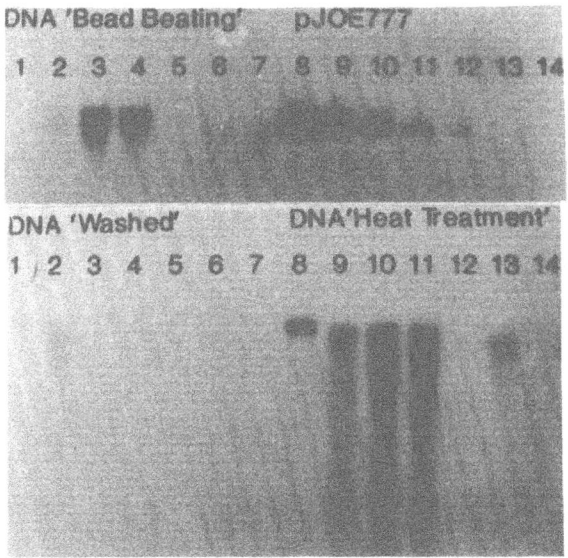

Figure 4. Detection of an amplified 5.7 kb DNA sequence in *Streptomyces lividans* growing in natural soil over 15 days. DNA bead-beating, spore and mycelium lysed; heat treatment, SDS-heat lysis of mycelium only; washed, DNA extraction without lysis steps; lane 1, uninoculated soil; 2, day 0; 3, day 1; 4, day 2; 5, day 5; 6, day 10; 7, day 15; lane 8, positive control pJOE77 containing 5.7 kb, 6 ng; DNA heat treatment lanes 9-14 days 0-15; pJOE77 DNA lane 8, 50 ng ; 9, 25 ng; 10, 12.5 ng; 11, 6.25 ng; 12, 3.12 ng.

1 kb

BI 66

Limed soil

Unlimed soil

S. griseus

S. lividans

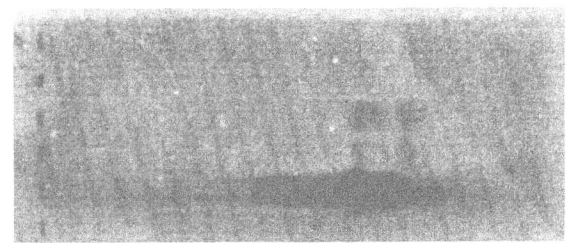

Figure 5. Detection of the *aphD* streptomycin resistance gene in limed and unlimed soil. Southern blot of PCR products following amplification of *aphD* using specific primers, probed with labelled PCR product.

genotype in soil and have been used extensively to track bacterial inoculants. Both the phenotype and the genotype can be determined *in situ* and light production under various conditions used to measure metabolic status of marked cells (Meikle *et al.*, 1992).

Differential DNA extractions can be used to locate where the gene is present in the soil community. For example, extractions omitting the cell lysis steps will extract extracellular DNA, while marked propagules such as spores, recalcitrant to SDS-heat lysis, can be detected in DNA extracted by bead-beating only (Cresswell *et al.*, 1991). Figure 4 illustrates the detection of an amplified sequence in *Streptomyces lividans* which codes for 5.7 kb and has been used to amplify marker genes on the chromosome (Altenbuchner and Cullum, 1987). Uninoculated soil did not give a signal when probed with a plasmid pJOE777 containing the 5.7 kb of DNA, whilst the marker was detected in soil inoculated with *S. lividans* both in the spores and mycelium growing or surviving in the soil. Extracellular marker was only detected at day 0, immediately after inoculation, when some lysis would have occurred. Free DNA was probably rapidly degraded after this time. Figure 4 also shows that cleaner DNA was recovered by bead-beating compared to the SDS-lysis procedure which tends to extract more phenolic components from the soil.

Detection of Indigenous Marker Genes

A range of genes are available for rapid identification and typing such as the genus-specific probes designed for detection of salmonellas. These are based on genes such as the *Hin* and *H-li* genes involved in the control of phase variation of *Salmonella* species and only present in these species and not in *E. coli* (Way *et al.*, 1993*)*. Insertion sequences such as IS*200* have also proved useful as genus-specific probes for salmonellas (Stanley *et al.*, 1992) where the insertion element was found to vary in number and its use as a probe of digested chromosomal DNA provides specific IS*200* fingerprints. This probe has not been applied to soil yet but the *Hin* and *H-Li* genes were used for specific detection of *Salmonella* species by multiplex PCR with specific primers allowing amplification of parts of these genes (Table 1).

Only a limited number of studies have focused on the detection of indigenous marker genes in soil but the potential for detection of plant pathogens and plant-beneficial bacteria is considerable. The bacterium responsible for crown gall, *Agrobacterium tumefaciens*, was detected in soil by PCR using primers specific to the *vir* region on the Ti plasmid. Other bacterial types capable of fixing nitrogen via the *nif* genes have been detected in soil although studies involved inoculants rather than detection of indigenous populations (Selenska and Klingmuller, 1991). Resistance genes to heavy metals such as mercury, *mer* gene (Bruce *et al.*, 1992) and antibiotics, *aphD* gene (Wellington *et al.*, 1993) have been detected directly in soil DNA following cell lysis *in situ*. In the latter study populations of bacteria in soil were examined for bioactive potential and specifically potential for streptomycin production which was implicated in the antibiosis of the inoculant *Bradyrhizobium japonicum*, used for nodulating soybeans. In addition to the direct detection of the streptomycin resistance gene, *aphD*, in soil, the genes for other parts of the streptomycin biosynthetic pathway were detected in soil isolates belonging to the *Streptomyces* genus. The practice of liming soil to increase the pH frequently results in increased numbers of streptomycetes and this was seen in the detection of *aphD* by PCR from soil DNA (Figure 5) where consistently more PCR product was obtained with limed soils. The resistance gene and the other genes in the streptomycin cluster were first detected in *S.griseus*, the type strain used for the industrial production of streptomycin.

DETECTION OF WHOLE CELLS IN SOIL USING GENE MARKERS

The detection of individual cells in soil has been achieved by the use of molecular markers, most notably *in situ* hybridizations to rRNA within fixed cells. This technique has been applied to the identification of specific bacteria in a range of habitats and studies with soil allowed detection of native populations (Hahn *et al.*, 1992). However, the method employs oligofluors for tagging of ribosomes and this may present problems in some soils as background fluorescence is common where organic matter is present. In addition, the metabolic status of the cell will determine the extent of fluorescence and bacteria in soil are frequently surviving in conditions of low nutrient status. Hahn *et al.*, (1992) noted that in order to detect indigenous bacteria in bulk soil a nutrient amendment was necessary. The large surface area of soil particles also causes problems of detection when microscopy is used so the limits of detection are frequently around 10^5, which precludes the use of highly specific probes for less numerous groups of soil bacteria.

Other marker genes such as *luxAB* and *lacZY* have been employed for detection of single cells but these rely on the phenotype of the marker so expression is essential for detection. In addition, only genetically marked bacteria can be detected. The *luxAB* genes have been used for the detection of single cells in soil slurries by use of charge coupled device-enhanced microscopy (Silcock *et al.*, 1992). Light production can be detected directly following the addition of the substrate n-decyl aldehyde. There is no background bioluminescence in soil, so the cells were clearly visible, but light production does depend on the metabolic status of the cell. It is likely that cells in bulk natural soil will also require some form of amendment to enable bioluminescence. Starved cells could produce light in the presence of an energy source (Meikle *et al.*, 1992).

Detection of indigenous populations of bacteria in soil has been achieved by using the *luxAB* genes by exploiting a bacteriophage as a delivery system (Turpin *et al.*, 1993). Cells infected *in situ* with a *lux*-marked phage will bioluminesce when the phage genome is transcribed prior to lysis. The amount of light produced can be related to cell numbers as the infection efficiency is relatively constant under standardized conditions of infection employing a 90 min incubation in the presence of nutrients and the substrate n-decyl aldehyde. Data on the detection limits for the use of a Plux22 bacteriophage for detection of *Salmonella typhimurium* in soil is presented in Table 2. For monitoring survival of salmonellas in soil the light output correlated with viable counts when nutrients were added during infection.

Table 2 Detection limits for *Salmonella typhimurium* LT2 enumerated in a range of samples using Plux22 bacteriophage

Sample	Treatment	Detection limit[1] (cells/ml or cells/g in original sample)
Liquid culture	(A) 10^{-1} dilution (0.5 ml in 4.5 ml Luria broth)	1×10^2
Water	(A) 10^{-1} dilution	1×10^2
	(B) 1L Membrane filtered : 0.2 μm pore	1
	: 0.45 μm pore	9
Sewage sludge	(A) 10^{-1} dilution : non sterile sludge	1.5×10^5
	(B) Flocculation of 150 ml	1.1×10^4
Soil	(A) 10^{-1} dilution	5×10^4
	(B) 0.5 g in 0.5 ml Luria broth, LKB luminometer	6×10^3
	(C) Extract & concentrate 5 g	5×10
	(D) Flocculation of 200 g	5×10
Grass	(A) 10^{-1} dilution	6×10^3
	(B) Stomach & concentrate 10 g	8.5×10^3

[1] Each 'detection limit' figure was derived from a calibration curve of a material inoculated with a range of *Salmonella typhimurium* LT2 inocula and calculated as the background reading plus the HSD (0.05) for that data set. Except for soil treatment B, all light readings were for a 5 ml volume in a Turner Designs luminometer. Turpin, Maycroft and Wellington, unpublished data.

Bacteriophage can be used to introduce various genes into a population of susceptible bacteria through the process of lysogeny (Marsh *et al.*, 1993). This is a useful technique for detecting susceptible indigenous hosts to specific phage. The actinophage KC301 carrying a thiostrepton resistance gene, *tsr*, was introduced into natural soil and lysogens selected for by screening for thiostrepton resistance. Once the phage infects a streptomycete host and enters the lysogenic cycle, the phage genome is stably maintained on the chromosome. Indigenous streptomycete lysogens were selected from the soil and the *tsr* gene detected on the chromosome of these strains.

CONCLUSIONS

The ecological applications of molecular markers are extensive and have provided a means of identifying bacteria and their specific potential activities *in situ*. Soil is a particularly difficult habitat to study so requires the application of a range of techniques to allow characterization of the bacterial diversity. Analysis of soil DNA and RNA has proved useful for tracking genes, detecting novel bacteria, analysing genetic diversity and identifying pathogens and important groups of plant-beneficial bacteria. Despite the extreme heterogeneity of soils, most methods developed have been applied to a range of soil types and there are now variations possible to adapt extraction procedures to highly organic soils or heavy clays.

REFERENCES

Aardema, B.W., Lorenz, M.G. and Krumbein, W.E. (1983) Protection of sediment-adsorbed transforming DNA against enzymic inactivation. *Appl. Environ. Microbiol.* 46,417-420.

Altenbuchner, J and Cullum, J. (1987) Amplification of cloned genes in Streptomyces. *Biotechnology* 5,1328-1329.

Amann, R. I., Krumholz, L. and Stahl, D. A. (1990a) Fluorescent-oligonucleotide probing of whole cells for determinative, phylogenetic, and environmental studies in microbiology. *J. Bacteriol.* 172,762-770.

Amann, R. I., Binder, B. J., Olson, R. J., Chisholm, S. W., Devereux, R. and Stahl, D. A. (1990b) Combination of 16S rRNA-targeted oligonucleotide probes with flow cytometry for analyzing mixed microbial populations. *Appl. Environ. Microbiol.* 56,1919-1925.

Amann, R. I., Springer, N., Ludwig, W., Gortz, H-D. and Schleifer, K-H. (1991) Identification *in situ* and phylogeny of uncultured bacterial endosymbionts. *Nature* 351,161-164.

Bramwell, P. (1992) The characterisation and detection of plant pathogenic streptomycetes in the natural environment. PhD Thesis, University of Warwick.

Bruce, K. D., Hiorns, W. D., Hobman, J. L., Osborn, A. M., Strike, P. and Ritchie, D. A. (1992) Amplification of DNA from native populations of soil bacteria by using the polymerase chain reaction. Appl. Environ. Microbiol. 58,3413-3416.

Cresswell, N., Saunders, V.A. and Wellington, E.M.H. (1991) Detection and quantification of *Streptomyces violaceolatus* plasmid DNA in soil. *Lett. Appl. Microbiol.* 13,193-197.

Cresswell, N. and Wellington, E.M.H. (1992) The detection of gene transfer in terrestrial environments, in: "Genetic Interactions Between Microorganisms in the Natural Environment",. (Wellington, E.M.H and Van Elsas, J.D. Eds.), pp. 59-82. Pergamon Press, Oxford.

Delong, E. F., Wickham, G. S. and Pace, N. R. (1989) Phylogenetic stains: ribosomal RNA-based probes for identification of single cells. *Science* 243,1360-1363.

Drahos, D.J., Hemming, B.C. and McPherson, S., (1986) Tracking recombinant organisms in the environment: ß-galactosidase as a selectable non-antibiotic marker for fluorescent pseudomonads. *Biotechnology* 4,439-444.

Fægri, A., Torsvik, V.L. and Goksoyr, J. (1977) Bacterial and fungal activities in soil: separation of bacteria and fungi by a rapid fractionated centrifugation technique, *Soil Biol. Biochem.* 9,105-112.

Greaves, M. P. and Wilson, M. J. (1969) The adsorption of nucleic acids by montmorillonite. *Soil Biol. Biochem.* 1,317-323.

Hahn, D., Amann, R. I., Ludwig, W., Akkermans, A. D. L. and Schleifer. K. H. (1992) Detection of micro-organisms in soil after *in situ* hybridization with rRNA-targeted, fluorescently labelled oligonucleotides. *J. Gen. Microbiol.* 138,879-887.

Hahn, D., Kester, R., Starrenburg, M. J. C. and Akkermans, A. D. L. (1990) Extraction of ribosomal RNA from soil for detection of *Frankia* with oligonucleotide probes. *Arch. Microbiol.* 154,329-335.

Herrick, J. B., Madsen, E. L., Batt, C. A. and Ghiorse, W. C. (1993) Polymerase chain reaction amplification of naphthalene-catabolic and 16S rRNA gene sequences from indigenous sediment bacteria. *Appl. Environ. Microbiol.* 59,687-694.

Herron, P.R. and Wellington, E.M.H. (1990) New method for extraction of streptomycete spores from soil and application to the study of lysogeny in sterile amended and nonsterile soil. *Appl. Environ. Microbiol.* 56,1406-12.

Holben, W.E., Jansson, J.K., Chelm, B.K. and Tiedje, J.M. (1988) DNA probe method for the detection of specific microorganisms in the soil bacterial community. *Appl Environ. Microbiol.* 54,703-11.

Holben, W. E. (1993) Isolation and purification of bacterial DNA from soil, in: " Methods of Soil Analysis" part 2 Chemical and Microbiological Properties -Agronomy Monograph no. 9 (3rd edition) American Society of Agronomy, Soil Science Society of America. In press.

Hughes, D. W. and Galau, G. (1988) Preparation of RNA from cotton leaves and pollen. *Plant Mol. Biol. Report* 6,253-257.

Jacobsen, C. S. and Rasmussen, O. F. (1992) Development and application of a new method to extract bacterial DNA from soil based on separation of bacteria from soil with cation-exchange resin. *Appl. Environ. Microbiol.* 58,2458-2462.

Karagouni, A.D., Vionis, A.P., Baker, P.W. and Wellington, E.M.H. (1993) The effect of soil moisture content on spore germination, mycelium development and survival of a seeded streptomycete in soil. *Microbial Releases* 2,47-51.

Liesack, W. and Stackebrandt, E. (1992) Occurrence of novel groups of the domain bacteria as revealed by analysis of genetic material isolated from an Australian terrestrial environment, *J. Bacteriol.* 174,5072-5078.

Lorenz, M.G. and Wackernagel, W. (1987) Adsorption of DNA to sand and variable degradation rates of adsorbed DNA. *Appl. Environ. Microbiol.* 53,2948-2956.

MacDonald, R.M. (1986) Sampling soil microfloras: Dispersion of soil by ion exchange and extraction of specific microorganisms from suspension by elutriation. *Soil Biol. Biochem.* 18,399-406.

Marsh, P., Toth, I., Meijer, M., Schilhabel, M.B. and Wellington, E.M.H. (1993) Survival of the temperate actinophage ØC31 and *Streptomyces lividans* in soil and the effects of competition and selection on the spread of lysogens. *FEMS Microbiol. Ecol.* 13, 13-21.

Meikle, M., Killham, K., Prosser, J. and Glover, L. A. (1992) Luminometric measurement of population activity of genetically modified *Pseudomonas fluorescens* in the soil. *FEMS Microbiol. Lett.* 99,217-220.

Moran, M. A., Torsvik, V.L., Torsvik, T. and Hodson, R.E. (1993) Direct extraction and purification of rRNA for ecological studies. *Appl. Environ. Microbiol.* 59,915-918.

Morgan, J.A.W., Winstanley, C., Pickup, R.W., Jones, J.G. and Saunders, J.R. (1989) Direct phenotypic and genotypic detection of a recombinant pseudomonad population released into lake water. *Appl. Environ. Microbiol.* 55,2537-2544.

Muyzer, G., De Waal, E. C. and Uitterlinden, A.G. (1993) Profiling of complex microbial populations by denaturing gradient gel electrophoresis analysis of polymerase chain reaction-amplified genes coding for 16S rRNA. *Appl. Environ. Microbiol.* 59,695-700.

Ogram, A., Sayler, G.S. and Barkay, T. (1987) The extraction and purification of microbial DNA from sediments. *J. Microbiol. Meth.* 7,57-66.

Ogram, A., Sayler, G. S., Gustin, D. and Lewis, R J. (1988) DNA adsorption to soils and sediments. *Environ. Sci. Technol.* 22,982-984.

Paul, J.H., Jeffrey, W.H. and Cannon, J.P. (1990) Production of dissolved DNA, RNA, and protein by microbial populations in a Florida river. *Appl. Environ. Microbiol.* 56,2957-2962.

Picard, C., Pononnet, C., Paget, E., Nesme, X. and Simonet, P. (1992) Detection and enumeration of bacteria in soil by direct DNA extraction and polymerase chain reaction. *Appl. Environ. Microbiol.* 58,2717-2722.

Saunders, J. R. and Saunders, V. A. (1993) Genotypic and phenotypic methods for the detection of specific released microorganisms, in: "Monitoring Genetically Manipulated Microorganisms in the Environment" (Edwards, C. Ed.) pp. 27-59. John Wiley & Sons Ltd., Chichester.

Sayler, G. S., Fleming J. T., Applegate, B. and Werner, C. (1992) Nucleic acid extraction and analysis: detecting genes and their activity in the environment, in "Genetic interactions between microorganisms in the natural environment" (Wellington E. M. H. and van Elsas, J D. Eds.) pp. 237-257. Pergamon press, Oxford.

Selenska, S. and Klingmuller, W. (1991) Direct detection of nif-gene sequence of Enterobacter agglomerans in soil. FEMS Microbiol. Letts. 80,243-246.

Silcock, D. E., Waterhouse, R. N., Glover, L. A., Prosser, J. I. and Killham, K. (1992) Detection of a single genetically modified bacterial cell in soil by using charge coupled device-enhanced microscopy. Appl. Environ. Microbiol. 58,2444-2448

Smalla, K., Cresswell, N., Mendonca-Hagler, L. C., Wolters, A. and van Elsas, J. D. (1993) Rapid DNA extraction protocol from soil for polymerase chain reaction-mediated amplification. *J. Appl. Bacteriol.* 74,78-85.

Stanley, J., Burnens, A., Powell, N., Chowdry, N. and Jones, C. (1992) The insertion sequence IS*200* fingerprints genotype and epidemiological relationships in *Salmonella heidelberg*. *J. Gen. Microbiol.* 138,2329-2336.

Steffan, R.J. & Atlas, R.M. (1988) DNA amplification to enhance detection of genetically engineered bacteria in environmental samples. *Appl. Environ. Microbiol.* 54,2185-91.

Steffan, R. J. and Atlas, R. M. (1991) Polymerase chain reaction applications in environmental microbiology. *Ann. Rev. Microbiol.* 45,137-161.

Torsvik, V., Goksyr, J. and Daae, F. L. (1990) High diversity in DNA of soil bacteria. *Appl. Environ. Microbiol.* 56,782-787.

Tsai, Y-L. & Olson, B.H. (1991) Rapid method direct extraction of DNA from soil and sediments. *Appl. Environ. Microbiol.* 57,1070-74.

Tsai, Y-L., Park, M. J. and Olson, B. H. (1991) Rapid method for direct extraction of mRNA from seeded soils. *Appl. Environ. Microbiol.* 57,765-768.

Turpin, P.E., Maycroft, K.A., and Wellington, E.M.H. (1993a) Viable but nonculturable Salmonellas in soil. *J. Appl. Bacteriol.* 74,421-427.

Turpin, P.E., Maycroft, K.A., Bedford, J. Rowlands, C.L. and Wellington, E.M.H. (1993b) A rapid luminescent-phage based MPN method for the enumeration of *Salmonella typhimurium* in environmental samples. *Lett. Appl. Microbiol.* 16,24-27.

van Overbeek, L.S., van Elsas, J. D., Trevors, J. T. and Starodub, M. E. (1990) Long-term survival of and plasmid stability in *Pseudomonas* and *Klebsiella* species and appearance of nonculturable cells in agricultural drainage water. *Microb. Ecol.* 19,239-249.

Way, J. S., Josephson, K. L., Pillai, S. D., Abbaszadegan, M., Gerba, C. P. and Pepper, I. L. (1993) Specific detection of *Salmonella* spp. by multiplex polymerase chain reaction. *Appl. Environ. Microbiol.* 59,1473-1479.

Wayne, L.G., Brenner, D.J., Colwell, R.R., Grimont, P.A.D., Kandler, O., Krichevsky, M.I., Moore, L.H., Moore, W.E.C., Murray, R.G.E., Stackebrandt, E., Starr, M.P. and Trüper, H.G. (1987) Report of the ad *hoc* committee on reconciliation of approaches to bacterial systematics, *Int. J. Syst. Bacteriol.* 37,463-464.

Wellington, E.M.H., Herron, P.R. and Cresswell, N. (1992) Genetic interactions between bacteria in soil. In: "Monitoring Genetically Manipulated Microorganisms in the Environment " (Edwards, C. Ed.), pp. 137-170. John Wiley & Sons Ltd., Chichester.

Wellington, E.M.H., Marsh, P., Toth, I., Cresswell, N., Huddleston, L. and Schilhabel, M. (1993) The selective effects of antibiotics in soils, in: "Trends in Microbial Ecology" (Guerrero R. and Pedros-Alio, C. Eds.), pp. 331-336. Spanish Society for Microbiology, Spain.

Wipat, A., Wellington, E.M.H. & Saunders, V.A. (1991) *Streptomyces* marker plasmids for monitoring survival and spread of streptomycetes in soil. *Appl. Environ. Microbiol.* 57,3322-3330.

PHYLOGENETIC DIVERSITY OF METHANOGEN ENDOSYMBIONTS OF ANAEROBIC CILIATES

T. Martin Embley[1] and Bland J. Finlay[2]

[1]Microbiology Group
The Natural History Museum
Cromwell Road
London SW7 5BD, UK

[2]Institute of Freshwater Ecology
Windermere Laboratory
Far Sawrey
Ambleside LA22 OLP, UK

INTRODUCTION

Anaerobic environments occur widely wherever the accumulation of organic material exceeds the rate at which oxygen becomes available for its aerobic decomposition. Examples include marine and freshwater sediments, anaerobic sewage deposits, landfill sites and the guts of many animals (Fenchel and Finlay, 1991a). In these situations protozoa and ciliates in particular, survive by feeding on rich and varied prokaryote communities. Ciliates which live exclusively in anaerobic environments lack mitochondria and their metabolism is based upon fermentation and substrate level phosphorylation (Müller, 1988; Fenchel and Finlay, 1991a). Most of the anaerobic ciliates contain microbodies some of which have been shown to be capable of oxidizing pyruvate to acetate with the production of ATP, CO_2 and H_2. In such cases they are called hydrogenosomes (Müller, 1988). The origins of hydrogenosomes are unresolved, but in some ciliates they contain cristae and a double membrane and thus they resemble mitochondria (Finlay and Fenchel, 1989).

When excited with short-wave ultraviolet light, many anaerobic ciliates are observed to contain large numbers of fluorescing bacteria. In at least one case (Van Bruggen et al., 1983) the fluorescent material has been extracted and shown to be coenzyme F_{420}. Large amounts of this compound are rarely found in bacteria but it is a key component in most methanogenic archaea where it functions as a low-potential electron carrier (Jones et al., 1987). The intracellular methanogens are metabolically active since methane production by ciliates containing fluorescing bacteria has been demonstrated (Van Bruggen *et al.*, 1986; Fenchel and Finlay, 1992). Since most methanogenic bacteria can use H_2 to reduce CO_2 and generate energy it is likely that hydrogen is transferred from hydrogenosome to symbionts. When endosymbiotic methanogenic activity is experimentally inhibited, the ciliates release hydrogen instead of methane (Fenchel and Finlay, 1992). The costs or benefits of the symbiosis to the host are difficult to identify but some large ciliates can certainly grow faster when symbionts are present (Fenchel and Finlay, 1991*b*).

The identities of the symbiotic methanogens in different ciliates have been difficult to establish with certainty. Endosymbiotic bacteria are notoriously difficult to culture and once a culture has been isolated it must be proven that it represents the endosymbiont and not a free-living contaminant. An experiment aimed at reinfecting an aposymbiotic strain of *Trimyema compressum* succeeded in establishing infection with methanogens isolated from *Metopus striatus* or *Pelomyxa palustris*, but the symbionts were quickly lost when the host growth rate was stimulated (Wagener *et al.*, 1990).

A simple and unambiguous approach for the identification of endosymbiotic bacteria would allow a number of fundamental questions concerning the nature of the symbiosis to be addressed, including :

1.   How specific are the interactions ?
2.   Do hosts contain only one, or several types of symbiont ?
3.   How common is the ability to colonise ciliates ? Is the ability to form associations restricted to a single methanogen lineage ?

Over the past few years molecular tools have been developed which allow the identities of bacteria to be determined easily. The key to this approach is twofold, the comparative analysis of ribosomal (r)RNA sequences to classify and identify bacteria (Woese, 1987); and the ease by which rRNA sequences can be recovered from uncultured bacteria using the polymerase chain reaction (Saiki *et al.*, 1988). Symbioses are attractive subjects for molecular methods because they normally comprise only a few partners whose identities can each be addressed in turn.

USE OF rRNA SEQUENCES TO CLASSIFY AND IDENTIFY ENDOSYMBIOTIC BACTERIA

The advantages of rRNA's for studying microbial phylogeny have been extensively discussed (Woese, 1987). Of particular relevance to the present discussion is that, on the basis of rRNA sequence comparisons, all cellular life can be divided into three domains; the *Bacteria*, *Archaea* and *Eucarya* (Woese *et al.*, 1990). Each domain contains some conserved sequence motifs which are absent from the other two. It is thus relatively simple to design PCR primers that will specifically amplify RNA sequences from particular phylogenetic groups directly from heterogeneous mixtures of templates (Giovannoni *et al.*, 1990; Amman *et al.*, 1991; Fry *et al.*, 1991). Judging from the data available, the methanogen phenotype appears to occur mainly in a phylogenetically coherent subdomain of the *Archaea* called the Euryarchaeota. PCR primers which specifically amplify rRNA sequences from the Euryarchaeota, without concomitant amplification of ciliate rRNA genes, have been designed (Embley *et al.*, 1992*a*). Using these primers it is not necessary to purify symbiont DNA prior to amplification and PCR can be done directly using DNA released from small numbers (typically 100-1000 washed cells) of heat-lysed ciliates (Embley *et al.*, 1992*a,b*; Finlay *et al.*, 1993). Once a sequence has been obtained the simplest way to prove it belongs to an endosymbiont is carry out an *in situ* hybridization experiment. Oligonucleotide probes to rRNA (which can be designed from the sequence of a symbiont PCR product) can be labelled with fluorescent dyes (DeLong *et al.*, 1989) and used to probe intact protozoa which have been made permeable by treatment with paraformaldehyde (Amman *et al.*, 1991; Embley *et al.*, 1992*a*). Ribosomal (r) RNA sequences are mosaics of conserved, semi-conserved and very variable sequences (Woese *et al.*, 1983; Gutell *et al.*, 1985) and it is relatively simple to design highly specific probes by targeting an appropriate region (Stahl and Amann, 1991). Probes to rRNA are very sensitive as ribosomes are abundant in actively growing cells (DeLong *et al.*, 1989).

Once a sequence has been shown to belong to an endosymbiont, its rRNA sequence can be compared to sequences from other *Archaea* (Larsen *et al.*, 1993) and its phylogenetic relationships to free-living methanogens can be established.

IDENTIFICATION OF THE ENDOSYMBIONTS OF ANAEROBIC CILIATES USING rRNA SEQUENCES

Metopus palaeformis

Metopus palaeformis is found in lake sediments and anaerobic municipal landfill material (Finlay and Fenchel, 1991*a*). It contains many rod-shaped methanogens which are spread throughout the host cytoplasm; they are closely associated with but not apparently attached to the hydrogenosomes. The growth of methanogens and host are closely coupled but the host obtains no measurable energetic advantage from the association (Finlay and Fenchel, 1992). PCR primers specific for Euryarchaeota were used to amplify a fragment of the symbiont 16S rRNA gene which was sequenced and compared to sequences from cultured methanogens (Embley *et al.*, 1992*a*). A fluorescent probe was then designed and used in whole cell hybridizations to prove that the amplification product was derived from the rod-shaped endosymbiont. Figure 1 is a phylogenetic tree which clearly demonstrates that the symbiont is part of the genus *Methanobacterium* but that its sequence is different from those of free-living methanobacteria.

Metopus striatus **and** *Pelomyxa palustris*

In separate experiments methanogens were isolated from washed cells of *Metopus striatus* and *Pelomyxa palustris*, both of which had been isolated from the same freshwater aquarium (Van Bruggen *et al.*, 1983, 1984; Van Bruggen, 1986). On the basis of morphology and DNA base composition (%GC) both methanogens were classified as *Methanobacterium formicicum*. When the 16S rRNA genes of these two strains were amplified and partially sequenced (Embley and Finlay, 1993) it became clear that the two strains were identical to each other, but different from *Methanobacterium formicicum* and also from the *Methanobacterium* identified in *Metopus palaeformis* (Figure 1). The level of sequence divergence between the isolates and *M. formicicum* indicates that they are closely related species.

Metopus contortus

Metopus contortus is a large ciliate which lives in sulphide-rich marine sands (Fenchel *et al.*, 1977). It has been the subject of two investigations (on different isolates) aimed at identifying its endosymbionts; one culture based (Van Bruggen *et al.*, 1986) and one using PCR and fluorescent probes (Embley *et al.*, 1992*b*).

Van Bruggen *et al.* (1986) isolated a polymorphic methanogen which they classified as *Methanoplanus endosymbiosus* on the basis of morphology and DNA base compostion. When the 16S rRNA of this strain was subsequently sequenced (Embley and Finlay, 1993) it was confirmed as a species of *Methanoplanus* and shown to be very closely related to the free-living species *Methanoplanus limicola*.

A second strain of *Metopus contortus* was investigated by Finlay and Fenchel (1991*b*). This strain of *Metopus contortus* contained a variety of different shapes of endosymbiont which were interpreted to be stages in the morphological transformation of a single species. The final stage of this morphological transformation was the intimate association of the enlarged symbiont with host hydrogenosomes. PCR amplifications recovered only a single sequence from the ciliate and a fluorescent probe was designed to bind to the homologous rRNA (Table 2; Embley *et al.*, 1992*b*). The probed symbionts in whole cells of *Metopus contortus* exhibited a variety of morphologies and some symbionts were approximately twice the size of others. These data support the hypothesis that a single species of symbiont undergoes a morphological transformation as part of the symbiotic process. Analysis of the symbiont rRNA sequence showed that it is closely related to the free-living species *Methanocorpusculum parvum* but only distantly related to *Methanoplanus endosymbiosus* (Figure 1).

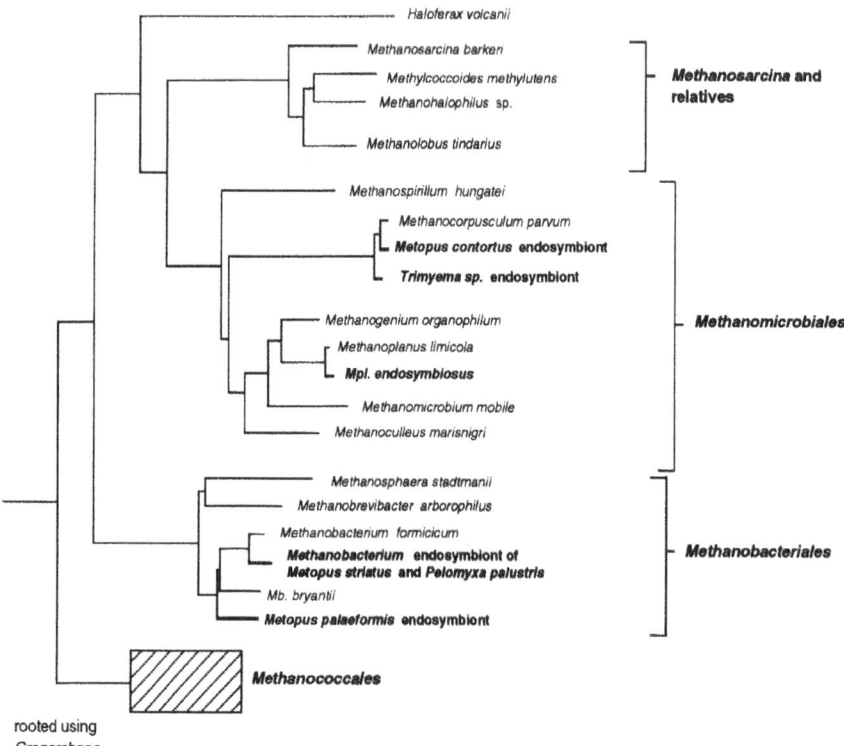

rooted using
Crenarchaea

Figure 1. Phylogenetic tree showing the relationships of endosymbiotic methanogens to free-living methanogens. Adapted from Embley and Finlay (1993). The tree was constructed using neighbour joining (Saitou and Nei, (1987). Reference sequences were obtained from the Ribosomal RNA Data Base Project (Larsen *et al.*, 1993)

Trimyema

The only other ciliate where there is an apparent morphological transformation of an endosymbiont is in a fresh-water strain of *Trimyema* (Finlay *et al.*, 1993). The smallest forms of the symbiont are disc-shaped and slightly less than 1 μm in length. They give rise to a continuous series of bacterial shapes by a transformation which involves a significant increase in size and a progressive invagination of the cell wall. In the final stages of this transformation, the methanogens appear stellate and almost totally enclosed by hydrogenosomes. Once again PCR amplifications revealed only a single sequence and a probe to the homologous rRNA detected a range of fluorescing particles of sizes consistent with morphological transformation of a single endosymbiont (Finlay *et al.*, 1993). Detailed analysis of the 16S rRNA sequence from the symbiont in *Trimyema* sp. showed that it is closely related to the *Methanocorpusculum*-like endosymbiont from *Metopus contortus*. The level of sequence divergence between the two (Figure 1) is very small but is greater than that between *Methanoplanus endosymbiosus* and *Methanoplanus limicola*.

Cyclidium porcatum

Cyclidium porcatum was the first freshwater anaerobic scuticociliate to be cultured and described. It contains a unique tripartite structure consisting of hydrogenosomes, interspersed with methanogens and a small number of unidentified Bacteria (Esteban *et al.*, 1993; Embley and Finlay, 1993). The detailed identities of the partners is currently unknown but Archaeal and Bacterial specific probes have confirmed the domain-level affiliations of each prokaryote.

DISCUSSION

Symbiotic associations between *Eucarya* and *Bacteria* are common and well documented (Gray, 1992). The most famous examples are chloroplasts and mitochondria: symbionts which have evolved to the stage where they are no longer capable of a free-living existence. Whereas associations between *Eucarya* and *Bacteria* are common, the only *Eucarya* to contain endosymbiotic *Archaea* are anaerobic protozoa. In this review we have presented the results of published investigations aimed at identifying the methanogenic *Archaea* living inside anaerobic protozoa (mainly ciliates) using rRNA sequences. These results allow us to answer some questions concerning the specificity of the symbiotic interactions.

The methanogenic symbionts living inside anaerobic protozoa are taxonomically very diverse. The methanogen phenotype is found in only three major Archaeal lineages and symbionts (from a very small sample of hosts) are drawn from two of these. Symbionts occur in the genera *Methanobacterium*, *Methanocorpusculum* and *Methanoplanus* (Figure 1). The symbiont sequences are always novel but they are closely related to, and separated by, sequences from free-living methanogens. The ability to colonise ciliates is thus widespread among methanogens and the symbioses with different hosts appear to have arisen independently.

On the basis of classical criteria of morphology and ultrastructure (Lynn and Corliss, 1991), the trichostomatid *Trimyema* and the heterotrich *Metopus contortus* are unrelated. Nevertheless, they contain *Methanocorpusculum*-like endosymbionts which are very closely related (Figure 1). Both endosymbionts show a morphological transformation. In contrast, the congeneric ciliates *Metopus palaeformis* and *M.*

contortus contain symbionts which are distantly related to each other and closely related to free-living species. These data show that at least one of these symbiotic associations is relatively recent and must postdate the speciation of *M. palaeformis* and *M. contortus*.

The sequence data for the strains which have been cultured from anaerobic protozoa paint a similar picture to the *in situ* molecular studies (with the caveat that there remains some doubt that culture-based studies have actually isolated the endosymbiont). The same *Methanobacterium* species was recovered from the ciliate *Metopus striatus* and the amoeba *Pelomyxa palustris*, whereas *Metopus contortus* contained a symbiont from a different order; *Methanoplanus endosymbiosus*. When the results of molecular and culture based studies on *Metopus contortus* are combined they show that different strains of this ciliate can contain endosymbionts from two completely different genera of methanogen.

The association between protozoa and methanogens is probably an ancient one. The hosts are taxonomically very diverse and include at least three orders of ciliate (Trichostomes, Heterotrichs and Scuticociliates) and the anaerobic amoeba *Pelomyxa*. A recent publication has now described a methanogen endosymbiont in the anaerobic flagellate *Psalteriomonas vulgaris* (Broers *et al.*, 1993). The data from the molecular studies described above suggest that associations can form very freely. However, some of the symbioses show very sophisticated interactions. The symbionts in *Metopus contortus* and *Trimyema* sp. display a morphological transformation which can be rationalised as facilitating hydrogen capture (Finlay and Fenchel, 1991; Finlay *et al.*, 1993). In *Cyclidium porcatum* three partners, hydrogenosomes, methanogens and anaerobic bacteria are organised into a single complex whose division is synchronised and which occupies about a third of the host cell (Esteban *et al.*, 1993). This last arrangement is all the more impressive when it is considered that the genus *Cyclidium* also contains aerobic species which contain mitochondria.

REFERENCES

Amann, R.I., Springer, N., Ludwig, W., Gortz, H-D. and Schleifer, K-H. (1991) Identification and *in situ* phylogeny of uncultured bacterial endosymbionts. Nature **351**, 161-164.

Broers, C. A. M., Meijers, H.H.M., Symens, J. C., Stumm, C. K., Vogels, G. D. and Brugerolle, G. (1993) Symbiotic association of *Psalteriomonas vulgaris* n. spec. with *Methanobacterium formicicum*. Europ. J. Protistol. **29**, 98-105.

DeLong, E. F., Wickham, G. S. and Pace, N. R. (1989) Phylogenetic stains : ribosomal RNA based probes for the detection of single cells. Science **243**, 1360-1363.

Distel, D. L., DeLong, E. F. and Waterbury, J. B. (1991) Phylogenetic characterisation and *in situ* localisation of the bacterial symbiont of shipworms (*Teredinidae*: *Bivalvia*) by using 16S rRNA sequence analysis and oligodeoxynucleotide probe hybridisation. Appl. Environ. Microbiol. **57**, 2376-2382.

Embley, T. M. and Finlay, B. J. (1993) Systematic and morphological diversity of endosymbiotic methanogens in anaerobic ciliates. Ant. van Leeuwen. (in press).

Embley, T.M., Finlay, B.J., Thomas, R.H. and Dyal, P.L. (1992a) The use of rRNA sequences and fluorescent probes to investigate the phylogenetic positions of the anaerobic ciliate *Metopus palaeformis* and its archaeobacterial endosymbiont. J. Gen. Microbiol. **138**, 1479-1487.

Embley T.M., Finlay, B.J. and Brown, S. (1992b) RNA sequence analysis shows that the symbionts in the ciliate *Metopus contortus* are polymorphs of a single methanogen species. FEMS Microbiol. Lett. **97**, 57-62.

Esteban, G., Guhl, B.E., Clarke, K.J., Embley, T.M. and Finlay B.J. (1993) *Cyclidium porcatum* n.sp.: a free-living anaerobic scuticociliate containing a stable complex of hydrogenosomes, eubacteria and archaeobacteria. Eur. J. Protistol. **29**, 262-270.

Fenchel, T. and Finlay, B.J. (1990) Anaerobic free-living protozoa : growth efficiencies and the structure of anaerobic communities. FEMS Microbiol. Ecol. **74**, 269-276.

Fenchel, T. and Finlay, B. J. (1991*a*) The biology of free-living anaerobic ciliates. Europ. J. Protistol. **26**, 201-215.

Fenchel, T. and Finlay, B.J. (1991b) Endosymbiotic methanogenic bacteria in anaerobic ciliates: significance for the growth efficiency of the host. J. Protozool. **38**, 18-22.

Fenchel, T. and Finlay, B.J. (1992) Production of methane and hydrogen by anaerobic ciliates containing symbiotic methanogens. Arch. Microbiol. **157**, 475-480.

Fenchel, T., Perry, T. and Thane, A. (1977). Anaerobiosis and symbiosis with bacteria in free-living ciliates. J. Protozool. **24**, 154-163.

Finlay, B.J. and Fenchel, T. (1989) Hydrogenosomes in some anaerobic ciliates resemble mitochondria. FEMS Microbiol. Lett. **65**, 311-314.

Finlay, B.J. and Fenchel, T. (1991*a*) An anaerobic protozoon, with symbiotic methanogens, living in municipal landfill material. FEMS Microbiol. Ecol. **85**, 169-180.

Finlay BJ and Fenchel T (1991b) Polymorphic bacterial symbionts in the anaerobic ciliated protozoon *Metopus contortus*. FEMS Microbiol. Lett. **79**, 187-190.

Finlay, B. J. and Fenchel, T. (1992) An anaerobic ciliate as a natural chemostat for the growth of endosymbiotic methanogens. Europ. J. Protistol. **28**: 127-137.

Finlay, B.J., Embley, T.M. and Fenchel, T. (1993) A new polymorphic methanogen, closely related to *Methanocorpusculum parvum*, living in stable symbiosis within the anaerobic ciliate *Trimyema* sp. J. Gen. Microbiol. **139**, 371-378.

Fry, N. K., Rowbotham, T. J., Saunders, N. A. and Embley, T. M. (1991) Direct amplification and sequencing of the 16S ribosomal DNA of an intracellular *Legionella* species recovered by amoebal enrichment from the sputum of a patient with pneumonia. FEMS Microbiol. Lett. **83**, 165-168

Giovannoni, S. J., Britschgi, T. B., Moyer, C. L. and Field, K. G. (1990) Genetic diversity in Sargasso Sea bacterioplankton. Nature **345**, 60-63

Gray, M. (1992). The endosymbiont hypothesis revisited. Int. Rev. Cytol. **141**, 233-357.

Gutell, R. R., Weiser, B., Woese, C. R. and Noller, H. F. (1985) Comparative anatomy of 16S-like ribosomal RNA. Progr. Nucl. Res. Molec. Biol. **32**, 155-216.

Jones, W. J., Nagle, D. P. and Whitman, W. B. (1987) Methanogens and the diversity of archaebacteria. Microbiol. Rev. **51**, 135-177.

Jukes, T.H. and Cantor, C.R. (1969) Evolution of protein molecules. In " *Mammalian Protein Metabolism*" (Munro, H.N. Ed.), pp. 21-132. Academic Press, New York.

Larsen, N., Olsen, G. J., Maidak, B. L., McCaughey, M. J., Overbeek, R., Macke, T. J., Marsh, T. L. and Woese, C. R. (1993). The ribosomal database project. Nucl. Acid. Res. **21**, 3021-3023.

Lynn, D. and Corliss, J. (1991). Ciliophora. In *"Microscopic Anatomy of Invertebrates, Volume 1 Protozoa"* (Harrison, F. W. and Corliss, J. O. Ed.), pp. 333-467. Wiley-Liss Inc., New York.

Müller, M. (1988) Energy metabolism in protozoa without mitochondria. Ann. Rev. Microbiol. **42**, 465-488.

Saiki, R. K., Gelfand, D.H., Stoffel, S., Scharf, S.J., Higuchi, R., Horn, G.T., Mullis, K.B. and Erlich, H.A. (1988) Primer-directed enzymatic amplification of DNA with a thermostable DNA polymerase. Science **239**, 487-491.

Saitou, N. and Nei, M. (1987) The neighbour joining method : a new method for constructing phylogenetic trees. Mol. Biol. Evol. **4**, 406-425.

Stahl, D.A. and Amann, R.I. (1991) Development and application of nucleic acid probes in bacterial systematics. In " *Nucleic acid techniques in bacterial systematics*". (Stackebrandt, E. and Goodfellow, M. Ed.), pp. 205-248. John Wiley, Chichester.

Van Bruggen, J. J. A. (1986). Methanogenic bacteria as endosymbionts of sapropelic protozoa. PhD. Thesis, University of Nijmegan.

Van Bruggen, J.J.A., Stumm, C.K. and Vogels, G.D. (1983) Symbiosis of methanogenic bacteria and sapropelic protozoa. Arch. Microbiol. **136**, 89-95.

Van Bruggen, J.J.A., Zwart, K.B., van Assema, R.M., Stumm, C.K. and Vogels, G.D. (1984) *Methanobacterium formicicum*, an endosymbiont of the anaerobic ciliate *Metopus striatus* McMurrich. Arch. Microbiol. **139**, 1-7.

Van Bruggen, J.J.A., Zwart, K.B., Hermans, J.G.F., van Hove, E.M., Stumm, C.K. and Vogels, G.D. (1986) Isolation and characterisation of *Methanoplanus endosymbiosus* sp. nov. an endosymbiont of the marine sapropelic ciliate *Metopus contortus* Quennerstedt. Arch. Microbiol. **144**, 367-374.

Wagener, S., Bardele, C.F. and Pfennig, N. (1990) Functional integration of *Methanobacterium formicicum* into the anaerobic ciliate *Trimyema compressum*. Arch. Microbiol. **153**, 496-501.

Woese, C. R. (1987) Bacterial evolution. Microbiol. Rev. **51**, 221-271.

Woese, C. R., Gutell, R., Gupta, R., and Noller, H. F. (1983) Detailed analysis of the higher order structure of 16S-like ribosomal ribonucleic acids. Microbiol. Rev. **47**, 621-669

Woese, C. R., Kandler, O. and Wheelis, M. L. (1990) Towards a natural system for organisms : proposal for the domains *Archaea*, *Bacteria* and *Eucarya*. Proc. Natl. Acad. Sci. USA **87**, 4576-4579.

DIVERSITY, DYNAMICS AND TOPOGRAPHIC ARRANGEMENT OF MICROORGANISMS ARE ESSENTIAL PARAMETERS THAT IDENTIFY A MICROBIAL CONSORTIUM

Everly Conway de Macario and Alberto J.L. Macario

Wadsworth Center for Laboratories and Research, New York State Department of Health, P.O. Box 509, and Department of Biomedical Sciences, School of Public Health, The University at Albany, Albany NY 12201-0509, USA

INTRODUCTION

Years ago it was already clear that elucidation of the microbial flora of complex ecosystems requires identification of microrganisms directly, in samples from the ecosystems, avoiding culture-isolation procedures (Colwell *et al.*, 1985; Macario and Conway de Macario, 1982; Zambon *et al.*, 1984; and references therein). The latter procedures have been, and will always be, extremely useful for the purification and characterization of the components of a microbial community, and to provide axenic cultures to the scientific community. However, culture-isolation is insufficient, and may even be misleading for studies in microbial ecology, inasmuch as it may reveal the microbes culturable in the media utilized, but not necessarily the most significant ones for the ecosystem under analysis. Therefore, for accurate characterization of microbial communities, culture-isolation must be complemented with direct identification and quantification, with minimal sample manipulation, to avoid artifactual distortion of the microbial flora.

Immunology can help in this regard. The application of antibodies and immunotechnology to the direct identification of microbes, and to their classification, has a long history. The reliability, usefulness, and contributions of the immunologic approach to microbiology itself, and to various related disciplines (medical, environmental, ecological) with a microbiologic component are well established (Colwell *et al.*, 1985; Dahle and Laake, 1982; Macario and Conway de Macario, 1982; Ørskov and Ørskov, 1992; Schmidt, 1973; Schmidt *et al.*, 1968; Zambon *et al.*, 1984). The immunologic approach remains valid, other newly developed strategies notwithstanding. Immunology can contribute to the advancement of microbiology and microbial ecology in ways no other approach can. A pertinent review has been published recently (Macario and Conway de Macario, 1993).

Bacterial Diversity and Systematics
Edited by F.G. Priest *et al.*, Plenum Press, New York, 1994

SCOPE

In this brief review of work done in our laboratories, we will focus on four main findings: (1). The microbial diversity in natural and manufactured ecosystems is greater than that suggested by the array of microbes in culture collections and by the results of early surveys; (2). The microbial community structure of a given ecosystem is the consequence of a set of conditions affecting the ecosystem and, therefore, it is (within certain limits) not distinctive of the ecosystem, but of a particular set of conditions; (3). The microbial community structure is a dynamic entity, which changes with time and in response to conditions affecting the ecosystem; and (4). Elucidation of the spatial, tridimensional arrangement of the microorganisms in a consortium using histological methods contributes essential information for the elucidation of the structure of communities that form aggregates, such as flocs, granules, or biofilms.

DIVERSITY AND DISTINCTIVENESS OF MICROBES AND SUBPOPULATION PATTERNS

When we identified methanogenic bacteria (methanogens) by immunologic means in samples from anaerobic methanogenic bioreactors, we found an array of species and immunotypes that demonstrated a diversity much larger than anticipated on the basis of traditional knowledge (Macario and Conway de Macario, 1988; Macario *et al.*, 1989*a*).

The array of species present in each bioreactor was called the pattern of methanogenic subpopulations and was distinctive of the set of conditions under which the bioreactor was operated, *e.g.*, bioreactor type, substrate, and temperature (Fig. 1) (Macario *et al.*, 1989*b*). This occurred despite the fact that all the different bioreactors studied, run under different sets of conditions (but maintained constant during the experiment) had been inoculated at the outset (started up) with aliquots of the same inoculum. It was concluded that the pattern of methanogenic subpopulations which became established was the result of selection by the operating conditions from the array of methanogens present in the original inoculum. Different sets of conditions selected different subpopulations and promoted typical degrees of expansion of the selected subpopulations.

These observations have been confirmed in all other similar studies. For example, in a bioreactor with wood (hybrid poplar) as substrate, an unique pattern of methanogenic subpopulations was found, different from all others described previously for bioreactors operating with other substrates (Macario *et al.*, 1991*a*). The most salient feature of the wood-fermenting bioreactor was an overwhelming predominance of a methanogenic subpopulation antigenically related to, and morphologically indistinguishable from, *Methanomicrobium mobile* BP (Balch *et al.*, 1979; Boone and Whitman, 1988).

BACTERIAL POPULATION DYNAMICS AND PATTERN CHANGES

The pattern of methanogenic subpopulations is typical of each set of conditions that operate in a given ecosystem; if one or more conditions change, the pattern changes (Macario *et al.*, 1989*a,b*). This is because each subpopulation either decreases and becomes undetectable by the available methods, or increases considerably above the initial levels. In a series of illustrative experiments with upflow anaerobic sludge blanket (UASB) granules, the time-course changes of the methanogenic subpopulations after shifting the temperature from 38°C to either 46, 55 or 64°C, were determined (Visser *et al.*, 1991). The subpopulations responded to the temperature shift with distinctive profiles of four types (Fig. 2, I-IV). It is clear from the results that the quantitative pattern of methanogenic subpopulations was

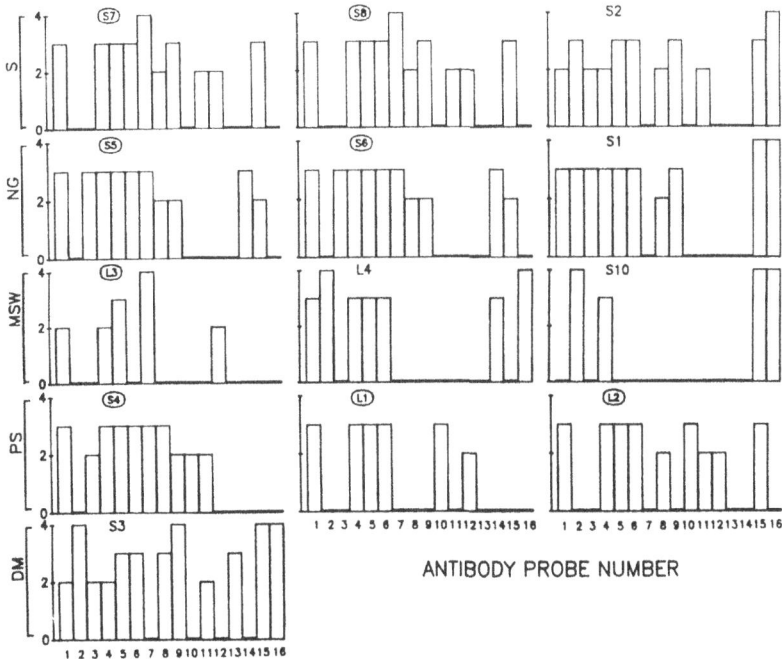

Figure 1. Patterns of methanogenic subpopulations of 13 bioreactors operated under controlled conditions for a comparative study. Each bar represents a methanogen identified by antigenic fingerprinting. The bar's height indicates the degree of antigenic relatedness with the reference methanogen shown at the bottom by a number: 1, *Methanobacterium formicicum* MF; 2, *Methanobacterium thermoautotrophicum* ΔH; 3 and 4, *Methanobacterium bryantii* MoH and MoHG, respectively; 5 and 6, *Methanobrevibacter smithii* ALI and PS, respectively; 7 and 8, *Methanobrevibacter arboriphilus* AZ and DC, respectively; 9, *Methanobrevibacter ruminantium* M1; 10, *Methanospirillum hungatei* JF1; 11, *Methanococcus vannielii* SB; 12, *Methanothrix soehngenii* Opfikon; 13, *Methanothrix* strain CALS-1; 14, *Methanosarcina mazei* S6; 15, *Methanosarcina barkeri* W; and 16, *Methanosarcina thermophila* TM-1. Bioreactors are designated by an S or an L followed by a number. They were run at 35°C (ovals) or at 55°C. Substrates were: sorghum (S), napier grass (NG), municipal solid waste (MSW), primary sludge (PS) and dairy manure (DM). Reproduced with permission from Macario *et al.*, 1989b.

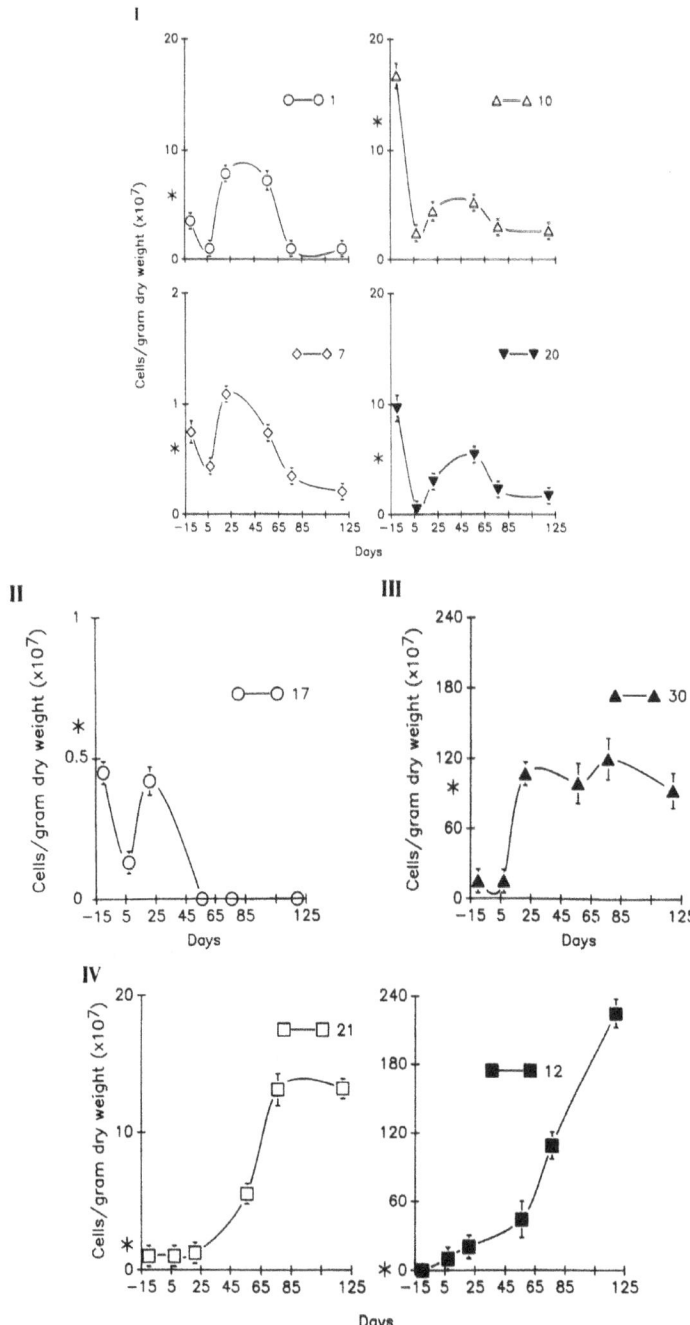

Figure 2. Time-course quantitative profiles of methanogenic subpopulations in UASB bioreactor granules starting 10 days (first point) before shifting the temperature from 38° to 55°C at day 1. The asterisk indicates the cell concentration of the granules in the original inoculum. I. 1, *Methanobrevibacter smithii* PS; 10, *Methanobrevibacter smithii* ALI; 7, *Methanospirillum hungatei* JF1; and 20, *Methanosarcina thermophila* TM-1; II. 17, *Methanogenium cariaci* JR1; III. 30, *Methanothrix soehngenii* Opfikon; and IV. 21, *Methanobrevibacter arboriphilus* AZ; and 12, *Methanobacterium thermoautotrophicum* ΔH. Reproduced with permission from Visser *et al.*, 1991.

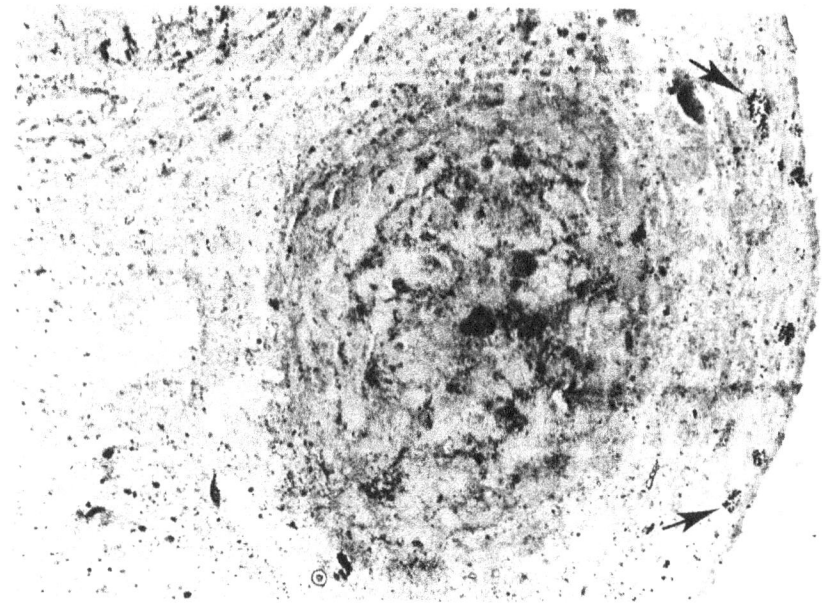

Figure 3. Section of a granule (Gram stain; X200) before the temperature shift showing the medulla and cortex and, in the latter, small circular areas lined by cocci (arrows).

typical of each time-point examined; it was different from the patterns found before or after, *i.e.*, the microbial community structure is a dynamic entity.

TRIDIMENSIONAL ARRANGEMENT OF MICROBES AS A DEFINING PARAMETER OF MICROBIAL COMMUNITY STRUCTURE

The application of histologic methods that are routinely employed to study animal and human tissues in normal and pathologic conditions, to the analysis of the microanatomy of microbial consortia has proven informative and useful (Macario *et al.*, 1991*b*). For example, we found that the modifications in the pattern of methanogenic subpopulations, and in their size, that occurred in the UASB granules as a response to a temperature shift mentioned above (Fig. 2, I-IV), were accompanied by changes in the architecture of the granules (Visser *et al.*, 1991; Macario *et al.*, 1991*b*).

The general architectural features of the granules were elucidated by applying standard histochemical methods. These methods also revealed a variety of microbial colonies with characteristic shapes, sizes and abundance. The colonies of the methanogenic subpopulations were identified using specific antibody probes and immunohistochemical procedures. A summary of the main findings follows: (1). The UASB granules maintained at 38°C, *i.e.*, before the temperature shift, showed a compact structure with a central zone, termed medulla, and a peripheral zone, or cortex, suggesting functional compartmentation (Fig. 3); (2). In the cortex, a series of circular or oval cavities lined by cocci, some of which were also in the lumen of the cavities were observed (Fig. 3). These circular areas represent the cross-section of tubules running through the cortex. Some of these tubules have their ends open to the outside on the granule's surface. The tubules might be an internal

165

communication system through which substrate, metabolic products, and cells circulate; (3). A firm peripheral structure, resembling a capsule formed by a compact layer of argyrophilic fibers, was seen enveloping the granules. This envelope might play a role (passive, if nothing else) in regulating what goes in (*e.g.*, nutrients) and, what comes out (*e.g.*, metabolic products) of the granular matrix; (4). The structural compactness and the tubules were considerably less evident just one week after the temperature shift from 38°C to 46°C or higher (55°C and 64°C). This tendency to a more spongy structure was well manifest by the end of the observation period (4 months) at high temperature, when the granules were found to contain many empty spaces like bubbles or crevices, but the medulla and cortex regions were still well defined (Fig. 4); (5). Concomitantly, the subpopulations of methanogens that had been present at 38°C were gradually replaced by others that became predominant by the end of the observation period. This was manifest not only from the cell counts, but also from the fact that the corresponding colonies increased in number and size. For example, a methanogen antigenically related to *Methanobacterium thermoautotrophicum* ΔH (thermophilic; Balch *et al.*, 1979; Boone and Whitman, 1988), which was virtually undetectable in the granules at 38°C, formed easily visible colonies just one week after the shift up to the thermophilic range (Fig. 5). These colonies reached much larger sizes and were more numerous by the end of the observation period (Macario *et al.*, 1991*b*). Similarly, a methanosarcina antigenically related to *Methanosarcina barkeri* W (which grows optimally at 37°C) was present in the granules before the temperature shift (Fig. 6), but was replaced by larger and more abundant colonies of a methanosarcina antigenically related to the thermophilic species (optimal growth temperature 45°C; Boone and Whitman, 1988) *Methanosarcina thermophila* TM-1 (Macario *et al.*, 1991*b*).

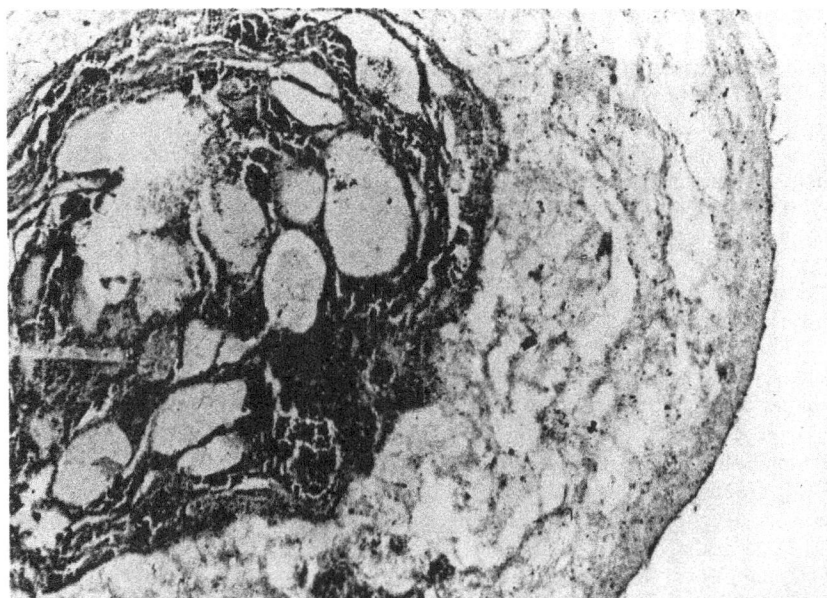

Figure 4. Granule section (Gram stain; X200) showing the medulla and cortex with a lax structure after 4 months at 55°C.

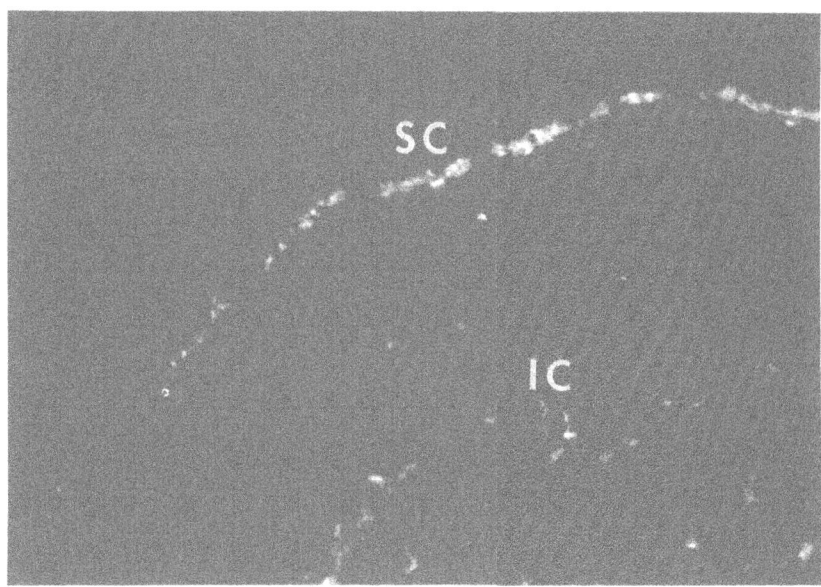

Figure 5. Section of a granule (X1,000) one week after the temperature shift. Shown are an incipient long, thin, surface colony (SC) and a small inner colony (IC) of a methanogen antigenically related to *Methanobacterium thermoautotrophicum* ΔH revealed by the antibody probe for this organism and indirect immunofluorescence.

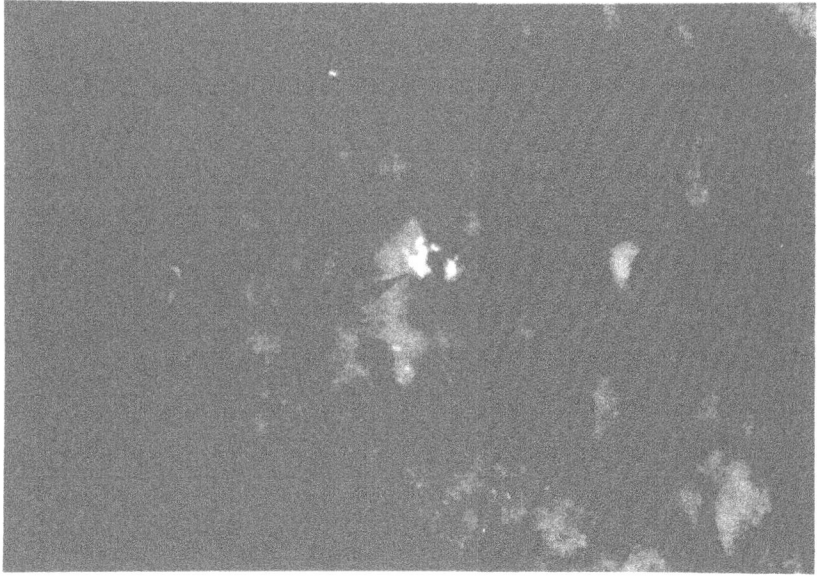

Figure 6. Section of a granule (X1,000) before the temperature shift. Shown is a small colony (arrow) of a methanosarcina antigenically related to *Methanosarcina barkeri* W revealed by the antibody probe for this organism and indirect immunofluorescence in the deep cortex.

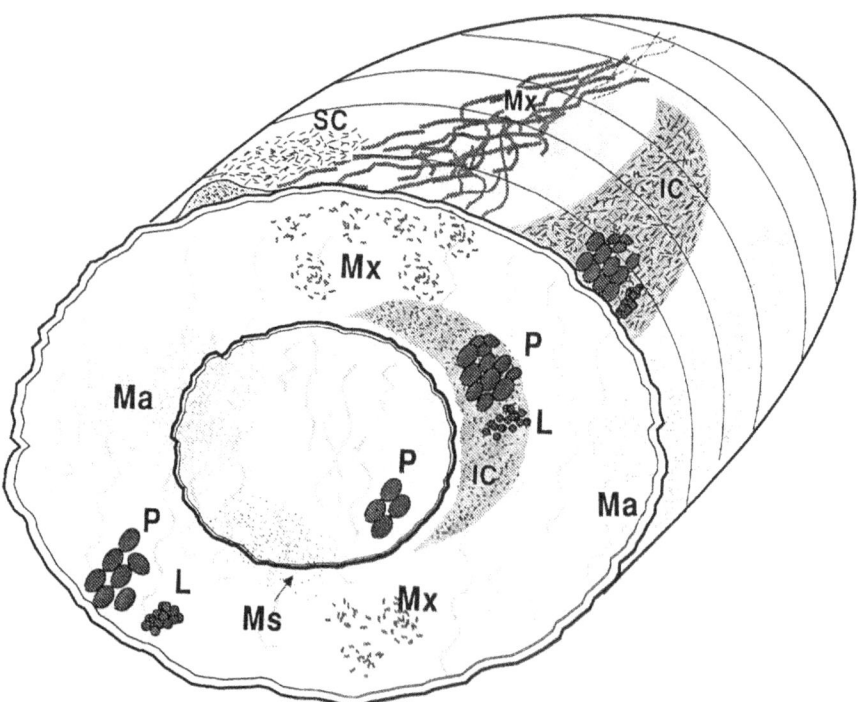

Figure 7. Diagram of an UASB granule showing methanogenic subpopulations, composed using the images provided by thin sections of granules at 55°C for four months. SC and IC, surface and inner colonies, respectively, of *Methanobacterium thermoautotrophicum*; P and L, packets and lamina, respectively, of *Methanosarcina thermophila*; Mx, *Methanothrix soehngenii* rods; Ma, *Methanobrevibacter arboriphilus*; and MS, *Methanobrevibacter smithii*. Reproduced with permission from Macario *et al.*, 1991*b*.

Based on these findings, and after examination of many sections of granules, a diagram of the tridimensional granular structure was generated (Fig. 7). It represents the general architecture of the granules and the location and shape of the methanogens' colonies. Interesting observations are: (1). The intimate association between two thermophilic methanogens, one acetoclastic (*i.e.*, a *Methanosarcina* sp.) and the other a H_2-consumer (*i.e.*, *M. thermoautotrophicum*). This close spatial association could very well be interpreted to represent the microanatomic substratum for biochemical exchanges between the two species; (2). The dual distribution of colonies of the same methanogen in separate locations (*i.e.*, deep inside the granule and on its surface) where the microenvironmental conditions are different. This implies a physiologic plasticity for methanogens, since the same species can thrive in the midst of different surroundings in microbiologic, biochemical, and structural terms; and (3). The occurrence of two distinct forms of the methanosarcina related to *M. thermophila* TM-1, one appearing as the classical globular packets, and the other as the recently described lamina (Mayerhofer *et al.*, 1992). This polymorphism brings to mind disparate physiologic characteristics.

The microanatomic modifications that accompanied the changes of the methanogenic subpopulations support the notion that there is a close functional link in the UASB granules between cells with various specializations, and between cells and supporting structures.

METHODS

The studies described in this review were done using polyclonal antibody probes derived from rabbit antisera raised by immunization with reference strains (Macario and Conway de Macario, 1985a). Methanogens were identified by the antigenic fingerprinting method (Macario and Conway de Macario, 1983) using antibody probes of pre-defined specificity spectra (Macario and Conway de Macario, 1985b). Indirect immunofluorescence and a quantitative slide immunoenzymatic assay (SIA) were applied as described (Macario and Conway de Macario, 1985a,b). Quantification of individual subpopulations was done by direct microscopic counting of microbes identified by indirect immunofluorescence, and by SIA (Macario and Conway de Macario, 1988). To elucidate the microanatomy of granular consortia, standard histologic and histochemical techniques were applied (Macario *et al.*, 1991b, and references therein). The methanogenic subpopulations were mapped within the granules by examining serial histologic sections stained with antibody probes using indirect immunofluorescence (Macario *et al.*, 1991b). Cell suspensions for identification and counting of individual organisms were prepared by gently disassembling of the granules (Visser *et al.*, 1991).

CONCLUDING REMARKS

The combination of histological and histochemical procedures with immunohistochemical techniques using specific antibody probes has advanced our knowledge of the UASB granules and, by implication, that of other granular consortia and biofilms. Identification of methanogens *in situ*, in the consortium of which they are a part, has allowed the physical mapping of the various species detected with regard to each other, and with respect to the supporting structures. This kind of approach will lead to a complete elucidation of the tridimensional structure of important microbial consortia. Methods and probes that can only identify individual cells without providing topographical information have limited potential. One needs to gain insights into the microanatomic features of an ecosystem that are functionally relevant to understand its ecophysiology. For this purpose, the strategies outlined here are helpful. Architectural information is also needed to design consortia, using selected or engineered strains, endowed with the anatomic features necessary for efficient performance.

The data obtained applying histologic and immunologic methods combined, demonstrated that the quali- and quantitative patterns of methanogenic subpopulations (and by inference those of other microbial populations in the same consortium) depend closely on the conditions of the ecosystem (*e.g.*, temperature). Patterns are dynamic. They change with time, particularly after a change in one or more environmental conditions affecting the consortium.

ACKNOWLEDGEMENTS

Work done in our laboratory was supported by grants No. 706-RIER-BEA-85 from NYSERDA and No. 870787 from NATO.

REFERENCES

Balch, W.E., Fox, G.E., Magrum, L.J., Woese, C.R. and Wolfe, R.S. (1979) Methanogens: Reevaluation of a unique biological group. Microbiol. Rev. 43, 260-296.

Boone, D.R. and Whitman, W.B. (1988) Proposal of minimal standards for describing new taxa of methanogenic bacteria. Int. J. Syst. Bacteriol. 38, 212-219.

Colwell, R.R., Brayton, P.R., Grimes, D.J., Roszak, D.B., Huq, S.A. and Palmer, L.M. (1985) Viable but non-culturable *Vibrio cholerae* and related pathogens in the environment: Implications for release of genetically engineered microorganisms. Biotechnology 3, 817-820.

Dahle, A.B. and Laake, M. (1982) Diversity dynamics of marine bacteria studied by immunofluorescent staining on membrane filters. Appl. Environ. Microbiol. 43, 169-176.

Macario, A.J.L. and Conway de Macario, E. (1982) Immunology of methanogens: A new development in microbial biotechnology. Immunology Today 3, 279-284.

Macario, A.J.L. and Conway de Macario, E. (1983) Antigenic fingerprinting of methanogenic bacteria with polyclonal antibody probes. Syst. Appl. Microbiol. 4, 451-458.

Macario, A.J.L. and Conway de Macario, E. (1985a) A preview of the uses of monoclonal antibodies against methanogens in fermentation biotechnology: Significance for public health, in "Monoclonal Antibodies Against Bacteria", (Macario, A.J.L. and Conway de Macario, E., Eds.), Vol. I, pp. 269-286. Academic Press, Orlando, Florida, U.S.A.

Macario, A.J.L. and Conway de Macario, E. (1985b) Monoclonal antibodies of predefined molecular specificity for identification and classification of methanogens and for probing their ecologic niches, in "Monoclonal Antibodies Against Bacteria", (Macario, A.J.L. and Conway de Macario, E., Eds.), Vol. II, pp. 213-247. Academic Press, Orlando, Florida, U.S.A.

Macario, A.J.L. and Conway de Macario, E. (1988) Quantitative immunologic analysis of the methanogenic flora of digestors reveals a considerable diversity. Appl. Environ. Microbiol. 54, 79-86.

Macario, A.J.L. and Conway de Macario, E. (1993) Manipulation and mapping of microbes with antibodies, in "Trends in Microbial Ecology", (Guerrero, R. and Pedrós-Alió, C., Eds.), pp. 505-510. Spanish Society for Microbiology, Barcelona, Spain.

Macario, A.J.L., Conway de Macario, E., Ney, U., Schoberth, S.M. and Sahm, H. (1989a) Shifts in methanogenic subpopulations measured with antibody probes in a fixed-bed anaerobic bioreactor treating sulfite evaporator condensate. Appl. Environ. Microbiol. 55, 1996-2001.

Macario, A.J.L., Earle, J.F.K., Chynoweth, D.P. and Conway de Macario, E. (1989b) Distinctive patterns of methanogenic flora determined with antibody probes in anaerobic digestors of different characteristics operated under controlled conditions. Syst. Appl. Microbiol. 12, 216-222.

170

Macario, A.J.L., Peck, M.W., Conway de Macario, E. and Chynoweth, D.P. (1991a) Unusual methanogenic flora of a wood-fermenting anaerobic bioreactor. J. Appl. Bacteriol. 71, 31-37.

Macario, A.J.L., Visser, F.A., van Lier, J.B., and Conway de Macario, E. (1991b) Topography of methanogenic subpopulations in a microbial consortium adapting to thermophilic conditions. J. Gen. Microbiol. 137, 2179-2189.

Mayerhofer, L.E., Macario, A.J.L. and Conway de Macario, E. (1992) Lamina, a novel multicellular form of *Methanosarcina mazei* S-6. J. Bacteriol. 174, 309-314.

Ørskov, F. and Ørskov, I. (1992) *Escherichia coli* serotyping and disease in man and animals. Can. J. Microbiol. 38, 699-704.

Schmidt, E.L. (1973) Fluorescent antibody techniques for the study of microbial ecology. Bull. Ecol. Res. Comm. (Stockholm) 17, 67-76.

Schmidt, E.L., Bankole, R.O. and Bohlool, B.B. (1968) Fluorescent-antibody approach to study of *Rhizobia* in soil. J. Bacteriol. 95, 1987-1992.

Visser, F.A., van Lier, J.B., Macario, A.J.L. and Conway de Macario, E. (1991) Diversity and population dynamics of methanogenic bacteria in a granular consortium. Appl. Environ. Microbiol. 57, 1728-1734.

Zambon, J.J., Huber, P.S., Meyer, A.E., Slots, J., Fornalik, M.S. and Baier, R.E. (1984) *In situ* identification of bacterial species in marine microfouling films by using an immunofluorescence technique. Appl. Environ. Microbiol. 48, 1214-1220.

CHEMOTAXONOMY AND THE IDENTIFICATION

OF THERMOPHILIC BACTERIA

Milton S. da Costa [1] and M. Fernanda Nobre [2]

[1] Departamento de Bioquímica and [2] Departamento de Zoologia
Universidade de Coimbra
3049 Coimbra Codex
Portugal

INTRODUCTION

Ever since Thomas D. Brock and his co-workers described several landmark thermophiles in the late 1960s and early 1970s, the interest in thermophilic organisms has never ceased. The fascination with these organisms ranges from the evolutionary implications of hyperthermophilic Archaea and Bacteria as ancient lineages, the secondary adaptations of moderate thermophiles to high-temperature environments, the physiology, metabolism and biochemistry, to the obvious biotechnological applications of thermophiles.

The interest in hyperthermophilic Archaea and Bacteria in the last decade has also increased steadily due to our innate need to test the limits of life on Earth and to discover the organisms capable of growth at the highest temperatures. The search for the most thermophilic organisms is somewhat reminiscent of the famous Bone Wars waged between O. C. Marsh and E. D. Cope who, during the latter part of the nineteenth century, competed with each other to describe the largest dinosaurs and the largest number of fossil animals species. Both became famous, not only for their discoveries, but for their memorable feud. These two gentlemen, rivalries aside, were responsible for a quantum jump in the number of fossil animals discovered and made notable contributions to vertebrate paleontology. Neither of them did, however, find the largest dinosaur.

Although the hyperthermophilic Archaea are currently the focus of a large amount of research, there is a continued interest in the "moderately" thermophilic bacteria and new organisms are frequently described and hundreds of strains are isolated every year. Most of these are never reported, but large numbers are screened for enzyme activities or other properties that may be of industrial interest. Even researchers that are only interested in particular properties of bacteria still find it necessary to know what organisms are being studied and have to identify them.

Rapid and reliable methods for identification of bacteria have been devised and new ones are continually being developed and tested. Chemotaxonomic parameters which, among others, include cell wall constituents, polar lipid, respiratory lipoquinones, polyamines, haponoids, fatty acid composition, computer-assisted whole-cell protein sodium dodecyl sulfate polyacrylamide gel electrophoresis (SDS-PAGE) and multilocus enzyme electrophoresis, have provided valuable tools for the identification and taxonomy of bacteria. These methods cannot, however, be considered superior to phenotypic (biochemical) characterization, or genetic-molecular studies. They should be viewed as additional methods available to a polyphasic approach to bacterial taxonomy, as adjuncts for the identification of strains and, in several cases, as the primary methods to identify bacteria rapidly and reliably.

Several chemotaxonomic parameters have been applied successfully to the taxonomy and identification of inumerous mesophilic bacteria and there is, of course, no reason why they cannot be applied to thermophilic bacteria. They have, in fact, been important to the classification of some thermophiles, but have not been extensively used. This review deals with some of the applications of chemotaxonomy to the identification and taxonomy of thermophilic bacteria.

OVERVIEW OF THERMOPHILIC BACTERIA

Several terms, such as, moderate thermophile, thermophile, extreme thermophile, caldoactive, hyperthermophile, have been used to define organisms growing at elevated temperatures. These terms, based on one single physiological parameter, are always difficult to apply to large numbers of different species and perhaps it is best to use Brock's (1986) definition of thermophile. This description states that "a thermophile is an organism capable of living at temperatures at or near the maximum for the taxonomic group of which it is a part". Brock (1986) also defined a thermophilic boundary for the Archaea and the Bacteria of 55°C to 60°C, based on ecological and evolutionary considerations. By this boundary, he stressed the relative rarity of natural environments with temperatures above 55°C-60°C and the fact that the known Eucarya have an upper temperature limit of about 60°C. More recently, Kristjansson and Stetter (1992) adopted the same general definition, although these authors, as a matter of expediency, only list the Bacteria and Archaea with maximum growth temperatures of 65°C or higher.

Intuitively, microbiologists know which organisms are thermophilic and a precise definition cannot be produced that satisfies everyone. Very recent developments have, in fact, shown that exceptions to the established rules are always creeping up. The Thermotogales, for example, were, for several years, the archetypal hyperthermophiles of the Bacteria. This year, however, Davey et al. (1993) described three species belonging to two genera within the Thermotogales with the absurdly low optimum growth temperatures of 45°C to 55°C. Many, perhaps, would not consider these organisms thermophiles, at all.

Almost all bacterial phyla contain thermophilic species and, perhaps, it came as no surprise when a thermophilic spirochete, *Spirochaeta thermophila*, was announced a few years ago (Aksenova et al.,1992). While mesophiles make up the vast majority of the species of bacteria, the most ancient lineages are exclusively or predominantly composed of thermophilic bacteria. Moreover, the most thermophilic bacteria, such as, *Aquifex* spp. and *Thermotoga* spp. belong to the two most ancient lines of descent leading to the conclusion that the ancestral bacteria were thermophiles (Achenbach-Richter et al., 1987; Woese, 1987; Kandler, 1993). It should also come as no surprise that the organisms of these ancient phyla have some of the most unusual chemotaxonomic characteristics of the Bacteria.

It is not the purpose of this review to list the thermophilic bacteria that have been described. Rather, we will concentrate on a few groups, namely the genus *Thermus*, which are actively being studied and to which chemotaxonomic methods can be applied to large number of strains. We will, however, give examples of several thermophilic bacteria for which chemotaxonomic parameters have been described and point out the usefulness of chemotaxonomy to the identification and to the taxonomy of these organisms.

Aquifex-Hydrogenobacter

The microaerophilic, hyperthermophilic bacterium designated *Aquifex pyrophilus*, with an optimum growth temperature of about 85°C, was described recently (Huber *et al.*, 1992). This organism is related to *Hydrogenobacter thermophilus* which has an optimum growth temperature of 72°C-75°C (Kawasumi *et al.*, 1984; Kristjansson *et al.*, 1985) and together represent the earliest lineage of the bacterial phylogenetic tree (Burggraf *et al.*, 1992). Both organisms are strictly chemolithoautotrophic, but do not, otherwise, share many physiological characteristics. With respect to chemotaxonomic parameters that could be useful in confirming this relatedness, very little information is yet available.

In *H. thermophilus* the major lipoquinone is 2-methylthio-3-VI, tetrahydroheptaphenyl-1, 4-naphthoquinone (methioquinone) that is unique to this organism (Ishii *et al.*, 1983), but the lipoquinones of *A. pyrophilus* have not been described. The polar lipid composition of *A. pyrophilus* features an aminophospholipid, a glycolipid and a phospholipid. The main core lipid is an alkylglycerol diether (1,2-di-*O*-alkylglycerol), which is also found in *Thermodesulfobacterium commune* (Langworthy *et al.*, 1983), and an alkylglycerol monoether, rather than the more common glycerolipids of bacteria (Huber *et al.*, 1992). The core lipids of *H. thermophilus* have not, to our knowledge, been described, but *n*-C18:0 and *n*-C20:1 are the main fatty acids of this organism (Ishii *et al.*, 1983). However, the fatty acids of *A. pyrophilus*, which constitute about 10% of the acid methanolysate, were not reported.

A large number of *Hydrogenobacter* spp. have been isolated from worldwide geothermal areas and five groups were detected based on DNA:DNA hybridization studies (Aragno, 1992). Since the physiological characteristics do not discriminate between these groups, chemotaxonomic parameters, particularly lipid analysis and whole-cell protein SDS-PAGE, could be useful to distinguish them.

The Thermotogales

The extremely thermophilic bacteria *Thermotoga maritima* (Huber *et al.*, 1986), *T. neapolitana* (Jannasch *et al.*, 1988), *T. thermarum* (Windberger *et al.*, 1989), *Thermosipho africanus* (Huber *et al.*, 1989), *Fervidobacterium nodosum* (Patel *et al.*, 1985) and *F. islandicum* (Huber *et al.*, 1990) represent the second most deeply branching phylum of the Bacteria. In addition to these organisms, three new species, with growth temperatures of 45°C to 55°C, designated *Geotoga petraea*, *G. subterranea* and *Petrotoga miotherma* from oil field brines, were recently added to the Thermotogales (Davey *et al.*, 1993). These organisms are all strictly anaerobic chemoorganotrophs and share several phenotypic characteristics.

The cell wall of the Thermotogales is sensitive to lysozyme; *meso*-diaminopimelic acid is not detected in the cell wall, but D-glutamate and D- and L-lysine are present. The lipid composition of these organisms constitutes a very interesting and unifying characteristic of this group. A novel ether core lipid, found in all *Thermotoga* spp., has been identified as 15, 16-dimethyl-30-glyceryloxytriacontanoic acid (De Rosa *et al.*, 1989). The fatty acids consist primarily of even-numbered straight-chain molecules (C14:0, C16:0, C18:0). In addition to the monocarboxylic fatty acids, the Thermotogales also possess long chain α, ω-dicarboxylic fatty acids with a single internal 15-methyl branch or vicinal 15, 16-dimethyl branches. The carbon lengths of these dicarboxylic acids varies between C28 and C34. Dicarboxylic acids like these, named diabolic acids, are also found in a mesophilic *Butyrivibrio* sp. (Klein *et al.*, 1979).

Although the respiratory lipoquinones of the Thermotogales have not been reported, the presence of diabolic acids and 15, 16-dimethyl-30-glyceryloxytriacontanoic acid are good taxonomic markers for these extreme thermophiles. The phylogenetic position of the

Butyrivibrio sp. of Klein *et al*. (1979) remains to elucidated; other *Butyrivibrio* spp. belong to the Gram-positive bacteria (Woese, 1987), although they stain Gram-negative. The lipid composition of the two *Geotoga* spp. and *Petrotoga miotherma* was not fully described, except that the main monocarboxylic fatty acids were C16:0, C16:1 and C18:1. Dicarboxylic acids and the core lipids were not investigated. At present, chemotaxonomic parameters, that could relate these newly described species to the other Termotogales, are not available.

The Green Non-Sulfur Bacteria

The green non-sulfur bacterial phylum is composed of the thermophilic bacteria *Chloroflexus aurantiacus* and *Thermomicrobium roseum*, and the mesophilic *Herpetosiphon* spp. (Oyaizu *et al*., 1987; Woese, 1987). The bacteria of this phylum form an ancient line of descent but share few, if any, phenotypic characteristics. The limited amount of information available also indicates that these organisms have few chemotaxonomic characteristics in common.

The phototrophic bacterium *Cloroflexus aurantiacus* has a polar lipid and fatty acid profile common for the Bacteria. The polar lipids of two strains of *C. aurantiacus* consist of phosphatidylglycerol, phosphatidylinositol, two glycolipids and a sulfoquinovosyldiglyceride, and *n*-C16:0 and *n*-C18:0 are the major fatty acids. Fatty alcohols, primarily C18:0 are also present (Knudsen *et al*., 1982). The chemoorganotrophic species *Herpetosiphon (giganteus) aurantiacus* has a fatty acid composition very similar to *C. aurantiacus*; the major fatty acids are C16:0 and C18:1 (Kleinig and Reichenbach, 1977; Reichenbach *et al*., 1978). The major lipoquinones of *C. aurantiacus* are menaquinone 8 (MK-8) and MK-10 (Hale *et al*., 1983) while in *H. aurantiacus* MK-6 is present with minor MK-7.

The species *Thermomicrobium roseum* is represented by one strain isolated in Yellowstone National Park, USA (Jackson *et al*., 1973). This organism is chemoorganotrophic, has an optimum growth temperature of about 70°C, a maximum growth temperature about 80°C and is obligately aerobic. Unlike most bacteria, the cell wall of *T. roseum* is composed of protein and lacks peptydoglycan (Merkel *et al*., 1980). Strains, initially designated *Thermomicrobium fosteri* (ATCC 29033) by Phillips and Perry (1976), have peptydoglycan and are, therefore, not related to *T. roseum*.

Glycerolipids are replaced by long chain 1, 2-diols in *T. roseum*. The major diol is the *n*-C21 diol, 1, 2-eicosonediol; smaller amounts of 13-methyl-1, 2-nonadecanediol (C20) and 1,2-nonadecanediol (C19) are also present. The polar head groups are linked to the terminal hydroxyl group of the diols and several phospholipids and glycolipids can be detected by TLC. The structure of the diols is analogous to the glycerolipids and fatty acids are ester-linked to the hydroxyl group of carbon 2 or amide-linked to the glycolipid polar head group. The major fatty acid of *T. roseum* is the internally-branched 12-methyl-C18:0; 10-methyl-C16:0, 10-methyl-C17:0, 12-methyl-C17:0, 12-methyl-C19:0 and 14-methyl-C20:0; *n*-C16:0, *n*-C18:0, *n*-C19:0 and *n*-C20:0 are also detected in lower relative proportions (Pond *et al*., 1986; Pond and Langworthy, 1987).

Internally-branched fatty acids are not restricted to *T. roseum*; fatty acids of this type are, in fact, common in Gram-positive bacteria of the genera *Brevibacterium*, *Corynebacterium*, *Nocardia*, *Mycobacterium*, *Actinomadura*, *Nocardiopsis*, among others (Fischer *et al*., 1983; Kroppenstedt, 1985), and Proteobacteria (Purple Bacteria), such as, the sulfate-reducing bacteria *Desulfobacter curvatus* and *Desulfobacterium autotrophicum* which contain low amounts of 10-methyl-C16:0 and 10-methyl-C17:0 (Vainshtein *et al*., 1992). In this study, the thermophilic sulfate-reducer *Thermodesulfobacterium commune* and *Desulfovibrio thermophilus* were found to be closely related to each other on the basis of fatty acid analysis, but had a fatty acid composition different from the other mesophilic sulfate-reducing bacteria belonging to the delta subdivision of the Proteobacteria (Woese, 1987). The principal fatty acids of these two organisms are *iso*-C16:0, *iso*-C17:0, *anteiso*-C17:0, *iso*-C18:0 and *n*-C18:0.

Alkylglycerol diethers constitute about 80% of the core lipids of *T. commune* and the remainder appears to be constituted by alkylglycerol monoethers. The monocarboxylic fatty acids are derived from the alkylglycerol monoethers (Langworthy *et al.*, 1983). As mentioned above, these core lipids are also found in *Aquifex pyrophilus*, but a phylogenetic relationship between the two organims has not been established, although *T. commune* is considered to form a deep line of descent of the bacterial phylogenetic tree (Woese, 1987). There is also some confusion concerning the taxonomy of *T. commune*, since this species was never validly described (Zeikus *et al.*, 1983) and the name *T. mobile* has been accepted for *D. thermophilus* (Vainshtein *et al.*, 1992). The major respiratory quinone of *T. commune* and *D. thermophilus* is menaquinone 7 (MK-7) which is also the principal lipoquinone of other unrelated sulfate-reducing bacteria (Collins and Widdel, 1986).

Some Thermophilic Gram-Positive Bacteria

Thermophilic *Bacillus* species are numerous and have been isolated from a large variety of thermal and non-thermal environments (Sharp *et al.*, 1992). These rod-shaped organisms share many biochemical phenotypic and morphological characteristics, paramount among which is the ability to form endospores. This characteristic is taken as an unifying characteristic in the description of this genus. Recent studies show that the formation of endospores is not always easily induced and that their presence does not necessarily signify that the organisms should be placed in this genus (Stackebrandt *et al.*, 1987; Rössler *et al.*, 1991). The identification of these organisms can be achieved with a large degree of success by the determination of the respiratory quinones, polar lipid patterns and the fatty acid composition. Some of these parameters are also very useful to the taxonomy of these organisms.

All bacteria of the genus *Bacillus* possess menaquinone 7 (MK-7) as the major respiratory quinone (Collins and Jones, 1981). Phosphatidylethanolamine (PE), phosphatidylglycerol (PG) and diphosphatidylglycerol (DPG) are found in most species and are easily detected by two-dimensional thin-layer chromatography with the utilization of specific spray reagents. AminoacylPG, lysoPE, glucosaminylPG, phosphatidic acid and glycolipids are often encountered in small quantities. The major fatty acids of some thermophilic species of *Bacillus*, for example, *B. stearothermophilus*, *B. flavothermus*, *B. thermoruber*, "*B. caldotenax*", "*B. caldovelox*", "*B. caldolyticus*" and *B. thermocloacae* are *iso*-C15:0, *iso*-C16:0, *iso*-C17:0 and *anteiso*-C17:0; straight-chain fatty acids are also found, generally, in lower amounts (O'Leary and Wilkinson, 1988; Demharter and Hensel, 1989). Unfortunately, to our knowledge, no study has been performed on a large number of strains, under standardized conditions, to complement the taxonomy of thermophilic *Bacillus* spp. or to use fatty acid analysis as a means to aid the identification of new isolates.

Three thermoacidophilic species previously classified in he genus *Bacillus* as *B. acidocaldarius* (Darland and Brock, 1971), *B. acidoterrestris* (Deinhard *et al*, 1987*a*) and *B. cycloheptanicus* (Deinhard *et al.*, 1987*b*) have a lipid composition which sets them appart from the other species of *Bacillus*. These three thermoacidophilic species possess ω-alicyclic fatty acids with a cyclohexyl or cycloheptyl ring at the end of the hydrocarbon chain. The presence of ω-alicyclic acids in these thermoacidophilic bacteria also suggests a taxonomic relationship between them. Based on the fatty acid composition, numerical taxonomic studies (Priest *et al.*, 1988) and 16S rRNA sequence analysis Wisotzkey *et al.* (1992) proposed the creation of the genus *Alicyclobacillus* to accommodate these three species. The predominant fatty acids of *A. acidocaldarius* and *A. acidoterrestris* are the ω-cyclohexyl acids, ω-cyclohexylundecanoic acid (C17) and ω-cyclohexyltridecanoic acid (C19) which reach between 50% and 90% of the total fatty acids (De Rosa *et al.*, 1972; Langworthy and Pond, 1986; Deinhard *et al.*, 1987*a*). The third species, *A. cycloheptanicus*, has ω-cycloheptyl fatty acids namely, ω-cycloheptylundecanoic acid (C18) and ω-cycloheptyltridecanoic acid (C20) comprising about 90% of the total fatty acids (Deinhard *et al.*, 1987*b*). The remainder of the fatty acids are *iso*- and *anteiso*-

branched. All three species have a sulfonolipid and, like all species of *Bacillus*, have MK-7 as the major respiratory quinone. The aberrant fatty acid composition of *A. acidocaldarius* was, initially, attributed to an adaptation to the extreme thermoacidophilic environment inhabited by this species. In the case of *A. acidocaldarius*, *A. acidoterrestris*, *A. cycloheptanicus* and other organisms, as well, there may be a relationship between extreme environmental conditions and the appearence of certain rare fatty acids (Kannenberg *et al.*, 1984; Langworthy and Pond, 1986). However, cyclohexyl fatty acids are also found in the Gram-positive, non-acidophilic, mesophile *Curtobacterium pusillum* where they reach about 80% of the total fatty acids (Suzuki *et al.*, 1981) and in *Mycobacterium* spp. and *Nocardia* spp. grown on *n*-alkyl-substituted cycloparaffins (Beam and Perry, 1974). The relationship of *Curtobacterium pusillum*, which contains MK-9, to the genus *Alicyclobacillus* is not known at present.

Several species of thermophilic clostridia have been described (Wiegel, 1992) of which, *Clostridium thermocellum* is probably the best studied in terms of useful chemotaxonomic parameters. All clostridia have menaquinone 7 (MK-7) as the major lipoquinone (Collins and Jones, 1981). *Iso-* and *anteiso-*fatty acids are the predominant fatty acids in *C. thermocellum* (Herrero *et al.*, 1982), *C. thermosulfurogines* and *C. thermohydrosulfuricum* (Langworthy and Pond, 1986). Fatty aldehydes and 1-*O*-alkylglycerols are also present and the hydrocarbon chains are also branched-chains as in the monocarboxylic fatty acids. Dicarboxylic fatty acids are, however, also found in *C. thermosulfurogines* and *C. thermohydrosulfuricum*. The main dicarboxylic acid in these species has 30 carbons with two internal methyl branches that are not vicinal. These dicarboxylic acids appear to be made up of two condensed *iso*-C15 fatty acids (Langworthy and Pond, 1986) and are, therefore, structurally different from the dicarboxylic acids of the Thermotogales. A phylogenetic relationship between these thermophilic clostridia and the Thermotogales need not be invoked on the basis of fatty acid composition. Recently, Lee *et al.* (1993) proposed the transfer of these two *Clostridium* spp. to the Gram-positive bacteria of the genera *Thermoanaerobacterium* and *Thermoanaerobacter*, respectively. It should be recalled that endospores were recently described in *Thermoanaerobacter* (*Thermoanaerobium*) *brockii* (Cook *et al.*, 1991) and that this genus is, probably, closely related to the genus *Clostridium*. The presence of C30 dicarboxylic acids in other species of *Thermoanaerobacter* would be of taxonomic significance and aid in their identification.

Rhodothermus marinus

The first strains of the halo-thermophilic species *Rhodothermus marinus* were isolated and described by Alfredsson *et al.* (1988) from marine hot springs in the northern coast of Iceland. A few years later, strains, similar to *R. marinus*, were isolated from hot springs on a sandy beach on the Island of S. Miguel in the Azores (Nunes *et al.*, 1992*b*). All organisms are orange-pigmented, have an optimum growth temperature of about 65°C and a maximum growth temperature of about 77°C. These organisms are slightly halophilic growing in media containing 0.5% NaCl to 6% or 7% NaCl.

The major respiratory lipoquinone is menaquinone 7 (MK-7). The polar lipids are constituted primarily by phospholipids, but small amounts of glycolipids are also detected. The major phospholipids identified are phosphatidylethanolamine (PE), diphosphatidyl-glycerol (DPG), and phosphatidylglycerol (PG). A major unidentified phospholipid, migrating close to the origin on two-dimensional TLC, is also present in all strains (Tindall, 1991; Nunes *et al.*, 1992*a*).

The fatty acids are predominantly branched-chain. Nunes *et al.* (1992*a*) found that, in the strains from Iceland and the Azores, the *iso*-odd (*iso*-C15:0, *iso*-C17:0) and the *anteiso*-odd (*anteiso*-C15:0, *anteiso*-C17:0) were the predominant fatty acids, while Tindall (1991) found that the *iso*-even (*iso*-C14:0, *iso*-C16:0, *iso*-C18:0) were the major fatty acids in the Icelandic strains. The discrepancies encountered in the two original studies resulted primarily from the medium composition; while Tindall (1991) used a medium containing yeast extract and glutamate, Nunes *et al.* (1992*a*) used a medium containing yeast extract and tryptone. In a joint study it was found that the incorporation of glutamate into the growth medium led to a large increase in *iso*-C16:0 and a concomitant decrease in *iso*-C15:0 and *iso*-C17:0. The most startling results occurred when glutamate was used as the sole source of carbon and nitrogen; under these conditions, the major fatty acid was *n*-C16:0. In this medium, the normal-chain fatty acids reached about 50% of the total fatty acids, while, in the other media, the normal-chain fatty acids never exceeded about 6% (Chung *et al.*, 1993).

THE GENUS *THERMUS*

Taxonomy of the Genus *Thermus*

Bacteria of the genus *Thermus* were originally isolated from hot springs at Yellowstone National Park, USA, by Brock and Freeze (1969), who designated the type species of the genus *T. aquaticus*. Since then, strains of this genus have been isolated from numerous natural geothermal and artificial thermal environments throughout the world (Table 1). Most isolates have been described from Yellowstone National Park (Brock and Freeze, 1969; Munster *et al.*, 1986), terrestrial hot springs in Iceland (Pask-Hughes and Williams, 1977; Kristjansson and Alfredsson, 1983; Kristjansson *et al.*, 1986; Hudson *et al.*, 1987*a*; Kristjansson *et al.*, 1994), New Zealand (Hudson *et al.*, 1986, 1987*b*, 1989), and Portugal (Prado *et al.*, 1988; Santos *et al.*, 1989; Manaia and da Costa, 1991). A few strains are also available from Japan (Saiki *et al.*, 1972; Oshima and Imahori, 1974; Taguchi *et al.*, 1982).

The vast majority of the strains are isolated from terrestrial hydrothermal areas, but a few *Thermus* isolates have originated from shallow marine hot springs off the coast of Iceland, the Azores and Fiji (Kristjansson *et al.*, 1986; Hudson *et al.*, 1989; Manaia and da Costa, 1991). In addition to natural hydrothermal areas, *Thermus* spp. have been isolated from hot tap water, domestic and industrial hot water systems (Brock and Boylen, 1973; Pask-Hughes and Williams, 1975) and thermally polluted streams (Ramaley and Hixson, 1970; Degryse *et al.*, 1978).

Three species of the genus *Thermus* have been validly described. The species designated *T. ruber* (Loginova *et al.*, 1984) has an optimum growth temperature of about 60°C and a maximum growth temperature just below 70°C and all known strains are red-pigmented. This species was originally described from isolates from the Kamchatka Peninsula in Russia, but other strains have been isolated from natural and artificial thermal environments throughout the world (Hensel *et al.*, 1986, 1988; Sharp and Williams, 1988; Sharp *et al.*, unpublished results).

In spite of the large number of high-temperature and generally yellow-pigmented *Thermus* strains with optimum growth temperatures of 70°C to 75°C and maximum growth temperatures slightly below 80°C, only two species have been validly described (Williams and da Costa, 1992). These are *T. aquaticus* (Brock and Freeze, 1969), described from strains isolated in Yellowstone National Park, USA, and *T. filiformis* (Hudson *et al.*, 1987*b*) based on strain Wai33 A1 (ATCC 43280; DSM 4687) from New Zealand. In contrast to all other *Thermus* strains described, which form rod-shaped cells of various lengths and short filaments on *Thermus* medium (Brock, 1978) and medium 162 of Degryse *et al.* (1978), *T. filiformis* forms long intertwining filaments and rod-shaped cells are never seen (Hudson *et al.*, 1987*b*). However, according to Georganta *et al.* (1993), all non-filamentous strains from New Zealand, subjected to DNA:DNA hybridization studies, also belong to *T. filiformis*.

Table 1. Species of the genus *Thermus*

Species and Type Strains	Other Strains in Species	Growth Temperature		
		Minimum	Optimum	Maximum
T. aquaticus YT-1 ATCC 25104ᵗ, DSM 625	Strains in groups 1a and 1b (Munster *et al.*, 1986)	40°C	70°C	78°C
T. filiformis Wai33 A1 ATCC 43280ᵗ, DSM 4687	Other strains from New Zealand (Georganta *et al.*, 1993)	37°C	73°C	80°C
"*T. thermophilus*" HB-8 ATCC 27634, DSM 579	AT-62; GK-24; B; RQ-1 (Williams, 1989)	40°C	73°C	85°C
"*T. brockianus*" YS 38 NCIMB 12676	Strains in cluster B (Munster *et al.*, 1986)	40°C	70°C	78°C
T. scotoductus SE-1	Isolates from Iceland and X-1 (Kristjansson *et al.*, 1994)	42°C	65-70°C	75°C
T. ruber BKMB-1258 ATCC 35948ᵗ, DSM 1279	Red-pigmented strains worldwide (Sharp *et al.*, unpublished)	37°C	60°C	69°C
"*T. ruber*"- like Vi-R1	Red-pigmented strains Vizela, Portugal (Tenreiro *et al.*, unpublished)	-	50°C	60°C

The high-temperature species, designated "*T. thermophilus*" (Oshima and Imahori, 1974) was validly described on the basis of strain HB-8, but was considered a synonym of *T. aquaticus* and was not included in the Approved Lists of Bacterial Names (Skerman *et al.*, 1980). It is, however, clear from phenotypic characteristics, such as, the ability to grow at temperatures up to about 85°C, growth in media with more than 2% NaCl (Santos *et al.*, 1989; Manaia and da Costa, 1991) and DNA:DNA hybridization studies (Williams, 1989) that several strains including HB-8, B, RQ-1, AT-62 and GK-24 belong to "*T. thermophilus*". Another species, based on several isolates from Yellowstone National Park, USA, (Munster *et al.*, 1986), also clearly distinct from *T. aquaticus* on the basis of biochemical tests and DNA homology studies, has tentatively been named "*T. brockianus*", but has not been validly described at this time (Williams, 1989).

Many non-pigmented high-temperature strains have been isolated worldwide (Williams and da Costa, 1992). The vast majority of these organisms originate from artificial thermal environments without sunlight. All of the strains from Vizela, Portugal, isolated over a three year period, are non-pigmented. This spring is located at the end of a 450 m tunnel that is kept dark, except for periodic repairs.

Several non-pigmented strains, designated *T. scotoductus*, with optimum growth temperatures of 65°C to 70°C were isolated from one geothermal hot water system in Selfoss, Iceland, over a period of several years (Kristjansson *et al.*, 1994). These strains produce a water-soluble brown pigment on medium 162 and were found to share a high DNA homology with the non-pigmented X-1 strain of Ramaley and Hixson (1970).

One of the major problems with the taxonomy of *Thermus* species is related to the supposedly high phenotypic diversity of these organisms. This diversity may, however, be due to different techniques used for accessing phenotypic characteristics, difficulties in analyzing the results or a true inherent variability of the organisms. The difficulty in accessing taxonomic relationships on the basis of the numerical taxonomy of phenotypic characteristics can be exemplified by the extreme diversity of phenotypes of a large number of *T. ruber* strains from worldwide sources which, nevertheless, share very high DNA homologies and constitute a homogeneous genomic species (Sharp and Williams, 1988; Sharp *et al.*, unpublished results).

Chemotaxonomy of the Genus *Thermus*

Due to the great diversity of phenotypic characteristics of these organisms it is, in fact, absolutely necessary to find other, more reliable, methods for the identification of the species of *Thermus*. The analysis of the polar lipids, respiratory lipoquinones, fatty acids, cell wall components and computer-assisted whole protein SDS-PAGE can be useful, both as a means to aid the taxonomy of the strains of *Thermus* and as rapid methods for the identification of many of the isolates

The peptidoglycan of all *Thermus* strains examined contains ornithine. This amino acid is also found in the murein of the mesophilic bacteria of the genus *Deinococcus* (Pask-Hughes and Williams, 1978; Hensel *et al.*, 1986; Embley *et al.*, 1987). The primary structure of the peptidoglycan of *T. ruber* is, in fact, identical to the peptidoglycan of strains of the genus *Deinococcus* (Schleifer and Kandler, 1972; Hensel *et al.*, 1986). The presence of ornithine in the peptidoglycan of the bacteria of the genera *Thermus* and *Deinococcus*, in addition to menaquinone 8 (MK-8), reinforces the results of 16S rRNA sequencing showing that these two genera are related to each other and form a separate phylum within the Bacteria (Hensel *et al.*, 1986; Weisburg *et al.*, 1989).

All *Thermus* strains, both low-temperature and high-temperature strains, have a very characteristic polar lipid pattern on single- and two-dimensional thin-layer chromatography (TLC). The polar lipid patterns, together with the presence of MK-8, provide an excellent and simple method for the identification of *Thermus* strains among large numbers of isolates of aerobic chemoorganotrophic thermophilic bacteria.

A major glycolipid (GL1) and a major phospholipid (PL2) are present in all high-temperature *Thermus* strains studied (Pask-Hughes and Shaw, 1982; Prado *et al.*, 1988; Donato *et al.*, 1990). At growth temperatures between 70°C and 75°C, two minor phospholipids and two minor glycolipids are also present in most strains. Other minor glycolipids and phospholipids are also consistently visualized in some strains and could serve as a means for their identification (Donato *et al.*, 1990; da Costa, 1994). There is, however, the chance that the minor polar lipids can be missed on TLC plates and therefore do not constitute a reliable method of identification of specific strains. There is no evidence that these minor polar lipids have taxonomic significance (Figure 1).

The major phospholipid has a constant migration on single-dimension TLC in all *Thermus* strains, while the major glycolipid exhibits small, but reproducible, differences in migration in some of the high-temperature strains. The different migration of the major glycolipid of some strains may be due to the different composition of the polar head group and could be of taxonomic significance (B. J. Tindall, personal communication). Nevertheless, several strains belonging to *Thermus aquaticus*, on the basis of DNA:DNA hybridization results, phenotypic properties and fatty acid analysis, have major glycolipids with different rates of migration on TLC.

Recently, after separating two colony types of one high-temperature *Thermus* strain, we observed that one of the organisms lacked GL1. Instead, the major glycolipid was replaced by very large amounts of GL2 which, in all strains examined, is a minor component of the polar lipid fraction (Prado *et al.*, 1988; Tenreiro *et al*, unpublished results). This organism is either, a spontaneous mutant or the original strain consists of a mixed culture. The fatty acid profiles of the two organisms are identical. At present there is no conclusive evidence that differences in glycolipid composition are taxonomically discriminating

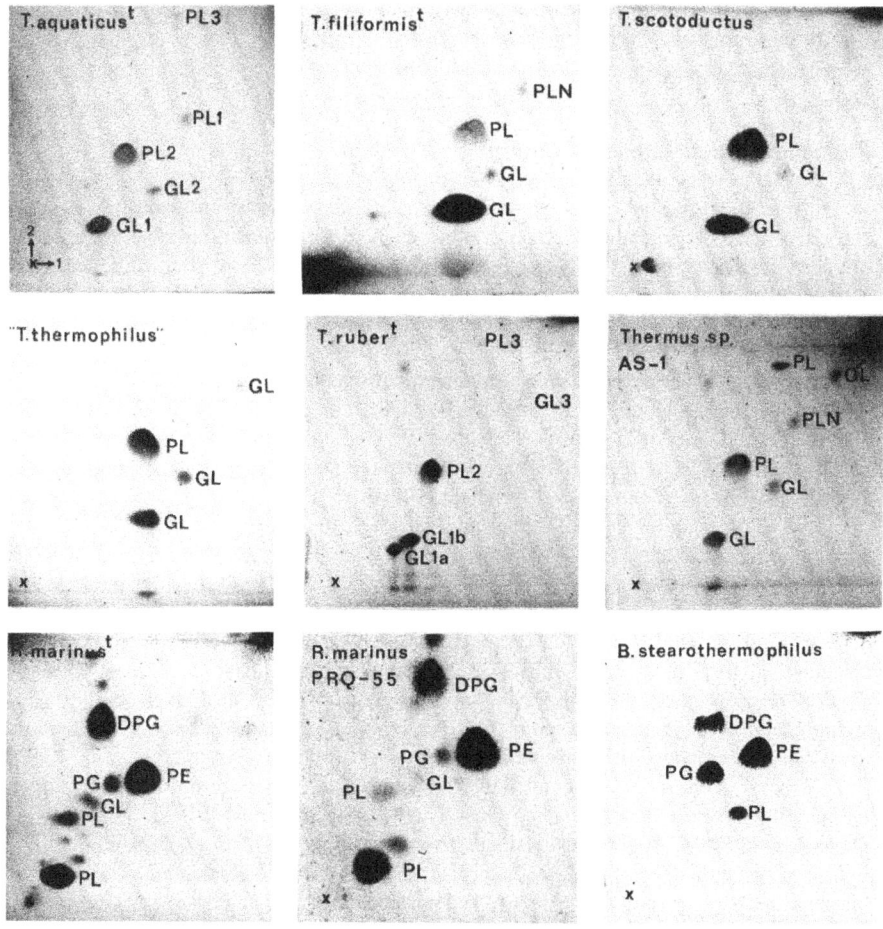

Figure 1. Two dimensional thin-layer chromatography of the polar lipids from *Thermus* spp., *Rhodothermus* sp. and *Bacillus stearothermophilus*.

Methods as in Donato *et al.*, (1990).

Abbreviations: GL, glycolipid; GL1, glycolipid 1; GL1a, glycolipid 1a; GL1b, glycolipid 1b; GL2, glycolipid 2; GLN, aminoglycolipid; PL, phospholipid; PL2, phospholipid 2; PL3, phospholipid 3; PLN, aminophospholipid; DPG, diphosphatidylglycerol; PE, phosphatidylethanolamine; PG, phosphatidylglycerol.

The structure of the major glycolipid (GL1) was first determined by Oshima and Yamakawa (1974) for "*T. thermophilus*" HB-8. The major glycolipids of other *Thermus* strains have now been studied and, while their precise structure is not known, they can be described as diglycosyl-(N-acyl)glycosaminyl-glucosyldiacylglycerols. One residue of glucose is found in the major glycolipid of all strains examined, but the two additional hexoses can be composed of two glucose residues, one glucose and one galactose residue or two galactose residues. With the exception of *T. aquaticus* YT-1, where galactosamine is present, all other high-temperature *Thermus* strains examined have glucosamine (Pask-Hughes and Shaw, 1982; Prado *et al.*, 1988; da Costa, 1994).

The Icelandic strains of *T. scotoductus* were initially believed to represent a new genus and species of thermophilic bacteria and were designated "*Scotothermus ductus*" (Kristjansson and Alfredsson, 1992). On the basis of the polar lipid pattern alone, it was imediately clear that these organisms belonged to the genus *Thermus* or were closely related to *Thermus*. All Icelandic strains had the canonical polar lipid pattern composed of one major glycolipid and one major phospholipid along with a few minor glycolipids and phospholipids. Moreover, the major respiratory lipoquinone was MK-8 (Tenreiro *et al.*, unpublished results).

The low-temperature strains related to *T. ruber* have a polar lipid pattern which is very similar to the high-temperature strains of the genus *Thermus*, except that there are two prominent glycolipids (GL1a and GL1b) that migrate at the position of the major glycolipid (GL1) of the high-temperature strains. In a study of *T. ruber* strains from worldwide sources, all organisms exhibited this pattern. There are, however, easily discernable differences in the relative concentration of each of the glycolipids following TLC which have been quantified in the type strain of *T. ruber* and the patent strain designated "*T. rubens*" (Donato *et al.*, 1991).

Several red-pigmented strains, with an optimum growth of about 50°C and a maximum growth temperature of about 60°C, were isolated from one hot spring in Vizela, Portugal, over a two year period. Yellow-pigmented low-temperature strains, incapable of growth at temperatures above about 60°C to 65°C, were also isolated recently from another hot spring in Portugal. All these organisms have MK-8 and exhibit a polar lipid pattern identical to the *T. ruber* strains. These organisms, therefore, appear to be closely related to *T. ruber*, but based on fatty acid composition, may represent new species within the *T. ruber* complex (Tenreiro *et al.*, unpublished results).

Branched-chain *iso-* and *anteiso*-fatty acids represent the major acyl chains in all *Thermus* strains. It was initially shown that *iso*-C17:0 was the predominant fatty acid in a small number of strains that included *T. aquaticus* YT-1, "*T. thermophilus*" HB-8 and "*T. flavus*" AT-62 followed by lower relative proportions of *iso*-C15:0. The accompanying *anteiso*-odd fatty acids were found in smaller relative proportions, while *iso*-C16:0 and *n*-C16:0 were also present, sometimes in large relative proportions (Ray *et al.*, 1971; Oshima and Yamakawa, 1974; Pask-Hughes and Shaw, 1982). With the availability of large numbers of high-temperature and low-temperature strains, it is becoming increasingly clear that there is a large diversity in the fatty acid profiles which may be related to the taxonomy of the genus *Thermus* and can, in many cases, be used to identify specific strains.

One of the most interesting examples of a strain specific fatty acid profile is found in the type strain of *T. filiformis* Wai33 A1. Unlike all other *Thermus* strains examined, this organism has very low relative proportions of *iso*-C15:0 and *iso*-C17:0 and very large relative proportions of the corresponding *anteiso*-fatty acids (Table 2). Of all the high-temperature strains studied, *anteiso*-C17:0-3OH was only found in the type strain of *T. filiformis* where it reaches about 7% of the fatty acids. Recently, several non-filamentous *Thermus* strains from New Zealand were classified as *T. filiformis* on the basis of DNA:DNA hybridization studies (Georganta *et al.*, 1993). However, on the basis of the fatty acid composition, these strains, with high *iso*-fatty acids and low *anteiso*-fatty acids, could be considered to be unrelated to the type strain of *T. filiformis* (Ferraz *et al.*, unpublished results). It should, nevertheless, be taken into consideration that the abnormal fatty acid composition of the type strain of *T. filiformis* may be related to the filamentous morphology and the extra wall layer, external to the outer layer found in other *Thermus* species (Hudson *et al.*, 1987*b*). Filamentous cultures of several non-filamentous *Thermus* strains were obtained by growing the organisms on the normal *Thermus* medium supplemented with several D-amino acids and glycine (Janssen *et al.* 1991). The filamentous morphology of these strains was only stable as long as the amino acid was included in the medium; upon growth in the medium without the specific amino acid, the organisms would revert to the rod-shaped morphology. The effect of changes in the growth

Table 2. Fatty acid composition (% of total) of *T. aquaticus* and *Thermus* strains from New Zealand, grown at 70°C

Strains								Fatty Acid Composition								
	i-C13	i-C14	n-C14	i-C15	a-C15	n-C15	i-C16	n-C16	i-C15 3-OH	i-C17	a-C17	n-C17	i-C16 3-OH	n-C16 3-OH	i-C17 3-OH	a-C17 3-OH
T. aquaticus	-	0.9	1.3	23.4	2.0	-	14.5	15.7	2.1	27.6	2.5	-	1.8	1.1	5.2	-
T. filiformis	-	1.4	-	4.6	23.3	-	9.2	3.9	-	6.2	37.2	-	0.6	-	1.7	7.2
HS7 A1	0.7	-	-	49.0	3.7	4.3	1.5	7.1	-	29.5	1.9	1.7	-	-	-	-
Ork A2	1.4	2.8	-	38.9	6.4	0.6	13.4	5.7	-	25.7	4.3	-	-	-	-	-
Ork2 A3	-	0.9	-	32.2	2.6	0.6	15.6	6.9	-	35.9	3.9	-	-	-	-	-
Rot8 A1	0.8	2.1	-	40.4	12.8	-	10.7	6.4	-	18.7	7.3	-	-	-	-	-
Rot34 A5	1.2	1.3	-	42.3	6.9	-	7.6	8.4	-	28.0	4.3	-	-	-	-	-
Rt4 A2	0.6	1.5	-	40.7	5.2	0.8	11.3	5.5	0.7	28.4	4.3	-	-	-	-	-
T351	0.6	0.9	-	35.6	5.8	0.7	9.4	6.7	-	33.5	6.1	-	-	-	-	-
Tau8 A1	0.8	0.8	-	33.8	3.3	-	9.5	6.7	-	38.9	4.8	-	-	-	-	-
Tok4 A2	0.6	0.8	-	39.6	5.4	0.5	7.9	8.5	-	31.6	4.9	-	-	-	-	-
Tok5 A2	1.1	1.4	-	48.9	9.8	4.3	4.7	5.5	-	19.9	4.4	-	-	-	-	-
Tok34 A4	1.3	1.8	0.6	41.5	9.2	-	9.2	9.1	-	22.1	5.2	-	-	-	-	-
Wai2 A1	-	1.0	-	39.0	6.4	-	8.2	5.9	-	32.6	5.4	-	-	-	-	-
Wam1 A7	0.9	0.8	-	43.1	7.9	3.9	3.7	8.6	0.5	24.2	4.3	1.0	-	-	-	-
Wam8 A1	-	0.9	-	34.5	7.1	-	10.0	7.4	-	29.9	8.2	-	-	-	-	-

Abbreviations for fatty acids: *i*, iso-branched; *a*, anteiso-branched; *n*, straight chain; 3-OH, hydroxyl group at position 3.
Fatty acid methyl esters (FAMES) were obtained by saponification and methylation as described by Kuykendall *et al.*, (1988). FAMES were separated using a Tracor model gas-chromatograph fitted with a 5% phenyl-methyl-silicone capillary column (0.2 mm x 25 m), a flame ionization detector (300°C), column split ratio, 100:1, column temperature 170°C to 270°C at 5°C/min, injection port temperature 300°C.

medium remains to be investigated. The fatty acid profile of this strain may, therefore, not be taxonomically discriminating.

In addition to *T. aquaticus*, a second species from Yellowstone National Park, USA, tentatively named "*T. brockianus*", has been identified on the basis of phenotypic characteristics (Munster *et al.*, 1986) and DNA homology studies (Williams, 1989). The analysis of the fatty acids confirms the existence of these two species. In both species the *iso*-odd fatty acids predominate over the corresponding *anteiso*-odd fatty acids, which are present in low relative proportions, and the relative proportion of *iso*-C17:0 is always higher than *iso*-C15:0. All strains that were included in the species *T. aquaticus* by Williams (1989) on the basis of DNA homology studies have small amounts of 3-hydroxy fatty acids, namely *iso*-C15:0-3OH, *iso*-C16:0-3OH, *n*-C16-3OH and *iso*-C17-3OH, the last of which attains about 5% to 6% of the total fatty acids. In contrast to the *T. aquaticus* strains, hydroxy fatty acids were vestigial or absent in the "*T. brockianus*" strains (Table 3).

The Icelandic strains of *T. scotoductus* have a fatty acid profile which distinguishes them from all other high-temperature strains examined. In these organisms the *iso*-odd and *anteiso*-odd fatty acids have practically the same relative proportions and *iso*-C16:0 is a minor component. On the basis of the fatty acid composition, these strains are distinct from five other yellow-pigmented Icelandic strains studied, all of which have high relative proportions of *iso*-C15:0 and *iso*-C17:0 and the corresponding *anteiso*-fatty acids are minor components. The *Thermus* X-1 strain, included in the species *T. scotoductus*, has a fatty acid profile which is, however, also different from the Icelandic strains attributed to the same species (Table 3). Since this strain shares high DNA sequence homology with the strains from Iceland, it can only be concluded that the different fatty acid composition of strain X-1 is due to geographical separation of the organisms. As in the case of the *T. filiformis* strains, fatty acid composition may not be appropriate to confirm some of the species of *Thermus*.

All *T. ruber* isolates examined to date have a very similar fatty acid profile. In these organisms, irrespective of the site of isolation and at the optimum growth temperature, *iso*-C15:0 is always the most abundant fatty acid followed by *iso*-C17:0. The *anteiso*-odd fatty acids are minor components, frequently exceeded in concentration by *iso*-C16:0 and/or *n*-C16:0 (Hensel *et al.*, 1986; Donato *et al.*, 1991; Nordström and Laakso, 1992). At lower growth temperatures, Nordström and Laakso (1992) found that *anteiso*-C17:1 became a major component of the fatty acids of two *T. ruber* strains from Iceland.

When the fatty acid composition of the red-pigmented low-temperature *Thermus* strains from Vizela, Portugal, was compared to several *T. ruber* strains grown at 55°C, there were considerable differences in the fatty acid profiles. In the Vizela organisms, the *iso*-C15:0 fatty acid and the *anteiso*-C15:0 fatty acid have similar relative proportions and the *anteiso*-C17:0 fatty acid of the Vizela organisms is always higher than the corresponding *anteiso*-fatty acid of the *T. ruber* strains. There are several minor fatty acids that are present in one group or the other, but not in both. These results, in conjunction with the very large differences in the cardinal temperatures, indicate the existence of another species closely related to *T. ruber*.

The loss of pigmentation, the inability to grow at temperatures above 70°C and very large alterations in the fatty acid composition of five high-temperature *Thermus* strains were attributed to the elimination of plasmids with ethidium bromide (Virtala *et al.*, 1993). These results pose an intriguing problem for the utilization of the fatty acid composition in chemotaxonomy of *Thermus* strains since, naturally occuring plasmid-less strains (presumably, non-pigmented) would be classified differently from plasmid-containing isolates. Non-pigmented *Thermus* strains are common (Williams and da Costa, 1992) and the new species *T. scotoductus* is based on non-pigmented strains. Some non-pigmented strains, however, have plasmids (Moreira *et al.*, unpublished results) and plasmids have not been isolated from some yellow-pigmented strains (Kristjansson, personal communication; Munster *et al.*, 1985). If the results of Virtala *et al.* (1993) are confirmed, caution will have to be

Table 3. Fatty acid composition (% of total) of *T. aquaticus*, "*T. thermophilus*", "*T. brockianus*" and *T. scotoductus* strains, grown at 70°C

Fatty Acid Composition

Strains	i-C13	Unl	i-C14	n-C14	i-C15	a-C15	n-C15	i-C16	n-C16	i-C15 3-OH	i-C17	a-C17	n-C17	i-C16 3-OH	n-C16 3-OH	i-C18	i-C17 3-OH
T. aquaticus																	
YT-1 ATCC 25104[t]	-	-	0.9	1.3	23.4	2.0	-	14.5	15.7	2.1	27.6	2.5	-	1.8	1.1	-	5.2
15028	-	1.7	1.0	-	30.5	1.7	0.8	13.3	7.3	2.3	32.9	1.9	-	0.6	-	-	5.6
"*T. thermophilus*"																	
HB-8 ATCC 27634	-	-	1.1	-	40.9	5.1	-	6.9	6.2	-	35.1	3.9	-	-	-	-	-
"*T. brockianus*"																	
15011	0.8	-	2.1	-	38.0	3.1	-	13.8	8.6	-	30.0	2.4	-	-	-	-	-
15038	0.8	-	1.6	-	38.6	2.6	0.5	11.2	8.6	-	32.9	2.4	-	-	-	-	-
T. scotoductus																	
X-1 ATCC 27978	-	-	1.7	-	28.6	5.3	0.8	17.6	7.2	-	30.4	5.3	-	-	-	1.0	-
ITI-153	-	-	-	-	18.6	19.0	1.5	2.2	6.3	-	25.6	24.6	1.1	-	-	-	-
ITI-154	-	-	-	-	20.5	21.4	1.3	1.6	4.8	-	24.8	24.3	0.8	-	-	-	-

Abbreviations for fatty acids: *i*, *iso*-branched; *a*, *anteiso*-branched; *n*, straight chain; 3-OH, hydroxyl group at position 3; Un, unknown fatty acid. Methods as described in Table 2.

exercised when comparing the fatty acid profiles of naturally occuring plasmid-less *Thermus* strains.

CONCLUDING REMARKS

This review is restricted, almost entirely, to a discussion of the lipid composition in the chemotaxonomy of thermophilic bacteria. Other chemotaxonomic parameters, that could be useful to the taxonomy of these organisms, aside from cell wall composition, have been essentially overlooked. Even the information available on the lipid composition is sparse in comparison to the overwhelming amount of data available on the lipid composition of mesophilic bacteria.

In many cases the number of strains of each thermophilic species is small and in some instances limited to one strain. In these cases, the utilization of chemotaxonomic parameters, useful at or below the species level, such as, fatty acid composition, whole-protein SDS-PAGE or multilocus enzyme electrophoresis, should add very little information to the description of the organisms or should be of little value to identification. Moreover, many thermophiles are easily identified by morphological and phenotypic characteristics and it is not necessary to resort to chemotaxonomic identification. There are, however, a number of species represented by many isolates from different geographical areas and environments that could easily be subjected to chemotaxonomic analysis. Many genera, composed of closely related species, could also be more easily classified by chemotaxonomy. For example, to our knowledge, whole-cell protein electrophoresis has been performed twice on thermophilic bacteria; Lee *et al.* (1993) used SDS-PAGE of soluble proteins to distinguish between the genera *Thermoanaerobacter* and *Thermoanaerobium*. Previously, Williams (1989) differentiated *T. aquaticus* from "*T. brockianus*".

Recently Nunes *et al.* (unpublished results) applied computer-assisted whole-cell protein SDS-PAGE to twenty-four Azorean and two Icelandic *Rhodothermus* strains using the methods described by Pot *et al.* (1989) and Vauterin and Vauterin (1992). This study showed that all strains have a similarity of 90% or over and confirmed the results of fatty acid composition (Nunes *et al.*, 1992*a*) and phenotypic characteristics (Nunes *et al.*, 1992*b*). While the results of the DNA:DNA hybridization are not yet available, the chemotaxonomic results clearly indicate that all strains from the two sites belong to the same species.

Whole-cell protein SDS-PAGE was also performed on a wide range of high-temperature *Thermus* strains belonging to all recognized species and genospecies, but the results do not always agree with the established taxonomy of Williams (1989). These results conflict with DNA:DNA hybridization studies suggesting that difficulties with this technique, arising from the thermophilic nature of the organisms, have yet to be resolved (Manaia and da Costa, unpublished results).

Chemotaxonomic parameters have some advantage over classical morphological and phenotypic characterization for identification, but do not replace it. Semi-automated fatty acid analysis and computer-assisted whole cell protein SDS-PAGE have the advantage of being applicable to large numbers of organisms, with speed and reliability. Other chemotaxonomic parameters can be used at several taxonomic levels (Goodfellow and Minnikin, 1985) and have proven useful to the taxonomy of thermophilic bacteria.

New molecular-genetic methods are being applied successfully to a large number of organisms and will become progressively more important as rapid methods are developed. Whatever techniques are employed, what really matters is the best identification, taxonomy and phylogeny possible of the organisms we study.

ACKNOWLEDGEMENTS

The authors wish to thank the Junta Nacional de Investigação Científica e Tecnológica (JNICT) for supporting the research on thermophilic bacteria in our laboratory. We would like to express our gratitude to Laura Carreto, Paula Chung, Ana Ferraz, Célia Manaia, Leonilde Moreira, Olga Nunes and Sandra Tenreiro for unpublished data. We also wish to thank Don Cowan (London, U.K.), Jakob Kristjansson (Reykjavik, Iceland), Hugh Morgan (Hamilton, New Zealand) and Richard Sharp (Salisbury, U.K.) for the gift of many strains.

REFERENCES

Achenbach-Richter, L., Gupta, R., Stetter, K. O. and Woese, C. R. (1987) Were the original eubacteria thermophiles? System. Appl. Microbiol. 9, 34-39.

Aksenova, H. Y., Rainely, F. A., Janssen, P. H., Zavarzin, G. A. and Morgan, H. W. (1992) *Spirochaeta thermophila* sp. nov., an obligately anaerobic, polysaccharolytic, extremely thermophilic bacterium. Int. J. System. Bacteriol. 42, 175-177.

Alfredsson, G. A., Kristjansson, J. K., Hjorleifsdottir, S. and Stetter, K. O. (1988) *Rhodothermus marinus*, gen. nov., sp. nov., a thermophilic, halophilic bacterium from submarine hot springs in Iceland. J. Gen. Microbiol. 134, 299-306.

Aragno, M. (1992) The aerobic, chemolithoautotrophic, thermophilic bacteria, in "Thermophilic Bacteria" (Kristjansson, J. K., Ed.), pp. 77-104. CRC Press, Boca Raton.

Beam, H. W. and Perry, J. J. (1974) Microbial degradation and assimilation of *n*-alkyl-substituted cycloparaffins. J. Bacteriol. 118, 394-399.

Brock, T. D. (1978) "Thermophilic Microorganisms and Life at High Temperatures". Springer-Verlag, Heidelberg.

Brock, T. D. (1986) Introduction: an overview of thermophiles, in "Thermophiles: General, Molecular, and Applied Microbiology". (Brock, T. D., Ed.), pp. 1-16. John Wiley and Sons, New York.

Brock, T. D. and Boylen, K. L. (1973) Presence of thermophilic bacteria in laundry and domestic hot-water heaters. Appl. Microbiol. 25, 72-76.

Brock, T. D. and Freeze, H. (1969) *Thermus aquaticus* gen. n. and sp. n., a non-sporulating extreme thermophile. J. Bacteriol. 98, 289-297.

Burggraf, S., Olsen, G. J., Stetter, K. O. and Woese, C. R. (1992) A phylogenetic analysis of *Aquifex pyrophilus*. System. Appl. Microbiol. 15, 352-356.

Chung, A. P., Nunes, O.C., Tindall, B. J. and da Costa, M. S. (1993) The effect of the growth medium composition on the fatty acids of *Rhodothermus marinus* and "*Thermus thermophilus*" HB-8. FEMS Microbiol. Lett. 112, 13-18.

Collins, M. D. and Jones, D. (1981) Distribution of isoprenoid quinone structural types in bacteria and their taxonomic implications. Microbiol. Rev. 45, 316-354.

Collins, M. D. and Widdel, F. (1986) Respiratory quinones of sulfate-reducing and sulfur-reducing bacteria: a systematic investigation. System. Appl. Microbiol. 8, 8-18.

Cook, G. M., Janssen, P. H. and Morgan, H. W. (1991) Endospore formation by *Thermoanaerobium brockii* HTD4. System. Appl. Microbiol. 14, 240-244.

da Costa, M. S. (1994) The cell wall and lipids of the bacteria of the genus *Thermus*, in "The Genus *Thermus*" (Sharp, R. and Williams, R. A. D., Eds.), Biotechnology Handbooks Series, Plenum Press, New York (in press).

Darland, G. and Brock, T. D. (1971) *Bacillus acidocaldarius* sp. nov., an acidophilic thermophilic spore-forming bacterium. J. Gen. Microbiol. 67, 9-15.

Davey, M. E., Wood, W. A., Key, R., Nakamura, K. and Stahl, D. A. (1993) Isolation of three species of *Geotoga* and *Petrotoga*: two new genera, representing a new lineage in the bacterial line of descent distantly related to the "*Thermotogales*". System. Appl. Microbiol. 16, 191-200.

De Rosa, M., Gambacorta, A., Minale, L. and Bu'Lock, J. D. (1972) The formation of ω- cyclohexyl-fatty acids from shikimate in an acidophilic thermophilic *Bacillus*. Biochem. J. 128, 751-754.

De Rosa, M., Gambacorta, A., Huber, R., Lanzotti, V., Nicolaus, B., Stetter, K. O. and Tricone, A. (1989) Lipid structures in *Thermotoga maritima*, in "Microbiology of Extreme Environments and its Potential for Biotechnology" (da Costa, M.S., Duarte, J. C. and Williams, R. A. D., Eds.), pp. 167-173. Elsevier, London.

Degryse, E., Glansdorff, N. and Piérard, A. (1978) A comparative analysis of extreme thermophilic bacteria belonging to the genus *Thermus*. Arch. Microbiol. 117, 189-196.

Deinhard, G., Blanz, P., Poralla, K. and Altan, E. (1987a) *Bacillus acidoterrestris* sp. nov., a new thermotolerant acidophile isolated from different soils. System. Appl. Microbiol. 10, 47-53.

Deinhard, G., Saar, J., Krischke, W. and Poralla, K. (1987b) *Bacillus cycloheptanicus* sp. nov., a new thermoacidophile containing ω-cycloheptane fatty acids. System. Appl. Microbiol. 10, 68-73.

Demharter, W. and Hensel, R. (1989) *Bacillus thermocloaceae* sp. nov., a new thermophilic species from sewage sludge. System. Appl. Microbiol. 11, 272-276.

Donato, M. M., Seleiro, E. A. and da Costa, M. S. (1990) Polar lipid and fatty acid composition of strains of the genus *Thermus*. System. Appl. Microbiol. 13, 234-239.

Donato, M. M., Seleiro, E. A. and da Costa, M. S. (1991) Polar lipid and fatty acid composition of strains of *Thermus ruber*. System. Appl. Microbiol. 14, 235-239.

Embley, T. M., O'Donnell, A. G., Wait, R. and Rostron, J. (1987) Lipid and cell wall amino acid composition in the classification of members of the genus *Deinococcus*. System. Appl. Microbiol. 10, 20-27.

Fischer, A., Kroppenstedt, R. M. and Stackebrandt, E. (1983) Molecular-genetic and chemotaxonomic studies on *Actinomadura* and *Nocardiopsis*. J. Gen. Microbiol. 129, 3433-3446.

Georganta, G., Smith, K. E. and Williams, R. A. D. (1993) DNA: DNA homology and cellular components of *Thermus filiformis* and other strains of *Thermus* from New Zealand hot springs. FEMS Microbiol. Lett. 107, 145-150.

Goodfellow, M. and Minnikin, D. E. (1985) Introduction to chemosystematics, in "Chemical Methods in Bacterial Systematics", (Goodfellow, M and Minnikin, D. E., Eds.), pp. 1-15. Academic Press, London.

Hale, M. B., Blankenship, R. E. and Fuller, R. C. (1983) Menaquinone is the sole quinone in the facultatively aerobic green photosynthetic bacterium *Chloroflexus aurantiacus*. Biochim. Biophys. Acta 723, 376-382.

Hensel, R., Demharter, W., Kandler, O., Kroppenstedt, R. M. and Stackebrandt, E. (1986) Chemotaxonomic and molecular-genetic studies of the genus *Thermus*: Evidence for a phylogenetic relationship of *Thermus aquaticus* and *Thermus ruber* to the genus *Deinococcus*. Int. J. System. Bacteriol. 36, 444-453.

Hensel, R., Demharter, W. and Hilpert, R. (1988) The microflora involved in aerobic-thermophilic sludge stabilization. System. Appl. Microbiol. 11, 312-319.

Herrero, A. A., Gomez, R. F. and Roberts, M. F. (1982) Ethanol-induced changes in the membrane lipid composition of *Clostridium thermocellum*. Biochim. Biophys. Acta 693, 195-204.

Huber, R., Langworthy, T. A., König, H., Thomm, M., Woese, C. R., Sletyr, U. B. and Stetter, K. O. (1986) *Thermotoga maritima*, sp. nov. represents a new genus of unique extremely thermophilic eubacteria growing up to 90°C. Arch. Microbiol. 144, 324-333.

Huber, R., Woese, C. R., Langworthy, T. A., Fricke, H. and Stetter, K. O. (1989) *Thermosipho africanus* gen. nov., represents a new genus of thermophilic eubacteria within the "*Thermotogales*". System. Appl. Microbiol. 12, 32-37.

Huber, R., Woese, C. R. Langworthy, T. A., Kristjansson, J. K. and Stetter, K. O. (1990) *Fervidobacterium islandicum* sp. nov., a new extremely thermophilic eubacterium belonging to the "*Thermotogales*". Arch. Microbiol. 154, 105-111.

Huber, R., Wilharm, T., Huber, D., Trincone, A., Burggraf, S., König, H., Rachel, R., Rockinger, I., Fricke, H. and Stetter, K. O. (1992) *Aquifex pyrophilus* gen. nov. sp. nov., represents a novel group of marine hyperthermophilic hydrogen-oxidizing bacteria. System. Appl. Microbiol. 15, 340-351.

Hudson, J. A., Morgan, H. W. and Daniel, R. M. (1986) A numerical classification of some *Thermus* isolates. J. Gen. Microbiol. 132, 532-540.

Hudson, J. A., Morgan, H. W. and Daniel, R. M. (1987a) Numerical classification of some *Thermus* isolates from Icelandic hot springs. System. Appl. Microbiol. 9, 218-223.

Hudson, J. A., Morgan, H. W. and Daniel, R. M. (1987*b*) *Thermus filiformis* sp. nov., a filamentous caldoactive bacterium. Int. J. System. Bacteriol. 37, 431-436.

Hudson, J. A., Morgan, H. W. and Daniel, R. M. (1989) Numerical classification of *Thermus* isolates from globally distributed hot springs. System. Appl. Microbiol. 11, 250-256.

Ishii, M., Kawasumi, T., Igarashi, Y., Kodama, T. and Minoda, Y. (1983) 2-Methylthio-1,4-naphtoquinone, a new quinone from an extremely thermophilic hydrogen bacterium. Agric. Biol. Chem. 47, 167-173.

Jackson, T. J., Ramaley, R. F. and Meinschein, W. G. (1973). *Thermomicrobium*, a new genus of extremely thermophilic bacteria. Int. J. System. Bacteriol. 23, 28-36.

Jannasch, H. W., Huber, R., Belkin, S. and Stetter, K. O. (1988) *Thermotoga neapolitana* sp. nov. of the extremely thermophilic eubacterial genus *Thermotoga*. Arch. Microbiol. 150, 103-104.

Janssen, P. H., Parker, L. E. and Morgan, H. W. (1991) Filament formation in *Thermus* species in the presence of some D-amino acids or glycine. Antonie van Leeuwenhoek. 59, 147-154.

Kandler, O. (1993) Archaea (Archeabacteria), in "Progress in Botany", vol. 54 (Benhke, H.-D., Lüttge, U., Esser, K., Kadereit, J. W. and Runge, M. , Eds.), pp. 1-24. Springer-Verlag, Berlin.

Kannenberg, E., Blume, A. and Poralla, K. (1984) Properties of ω-cyclohexane fatty acids in membranes. FEBS Lett. 172, 331-334.

Kawasumi, T., Igarashi, Y., Kodama, T. and Minoda, Y. (1984) *Hydrogenobacter thermophilus* gen. nov., sp. nov., an extremely thermophilic, aerobic, hydrogen-oxidizing bacterium. Int. J. System. Bacteriol. 34, 5-10.

Klein, R. A., Hazelwood, G. P., Kemp, P. and Dawson, R. M. C. (1979) A new series of long-chain dicarboxylic acids with vicinal dimethyl branching found as major components of the lipids of *Butyrivibrio* spp. Biochem. J. 183, 691-700.

Kleinig, H. and Reichenbach, H. (1977) Carotenoid glucosides and menaquinones from the gliding bacterium *Herpetosiphon giganteus* Hpa 2. Arch. Microbiol. 112, 307-310.

Knudsen, E., Jantzen, E., Bryn, K., Ormerod, J. G. and Sirevag, R. (1982) Quantitative and structural characteristics of lipids in *Chlorobium* and *Chloroflexus*. Arch. Microbiol. 132, 149-154.

Kristjansson, J. K. and Alfredsson, G. A. (1983) Distribution of *Thermus* spp. in Icelandic hot springs and a thermal gradient. Appl. Environ. Microbiol. 45, 1785-1789.

Kristjansson, J. K. and Alfredsson, G. A. (1992) The heterotrophic, thermophilic genera *Thermomicrobium*, *Rhodothermus*, *Saccharococcus*, *Acidothermus* and *Scotothermus*, in "Thermophilic Bacteria" (Kristjansson, J. K. , Ed.), pp. 63-76. CRC Press, Boca Raton.

Kristjansson, J. K. and Stetter, K. O. (1992) Thermophilic bacteria, in "Thermophilic Bacteria" (Kristjansson, J. K. , Ed.), pp. 1-18. CRC Press, Boca Raton.

Kristjansson, J. K., Ingasson, A. and Alfredsson, G. A. (1985) Isolation of thermophilic obligately autotrophic hydrogen-oxidizing bacteria, similar to *Hydrogenobacter thermophilus*, from Icelandic hot springs. Arch. Microbiol. 140, 321-325.

Kristjansson, J. K., Hreggvidsson, G. O. and Alfredsson, G. A. (1986) Isolation of halotolerant *Thermus* spp. from submarine hot springs in Iceland. Appl. Environ. Microbiol. 52, 1313-1316.

Kristjansson, J. K., Hjörleifsdottir, S., Marteinsson, V. T. and Alfredsson, G. A. (1994) *Thermus scotoductus*, sp. nov., a pigment-producing thermophilic bacterium from hot tap water in Iceland and including *Thermus* sp. X-1. System. Appl. Microbiol. (in press).

Kroppenstedt, R. M. (1985) Fatty acids and menaquinone analysis of Actinomycetes and related organisms, in " Chemical Methods in Bacterial Systematics". (Goodfellow, M. , Ed.), pp.173-200. Academic Press, London.

Kuykendall, L. D., Roy, M. D., A' Neill, J. J. and Devine, T. E. (1988) Fatty acids, antibiotic resistance, and deoxyribonucleic acid homology groups of *Bradyrhizobium japonicum*. Int. J. System. Bacteriol. 38, 358-361.

Langworthy, T. A. and Pond, J. L. (1986) Membranes and lipids of thermophiles, in "Thermophiles: General, Molecular and Applied Microbiology" (Brock, T. D. , Ed.), pp. 107-135. John Wiley and Sons, New York.

Langworthy, T. A., Holzer, G., Zeikus, J. G. and Tornabene, T. G. (1983) Iso- and anteiso-branched glycerol diethers of the thermophilic anaerobe *Thermodesulfotobacterium commune*. System. Appl. Microbiol. 4, 1-17.

Lee, Y-E., Mahendra, K. J., Chanyong, L., Lowe, S. E. and Zeikus, J. G. (1993) Taxonomic distinction of saccharolytic thermophilic anaerobes: description of *Thermoanaerobacterium xylanolyticum* gen. nov., sp. nov., and *Thermoanaerobacterium saccharolyticum* gen. nov., sp. nov.; reclassification of *Thermoanaerobium brockii, Clostridium thermosulfurogenes*, and *Clostridium thermohydrosulfuricum* E100-69 as *Thermoanaerobacter brockii* comb. nov., *Thermoanaerobacterium thermosulfurigenes* comb. nov., and *Thermoanaerobacter thermohydrosulfuricus* comb. nov., respectively; and transfer of *Clostridium thermohydrosulfuricum* 39E to *Thermoanaerobacter ethanolicus*. Int. J. System. Bacteriol. 43, 41-51.

Loginova, L. G., Egorova, L. A., Golovacheva, R. S. and Seregina, L. M. (1984) *Thermus ruber* sp. nov., nom. rev. Int. J. System. Bacteriol. 34, 498-499.

Manaia, C. M. and da Costa, M. S. (1991) Characterization of halotolerant *Thermus* isolates from shallow marine hot springs on S. Miguel, Azores. J. Gen. Microbiol. 137, 2643-2648.

Merkel, G. J., Durham, D. R. and Perry, J. J. (1980) The atypical cell wall composition of *Thermomicrobium roseum*. Can. J. Microbiol. 26, 556-559.

Munster, M. J., Munster, A. P. and Sharp, R. J. (1985) Incidence of plasmids in *Thermus* spp. isolated in Yellowstone National Park. Appl. Env. Microbiol. 50, 1325-1327.

Munster, M. J., Munster, A. P., Woodrow, J. R. and Sharp, R. J. (1986) Isolation and preliminary taxonomic studies of *Thermus* strains isolated from Yellowstone National Park, USA. J. Gen. Microbiol. 132, 1677-1683.

Nordström, K. M. and Laakso, S. M. (1992) Effect of the growth temperature on the fatty acid composition of ten *Thermus* strains. Appl. Environ. Microbiol. 58, 1656-1660.

Nunes, O. C., Donato, M. M., Manaia, C. M. and da Costa, M. S. (1992a) The polar lipid and fatty acid composition of *Rhodothermus* strains. System. Appl. Microbiol. 15, 59-62.

Nunes, O. C., Donato, M. M. and da Costa, M. S. (1992b) Isolation and characterization of *Rhodothermus* strains from S. Miguel, Azores. System. Appl. Microbiol. 15, 92-97.

O' Leary, W. M. and Wilkinson, S. G. (1988) Gram-positive bacteria, in "Microbial Lipids" Vol.1. (Ratledge, C. and Wilkinson, S. G. , Eds.), pp.117-201. Academic Press, London .

Oshima, M. and Yamakawa, T. (1974) Chemical structure of a novel glycolipid from an extreme thermophile, *Flavobacterium thermophilum*. Biochemistry 13, 1140-1146.

Oshima, T. and Imahori, K. (1974) Description of *Thermus thermophilus* (Yoshida and Oshima) comb. nov., a non-sporulating thermophilic bacterium from a Japanese thermal spa. Int. J. System. Bacteriol. 24, 102-112.

Oyaizu, H., Debrunner-Vossbrinck, B., Mandelco, L., Studier, J. A. and Woese, C. R. (1987) The green non-sulfur bacteria: a deep branching in the eubacterial line of descent. System. Appl. Microbiol. 9, 47-53.

Pask-Hughes, R. A. and Shaw, N. (1982) Glycolipids from some extreme thermophilic bacteria belonging to the genus *Thermus*. J. Bacteriol. 149, 54-58.

Pask-Hughes, R. A. and Williams, R. A. D. (1975) Extremely thermophilic Gram-negative bacteria from hot tap water. J. Gen. Microbiol. 88, 321-328

Pask-Hughes, R. A. and Williams, R. A. D. (1977) Yellow-pigmented strains of *Thermus* spp. from Icelandic hot springs. J. Gen. Microbiol. 102, 375-383.

Pask-Hughes, R. A. and Williams, R. A. D. (1978) Cell envelope components of strains belonging to the genus *Thermus*. J. Gen. Microbiol. 107, 65-72.

Patel, B. K. C. , Morgan, H. W. and Daniel, R. M. (1985) *Fervidobacterium nodosum* gen. nov. and spec. nov., a new chemoorganotrophic, caldoactive, anaerobic bacterium. Arch. Microbiol. 141, 63-69.

Phillips, W. E. and Perry, J. J. (1976) *Thermomicrobium fosteri* sp. nov., a hydrocarbon-utilizing obligate thermophile. Int. J. System. Bacteriol. 26, 220-225.

Pond, J. L. and Langworthy, T. A. (1987) Effect of growth temperature on the long-chain diols and fatty acids of *Thermomicrobium roseum*. J. Bacteriol. 169, 1328-1330.

Pond, J. L., Langworthy, T. A., Holzer, G. (1986) Long-chain diols: a new class of membrane lipids from a thermophilic bacterium. Science 231, 1134-1136.

Pot, B., Gillis, M., Hoste, B., van de Velde, A., Bekaert, F., Kersters, K. and De Ley, J. (1989) Intra- and intergeneric relationships of the genus *Oceanospirillum*. Int. J. System. Bacteriol. 39, 23-34.

Prado, A., da Costa, M. S. and Madeira, V. M. C. (1988) Effect of growth temperature on the lipid composition of two strains of *Thermus* sp. J. Gen. Microbiol. 134: 1653-1660.

Priest, F. G., Goodfellow, M. and Todd, C. (1988) A numerical classification of the genus *Bacillus*. J. Gen. Microbiol. 134, 1847-1882.

Ramaley, R. F. and Hixson, J. (1970) Isolation of a non-pigmented, thermophilic bacterium similar to *Thermus aquaticus*. J. Bacteriol. 103, 527-528.

Ray, P. H., White, D. C. and Brock, T. D. (1971) Effect of temperature on the fatty acid composition of *Thermus aquaticus*. J. Bacteriol. 106, 25-30.

Reichenbach, H., Boyer, P. and Kleinig, H. (1978) The pigments of the gliding bacterium *Herpetosiphon giganteus*. FEMS Microbiol. Lett. 3, 155-156.

Rössler, D., Ludwig, W., Schleifer, K. H., Lin, C., McGill, T. J., Wisotzkey, J. D., Jurtshuk, Jr., P. and Fox, G.E. (1991) Phylogenetic diversity in the genus *Bacillus* as seen by 16S rRNA sequencing studies. System. Appl. Microbiol. 14, 266-269.

Saiki, T., Kimura, R. and Arima, K. (1972) Isolation and characterization of extremely thermophilic bacteria from hot springs. Agric. Biol. Chem. 36, 2357-2366.

Santos, M. A., Williams, R. A. D. and da Costa, M. S. (1989) Numerical taxonomy of *Thermus* isolates from hot springs in Portugal. System. Appl. Microbiol. 12, 310-315.

Schleifer, K. H. and Kandler, O. (1972) Peptidoglycan types of bacterial cell walls and their taxonomic implications. Bacteriol. Rev. 36, 407-477.

Sharp, R. J. and Williams, R. A. D. (1988) Properties of *Thermus ruber* strains isolated from Icelandic hot springs and DNA:DNA homology of *Thermus ruber* and *Thermus aquaticus*. Appl. Environ. Microbiol. 54, 2049-2053.

Sharp, R. J., Riley, P. W. and White, D. (1992) Heterotrophic thermophilic *Bacilli*, in "Thermophilic Bacteria" (Kristjansson, J. K. , Ed.), pp. 19-50. CRC Press, Boca Raton.

Skerman, V. B. D., McGowan, V. and Sneath, P. H. A. (1980) Approved lists of bacterial names. Int. J. System. Bacteriol. 30, 225-420.

Stackebrandt, E., Ludwig, W., Weizenegger, M., Dorn, S., McGill, T. J., Fox, G. E., Woese, C. R., Schubert, W. and Schleifer, K. H. (1987) Comparative 16S rRNA oligonucleotide analysis and murine type of round-spore-forming bacilli and non-sporeforming relatives. J. Gen. Microbiol. 133, 2523-2529.

Suzuki, K. I., Saito, K., Kawaguchi, A., Okuda, S. and Komagata, K. (1981) Occurence of ω-cyclohexyl fatty acids in *Curtobacterium pusillum*. J. Gen. Appl. Microbiol. 27, 261-266.

Taguchi, H., Yamashita, M., Matsuzawa, H. and Ohta, T. (1982) Heat-stable and fructose 1,6-bisphosphate-activated L-lactate dehydrogenase from an extremely thermophilic bacterium. J. Biochem. 9, 1343-1348.

Tindall, B. J. (1991) Lipid composition of *Rhodothermus marinus*. FEMS Microbiol. Lett. 80, 65-68.

Vainshtein, M., Hippe, H. and Kroppenstedt, R. M. (1992) Cellular fatty acid composition of *Desulfovibrio* species and its use in the classification of sulfate-reducing bacteria. System. Appl. Microbiol. 15, 554-566.

Vauterin, L. and Vauterin, P. (1992) A new system for the standardized and objective comparison of electrophoresis patterns: applications in bacterial identification, in " Proceedings of the Conference on Taxonomy and Automated Identification of Bacteria. (Schindler, J. , Ed.), pp. 19-22. Prague.

Virtala, M. K., Nordström, K. M., Karp, M. T. and Laakso, S. V. (1993) Alterations in growth temperature range and fatty acid composition of *Thermus* as a result of plasmid elimination. Arch. Microbiol. 160, 12-17.

Weisburg, W. G., Giovannoni, S. J. and Woese, C. R. (1989) The *Deinococcus-Thermus* phylum and the effect of rRNA composition on phylogenetic tree construction. System. Appl. Microbiol. 11, 128-134.

Wiegel, J. (1992) The obligately anaerobic thermophilic bacteria, in "Thermophilic Bacteria" (Kristjansson, J. K. , Ed.), pp. 105-184. CRC Press, Boca Raton.

192

Williams, R. A. D. (1989) Biochemical taxonomy of the genus *Thermus*, in "Microbiology of Extreme Environments and its Potential for Biotechnology" (da Costa, M.S., Duarte, J. C. and Williams, R. A. D., Eds.), pp. 82-97. Elsevier, London.

Williams, R. A. D. and da Costa, M. S. (1992) The genus *Thermus* and related microorganisms, in "The Prokaryotes, 2nd Edition, A Handbook on the Biology of Bacteria: Ecophysiology, Isolation, Identification, Applications" (Balows, A., Trüper, H. G., Dworkin, M., Harder, W. and Schleifer, K.-H. , Eds.), pp. 3745-3753. Springer-Verlag, New York.

Windberger, E., Huber, R., Trincone, A., Fricke, H. and Stetter, K. O. (1989) *Thermotoga thermarum* sp. nov. and *Thermotoga neapolitana* occuring in African continental solfataric springs. Arch. Microbiol. 151, 506-512.

Wisotzkey, J. D., Jurtshuk, Jr., P., Fox, G. E., Deinhard, G. and Poralla, K. (1992) Comparative sequence analysis on the 16S rRNA (rDNA) of *Bacillus acidocaldarius*, *Bacillus acidoterrestris*, and *Bacillus cycloheptanicus* and proposal for creation of a new genus, *Alicyclobacillus* gen. nov. Int. J. System. Bacteriol. 42, 263-269.

Woese, C. R. (1987) Bacterial evolution. Microbiol. Rev. 51, 221-271.

Zeikus, J. G., Dawson, M. A., Thompson, T. E., Ingvorsen, K. and Hatchikian, E. C. (1983) Microbial ecology of volcanic sulphidogenesis: isolation and characterization of *Thermodesulfobacterium commune* gen. nov. and sp. nov. J. Gen. Microbiol. 129, 1159-1169.

ALKALIPHILES: DIVERSITY AND IDENTIFICATION

Brian E. Jones[1], William D. Grant[2], Nadine C. Collins[2], and Wanjiru E. Mwatha[3]

[1]Gist-brocades B.V., R&D, Postbus 1, 2600 MA Delft, The Netherlands
[2]Department of Microbiology, University of Leicester, Leicester LE1 9HN, U.K.
[3]Department of Botany, Kenyatta University, Nairobi, Kenya

INTRODUCTION

Micro-organisms that grow in hostile or extreme environments are currently a popular subject for study. The fascination with so-called 'extremophiles' reflects a perceived microbial biotechnological importance and a desire to investigate possible early conditions on Earth and the origins of life. Extremes of pH represent a particularly hostile regime for microbial life, especially when combined with extremes of salinity or temperature. This review deals with some of the aspects of micro-organisms that inhabit environments in the alkaline range of the pH spectrum.

Organisms that grow optimally at a pH greater than pH 8 are properly defined as alkaliphiles. Although the term alkalophile is also frequently used, alkaliphile is probably etymologically more correct (from the Arabic root, al qaliy = alkali). Obligate alkaliphiles generally have a pH optimum for growth between pH 9 and pH 10 and are incapable of growth at neutral pH. Organisms that are capable of growth at alkaline pH values, but with optima in the neutral to acid range are termed alkalitolerant. Most of the organisms described as growing under very alkaline conditions are prokaryotes, although in some environments a substantial component of the phototroph population consists of eukaryotic algae. Very few isolates have been subjected to detailed study and often it is not rigorously established whether or not an organism is truly alkaliphilic.

Alkaliphiles and alkalitolerant bacteria are widely distributed and may be isolated from many differing environments including environments made alkaline by human industrial activity. Processes such as cement manufacture and casting or the disposal of blast furnace slag result in the generation of $Ca(OH)_2$, while paper and board manufacture, electroplating and food processing produce NaOH-containing effluents. It has been shown that some of these man-made environments are too toxic to support a microbial population (Grant, unpublished results, Grant and Tindall, 1986) but in the main there is little microbiological information available on these types of environment. Exceptions are the KOH-containing effluents from the lye treatment of potatoes (skin removal) which are colonized by an orange pigmented, non-spore forming, Gram positive bacterium defined as *Exiguobacterium aurantiacum* (Gee *et al.*, 1980; Collins *et al.*, 1983) and the alkaliphilic *Bacillus* sp. isolated from the traditional Japanese indigo fermentation (Horikoshi, 1991). A Gram-negative, alkaliphilic rod described as an *Ancylobacter* sp. has been isolated from effluents from the kraft pulping process used in paper manufacture (Strand *et al.*, 1984).

Naturally occurring stable alkaline environments are not common (Grant, 1992), but even so, alkaliphiles can be isolated from a whole range of environments including soils, which on the basis of bulk pH measurements would not be considered particularly alkaline (Horikoshi, 1991). The types of alkaliphilic bacteria in soil environments, where alkalinity is probably transient and highly localized, are rather restricted (Horikoshi, 1991) and many of the isolates should perhaps be more correctly defined as alkalitolerant. On the other hand, a more diverse variety of alkaliphilic types, particularly of obligate alkaliphiles can be found in naturally occurring stable alkaline environments. These are of two types: (a) calcium hydroxide dominated ground waters, and (b) sodium carbonate (low Ca^{2+}) dominated soda lakes and deserts (Grant and Tindall, 1986; Grant and Horikoshi, 1989; Grant *et al.*, 1990; Grant, 1992; Grant and Horikoshi, 1992).

DIVERSITY IN THE ALKALINE ENVIRONMENT

Alkaliphile Diversity in 'Normal' Soils and Waters

Most soils, even those considered to be alkaline, rarely achieve pH > 8.5 due to the buffering effect of CO_2. Due to normal microbial activity the interstitial pCO_2 values in soil are higher than atmospheric. However, Gram-positive endospore-forming, rod-shaped isolates, often pigmented shades of yellow or red which grow over the pH range 8-11.5 can be easily obtained from soils by enrichment or direct plating on nutrient-rich media containing Na_2CO_3 as described by Grant and Tindall (1980) or Horikoshi (1991). The widespread occurrence of such organisms represents something of a conundrum and it can only be assumed that in the heterogeneous soil environment localized and transient alkalinity is generated by biological processes such as ammonification, sulphate reduction and oxygenic photosynthesis (Langworthy, 1978). Very many of these Gram-positive endospore forming bacteria have been isolated, particularly by Horikoshi and co-workers, mainly because of the biotechnological interest in alkali-stable enzymes. Most of the isolates are classified as *Bacillus* sp. and some bear names that have not been validated, and therefore have no taxonomic standing. Vedder (1934) was the first to isolate an alkaliphilic *Bacillus* of this type, which was described as *Bacillus alcalophilus*. Gordon and Hyde (1982) examined a large number of these isolates and assigned them to the *B. firmus* - *B. lentus* complex and distinguished five sub-groups. *Bacillus alcalophilus* was placed as a *B. firmus* - *B. lentus* intermediate. These isolates, together with several others, including some strains isolated from soda lakes, were re-examined by Fritze *et al.* (1990) on the basis of DNA base composition. Six phenotypic groups were recognized which correlated with three clusters on the basis of G+C content, but there was considerable heterogeneity within the homology groups. Vedder's strains, including the type strain of *B. alcalophilus*, exhibited no close relationship to the other alkaliphilic *Bacillus* strains included in the study. Ash *et al.* (1991) reached a similar conclusion from a comparative 16S rRNA sequence analysis which indicated that *B. alcalophilus* represented a separate line of descent and was unrelated to any other *Bacillus* group. Recently, a number of strains belonging to "homology group A" (Fritze *et al.*, 1990) were re-assessed (Spanka and Fritze, 1993). This study also used some new isolates from the Wadi Natrûn soda lakes and from horse meadow soils where ammonification is the likely source of alkalinity. On the basis of phenotypic properties and DNA-DNA homology values, the new strains isolated from horse faeces and soils formed a closely related group of obligately alkaliphilic bacilli for which the name *Bacillus cohnii* has been proposed. The soda lake strains were clearly quite different (see later). All these results served to illustrate the considerable heterogeneity among the alkaliphilic bacilli which

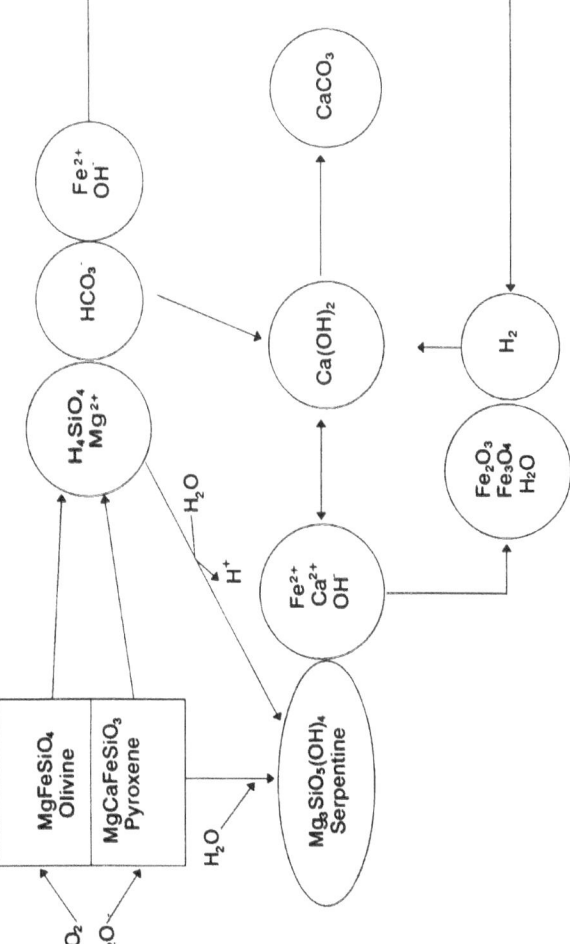

Figure 1. A simplified scheme for the hydrogeochemical generation of alkalinity in Oman ground waters.

probably represent several major taxa at the species level, especially since Horikoshi (1991) reports that strains isolated in his laboratory cover the range for G+C content from 35.5 to 52.4 mol%. The analysis of more isolates and a considerable amount of work will be required to unravel the true nature of the systematics of this group.

Compared to the alkaliphilic bacilli, non-spore-forming alkaliphilic strains from soils and water appear to be isolated less frequently. There are, however reports of alkaliphilic strains of *Pseudomonas* (Watanabe *et al.*, 1977), *Paracoccus* (Urakami *et al.*, 1989), *Micrococcus* (Akiba and Horikoshi, 1978; Kimura and Horikoshi, 1990), *Aeromonas* (Ohkoshi *et al.*, 1985), *Corynebacterium* (Kobayashi and Horikoshi, 1980) and streptomycetes (Tsujibo *et al.*, 1988), (see Horikoshi, 1991 for review). Strict anaerobes have apparently received less attention, although there is a recent report of a moderately thermophilic, alkaliphilic anaerobe, *Clostridium paradoxum*, isolated from an aeration pond of an anaerobic waste digestor (Li *et al.*, 1993). It appears that Gram-positive alkaliphilic types are most common in this environment.

High Ca^{2+} Alkaline Environments

Calcium (and magnesium) is a plentiful cation in most environments and combination with common anions (CO_3^{2-}, SO_4^{2-}, SiO_2, etc.) usually results in the precipitation of insoluble minerals at alkaline pH. Therefore, even though highly alkaline (pH > 11) calcium-bearing ground waters are apparently extremely rare, such conditions have been identified in various parts of the world. These include locations in California (Barnes *et al.*, 1972), Oman, Yugoslavia (Barnes *et al.*, 1978), Cyprus (Pantazis, 1976) and Jordan (Barnes *et al.*, 1982). Microbiological analysis of these environments is very scarce but because they are analogous to the effluent from cement manufacture they are of interest as model systems for investigating chemical and biological interactions in cement pore waters (Bath *et al.*, 1987; Grant *et al.*, 1990; Grant 1992).

The alkalinity is generated by a specific geological process, namely the low-temperature weathering of the calcium and magnesium containing silicate minerals olivine and pyroxene. The process, known as serpentinization, is depicted schematically in Figure 1. Carbon dioxide-charged surface waters decompose the silicate minerals releasing Ca^{2+} and OH^- into solution. Magnesium (Mg^{2+}) is immobilized as serpentine or precipitated as $Mg(OH)_2$ (brucite), $MgCO_3$ (magnesite), $CaMg(CO_3)_2$ (dolomite) or $Mg_3Ca(CO_3)_4$ (huntite). Carbonate is further removed as calcite ($CaCO_3$) but Ca^{2+} is in vast excess, resulting in a $Ca(OH)_2$-dominated brine in which solid-phase $Ca(OH)_2$ is in equilibrium with soluble phase Ca^{2+} and OH^- maintaining an alkaline environment around pH 11. The ionic composition of these waters is very dilute (Table 1), unlike the situation in soda lakes, since the solubility of $Ca(OH)_2$ is very low (about 10 mM). The buffering capacity of such waters on exposure to atmospheric CO_2 (or microbial activity?) is limited when separated from solid-phase $Ca(OH)_2$, again a situation quite unlike the CO_3^{2-}-dominated soda lake environment. The serpentinization process also produces extremely reducing conditions due to the release of Fe^{2+} and the production of hydrogen from intermediate iron hydroxides.

Alkaliphile Diversity in the Calcium Hydroxide Environment

Systematic microbiological analyses of a somewhat preliminary nature have been carried out in a number of $Ca(OH)_2$ springs in Oman where the ground waters evolve from formations undergoing active serpentinization (Bath *et al.*, 1987). The waters were very dilute, the major ions (Cl^-, Na^+, OH^-, Ca^{2+} and K^+) being present at concentrations less than 10 mM (Table 1). Furthermore, the very low carbon, nitrogen and phosphorous levels

Table 1. Hydrochemistry of alkaline springs in Oman
[from Bath *et al.*, 1987]

	Nizwa Jill	Karku	Jebel Awq	Nidab	Bahla
Na^+	9.5	11.2	26.2	5.7	8.2
K^+	0.24	0.29	0.71	0.09	0.22
Ca^{2+}	1.37	1.80	1.38	1.59	1.56
Mg^{2+}	<0.01	<0.01	<0.01	<0.01	<0.01
OH^-	4.9	6.1	6.1	3.3	4.9
CO_3^{2-}	0	0	0	0	0
Cl^-	8.2	9.9	24.2	5.4	7.7
SO_4^{2-}	0.01	0.02	0.36	0.06	<0.01
S^{2-}		0.8		0.5	
NO_3^-	<0.01	<0.01	0.5	<0.01	<0.01
NH_4^+	<0.01	0.02	<0.01	0.09	0.03
Total P^a	<0.015	<0.015	<0.015	<0.015	<0.015
TOC^b	2.5	2.1	1.5	0.5	2.3
pH	11.2	11.4	11.4	11.2	11.4
T °C	33	36	31	35	34

Figures are mM except [a] total phosphorous (mg/l); [b] total organic carbon (mg/l).

indicated ultra-oligotrophic waters. The microbial population was low; the viable counts of organotrophs were in the range 10^1 to 10^4 cfu/ml. Most of the isolates were facultative anaerobes, reflecting the reducing environment and although most of the strains were able to grow at pH 10 or above, only a few were obligately alkaliphilic. The enrichment of nitrifying, denitrifying, sulphur-oxidizing and methanogenic bacteria were all negative. On the other hand, enrichments for sulphate-reducing bacteria at pH 10.2 were positive from several of the spring water sites. Although strains of sulphate-reducing bacteria were obtained growing best at pH 8.5 to pH 9.5 these were not further characterized.

An attempt was made to identify the organotrophic bacteria (Table 2) with the conclusion that the genera found in these alkaline springs were similar to those found in less extreme soil and water environments. It is not at all clear if any of these organisms isolated are actually indigenous to the ground- and spring-waters, or whether they are merely surface contaminants. Certainly some of the bacteria could be attributed to introduction by animals. It would appear that the extreme pH is not the main limitation to microbial activity in these nutrient limited waters. Such environments may support oligotrophic alkaliphilic anaerobes, but any evidence is lacking since no cultures were made under conditions comparable to that of the chemistry of the springs.

It seems likely that the alkaline spring in California investigated by Souza *et al.* (1974) had similar chemistry to the Oman springs (Table 3). An obligately alkaliphilic,

Table 2. Tentative identification of heterotrophic bacteria isolated from spring waters in Oman
[from Bath et al., 1987]

| | Number of Isolates | |
Genus/Group	alkalitolerant	alkaliphilic
Actinobacillus	4	0
Bacillus	3	1
Hafnia	3	0
Vibrio	2	0
Clostridium	2	0
Caulobacter	0	1
Serratia	1	0
Enterobacteriaceae	2	0
Flavobacterium	1	1
Hyphomicrobium	1	0
Pseudomonas	1	0
Coryneform	1	0
Unknown	4	0

orange pigmented bacterium (strain A-1) isolated from enrichment cultures of the spring sediments grew optimally at pH 9-9.5. Although the strain was well characterized and found to resemble the coryneform group, it could not be assigned to any known genus (Souza and Deal, 1977). Unfortunately, the strain appears now to have been lost (Hochstein, personal communication to Grant).

Table 3. Major ion concentrations (mM) of Blackbird Valley Spring, Stanislaus County, California
[from Souza et al., 1974]

Ca^{2+}	1.20
Na^+	1.74
Cl^-	0.90
OH^-	3.06
pH	11.78

Although aquatic, highly alkaline (pH > 9), high Ca^{2+} environments appear to be very rare, there are isolated reports of exceptions (Waring et al., 1965; Da Costa, personal communication). These may be geographically very isolated and may reflect very localized

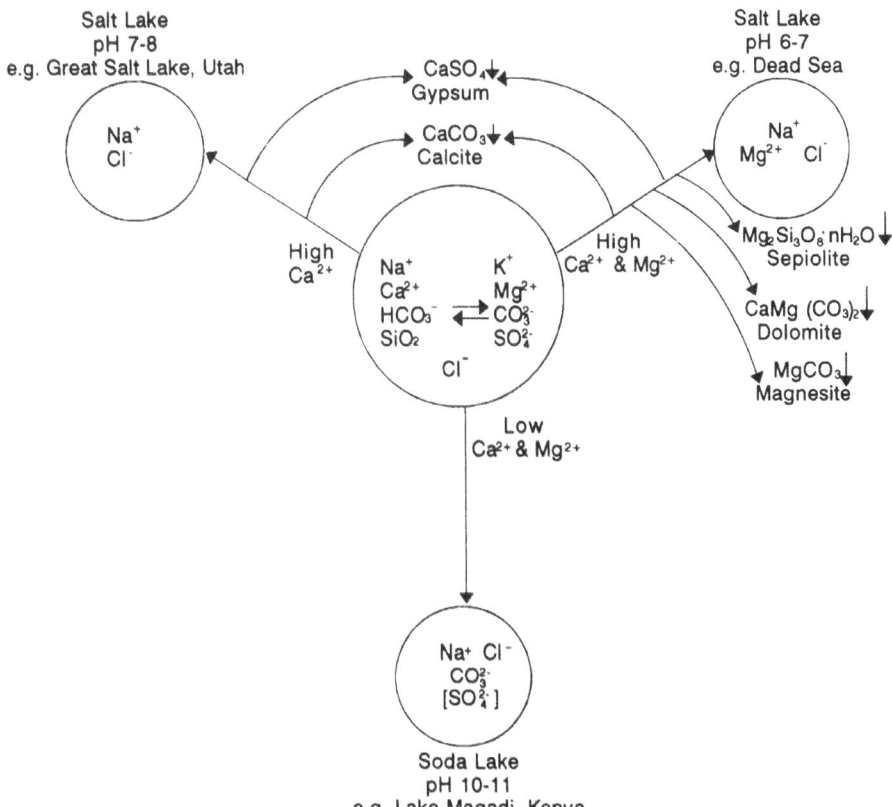

Figure 2. Schematic representation of the genesis of saline and alkaline lakes.

variations in the local geology. Such waters, if they do indeed exist, may repay microbiological investigation.

Soda Lakes and Soda Deserts - A Low Ca^{2+} Environment

Soda lakes probably represent the highest and most stable pH environments on earth, where alkalinity generates pH values > 11.5. In a world-wide sense, alkaline soda lakes and soda deserts are uncommon. Although quite abundant in certain regions, their non-random distribution gives some clue as to the conditions necessary for their formation. They often lie in isolated and inaccessible areas and consequently they are seldom explored environments, especially from a microbiological point of view. Lists of soda lakes and deserts and their locations may be found in Grant and Tindall (1986), Tindall (1988), Grant (1992).

The formation of soda lakes has much in common with athalassohaline (not derived from sea water) salt lakes, but with the major difference that soda lakes contain carbonate (or carbonate complexes) as a predominant anion (Figure 2) (Eugster and Hardie, 1978). Like salt lakes, soda lakes also require an unusual combination of geographical, topological and climatic circumstances for their formation, namely;

- a closed drainage basin with a restricted outflow;
- a high marginal relief which ensures enough rainfall to produce perennial streams of sufficient volume entering the basin to sustain a standing body of water;
- rates of evaporation that exceed inflow so that salts may accumulate by evaporative concentration.

Such conditions are found in the arid and semi-arid zones of tropical or sub-tropical rainshadow deserts, for example east of the Cordilleras of North America or in continental interiors. Other examples are found in areas of tectonic rifting such as the Great Rift Valley. A further important condition necessary for the formation of a soda lake is that significant amounts of Ca^{2+} and Mg^{2+} must be absent in the lithology of the catchment area, so that HCO_3^--containing ground waters are produced where the molar concentrations of HCO_3^- /CO_3^{2-} greatly exceed those of Ca^{2+} and Mg^{2+}. Through evaporative concentration, waters rapidly achieve saturation with respect to alkaline earth cations which precipitate as insoluble carbonates, leaving Na^+, Cl^-, HCO_3^-/CO_3^{2-} as the major ions in solution (Figure 2). As a general mechanism this requires bed rocks of a volcanic or metamorphic origin and a lack of sedimentary deposits. Alkalinity develops by a shift in the CO_2/HCO_3^-/CO_3^{2-} equilibrium, i.e.

$$2HCO_3^- \longrightarrow CO_3^{2-} + CO_2\uparrow + H_2O$$

In the course of the formation of alkalinity other ions also concentrate, especially Na^+ and Cl^-, making the lakes also somewhat saline, but each lake is unique presenting a bewildering variety of ion compositions and concentration ranges dictated by local geological conditions. This can give rise to circumstances where some ions which are normally only present in trace amounts achieve significant proportions (e.g. F^-, PO_4^{3-}, etc.) and this may be reflected in their microbiology. In contrast to the $Ca(OH)_2$ environment, chemical analysis of the East African soda lakes indicates a greater concentration of solutes with the major ions present in molar amounts (Table 4) (Jones, unpublished data). These figures are in broad agreement with earlier analyses (Grant and Tindall, 1986).

The possible mechanisms contributing to the formation of alkaline soda lakes have been well researched especially with regard to Lake Magadi (Baker, 1958; Eugster, 1970; Hardie and Eugster, 1970; Jones et al., 1977; Eugster, 1986) and Lake Bogoria (Renaut et al., 1986) in the Gregory Rift Valley (Kenya). Lake Magadi, at the lowest point in the Rift Valley, represents the final stage of maximum evaporite (trona, $NaHCO_3 \cdot Na_2CO_3 \cdot 12H_2O$) productivity, whereas the more dilute Lake Bogoria represents an earlier evolutionary stage. Figure 3 gives an idealized scheme of the processes taking place at Magadi. In the Rift

Table 4. Chemical analysis of Kenyan soda lake waters (October 1988: Jones, unpublished data)

Lake	Na^+	K^+	Ca^{2+}	Mg^{2+}	SiO_2	PO_4^{3-}	Cl^-	SO_4^{2-}	CO_3^{2-}
Elmenteita	195.7	3.6	0.07	<0.004	2.9	0.03	65.1	2.0	68.0
Nakuru	326.1	5.6	0.15	<0.004	3.3	0.15	57.5	0.5	198.3
Bogoria (north)	734.8	5.5	0.21	0.008	2.2	0.09	100.9	1.0	476.7
Bogoria (south)	795.7	6.8	0.19	0.008	2.0	0.17	115.5	1.1	516.7
Sonachi	140.4	9.0	0.05	0.008	2.1	0.04	12.4	0.8	90.0
Oloidien	8.7	1.8	0.28	0.65	1.0	0.003	4.8	0.5	<10.0
Magadi (lake brines)	7000.0	57.0	<0.01	<0.01	14.9	1.82	3154.9	17.5	3900.0
(lagoon brines)	2826.1	26.1	0.03	<0.01	7.1	0.23	1123.9	12.8	1816.7
Little Magadi	4626.1	61.1	0.02	0.03	7.5	0.31	1856.3	13.1	2433.3
Natron	4521.7	43.7	0.04	0.03	3.1	4.21	1464.8	1.7	2666.7

All concentrations given in mM.

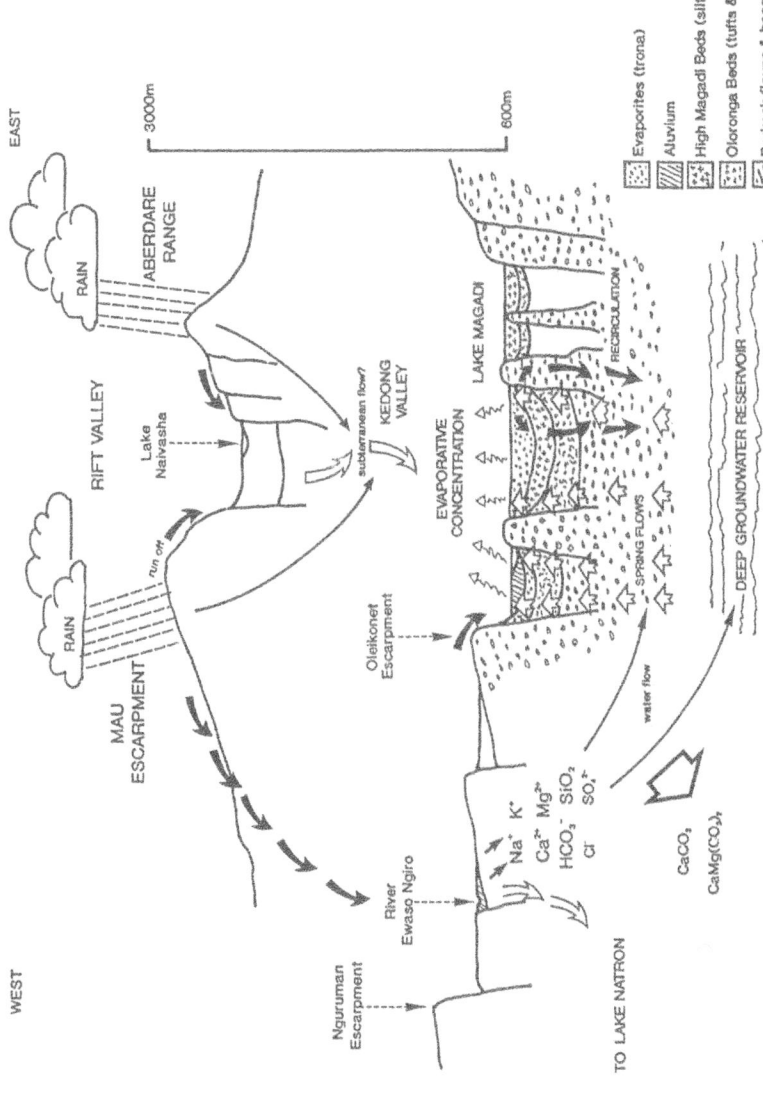

Figure 3. A diagrammatic cross-section of the southern section of the Gregory Rift Valley showing the possible processes involved in the formation of Lake Magadi - an alkaline, saline lake.

Valley, block faulting and lava flows have resulted in a series of closed basins in a rainfall catchment area characterized by bedrock composed of Pleistocene alkaline trachyte lavas (high Na^+, low Ca^{2+} and Mg^{2+}) (Baker, 1986). It is the rapid weathering of the volcanic rocks, particularly the hydrolysis of silicate minerals that produces HCO_3^--rich waters high in silica. A small proportion of the alkalinity may be derived through the leaching by CO_2-charged surface waters of natrocarbonatite ash deposits from the volcano Ol Doinyo Lengai in Tanzania (100 km south of Magadi). Although the very complex geochemical and hydrological processes leading to the selective removal of ions and crystalization of minerals with the genesis of alkalinity appear well understood, microbiological processes undoubtably play a role (Norton and Grant, 1988) that has yet to be fully assessed. The precipitation of trona (an active process at Lake Magadi) is dependent on the kinetics of CO_2 addition which must be influenced by microbial activity. If it were simply a matter of rapid evaporation, then thermonatrite ($Na_2CO_3 \cdot H_2O$) or natron ($Na_2CO_3 \cdot 10H_2O$), not trona, would crystallize preferentially. There is evidence in Lake Van, an alkaline, saline lake in eastern Turkey, of cyanobacterial involvement in the precipitation of calcareous minerals (aragonite, $CaCO_3$) in stromatolite-like reefs (Kempe et al., 1991). The travertine terraces and orifice mounds around the alkaline hot springs on the shores of Lake Bogoria are thickly colonized by (cyano)bacterial tufts (Jones, personal observations) which may be directly involved in the mineralization process (Casanova, 1986; Renaut et al., 1986). Diatoms may influence the removal of SiO_2 and the influence of clays, zeolites and silicate gels (e.g. Magadiite, $NaSi_2O_{13}(OH)_3 \cdot 3H_2O$) on microbiological processes in these environments has probably been underestimated. In one system at least, the soda lakes of the Wadi Natrûn depression in Eygpt, the action of sulphate-reducing bacteria in surrounding swamps formed by ground water seepage from the River Nile, has been proposed as the mechanism primarily responsible for the formation of alkalinity (Abd-el-Malek and Rizk, 1963), but this is unlikely to be a general mechanism.

Alkaliphile Diversity in Soda Lakes and Soda Deserts

Even to the casual observer, one of the striking features of some soda lakes can be their colour. Depending on the prevailing weather conditions and the chemical composition of the individual lakes, they may be coloured various hues of green or even pink, red or orange, due to massive permanent or seasonal blooms of micro-organisms. Soda lakes are regarded as being among the world's most productive naturally occurring aquatic habitats.

In these waters highly charged with carbonate, carbon in the form of CO_2 is virtually unlimited. Combine this with the high daily light intensities in these tropical and sub-tropical regions and productivity rates in excess of 10 g C/m^2/day can be achieved (Talling et al., 1973; Melack and Kilham, 1974; Grant et al., 1990). Only in polluted environments subject to interference by human activity are similar rates of productivity observed. Photosynthetic primary producers are the driving force behind all life in these lakes. The lakes of the East African Rift Valley are typical and are dominated by dense blooms of cyanobacteria especially in the more dilute lakes. Spirulina species are particularly prevalent at higher conductivities but there are also contributions from Cyanospira (Anabaenopsis) (Florenzano et al., 1985; Mwatha, 1991) and unicellular forms which may be Chroococcus spp. (Mwatha, 1991). Spirulina is the principle food of the vast flocks ($> 10^6$ individuals) of Lesser Flamingo (Phoeniconaias minor) that inhabit the Rift Valley soda lakes. Their role in nutrient recycling is likely to be very substantial. At lower conductivities, phototrophic eukaryotes of the diatom genera Nitzschia and Navicula are also common but their wider significance in this ecosystem, if any, has received only a little attention (Mwatha, 1991; Hecky and Kilham, 1973). There is also a substantial contribution to primary productivity made by anoxygenic phototrophic bacteria of the genus Ectothiorhodospira but this is as yet unquantified. Field observations suggest that these may play a role at lower conductivities, although there are several species which may each play a role at different salinities (Table 5).

Table 5. Optimal salinity ranges for *Ectothiorhodospira* spp.

E. mobilis	3-10%
E. shaposhnikovii	1-3%
E. vaculata	1-6%
E. marismortui	3-8%
E. halophila	12-30%
E. halochloris	14-20%
E. abdelmalekii	12-20%

Ectothiorhodospira spp. require reduced sulphur compounds and play a role in the sulphur cycle. Conventional wisdom would suggest that in soda lakes the importance of sulphate-reducing bacteria is likely to be considerable. Geochemists have often implied the presence of microbially mediated sulphate reducing activity in soda lakes in order to explain the appearance of the black lacustrine sediments or the relative depletion of sulphate in the brines of Lake Magadi for example (Jones *et al.*, 1977), but microbiological evidence is often lacking. The black colour of the sediments of the Kenyan soda lakes would indicate the presence of sulphide (Table 6) but there is often no detectable smell of H_2S (Jones and Owenson, personal observations), presumably because the high pH ($>$pH 11) retards the escape of S^{2-} as volatile H_2S. Analysis of sulphate levels (Table 4) suggests that sulphate

Table 6. Sulphide concentrations in soda lakes and salt lakes

Lake	mM S^{2-}
Magadi	6.2
Mono	15 - 40
Big Soda	12.8
Wadi Natrûn	0 - 0.04
Great Salt Lake	0.004 - 0.2
Dead Sea	0.01 - 0.02

Data from Javor (1989), except Magadi (Jones and Meijer, unpublished results).

availability is not a limiting factor. Alkaline enrichments of sulphate-reducing bacteria have been obtained from the lakes of the Wadi Natrûn (Abd-el-Malek and Rizk, 1963), Owens Lake and Searles Lake in California (Nehrkorn and Schwartz, 1961) and Lake Magadi (Jones, unpublished results) but no strain has ever been obtained in pure culture. Judging by the recent isolation of extreme halophilic sulphate-reducing bacteria from neutral, hypersaline environments; *Desulfovibrio halophilus* (Caumette *et al.*, 1991) and *Desulfohalobium retbaense* (Ollivier *et al.*, 1991), the problem with the alkaline environment appears to be technical rather than the lack of bacteria. Recently we appear to have overcome most of the technical difficulties, but the strains of alkaliphilic sulphate-reducing bacteria obtained still remain to be characterized. (Grant and Owenson, unpublished results).

Methanogenic bacteria occupy similar anoxic niches to sulphate-reducing bacteria and in neutral freshwater environments with high sulphate concentrations, methanogens compete unsuccessfully with sulphate-reducing bacteria for the same substrates. Usually, when sulphate is present, methanogenesis is suppressed since SO_4^{2-} is preferred to CO_2 as a terminal electron acceptor (Winfrey and Zeikus, 1977; Abram and Nedwell, 1978). However, in saline environments where sulphate is in excess, methanogens and sulphate-reducing bacteria may co-exist through 'non-competitive catabolism' (Whitman et al., 1992).

Biogenic methanogenesis in some alkaline, saline lakes has been demonstrated by enrichment and isolation studies, for example, Big Soda Lake (Nevada) (Oremland et al., 1982; Iversen et al., 1987), Mono Lake (California) (Oremland et al., 1987) and the lakes of the Wadi Natrûn (Egypt) (Boone et al, 1986; Mathrani et al, 1988) (Table 7). In these alkaline environments, methylotrophic methanogens use alternative substrates such as methanol and methylamines which are less easily utilized by sulphidogenic bacteria (Whitman et al., 1992). Methylated amines when present at high enough concentrations also serve as "non-competitive" substrates in saline environments, where they are formed from glycine betaine and other related osmoprotectants (King, 1988) produced by halophilic bacteria to regulate cytoplasmic osmolarity. Methylated reduced sulphur compounds such as dimethylsulphide and methane thiol are formed by the anaerobic degradation of microbial mats (Zinder et al., 1977) and may serve as additional substrates for methylotrophic methanogens.

Table 7. Methanogenesis in soda lakes

Lake	Situation	pH	Salinity	Reference
Big Soda Lake	Nevada, U.S.A.	9.7	8.8% [a] 2.6% [b]	Iversen et al., 1987 Oremland et al., 1982
Mono Lake	California, U.S.A.	9.8	9 - 10%	Oremland et al., 1987
West Alkali Lake	Oregon, U.S.A.	10.0	10.0%	Liu et al., 1990
Wadi Natrûn	Egypt	7.8-9.3	2.4%	Boone et al., 1986
Bosa		9.7	24.5%	Mathrani et al., 1988
Gaar		10.9	37.4%	
Zugm		11.0	39.4%	
Hamra		11.0	23.8%	Imhoff et al., 1979
Gabara		10.9	9.2%	
Muluk			15.9%	
Rizunia		11.2	38.9%	

a. in monimolimnion
b. in mixolimnion

Of the few alkaliphilic or alkalitolerant methanogens isolated and characterized (Table 8), only Methanohalophilus zhilinae (Mathrani et al., 1988) and Methanohalophilus oregonense (Liu et al., 1990) can be considered as obligately alkaliphilic. Since these species disproportionate methyl-containing compounds in the energy producing production of methane, they have been tentatively assigned to the family Methanosarcinaceae. These are obligately methylotrophic species and cannot use acetate or H_2-CO_2 as energy yielding

Table 8. Properties of some alkaliphilic methanogens

Organism	Source	pH optimum	pH range	Temperature°C optimum	Temperature°C range	Salinity (M) optimum	Salinity (M) range	Substrates	G+C mol%	Ref.
Methanohalophilus oregonense	West Alkali Lake, Oregon, USA	8.6	7.6-9.4	35	25-42	0.5	0.1-1.4	Trimethylamine Methanol	41	a.
Methanohalophilus zhilinae	Bosa Lake Wadi Natrûn Egypt	9.2	8.2-10.3	45	25-55	0.7	0.2-2.1	Methylamines Methanol	38	b.
Methanohalophilus mahii	Great Salt Lake, Utah, USA	7.5	6.5-8.5	35	25-44	2.0	0.5-3.5	Methylamines Methanol	48.5	c.
Methanohalobium evestigatum	Arabat, Crimea	7-7.5	6.0-8.3	50		4.3	2.6-5.1	Trimethylamine	37	d.
Methanobacterium alcaliphilum	Wadi Natrûn Egypt	8.4	6.5-9.9	37	25-40	<0.17	0-0.85	H_2-CO_2	57	e.
Methanobacterium ('thermoalcaliphilum') thermoautotrophicum	Biogas plant	7.5-8.5	6.5-9.5	60	40-69	0.01	0-0.43	H_2-CO_2	38.8	f. g.
'Methanohalococcus alcaliphilum' NY218	Salton Sea California	8.0-8.5	6.5-9.0	40	15-55	2.5-3	1.0-4.0	Methylamines Methanol	42	h.

a. Liu *et al.*, 1990; b. Mathrani *et al.*, 1988; c. Paterek and Smith, 1988; d. Zhilina and Zavarzin, 1990; e. Worakit *et al.*, 1986; f. Blotevogel *et al.*, 1985; g. Kotelnikova *et al.*, 1993; h. Nakatsugawa and Horikoshi, 1989; Nakatsugawa, 1991.

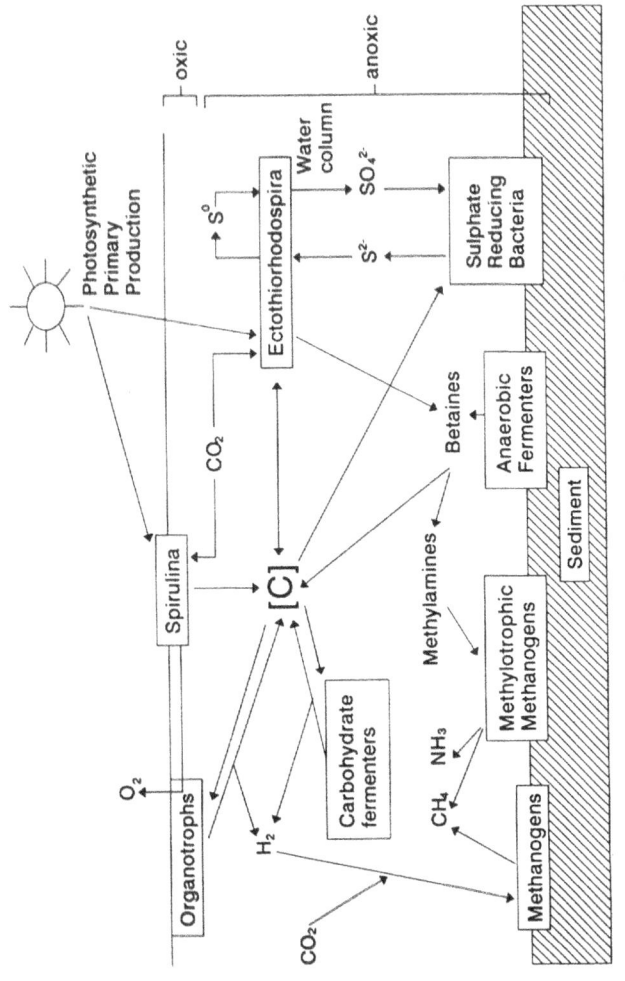

Figure 4. Proposed model for nutrient cycles in soda lakes.

209

substrates. DNA-DNA hybridization and 5S rRNA analyses demonstrated that these and other halophilic methylotrophic methanogens form a coherent group that are closely related to the methanosarcinas (Wilharm *et al.*, 1991; Whitman *et al.*, 1992). *Methanohalobium evestigatum* differs distinctly from the members of the genus *Methanohalophilus* confirming its separate taxonomic status (Zhilina and Zavarzin, 1990). The alkali-tolerant strain NY-218 isolated from Salton Sea, California (pH 8.3, NaCl 15%) (Nakatsugawa and Horikoshi, 1989; Nakatsugawa, 1991) has been given the name "*Methanohalococcus alcaliphilum*" in a patent (Nakatsugawa and Horikoshi, 1991) but the name has no taxonomic standing.

The two alkali-tolerant species assigned to the genus *Methanobacterium* (Table 8) are different in using solely H_2-CO_2 as substrate for growth and methanogenesis; any major carbon sources are unknown. *Methanobacterium thermoalcaliphilum* was isolated from a cattle manure digestor (Blotevogel *et al.*, 1985), but has very recently been reclassified as *Mb. thermoautotrophicum* (Kotelnikova *et al.*, 1993). This species is clearly not alkaliphilic, although some strains may be more alkalitolerant than others. *Methanobacterium alcaliphilum* is clearly alkaliphilic and an inhabitant of the soda lake environment having been isolated from several lakes of the Wadi Natrûn with low salt concentrations (total dissolved solids 24 g/l, pH 7.8 - 9.3) (Boone *et al.*, 1986; Worakit *et al.*, 1986). *Methanobacterium alcaliphilum* does not utilize acetate, trimethylamine, methanol or formate and has not been detected in other alkaline environments. However, a strictly autotrophic behaviour may be advantageous in the soda lake environment where CO_2 is in infinite excess. A search for obligately alkaliphilic strains of autotrophic methanogens in high pH soda lake environments may well repay the effort.

The presence in soda lake sediments of methanogenic bacteria utilizing H_2/CO_2 or methylamines implies the presence of further trophic groups of bacteria which for simplicity we can call "anaerobic fermenters" and "carbohydrate fermenters" (Figure 4). Our rather limited efforts to isolate alkaliphilic fermentative bacteria from both the less concentrated and the hypersaline Rift Valley soda lakes have so far only yielded facultative anaerobic types (Jones and Cohen, unpublished results). However, these represented a wide range of morphological types including, cocci, vibrios, spore-forming and non spore-forming rods. The isolates were almost exclusively obligately alkaliphilic, growing optimally at pH 8 to pH 10, and occasionally up to pH 11. A range of NaCl tolerances were exhibited. Some strains tolerated <10% NaCl, others tolerated >20%, but only a very few isolates were obligately halophilic. At least six different groups of bacteria were indicated by a simple numerical taxonomy study which also clearly showed the influence of the isolation procedure (direct plating, batch or continous culture) on the types of bacteria recovered. Japanese workers have characterized some similar non-spore-forming rods from lakes Magadi and Nakuru in Kenya and Owens Lake in the USA (Shiba and Horikoshi, 1989) for which the genus "*Haloalcalibium*" has been unofficially proposed (Shiba, 1989).

If we again take our cue from the rather more completely explored neutral hypersaline environment, where obligately anaerobic bacteria such as *Halobacteroides*, *Haloanaerobium* (Oren, 1986), *Haloincula* (Zhilina *et al.*, 1992), *Halocella* (Bolobova *et al.*, 1992), *Acetohalobium* (Kevbrin and Zavarzin, 1992) and clostridia (Fendrich *et al.*, 1990) perform the intermediary functions in nutrient recycling by fermenting carbohydrates and betaines, then we may credibly postulate the presence of specific alkaliphilic versions in the soda lake environment. Indeed, Shiba and Horikoshi (1989) have already reported sketchy details of particular strictly anaerobic bacteria from Kenyan and Californian soda lakes growing optimally at an alkaline pH.

Soda lakes also maintain dense populations of aerobic organotrophic bacteria supported by the primary productivity. Although there is a considerable body of limnological data on the Rift Valley soda lakes (Beadle, 1932; Jenkin, 1932; Rich, 1933; Hecky and Kilham, 1973; Talling *et al.*, 1973; Melack and Kilham, 1974; Melack, 1981), rather little attention had been paid until recently to the populations of non-phototrophic, organotrophic bacteria. These bacteria are easily isolated on nutrient-rich media poised at the appropriate pH (Na_2CO_3) and NaCl concentration (Grant and Tindall, 1980; Horikoshi, 1991). In contrast to the situation in soil, the soda lake populations are composed principally of Gram-negative bacteria. In the more dilute lakes (Bogoria, Nakuru, Elmenteita, etc.) with total salinities between 5% and 15% w/v, total counts using epifluoresence microscopy indicate bacterial populations of 10^7-10^8 per ml (Kilham, 1981). Viable counts of aerobic organotrophs ('copiotrophs') from these waters indicated 10^5-10^6 cfu/ml (Grant and Mwatha, 1989; Grant et al., 1990; Mwatha, 1991). It may seem suprising that these rather shallow (Nakuru is 3.3 m and Elmenteita 1.9 m deep, Bogoria is somewhat deeper), eutrophic lakes should contain so many aerobes. However, very high day-time rates of oxygen production (>2 g $O_2/m^2/h$ (Melack and Kilham, 1974) from Spirulina sp., which are gas vacuolated and rise to the surface in the early morning, together with strong regional winds producing sufficient wave action and mixing of a substantial proportion of the water column provides an adequate explanation (Melack and Kilham, 1974; and personal observations).

Monthly monitoring in the period 1988 - 1989 (usually two rainy seasons) of bacterial numbers in relation to physical parameters and algal numbers, etc., revealed a remarkably stable population of about 10^5 cfu/ml of aerobic organotrophs, in spite of considerable fluctuations in conductivity of the water (Figure 5). The cyanobacteria (as measured by chlorophyll a) appeared to have more difficulty adapting to fluctuations in salinity and alkalinity, but there was no evidence of any greater secondary productivity (of aerobic organotrophs) concomitant with or following the cyanobacterial bloom (Mwatha, 1991). The lakes experience two algal blooms per year with numbers and species composition dependant on the conductivity and alkalinity of the waters. Abstract mapping procedures using UNIRAS(UNIMAP) programmes revealed that the larger cyanobacterial bloom was nitrogen-limited in Lakes Nakuru and Elmenteita, but phosphate-limited in Lake Bogoria. Analyses of the bacterial data in relation to conductivity, alkalinity, phosphate and nitrogen levels indicated several distinct bacterial populations limited by an intricate combination of these variables. Since the bacteria are alkaliphilic their numbers were largely immune to any slight changes in alkalinity, but their growth was limited by available nitrogen and phosphate through competition by phototrophs. Cyanobacterial numbers were more susceptible to changes in alkalinity (Figure 5) and as their numbers varied in response to changing conditions, different poulations of organotrophs were able to dominate at a particular time (Mwatha, 1991). Although many of the same types of bacteria can be isolated from all of the lakes, it did appear that each lake had a unique population dominated by a few distinct types (Mwatha, 1991). This is further underlined when lake waters were examined using the BIOLOG system (Biolog, Inc., Hayward, California) which contains 95 carbon sources. When fresh samples of lake waters were used as inoculum, each lake produced a unique pattern (Jones, unpublished results). The work of Fritze et al. (1990) and Spanka and Fritze (1993) on the taxonomy of the alkaliphilic bacilli (including isolates from Mono Lake and the Wadi Natrûn) also serves to illustrate that soda lakes harbor unique types of bacteria even within recognized taxa.

A different population of prokaryotes is found in the concentrated brines and trona deposits of the strongly hypersaline (>20% NaCl), alkaline lakes such as Lake Magadi and Lake Natron in the Rift Valley, Owens Lake in California and some of the lakes of the Wadi Natrûn depression in Egypt. No cyanobacterial blooms are seen in this environment except rarely when flooding causes dilution of the surface brines. The origin of primary production is unclear but is possibly due to other Ectothiorhodospira spp. - Ectothiorhodospira halophila can be isolated at Magadi (Grant and Tindall, 1986 and unpublished observations) and E. halochloris and E. abdelmalekii impart a green colour to some of the Wadi Natrûn lakes (Imhoff et al., 1979). Secondary productivity is represented by obligately halophilic and alkaliphilic archaea, which are present in such abundance (10^6-10^7/ml) (Imhoff et al., 1979; Grant and Tindall, 1986) that the brines and trona surface are coloured red. These

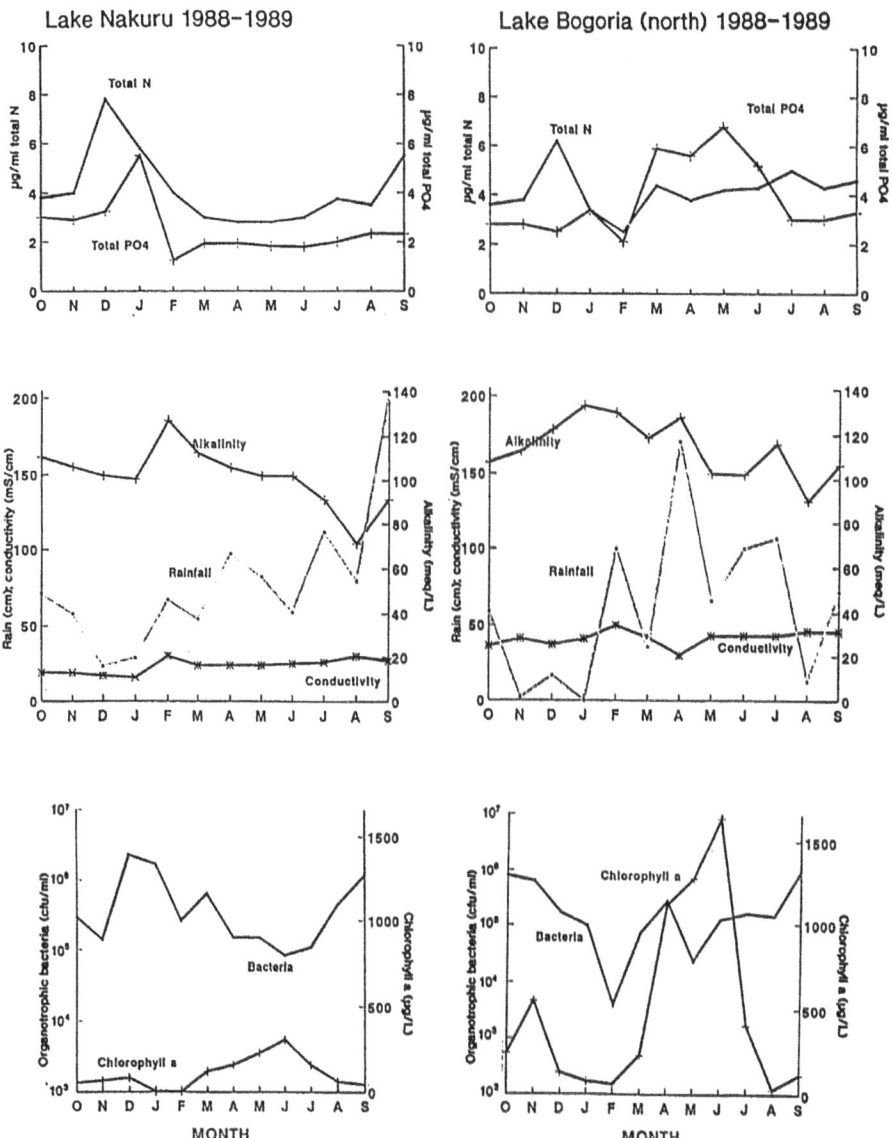

Figure 5. Monthly monitoring of two Rift Valley soda lakes.

organisms are currently classified into two genera (Grant and Larsen, 1989; Tindall, 1992): the coccoid types in the genus *Natronococcus* containing a single species *N. occultus* (Tindall *et al.*, 1984) and the rod-shaped isolates in the genus *Natronobacterium* with three recognized species - *N. pharaonis*, *N. gregoryi* and *N. magadii* (Tindall *et al.*, 1980; Soliman and Trüper, 1982; Tindall *et al.*, 1984). Recently we have described a new gas vacuolated isolate from Lake Magadi for which the name *Natronobacterium vacuolata* has been proposed (Mwatha and Grant, 1993). This organism may have the advantage of being able to adjust its buoyancy to the oxygen richer surface waters in brines largely devoid of oxygen. Natronococci seem to have a rather restricted distribution. Natronobacteria however, are widely distributed in soda lakes and organisms of this type have been isolated from the Wadi Natrûn (Soliman and Trüper, 1982); Sambhar salt lake in India (Upasani and Desai, 1990); Chagonnor, Chahannao and Wudunao soda lakes in China (Wang and Tang, 1989); and soda soils of Armenia (Zvyagintseva and Tarasov, 1988). Also, there is good evidence that these organisms were isolated from Californian soda lakes more than 30 years ago, before their significance was realized (Nehrkorn and Schwartz, 1961). Besides having some unusual lipids (De Rosa *et al.*, 1983; 1988), some of which may be species specific (Morth and Tindall, 1985; Lanzotti *et al.*, 1989) there are few diagnostic features to distinguish the recognized species (Grant, 1989). However, these genera are clearly halobacteria, representing two distinct lineages within this group (Lodwick *et al.*, 1991; McGenity and Grant, 1993).

The numerous warm springs (some $>80°C$) and 'soda' seeps in the Magadi-Natron basin represent a separate habitat within the soda lake environment. These are characterized by thick green or orange algal mats which appear to be the primary producers in quite a complex ecosystem which includes a cichlid fish (Tilapia, classified as *Oreochromis alcalicus grahami*) (Randall *et al.*, 1989), a 'brine fly' and a spider (Jones and Grant, personal observations). At least some of the green/orange algal mats are known to be the *Chloroflexus-Synechococcus* association which have been noted in thermal environments in other parts of the world (Grant and Tindall, 1986; and unpublished observations). These springs are more dilute than the lakes which they feed and have conductivities comparable with the 'northern' lakes (Bogoria, etc.). Although these springs and their overflow streams contain alkaliphilic organotrophs also found in the 'northern' lakes, they also contain some distinct types (Jones, unpublished observations). This environment provides some interesting pH and salinity gradients as well as thermal gradients which would be fruitful subjects of investigation - if they were more accessible. Recently we have succeeded in isolating some moderately thermophilic alkaliphiles from these sites (Jones and Meijer, unpublished results), but their true nature awaits investigation.

Diversity of Alkaliphilic Organotrophs from Kenyan Soda Lakes

Several hundreds of strains of aerobic, non-photosynthetic, organotrophic bacteria and archaea have been isolated from samples of soil, water and sediments in and around the soda lakes of the Kenyan-Tanzanian Rift Valley. The microorganisms were isolated on the basic media described by Grant and Tindall (1980) with a variety of carbon sources such as glucose, peptone, yeast extract, lactalbumin, casein, starch and olive oil. A further variety was introduced by the use of antibiotics (e.g. novobiocin, penicillin) and surfactants, including oxbile, in the isolation medium. These organisms are probably not exclusive to Kenyan soda lakes since a similar sort of bacterium has been reported from Cahannor soda lake in China (Ma *et al.*, 1992). Many of the isolates were biochemically reactive, presumably reflecting the nutrient-rich environment they inhabit and their manner of isolation; other isolates were nutritionally more fastidious. The vast majority of the isolates were obligately alkaliphilic with many strains exhibiting no growth below about pH 8. Although most of the isolates could tolerate up to about 12% NaCl, only a small number were obligately halophilic. An analysis of the G+C content of the DNA of 29 strains revealed a wide range of 34-67 mol%, indicating a variety of taxa. A random selection of 115 isolates representing all the soda lakes sampled, was divided into two categories on the basis of cell wall chemistry; namely, Gram-negative and Gram-positive types. Organisms growing at > 15% NaCl were assigned to a separate group of haloalkaliphiles. Normal Gram staining procedures including the modification originally devised for halophiles (Dussault, 1955) were unsatisfactory since most isolates stained Gram-negative regardless of cell wall chemistry. The isolates were instead assigned to cell wall groups on the basis of the KOH sensitivity test (Gregersen, 1978; Halebian et al., 1981), the aminopepidase reaction (Cerny, 1976; 1978) and a quinone analysis (Collins and Jones, 1981).

Taxonomic Analysis of Gram-Negative Alkaliphilic Isolates. Seventy isolates were tested for 104 character states using commercially available systems such as ATB 32GN and APIZYM (API-bioMérieux) where possible and the data compared with those from 20 known species representing 17 genera of Gram-negative bacteria. The data were analysed using the TAXPAK programme package (Sackin, 1987). Taxonomic resemblance was estimated by similarity matrices constructed using the Gower (S_G) and Jaccard (S_J) coefficients and cluster analysis performed using the UPGMA algorithm. The dendrogram resulting from the S_G/UPGMA analysis exhibited six clusters (labelled 1 to 6), at the 73% similarity level (Figure 6) (Jones et al., 1992). Cluster 3 could be divided at higher levels of similarity into four sub-clusters (labelled A to D). Five of the isolates did not fall into any of the clusters. Clusters 1 and 2 were the only phenons to show an association with any of the known organisms, i.e. cluster 1 incorporated *Comamonas terrigena* and *Pseudomonas putida*, and cluster 2, *Pseudomonas stutzeri*. Both of the *Pseudomonas* species clearly lay at the very boundary of phenetic hyperspace defining the clusters and since both species failed to associate when the S_J/UPGMA analysis was applied, a close taxonomic relationship can be discounted. In general, all the clusters generated by the S_G/UPGMA method were recovered in the dendrogram produced by the S_J/UPGMA analysis, except that the S_J transformation combined clusters 1 and 5. Since both of these clusters were characterized by biochemically fairly unreactive strains, this was perhaps not unexpected. The S_J transformation also resulted in cluster 4 becoming embedded within cluster 3.

The conclusions drawn from the numerical analysis were further supported by chemotaxonomic data (Table 10). In particular, ubiquinone-6 is diagnostic for all strains of cluster 3. Polar lipid patterns obtained by 2D TLC showed that all strains contained phosphatidylglycerol (PG), diphosphatidylglycerol (DPG), phosphatidylglycerol phosphate (PGP) and phosphatidylethanolamine (PE). In addition, most of the strains of sub-clusters 3B and 3D also contained phosphatidylglycerol sulphate (PGS), which may be useful as a non-exclusive taxonomic marker. Most of the bacteria also contained a glycolipid, possibly common to all these Gram-negative alkaliphiles. Fatty acid patterns obtained by GC and GC/MS analyses showed some discriminatory features and Table 10 shows unique profiles for each of the clusters. Clusters 1, 2, 3 and 4 are fairly typical of the majority of Gram-negative bacteria with C16:0 (25-42%) as the major saturated fatty acid, with lesser amounts of C14:0 and C18:0. The major unsaturated fatty acid in these four clusters is 11-*cis* C18:1 with lesser amounts of C16:1 in sub-clusters 3A and 3B. Again this is typical of Gram-negative bacteria, as is the lack of fatty acids with odd numbers of carbon atoms (Wilkinson, 1988). The minor amounts of C17:0 and C19:0 cyclopropane fatty acids are not atypical but appear diagnostic for cluster 2 and sub-cluster 3D. No hydroxy-fatty acids typical of *Pseudomonas* species were detected. The fatty acid profiles for the strains from clusters 5 and 6 were remarkable in containing major amounts of C16:0 (12-40%), and other major fatty acids with odd-numbered branched chain acids (40-85%). Also, these strains lacked significant amounts of C18:1 or any of the other unsaturated acids which were present in appreciable amounts in the alkaliphilic strains of clusters 1, 2, 3 and 4. The presence of

214

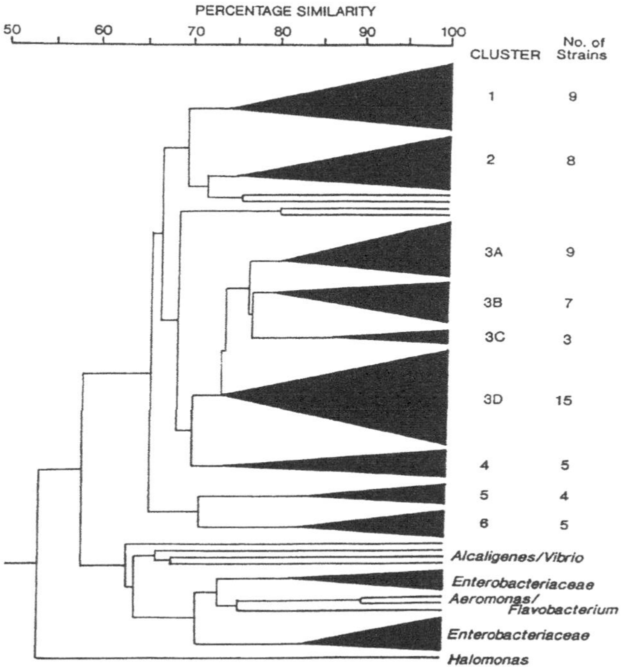

Table 9. A probability matrix for the identification of Gram-negative alkaliphiles

Figures are percentage distribution of positive discriminatory characters which define the clusters at the 73% level (S_G)

TEST	CLUSTER					
	1	2	3	4	5	6
N-acetylglucosamine(1)	13	0	26	20	0	100
Saccharose(1)	25	0	74	20	25	100
Maltose(1)	25	0	68	60	50	100
Lactate(1)	38	50	100	100	0	40
Propionate(1)	0	100	91	60	0	80
Valerate(1)	13	100	97	80	0	40
Citrate (1)	13	50	94	20	50	100
Histidine(1)	0	0	71	0	0	80
Glycogen(1)	0	13	26	20	25	100
3-hydroxybutyrate(1)	13	25	94	100	0	100
4-hydroxybenzoate(1)	0	0	71	80	0	0
Leucine arylamidase(2)	88	63	94	60	50	0
Valine arylamidase(2)	88	25	65	100	25	0
Phosphohydrolase(2)	88	13	3	20	75	40
α-galactosidase(2)	0	0	3	20	75	0
Ampicillin	50	63	56	0	100	80
Fusidic acid	25	3	3	20	100	100
Methicilin	50	13	50	20	100	100
Polymixin	88	50	81	100	0	0
Vancomycin	13	13	3	20	100	75
Yellow colony	0	0	0	0	100	0
Translucent colony	0	100	3	0	100	0
Lipase	0	100	21	0	0	0
Oxidase	25	88	6	0	0	0

(1): ATB 32GN system (bioMérieux)   (2): APIZYM test strips

Figure 6. Simplified dendrogram showing the relationship between clusters of Gram-negative alkaliphilic bacteria revealed by the S_G/UPGMA analysis.

Table 10. Chemotaxonomic characteristics of the Gram-negative taxa defined by the S_G/UPGMA analysis

CLUSTER	1	2	3A	3B	3C	3D	4	5	6
Polar lipids									
PG	+	+	+	+	+	+	+	+	+
DPG	+	+	+	+	+	+	+	+	+
PGP	+	+	+	+	+	+	+	+	+
PE	+	+	+	+	+	+	+	+	+
PGS	–	–	–	+	–	+	–	–	–
Ubiquinones	Q9	Q9	Q6(Q9)	Q6(Q9)	Q6(Q9)	Q6	Q9	Q9	Q9
Major fatty acids (>10%)	C16:0 11-*cis* C18:1	C16:0 C18:0 11-*cis* C18:1	C16:0 C16:1 11-*cis* C18:1	C16:0 C16:1 11-*cis* C18:1	C16:0 11-*cis* C18:1	C16:0 11-*cis* C18:1	C16:0 11-*cis* C18:1	*a*-C15:0 C16:0	*br*-C15:0 C16:0
Trace markers		C17/19 cyclo				C17/19 cyclo		*a*-C17:0	
G+C mol%	51.1–55.2	62.7	63.0	64.1		64.8–67.1	58.1–65.3	43.8–50.0	37.6–41.0

large amounts of C15:0 and C17:0 *iso* and *anteiso* acids is characteristic of only a very few groups of Gram-negative bacteria, notably species from exotic environments such as *Thermus*, or poorly defined taxa such as *Flavobacterium* (Wilkinson, 1988), further emphasizing the novelty of these soda lake isolates. It is possible that the fatty acid compositions on their own may permit an effective identification of these alkaliphilic bacteria.

The clusters were further analysed by determination of the most representative strain characterizing the cluster. The centroid of each individual cluster generated by the S_G/UPGMA method was computed using the RGROUPS sub-routine in TAXPAK. The centroid of a cluster of points representing real organisms projected into hyperspace represents a hypothetical average organism. The centroid rarely, if ever represents a real organism. Therefore, the Euclidean distances of each of the members of the cluster from the centroid of the cluster were calculated in order to establish which strain was closest to the hypothetical average organism. The strain closest to the centroid was designated the "centrotype" (indicated with the superscript "CT"). The centrotype organism can be thought of as the "Reference Strain" which most closely represents the essential and discriminating features of each cluster as a whole. DNA-DNA homology studies performed between DNA from the centrotype strains and DNA from other strains indicated higher levels of homology between strains of the same cluster than between strains from a different cluster, again emphasizing that the clusters are likely to represent discrete taxa (Collins, Grant & Jones, unpublished results).

So far, three of the centrotype strains (1E.1[CT], 28N.1[CT] and 64B.4[CT]) representing clusters 1, 3 and 6 have been subjected to direct automated DNA sequence analysis of the 16S rRNA gene, using a PCR method adapted from Weisburg *et al.* (1991) (Van Steenbergen, Roozen & Jones, unpublished results). Phylogenetic analysis revealed that all 3 strains belong to the gamma-3 subdivision of the Proteobacteria (Woese *et al.*, 1985) (Figure 7) (Van Steenbergen, McGenity & Jones, unpublished results). Strains 28N.1[CT] and 64B.4[CT] are clearly more closely related to each other than to strain 1E.1[CT]. Preliminary results indicate that strain 1E.1[CT] belongs to the Enteric-*Aeromonas*-Vibrio branch of the gamma-3 subdivision and probably deserves separate generic status. For strain 1E.1[CT] we suggest the name *"Magatibacter afermentans"* - the "non-fermentative soda-rod" (from "Magati" meaning "soda" in the language of the Maasai, a tribe of the Kenyan Rift Valley, and "afermentans" reflecting the restricted range of substrates for growth), prior to the further analysis required before a formal proposal can be made. Strains 28N.1[CT] and 64B.4[CT] are clearly members of the *Halomonas* group, which includes *Deleya*, a deep branch of the gamma-3 subdivision (Woese *et al.*, 1985). Although the percentage dissimilarity between strain 64B.4[CT], *Halomonas elongata* ATCC 33173 and *Halomonas halmophilum* ATCC 19717 (formally *Flavobacterium halmophilum*) based on analysis of 16S rRNA gene sequences is rather small, 4.9% and 4.8% respectively (McGenity and Grant, unpublished results), the significantly lower G+C content of the DNA of strain 64B.4[CT] (41 mol%) compared to the 59-63 mol% range for *Halomonas* (Vreeland, 1992) indicates a fundamental difference. Since strain 28N.1[CT] is no more closely related to strain 64B.4[CT] than 64B.4[CT] is related to *Halomonas* (Figure 8) and 28N.1[CT] is more distantly related to *Halomonas* than 64B.4[CT], we believe a separate generic status for the two obligately alkaliphilic strains is indicated, supported by G+C content, DNA-DNA reassociation studies and chemotaxonomic evidence (Figure 8). Conclusive evidence for the separation of 28N.1[CT] and 64B.4[CT] from *Halomonas* is seen in the lack of 3 or 4 of the 6 unique RNA signature sequences common to *Halomonas* (Figure 8).

A distinction from *Deleya* is more difficult to assess since only partial 16S rRNA gene sequences are available for *D. marina* (Kita-Tsukanoto *et al.*, 1993), but the approximate dissimilarity values (Table 11) (McGenity and Grant, unpublished results) and the alkaliphilic phenotype are indicative of separate genera. For strain 64B.4[CT] we are

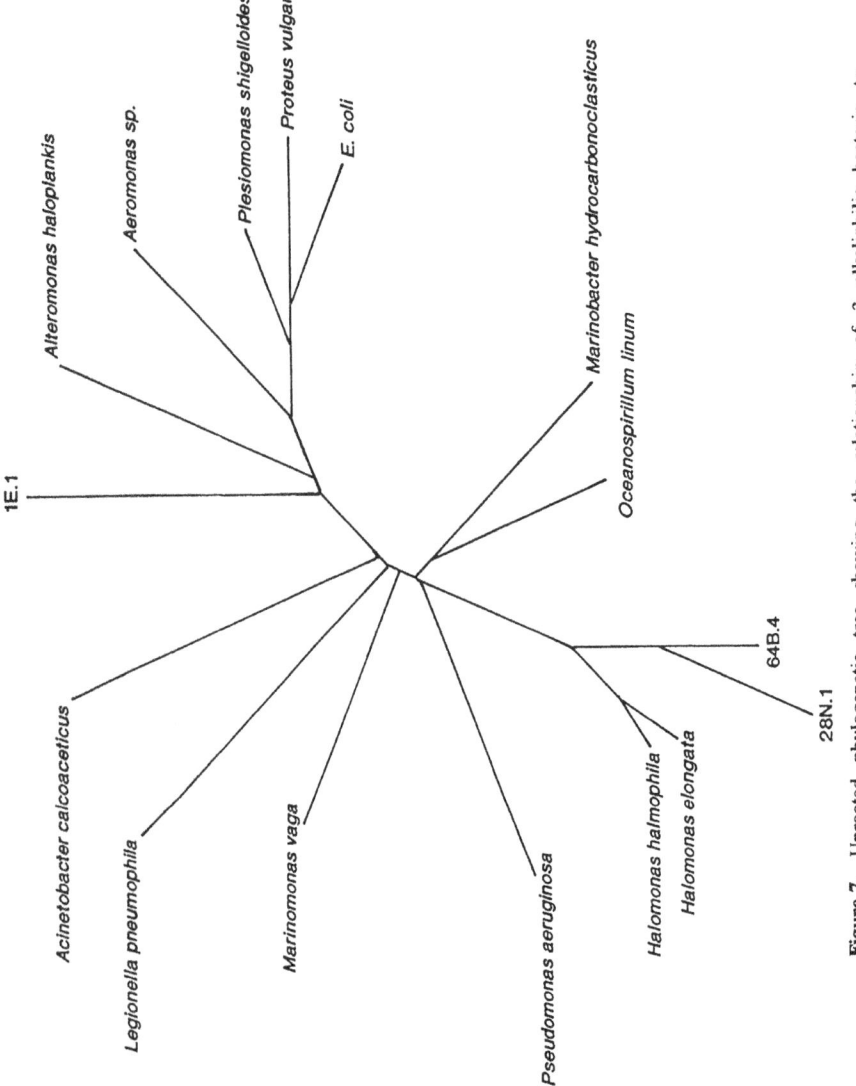

Figure 7. Unrooted phylogenetic tree showing the relationship of 3 alkaliphilic bacteria to representatives of the gamma-proteobacteria.

Relative Phylogenetic Distance

	H.h	H.e	64B.4
H.h			
H.e			
64B.4	54	66	
28N.1	71	75	54

H.h = *Halomonas halmophilum*, H.e = *Halomonas elongata*

G + C content mol%

Halomonas	59-63
28N.1	64
64B.4	41

DNA-DNA Hybridization

Strain	28N.1	64B.4
28N.1	100	30
64B.4	35	100

Major Respiratory Quinone

Halomonas	Q9
28N.1	Q6
64B.4	Q9

Major Fatty Acids

Halomonas	C12:0, 3-OH-C12:0, iso-C15:0, C16:0, C16:1, C17:0, 9-*trans*-C18:1
28N.1	C16:0, C16:1, 11-*cis*-C18:1
64B.4	iso-C15:0, anteiso-C15:0, C16:0

Polar Lipid Composition

Halomonas	PG DPG PE
28N.1	PG DPG PGP PE PGS
64B.4	PG DPG PGP PE

RNA Signature Sequences

	Halomonas	28N.1	64B.4
CCUAACUUCG	+	– CCUAACCUUCG	– CCUAACCUUCN
UUAAUACCCG	+	+	– UUAAUACCCU
AUAACUUG	+	– AUAACCUG	– UNANCGUG
CCCUCG	+	– CCUUCG	– CCUUCG
UCUCAG	+	+	+
UUAACG	+	+	+

Figure 8. Comparison of *Halomonas* with alkaliphilic strains 28N.1 and 64B.4.

proposing the name *"Igatibacter hanningtonii"* from "igati" meaning soda in the language of the Kikuyu, a tribe in the central Rift Valley of Kenya and "bacter" meaning rod. The species epithet is derived from Bishop Hannington who in 1885 first described Lake Bogoria the source of strain 64B.4[CT]. For strain 28N.1[CT] the genus *"Sodabacter"* is proposed with a new species *"Sodabacter nakuruensis"* since Lake Nakuru was where the organism was found.

Table 11. Percent Dissimilarity

	64B.4[CT]	*Deleya marina*	*Halomonas elongata*
64B.4[CT]	-		
Deleya marina	6.5	-	
Halomonas elongata	5.9	8.0	-
Halomonas halmophilum			2.9

Identification of Gram-Negative Alkaliphiles. One of the purposes of a numerical classification is to use the phenetic data defining the taxa for the assignment or identification of unknown strains. The classification test data can be analysed to determine the minimum set of tests which are required to define the clusters at the 73% (S_G) level, and identify those characters which are most diagnostic for each individual cluster. In other words, the minimum number of tests required to assign an unknown organism to a pre-determined cluster (taxon) with a high degree of predictability. From the minimum discriminatory tests, a probability matrix can be constructed for the identification of unknown strains. This analysis was performed using the TAXPAK programme package, namely the CHARSEP, DIACHAR and MCHOICE sub-routines with an evaluation of the identification matrix provided by the MOSTTYP and OVERMAT programmes. The mechanics and theory of the exercise are too complex for a full discussion here, but practical examples of the use of these programmes for the probabilistic identification of bacteria can be found in Williams *et al.* (1983), Priest and Alexander (1988) and Alexander and Priest (1990). Table 9 shows the set of 24 tests, the minimum required to define the six clusters, plus the identification matrix of percentage positive characters computed by the IDMAT programme. As a further stringent evaluation of the validity of the identification, matrix all 70 of the Gram-negative alkaliphilic bacteria used in the original numerical taxonomic analysis were subjected to cluster analysis using the S_G/UPGMA method using only the data for the derived set of 24 minimum discriminatory tests. Although there was some rearrangement of the positions of the clusters, their composition was largely unchanged at a 68% similarity level. The robustness of the classification was further illustrated by the successful assignment to clusters, with a high level of confidence, of unknown strains of Gram-negative, alkaliphilic bacteria subsequently isolated from Mono Lake, a hypersaline, alkaline lake in California (Javor, 1988) using only the battery of 24 tests and analysis by the MATIDEN programme.

Taxonomic Analysis of Gram-Positive Alkaliphilic Isolates. Of the 115 randomly selected isolates for taxonomic analysis, only 20 strains could be assigned to the Gram-positive type, presumably reflecting the relative proportions found in the soda lake environment. The isolates were analysed in much the same way as the Gram-negative isolates except that the inclusion of the BIOLOG system (Biolog, Inc., Hayward, CA) with its 95 substrates brought the number of character states up to 200. As controls, 17 known Gram-positive species were subject to the same analysis using the same conditions where appropriate. These reference bacteria included genera that are known to include facultative or obligate alkaliphilic species, many of which have been assigned, rightly or wrongly, to the genus *Bacillus*. Numerical analysis was performed using the S_G, S_J and simple matching (S_{SM}) coefficients and the UPGMA algorithm. The dendrogram resulting from the S_G/UPGMA transformation revealed six clusters (labelled 1 to 6), at the 79% similarity level (Figure 9) (Jones *et al.*, 1993). At the 79% level, clusters 1, 3 and 4 contained exclusively alkaliphilic bacteria representing 13 of the 20 newly-isolated strains from soda lakes. Five of the soda lake isolates failed to cluster. Clusters 2, 5 and 6 contained most of the *Bacillus* species, *Exiguobacterium aurantiacum* and two of the soda lake isolates. The main effect of the S_J/UPGMA and S_{SM}/UPGMA transformations was to gather all the *Bacillus* strains in a single cluster but still retaining three separate clusters of soda lake isolates. The significance of the clustering and the overlap characteristics were analysed by the TESTDEN programme and found to be satisfactory (P = 0.95-0.99). Chemotaxonomic data were not clearly discriminative (Table 12). No clear polar lipid pattern emerged which was distinct for any one cluster. The respiratory quinones likewise appeared to be of little value for a circumscription of the clusters, although the alkaliphilic soda lake strains tended to contain more of the shorter molecules, especially MK-4 and MK-6, as compared to *Bacillus* spp. Fatty acid compositions appear to be potentially more useful as markers, although they do point to a less homogeneous cluster 4 than the numerical analysis would suggest. The profiles for clusters 1, 3 and 4, which contain the soda lake isolates, are fairly typical for some members of the coryneform-mycobacterium-nocardioform group of the actinomycetes (Brennan, 1988), an association supported by their characteristic cell habit (irregular packets of cocci or irregular rods in pallisade formation), possession of dihydromenaquinones and their bright pigmentation (yellow, orange and red). The G+C compositions of the DNA from the soda lake strains also clearly sets them apart from the alkaliphilic *Bacillus* spp. but further work is required before the exact taxonomic status of the new isolates can be established fully.

Identification of Gram-Positive Alkaliphiles. As an aid to the identification of unknown soda lake isolates, the minimum number of tests required to define the clusters was computed in exactly the same manner as for the Gram-negative isolates. Table 14 sets out the 32 tests and the identification matrix. When all of the strains in this numerical taxonomic study were subjected to cluster analysis using only the data for the derived set of 32 minimum discriminatory tests, the reconstructed dendrogram exhibited a clear separation between the alkaliphilic soda lake strains and the alkaliphilic *Bacillus* species. A statistical analysis of cluster overlap, reliability of identification of the hypothetical median organisms and the identification scores for the re-assignment of centrotype strains to the correct cluster, all further confirmed the robustness of the identification matrix.

Taxonomic Analysis of Haloalkaliphilic Isolates. Twenty-five isolates were assigned to the category haloalkaliphile on the basis of their ability to grow in a NaCl concentration of 15% or more and at greater than pH 10. All the isolates were obtained from the environs of the most concentrated lakes with the highest pH and conductivity values, i.e. Lake Magadi, Little Lake Magadi and Lake Natron. The 25 strains were tested for 107 characters, which for practical purposes were divided into 123 character states for numerical analysis. Many of the strains were slow growing, which sometimes frustrated efforts for obtaining complete data. As controls, five known haloalkaliphilic archaea belonging to the genera *Natronobacterium* and *Natronococcus* were subjected to the same analysis using the same conditions. Numerical analysis was performed using the S_{SM}/UPGMA and S_J/UPGMA

Table 12. Chemotaxonomic characteristics of the Gram-positive alkaliphilic taxa defined in the S_G/UPGMA analysis

CLUSTER	1	2	3	4	5	6
Polar lipids						
PG	+	+	+	+	+	+
DPG	+	(+)	+	+	+	+
PGP	+	(+)	+	+	+	+
PI	(+)	(+)	+	+	+	+
PE	(+)	(+)	+	-	+	-
GL	(+)	+	-	+	(+)	+
Menaquinones	(MK-3) MK-4 (MK-5) (MK-5H$_2$) (MK-6) (MK-7) (MK-8)	(MK-4) (MK-6) MK-7 (MK-8) (MK-9)	(MK-4)(MK-6 (MK-7) (MK-11)	MK-4 (MK-5) (MK-6) (MK-7) (MK-8) (MK-8H$_2$) (MK-9)	(MK-4) (MK-6) MK-7 (MK-8)	MK-4 MK-5 MK-8
Major fatty acids (>10%)	a-C15:0 C16:0 a-C17:0 C18:0	C16:0 C18:0	C16:0 C18:0 C20:0	4A a-C15:0 C16:0 a-C17:0 C18:0 4B C16:0 C18:0	a-C15:0 br-C15:0 i-C15:0 C16:0 C16:0 a-C17:0 C18:0 C18:0	br-C15:0 C16:0 C18:0
Additional markers			br-C20:0	br-C16:1 br-C19:0 br-C20:0 br-C20:0		
G + C mol%	52.3-63.1	35.0-39.6	34.1-49.3	63.0-63.5	36.1-42.8	36.5-43.6

Data in brackets indicates a variable character

222

Table 13. A probability matrix for the identification of Gram-positive alkaliphiles. Figures are percentage distribution of positive discriminatory characters which define the clusters at the 79% level (S_G)

TEST	CLUSTER					
	1	2	3	4	5	6
Gelatin(1)	100	100	100	0	100	100
Fumarate	20	25	75	0	83	50
Fructose	60	100	75	0	83	100
Galactose	20	25	100	0	17	0
N-acetylglucosamine(2)	0	0	50	25	100	100
Saccharose(2)	0	75	25	0	100	100
Maltose(2)	20	100	0	25	100	100
Acetate(2)	20	25	0	75	100	100
Glucose(2)	20	75	0	25	100	100
Salicin(2)	0	100	50	25	67	100
Melibiose(2)	0	50	0	0	50	100
Propionate(2)	0	0	0	75	83	100
Valerate(2)	20	0	0	50	83	50
Glycogen(2)	40	75	0	0	100	100
L-serine(2)	40	0	25	0	17	100
Chymotrypsin(3)	40	25	75	25	100	0
β-glucosidase(3)	20	100	100	0	33	100
Serine	0	100	0	75	50	0
Arginine	0	50	100	100	33	0
Methionine	0	100	nc	100	33	25
Penicillin G	100	75	0	50	83	0
Methicillin	100	100	0	100	100	25
Streptomycin	40	100	50	75	67	0
Tetracyclin	80	75	0	100	100	25
Bacitracin	100	100	100	100	83	0
N-acetyl-D-glucosamine(4)	40	25	100	0	50	50
Cellobiose(4)	0	50	100	0	50	0
Turanose(4)	40	100	25	0	100	100
Methylpyruvate(4)	60	75	0	100	33	75
Monomethylsuccinate(4)	40	25	0	100	0	0
Thymidine(4)	80	25	50	50	83	0
Glycerol(4)	80	0	50	25	33	100

(1); "chargels" (Oxoid): (2); ATB 32GN test strip: (3); APIZYM test strip: (4); BIOLOG test system.

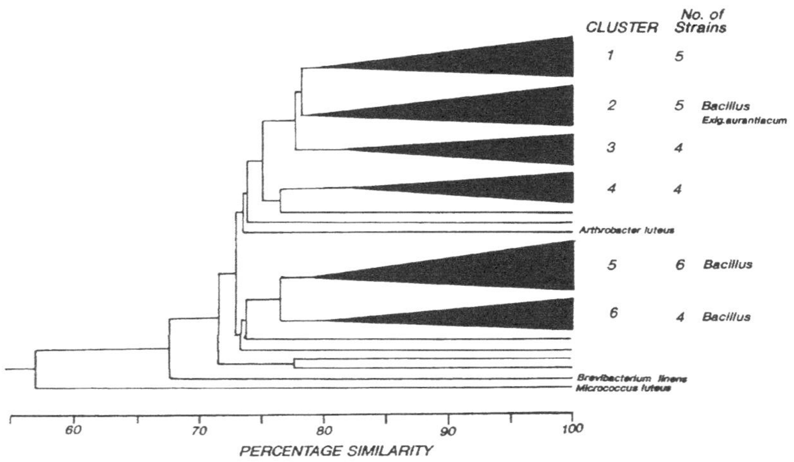

Figure 9. Simplified dendrogram showing the clusters of Gram-positive alkaliphilic bacteria revealed by the S_G/UPGMA analysis.

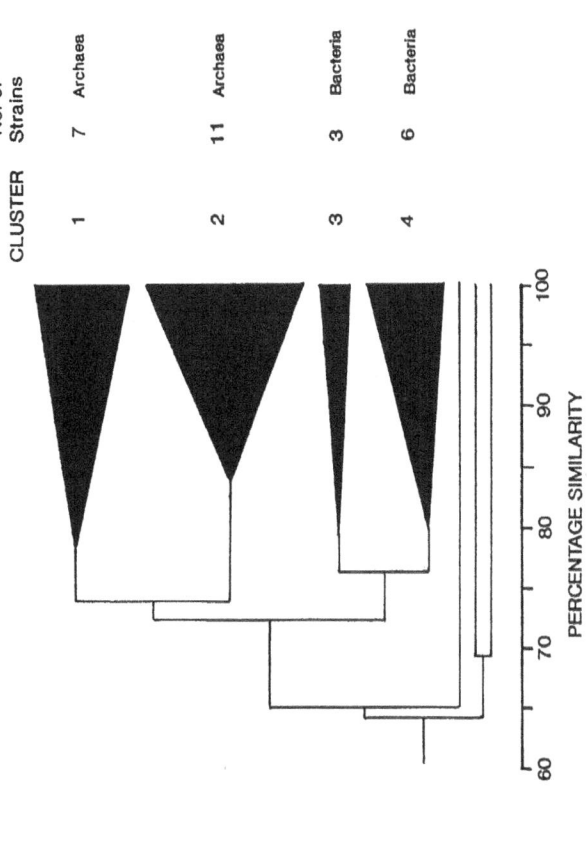

CLUSTER	No. of Strains	
1	7	Archaea
2	11	Archaea
3	3	Bacteria
4	6	Bacteria

PERCENTAGE SIMILARITY

Figure 10. Simplified dendrogram showing the relationships between clusters of haloalkaliphilic bacteria revealed by the S_G/UPGMA analysis.

methods. The dendrogram resulting from the S_{SM}/UPGMA transformation provided four clusters (labelled 1 to 4), at the 78.5% simililarity level (Figure 10) (Jones and Grant, 1993). The group membership of the clusters was identical when the S_J/UPGMA transformation was applied. Clusters 1 and 2 exclusively contained archaea, which was confirmed by insensitivity to penicillin and possession of membrane lipids containing glycerol diether moieties characteristic of the archaea (Kandler, 1993). On the basis of antibiotic sensitivity and membrane lipids based on ester linkages the nine soda lake isolates comprising clusters 3 and 4 can be assigned to the (eu)bacteria. While further work will be required to establish their exact taxonomic status, their unique phenotype (growth in 15-30% NaCl at pH 10) implies these may represent new taxa and reveals an unsuspected diversity of prokaryotic types in the most hypersaline of the soda lakes.

CONCLUSION

Of the three types of environment from which alkaliphilic bacteria may be isolated, the stable high pH environment of the soda lake appears to be the most bountiful when it comes to isolating new varieties. It is obvious that we have only just scratched the surface in revealing the diversity that this exotic and extreme environment may have to offer. A comparison with the rather better explored world of the neutral, saline environment is compelling and the trophic chains outlined in Figure 4 owe a debt to this work. In addition to the scientific challenge involved in putting such pieces of the puzzle together, there is an altogether different challenge to the adventurous microbiologist to get out and explore some of the more inhospitable and inaccessible regions of the world.

ACKNOWLEDGEMENTS
We should like to thank Dr. M.J. Sackin and Jonathan Radbourne for help with the TAXPAK programmes and computing; the Magadi Soda Company and Mr. Pravin Bowry of Nakuru for kind hospitality and assistance with field work; Mr. A.J. Engelen and colleagues of the Intracompany Service Laboratory for analyses; Dr. D. Jones for discussions and help with taxonomy and bacterial nomenclature; The British Council for a grant to W.E. Mwatha; and the directors of Gist-brocades for financial help. Part of this work was performed within the BIOTECH project "biotechnology of extremophiles" of the European Community, contract no. BIO2-CT93-0274.

REFERENCES

Abd-el-Malek, Y. and Rizk, S.G. (1963) Bacterial sulphate reduction and the development of alkalinity. III. Experiments under natural conditions in the Wadi Natrûn. J. Appl. Bact. 26, 20-26.

Abram, J.W. and Nedwell, D.B. (1978) Inhibition of methanogenesis by sulphate-reducing bacteria competing for transferred hydrogen. Arch. Microbiol. 117, 89-92.

Akiba, T. and Horikoshi, K. (1978) Localization of α-galactosidase in an alkalophilic strain of Micrococcus. Agric. Biol. Chem. 44, 2741-2742.

Alexander, B. and Priest, F.G. (1990) Numerical classification and identification of Bacillus sphaericus including some strains pathogenic for mosquito larvae. J. Gen. Microbiol. 136, 367-376.

Ash, C., Farrow, J.A.E., Wallbanks, S. and Collins, M.D. (1991) Phylogenetic heterogeneity of the genus Bacillus revealed by comparative analysis of small-subunit ribosomal RNA sequences. Lett. in Appl. Microbiol. 13, 202-206.

Baker, B.H. (1958) Geology of the Magadi area. Geological Survey of Kenya, Rep. No. 42, Government Printer, Nairobi.

Baker, B.H. (1986) Tectonics and volcanism of the southern Kenya Rift Valley and its influence on rift sedimentation, in "Sedimentation in the African Rifts" (Frostick, L.E., Renaut, R.W., Reid, I. and Tiercelin, J.J., Eds.), Geol. Soc. Spec. Publ., No. 25, Blackwell Scientific, Oxford.

Barnes, I., Rapp, J.B., O'Neil, J.R., Sheppard, R.A. and Gude, A.J. (1972) Metamorphic assemblages and the direction of flow of metamorphic fluids in four instances of serpentinization. Contrib. Mineral. Petrol. 35, 263-276.

Barnes, I., O'Neil, J.R. and Trescares, J.J. (1978) Present day serpentinization in New Caledonia, Oman and Yugoslavia. Geochim. Cosmochim. Acta 42, 144-145.

Barnes, I., Presser, T.S., Saines, M., Dickson, P. and Van Goos, A.F.K. (1982) Geochemistry of highly basis calcium hydroxide groundwater in Jordan. Chem. Geol. 35, 147-154.

Bath, A.H., Christofi, N., Neal, C., Philp, J.C., Cave, M.R., McKinley, I.G. and Berner, U. (1987) Trace element and microbiological studies of alkaline groundwaters in Oman, Arabian Gulf: a natural analogue for cement pore waters. Rep. Fluid Processes Res. Group, Brit. Geol. Surv. FLPU 87-2.

Beadle, L.C. (1932) Scientific results of the Cambridge Expedition to the East African Lakes 1930. IV. The water of some East African lakes in relation to their flora and fauna. J. Linn. Soc. Zool. **38**, 157-211.

Blotevogel, K-H., Fischer, U., Mocha, M. and Jannsen, S. (1985) *Methanobacterium thermoalcaliphilum* spec. nov., a new moderately alkaliphilic and thermophilic autotrophic methanogen. Arch. Microbiol. **142**, 211-217.

Bolobova, A.V. and Siman'kova, M.V. (1992) Cellulase complex of a new halophilic bacterium *Halocella cellulolytica*. Mikrobiologiya (Eng. translation), **61**, 557-562.

Boone, D.R., Worakit, S., Mathrani, I.M. and Mah, R.A. (1986) Alkaliphilic methanogens from high-pH lake sediments. Syst. Appl. Microbiol. **7**, 230-234.

Brennan, P.J. (1988) Mycobacterium and other actinomycetes, in "Microbial Lipids" vol. 1 (Ratledge, C. and Wilkinson, S.G., Eds.), pp.204-298, Academic Press, London.

Casanova, J. (1986) East African rift stromatolites, in "Sedimentation in the African Rifts" (Frostick, L.E., Renaut, R.W., Reid, I. and Tiercelin, J.J., Eds.), Geol. Soc. Spec. Publ., No. 25, Blackwell Scientific, Oxford.

Caumette, P., Cohen, Y. and Matheron, R. (1991) Isolation and characterization of *Desulfovibrio halophilus* sp. nov., a halophilic sulfate-reducing bacterium isolated from Solar Lake (Sinai). Syst. Appl. Microbiol. **14**, 33-38.

Cerny, G. (1976) Method for the distiction of Gramnegative from Grampositive bacteria. Eur. J. Appl. Microbiol. **3**, 223-225.

Cerny, G. (1978) Studies on the aminopeptidase test for the distinction of Gram-negative from Gram-positive bacteria. Eur. J. Appl. Microbiol. **5**, 113-122.

Collins, M.D. and Jones, D. (1981) Distribution of isoprenoid quinone structural types in bacteria and their taxonomic implications. Microbiol. Rev. **45**, 316-354.

Collins, M.D., Lund, B.M., Farrow, J.A.E. and Schleifer, K.H. (1983) Chemotaxonomic study of an alkalophilic bacterium, *Exiguobacterium aurantiacum* gen. nov., sp. nov. J.Gen. Microbiol. **129**, 2037-2042.

De Rosa, M., Gambacorta, A., Nicolaus, B. and Grant, W.D. (1983) A C_{25},C_{25} diether core lipid from archaebacterial haloalkaliphiles. J. Gen. Microbiol. **129**, 2333-2337.

De Rosa, M., Gambacorta, A., Grant, W.D., Lanzotti, V. and Nicolaus, B. (1988) Polar lipids and glycine betaine in haloalkaliphilic archaebacteria. J. Gen. Microbiol. **134**, 205-211.

Dussault, H.P. (1955) An improved technique for staining red halophilic bacteria. J. Bacteriol. **70**, 484-485.

Eugster, H.P. (1970) Chemistry and origins of the brines of Lake Magadi, Kenya. Mineral. Soc. Amer., Spec. Pap. 3, 213-235.

Eugster, H.P. (1986) Lake Magadi, Kenya: a model for rift valley hydrochemistry and sedimentation?, in "Sedimentation in the African Rifts" (Frostick, L.E., Renaut, R.W., Reid, I. and Tiercelin, J.J., Eds.), Geol. Soc. Spec. Publ., No. 25, Blackwell Scientific, Oxford.

Eugster, H.P. and Hardie, L.A. (1978) Saline lakes, in "Lakes: Chemistry, Geology and Physics" (Lermann, A., Ed.), pp.237-293, Springer-Verlag, New York.

Fendrich, C., Hippe, H. and Gottschalk, G. (1990) *Clostridium halophilium* sp. nov. and *C. litorale* sp. nov., an obligate halophilic and marine species degrading betaine in the Stickland reaction. Arch. Microbiol. **154**, 127-132.

Florenzano, G., Sili, C., Pelosi, E. and Vincenzini, M. (1985) *Cyanospira rippkae* and *Cyanospira capsulatus* (gen. nov. sp. nov.), a new filamentous heterocystous cyanobacterium from Lake Magadi (Kenya). Arch. Microbiol. **140**, 301-307.

Fritze, D., Flossdorf, J. and Claus, D. (1990) Taxonomy of alkaliphilic *Bacillus* strains. Int. J. Syst. Bacteriol. **40**, 92-97.

Gee, J.M., Lund, B.M., Metcalf, G. and Peel, J.L. (1980) Properties of a new group of alkalophilic bacteria. J.Gen. Microbiol. **117**, 9-17.

Gordon, R.E. and Hyde, J.L. (1982) The *Bacillus firmus* - *Bacillus lentus* complex and pH 7.0 variants of some alkalophilic strains. J. Gen. Microbiol. **128**, 1109-1116.

Grant, W.D. (1989) *Natronobacterium, Natronococcus*, in "Bergey's Manual of Systematic Bacteriology", vol. 3 (Staley, J.T., Bryant, M.P., Pfennig, N and Holt, J.G., Eds.), pp. 2230-2233, Williams and Williams, Baltimore.

Grant, W.D. (1992) Alkaline environments in "Encyclopedia of Microbiology", volume 1 (Lederberg, J., Ed.), Academic Press, San Diego.

Grant, W.D. and Horikoshi, K. (1989) Alkaliphiles, in "Microbiology of Extreme Environments and its Potential for Biotechnology" (Da Costa, M.S., Duarte, J.C. and Williams, R.A.D., Eds.), pp.346-366, Elsevier, London and New York.

Grant, W.D. and Horikoshi, K. (1992) Alkaliphiles: ecology and biotechnological applications, in "Molecular Biology and Biotechnology of Extremophiles" (Herbert, R.A. and Sharpe, R.J., Eds.), pp.143-162, Blackie, Glasgow and London.

Grant, W.D. and Larsen, H. (1989) Order *Halobacteriales*, in "Bergey's Manual of Determinative Bacteriology", vol. 3 (Staley, J.T., Bryant, M.P., Pfennig, N. and Holt, Eds.), pp.2216-2234, Williams and Wilkins, Baltimore.

Grant, W.D. and Mwatha, W.E. (1989) Bacteria from alkaline, saline environments, in "Recent Advances in Microbial Ecology" (proceedings of the 5th International Symposium on Microbial Ecology, Kyoto, Japan)(Hattori, T., Ishida, Y., Maruyama, Y., Morita, R.Y. and Uchida, A., Eds.), pp.64-67, Japan Scientific Societies Press, Tokyo.

Grant, W.D. and Tindall, B.J. (1980) The isolation of alkalophilic bacteria, in Microbial Growth and Survival in Extremes of Environment (Gould, G.W. and Corry, J.G.L., Eds.), pp.27-36, Academic Press, London.

Grant, W.D. and Tindall, B.J. (1986) The alkaline, saline environment, in "Microbes in Extreme Environments" (Herbert, R.A. and Codd, G.A., Eds.), pp.22-54, Academic Press, London.

Grant, W.D., Mwatha, W.E. and Jones, B.E. (1990) Alkaliphiles: ecology, diversity and applications. FEMS Microbiol. Rev. **75**, 255-270.

Gregersen, T. (1978) Rapid method for distinction of Gram-negative fron Gram-positive bacteria. Eur. J. Appl. Microbiol. Biotechnol. **5**, 123-127.

Halebian, S., Harris, B., Finegold, S.M. and Rolfe, R.D. (1981) Rapid method that aids in distinguishing Gram-positive from Gram-negative anaerobic bacteria. J. Clin. Microbiol. **13**, 444-448.

Hardie, L.A. and Eugster, H.P. (1970) The evolution of closed basin brines. Mineral. Soc. Amer., Spec. Pap. 3. 273-296.

Hecky, R.E. and Kilham, P. (1973) Diatoms in alkaline, saline lakes: ecology and geochemical implications. Limnol. Oceanogr. **18**, 53-71.

Horikoshi, K. (1991) "Microorganisms in Alkaline Environments", Kodansha, Tokyo and VCH Verlagsgesellschaft mbH, Weinheim, New York, Cambridge, Basel.

Imhoff, J.F., Sahl, H.G., Soliman, G.S.H. and Trüper, H.G. (1979) The Wadi Natrûn: chemical composition and microbial mass developments in alkaline brines of eutrophic desert lakes. Geomicrobiol. J. **1**, 219-234.

Iversen, N., Oremland, R.S. and Klug, M.J. (1987) Big Soda Lake (Nevada). 3. Pelagic methanogenesis and anaerobic methane oxidation. Limnol. Oceanogr. **32**, 804-814.

Javor, B. (1989) "Hypersaline Environments", Brock/Springer, Berlin and New York.

Jenkin, P.M. (1932) Reports of the Percy Sladen Expedition to some Rift Valley Lakes in Kenya in 1929. VII. Summary of the ecological results with special reference to the alkaline lakes. Ann. Mag. Nat. Hist. Ser. X, **8**, 133-181.

Jones, B.E. and Grant, W.D. (1993) Haloalkaliphilic microorganisms. European PatentApplication 0540 127 A1.

Jones, B.E., Grant, W.D. and Collins, N.C. (1992) Gram-negative alkaliphilic microorganisms. European Patent Application 0473 217 A1.

Jones, B.E., Grant, W.D. and Collins, N.C. (1993) Gram-positive alkaliphilic microorganisms. European Patent Application 0523 769 A1.

Jones, B.F., Eugster, H.P. and Rettig, S.L. (1977) Hydrochemistry of the Lake Magadi basin, Kenya. Geochim. Cosmochim. Acta. **41**, 53-72.

Kandler, O. (1993) Archaea (archaebacteria). Progress in Botany, **54**, 1-24.

Kempe, S., Kazmierczak, J., Landmann, G., Konuk, T., Reimer, A. and Lipp, A. (1991) Largest known microbialites discovered in Lake Van, Turkey. Nature **349**, 605-608.

Kevbrin, V.V. and Zavarzin, G.A. (1992) Effect of sulfur compounds on the growth of the halophilic homacetic bacterium *Acetohalobium arabaticum*. Mikrobiologiya (Eng. translation), **61**, 563-567.

Kiene, R.P., Oremland, R.S., Catena, A., Miller, L.G. and Capone, D.G. (1986) Metabolism of reduced methylated sulfur compounds in anaerobic sediments and by a pure culture of an estuarine methanogen. Appl. Environ. Microbiol. **52**, 1037-1045.

Kilham, P. (1981) Pelagic bacteria: extreme abundance in African saline lakes. Naturwissenschaften **68**, 380-381.

Kimura, T. and Horikoshi, K. (1990) The nucleotide sequence of an α-amylase gene from an alkalopsychrotrophic *Micrococcus* sp. FEMS Microbiol. Lett. **71**, 35-42.

King, G.M. (1988) Methanogenesis from methylated amines in a hypersaline algal mat. Appl. Environ. Microbiol. **54**, 130-136.

Kita-Tsukamoto, K., Oyaizu, H., Nanbak and Shimizu, U. (1993) Phylogenetic relationships of marine bacteria, mainly members of the family Vibrionaceae, determined on the basis of 16S rRNA sequences. Int. J. Syst. Bacteriol. **43**, 8-19.

Kobayashi, Y. and Horikoshi, K. (1980) Identification and growth characteristics of alkalophilic *Corynebacterium* sp. which produces NAD(P)-dependent maltose dehydrogenase and glucose dehydrogenase. Agric. Biol. Chem. **44**, 2261-2269.

Kotelnikova, S.V., Obraztsova, A.Y., Blotevogel, K-H. and Popov, I.N. (1993) Taxonomic analysis of thermophilic strains of the genus *Methanobacterium*: reclassification of *Methanobacterium thermoalcaliphilum* as a synonym of *Methanobacterium thermoautotrophicum*. Int. J. Syst. Bacteriol. **43**, 591-596.

Langworthy, T.A. (1978) Microbial life in extreme pH values, in "Microbial life in Extreme Environments" (Kushner, D.J., Ed.), pp.318-368, Academic Press, London and New York.

Lanzotti, V., Nicolaus, B., Trincone, A., De Rosa, M., Grant, W.D. and Gambacorta, A. (1989) A complex lipid with a cyclic phosphate from the archaebacterium *Natronobacterium occultus*. Biochim. Biophys. Acta **1001**, 31-34.

Li, Y., Mandelco, L. and Wiegel, J. (1993) Isolation and characterization of a moderately thermophilic anaerobic alkaliphile, *Clostridium paradoxum* sp. nov. Int. J. Syst. Bacteriol. **43**, 450-460.

Liu, Y., Boone, D.R. and Choy, C. (1990) *Methanohalophilus oregonense* sp. nov., a methyltrophic methanogen from an alkaline, saline aquifer. Int. J. Syst. Bacteriol. **40**, 111-116.

Lodwick, D., Ross, H.N.M., Walker, J.A., Almond, J.W. and Grant, W.D. (1991) Nucleotide sequence of the 16S ribosomal RNA gene from the haloalkaliphilic archaeon (archaebacterium) *Natronobacterium magadii*, and the phylogeny of halobacteria. Syst. Appl. Microbiol. **14**, 352-357.

Ma, Y., Tian, X., Zhou, P. and Wang, D. (1992) A new alkaliphilic bacterium. Abs. 6th Int. Symp. Microbial Ecol., Barcelona, Spain.

Mathrani, I.M., Boone, D.R., Mah, R.A., Fox, G.E. and Lau, P.P. (1988) *Methanohalophilus zhilinae* sp. nov., an alkaliphilic, halophilic, methylotrophic methanogen. Int. J. Syst. Bacteriol. **38**, 139-142.

McGenity, T.J. and Grant, W.D. (1993) The haloalkaliphilic archaeon (archaebacterium) *Natronococcus occultus* represents a distinct lineage within the *Halobacteriales*, most closely related to the other haloalkaliphilic lineage (*Natronobacterium*). Syst. Appl. Microbiol. **16**, 239-243.

Melack, J.M. (1981) Photosynthetic activity of phytoplankton in tropical African soda lakes. Hydrobiologicia **81**, 71-86.

Melack, J.M. and Kilham, P. (1974) Photosynthetic rates of phytoplankton in East African alkaline, saline lakes. Limnol. Oceanogr. **19**, 743-755.

Morth, S. and Tindall, B.J. (1985) Variation of polar lipid composition within haloalkaliphilic archaebacteria. Syst. Appl. Microbiol. **6**, 247-250.

Mwatha, W.E. (1991) Microbial Ecology of Kenyan Soda Lakes. Ph.D. Thesis, University of Leicester.

Mwatha, W.E. and Grant, W.D. (1993) *Natronobacterium vacuolata* sp. nov., a haloalkaliphilic archaeon isolated from Lake Magadi, Kenya. Int. J. Syst. Bacteriol. **43**, 401-404.

Nakatsugawa, N. (1991) Novel methanogenic archaebacteria which grow in extreme environments, in "Superbugs: Microorganisms in Extreme Environments" (Horikoshi, K. and Grant, W.D., Eds.) pp.212-220, Japan Scientific Societies Press, Tokyo and Springer-Verlag, Berlin, Heidelberg and New York.

Nakatsugawa, N. and Horikoshi, K. (1989) Isolation and charcterisation of two novel methanogens, a new haloalkaliphilic methanogen and a new alkaliphilic *Methanosarcina*, in "Microbiology of Extreme Environments and its Potential for Biotechnology" (Da Costa, M.S., Duarte, J.C. and Williams, R.A.D., Eds.), p.415, Elsevier, London and New York.

Nakatsugawa, N. and Horikoshi, K. (1991) Extremely halophilic methanogenic archaebacteria. U.S. Patent 5,055,406.

Nehrkorn, A. and Schwartz, W. (1961) Untersuchungen über lebensgemeinschaften halophiler Mikroorganismen. 1. Mikoorganismen aus Salzseen der californischen Wüstengebiete und aus einer Natriumchloride-Sole. Zeitschrift für Allg. Mikrobiol. **1**, 121-141.

Norton, C.F. and Grant, W.D. (1988) Survival of halobacteria within fluid inclusions in salt crystals. J. Gen. Microbiol. **134**, 1365-1373.

Ohkoshi, A., Kudo, T., Mase, T. and Horikoshi, K. (1985) Purification of three types of xylanases from an alkalophilic *Aeromonas* sp. Agric. Biol. Chem. **49**, 3037-3038.

Ollivier, B., Hatchikian, C.E., Prensier, G., Guezennec, J and Garcia, J.-L. (1991) *Desulfohalobium retbaense* gen. nov., sp. nov., a halophilic sulfate-reducing bacterium from sediments of a hypersaline lake in Senegal. Int. J. Syst. Bacteriol. **41**, 74-81.

Oremland, R.S., Marsh, L. and Des Marais, D.J. (1982) Methanogenesis in Big Soda Lake, Nevada: an alkaline, moderately hypersaline desert lake. Appl. Environ Microbiol. **43**, 462-468.

Oremland, R.S., Miller, L.G. and Whiticar, M.J. (1987) Sources and flux of natural gases from Mono Lake, California. Geochim. Cosmochim. Acta, **51**, 2915-2929.

Oren, A. (1986) The ecology and taxonomy of anaerobic halophilic eubacteria. FEMS Microbiol. Rev. **39**, 23-29.

Pantazis, T.M. (1976) Thermal mineral waters of Cyprus, in "Proceedings International Congress on Thermal Waters, Geothermal Energy and Volcanism, Mediterranean area", Vol. 2. pp.367-386, Athens.

Paterek, J.R. and Smith, P.H. (1988) *Methanohalophilus mahii* gen. nov., sp. nov., a methylotrophic halophilic methanogen. Int. J. Syst. Bacteriol. **38**, 122-123.

Priest, F.G. and Alexander, B. (1988) A frequency matrix for probabilstic identification of some bacilli. J. Gen. Microbiol. **134**, 3011-3018.

Randall, D.J., Wood, C.M., Perry, S.F., Bergman, H., Maloiy, G.M.O., Mommsen, T.P. and Wright, P.A. (1989) Urea excretion as a strategy for survival in a fish living in a very alkaline environment. Nature **337**, 165-166.

Renaut, R.W., Tiercelin, J.J. and Owen, R.B. (1986) Mineral precipitation and diagenesis in the sediments of the Lake Bogoria basin, Kenya Rift Valley, in "Sedimentation in the African Rifts" (Frostick, L.E., Renaut, R.W., Reid, I. and Tiercelin, J.J., Eds.), Geol. Soc. Spec. Publ., No. 25, Blackwell Scientific, Oxford.

Rich, F. (1933) Scientific results of the Cambridge Expedition to the East African lakes 1930. VII. The algae. J. Linn. Soc. Zool. **38**, 249-275.

Sackin, M.J. (1987) Programmes for classification and identification, in "Methods in Microbiology", vol. 19 (Colwell, R.R. and Grigorova, R., Eds.), pp.459-494, Academic Press, London.

Shiba, H. (1989) *Haloalcalibium sporogenum* gen. nov., sp. nov. and *Haloalcalibium grantii* sp. nov., a new type of facultatively anaerobic, haloalkaliphilic eubacteria. Abs. Superbugs Symp. "Microbial Life in Extreme Environments", Tokyo.

Shiba, H. and Horikoshi, K. (1989) Isolation and characterization of novel anaerobic, halophilic eubacteria from hypersaline environments of western America and Kenya, in "Microbiology of Extreme Environments and its Potential for Biotechnology" (Da Costa, M.S., Duarte, J.C. and Williams, R.A.D., Eds.), pp.371-374, Elsevier, London.

Soliman, G.S.H. and Trüper, H.G. (1982) *Halobacterium pharaonis* sp. nov., a new, extremely haloalkaliphilic archaebacterium with low magnesium requirement. Zbl. Bakt. Hyg., I. Abt. Orig. C **3**, 318-329.

Souza, K.A. and Deal, P.H. (1977) Characterization of a novel extremely alkalophilic bacterium. J. Gen. Microbiol. **101**, 103-109.

Souza, K.A., Deal, P.H., Mack, H.M. and Turnbull, C.E. (1974) Growth and reproduction of microorganisms under extremely alkaline conditions. Appl. Microbiol. **28**, 1066-1068.

Spanka, R. and Fritze, D. (1993) *Bacillus cohnii* sp. nov., a new, obligately alkaliphilic, oval-spore-forming *Bacillus* species with ornithine and aspartic acid instead of diaminopimelic acid in the cell wall. Int. J. Syst. Bacteriol. **43**, 150-156.

Strand, S.E., Dykes, J. and Chiang, V. (1984) Aerobic microbial degradation of glucoisosaccharinic acid. Appl. Environ. Microbiol. **47**, 268-271.

Talling, J.F., Wood, R.B., Prosser, M.V. and Baxter, R.M. (1973) The upper limit of photosynthetic productivity by phytoplankton: evidence from Ethiopian soda lakes. Freshwater Biol. **3**, 53-76.

Tindall, B.J. (1988) Prokaryotic life in the alkaline, saline, athalassic environment, in "Halophilic Bacteria" vol. 1 (Rodriguez-Valera, F., Ed.), pp.31-67, CRC Press, Inc., Boca Raton, Fl.

Tindall, B.J. (1992) The Famlily Halobacteriaceae, in "The Prokaryotes", 2nd edition, vol. 1 (Balows, A., Trüper, H.G., Dworkin, M., Harder, W. and Schleifer, K-H., Eds.), pp.768-808, Springer-Verlag, New York.

Tindall, B.J., Mills, A.A. and Grant, W.D. (1980) An alkalophilic red halophilc bacterium with a low magnesium requirement from a Kenyan soda lake. J. Gen. Microbiol. **116**, 257-260.

Tindall, B.J., Ross, H.N.M. and Grant, W.D. (1984) *Natronobacterium* gen. nov. and *Natronococcus* gen. nov., two new genera of haloalkaliphilic archaebacteria. Syst. Appl. Microbiol. **5**, 41-57.

Tsujibo, H., Sato, T., Inui, M., Yamamoto, H. and Inamori, Y. (1988) Intracellular accumulation of phenazine antibiotics, production by an alkalophilic actinomycete. Agric. Biol. Chem. **52**, 301-306.

Upasani, V. and Desai, S. (1990) Sambhar salt lake: chemical composition of the brines and studies on haloalkaliphilic archaebacteria. Arch. Microbiol. **154**, 589-593.

Urakami, T., Tamaoka,J., Suzuki, K-I. and Komagata, K. (1989) *Paracoccus alcaliphilus* sp. nov., an alkaliphilic and facultatively methylotrophic bacterium. Int. J. Syst. Bacteriol. **39**, 116-121.

Vedder, A. (1934) *Bacillus alcalophilus* n. sp.; benevens enkele ervaringen met sterk alcalische voedingsbodems. Ant. van Leeuwen. **1**, 141-147.

Vreeland, R.H. (1992) The family Halomonadaceae, in "The Prokaryotes", 2nd edition, vol. 4 (Balows, A., Trüper, H.G., Dworkin, M., Harder, W. and Schleifer, K-H., Eds.), pp.3181-3188, Springer-Verlag, New York.

Wang, D. and Tang, Q. (1989) Natronobacterium from soda lakes of China, in "Recent Advances in Microbial Ecology" (proceedings of the 5th International Symposium on Microbial Ecology, Kyoto, Japan) (Hattori, T., Ishida, Y., Maruyama, Y., Morita, R.Y. and Uchida, A., Eds.), pp.68-72, Japan Scientific Societies Press, Tokyo.

Waring, G.A., Blankenship, R.R. and Bentall, R. (1965) Thermal Springs of the United States and Other Countries of the World - a Summary. Geol. Surv. Prof. Paper 492, US Government Printing Office, Washington.

Watanabe, N., Ota, Y., Minoda, Y. and Yamada, K. (1977) Isolation and identification of some alkaline lipase producing microorganisms, culture conditions and some properties of crude enzymes. Agric. Biol. Chem. 41, 1353-1358.

Weisburg, W.G., Barns, S.M., Pelletier, D.A. and Lane, D.J. (1991) 16S ribosomal DNA amplification for phylogenetic study. J. Bacteriol. 173, 697-703.

Whitman, W.B., Bowen, T.L. and Boone, D.R. (1992) The methanogenic bacteria, in "The Prokaryotes" 2nd edition, vol. 1 (Balows, A., Trüper, H.G., Dworkin, M., Harder, W. and Schleifer, K-H., Eds.), pp.719-767, Springer-Verlag, New York.

Wilharm, T., Zhilina, T.N. and Hummel, P. (1991) DNA-DNA hybridization of methylotrophic halophilic methanogenic bacteria and transfer of *Methanococcus halophilus*^{T} to the genus *Methanohalophilus* as *Methanohalophilus halophilus* comb. nov. Int. J. Syst. Bacteriol. 41, 558-562.

Wilkinson, S.G. (1988) Gram-negative bacteria, in "Microbial Lipids" vol. 1 (Ratledge, C. and Wilkinson, S.G., Eds.), pp.299-488, Academic Press, London.

Williams, S.T., Goodfellow, M., Wellington, E.M.H., Vickers, J.C., Alderson, G., Sneath, P.H.A., Sackin, M.J. and Mortimer, A.M. (1983) A probability matrix for identification of some Streptomycetes. J. Gen. Microbiol. 129, 1815-1830.

Winfrey, M.R. and Zeikus, J.G. (1977) Effect of sulfate on carbon and electron flow during microbial methanogenesis in freshwater sediments. Appl. Environ. Microbiol. 33, 275-281.

Woese, C.R., Weisburg, W.G., Hahn, C.M., Paster, B.J., Zablen, L.B., Lewis, B.J., Macke, T.J., Ludwig, W. and Stackebrandt, E. (1985) The phylogeny of purple bacteria: the gamma subdivision. Syst. Appl. Microbiol. 6, 25-33.

Worakit, S., Boone, D.R., Mah, R.A., Abdel-Samie, M-E. and El-Halwagi, M.M. (1986) *Methanobacterium alcaliphilum* sp.nov., an H_2-utilizing methanogen that grows at high pH values. Int. J. Syst. Bacteriol. 36, 380-382.

Zhilina, T.N. and Zavarzin, G.A. (1990) Extremely halophilic, methylotrophic, anaerobic bacteria. FEMS Microbiol. Rev. 87, 315-322.

Zhilina, T.N., Zavarzin, G.A., Bulygina, E.S., Kevbrin, V.V., Osipov, G.A. and Chumakov, K.M. (1992) Ecology, physiology and taxonomy studies on a new taxon of *Haloanaerobiaceae*, *Haloincula saccharolytica* gen. nov., sp. nov. Syst. Appl. Microbiol. 15, 275-284.

Zinder, S.H., Doemel, W.N. and Brock, T.D. (1977) Production of volatile sulfur compounds during decomposition of algal mats. Appl. Environ. Microbiol. 34, 859-860.

Zvyagintseva, I.S. and Tarasov, A.L. (1988) Extreme halophilic bacteria from saline soils. Microbiologiya (Eng. Translation), 56, 664-669.

TAXONOMY AND PHYLOGENY OF MODERATELY HALOPHILIC BACTERIA

Antonio Ventosa

Department of Microbiology and Parasitology
Faculty of Pharmacy
University of Sevilla
41012 Sevilla, Spain

INTRODUCTION

Halobacteria are a group of procaryotes very well adapted to live in environments with high salinities. In fact, they grow optimally in media with 15 to 25% NaCl and are able to grow at saturated salt concentrations (Rodriguez-Valera, 1988; Kushner and Kamekura, 1988). They are grouped in the order Halobacteriales, in the family Halobacteriaceae, and currently species are included in six different genera: *Halobacterium, Haloarcula, Haloferax, Halococcus, Natronobacterium* and *Natronococcus* (Grant and Larsen, 1989). Phylogenetically, halobacteria are included within the archaebacteria (Domain Archaea), most closely related to methanogens (Woese, 1987; Woese *et al.*, 1990). However, in hypersaline environments, in addition to halobacteria, several other procaryotes which show haloadaptation can be observed. They are usually found in media of intermediate salinities, and are designated as moderate halophiles.

Moderately halophilic bacteria are defined as microorganisms growing optimally in media containing from 5 to 15% salt (Kushner and Kamekura, 1988). However, their salt response is very euryhaline, being able to grow over a wide range of salinities; for the majority of the species studied, the range is from seawater to extremely highly concentrated brines with 25 to 30% salt (Quesada *et al.*, 1982; Ventosa *et al.*, 1984; Rodriguez-Valera *et al.*, 1985; Rodriguez-Valera, 1988). Moderately halophilic bacteria are inhabitants of saline environments, such as lakes, salterns, saline soils and a variety of salted foods including salted cured meats and fish and brine fermented vegetables (Ventosa, 1988; Javor, 1989). The majority of recent studies concerning moderately halophilic bacteria have been carried out in hypersaline lakes (Dead Sea, Great Salt Lake, several Antarctic lakes and some African alkaline lakes) as well as in Mediterranean salterns in Spain and France. In contrast, the early studies on this group of microorganisms were based on strains isolated from salted foods, due to the fact that they could contaminate or spoil these products.

Bacterial Diversity and Systematics
Edited by F.G. Priest *et al.*, Plenum Press, New York, 1994

TAXONOMY OF MODERATELY HALOPHILIC BACTERIA

The current taxonomic status of the moderately halophilic bacteria contrasts with our knowledge about this group of bacteria a decade ago. In fact, in 1980 only six moderately halophilic species were accepted as validly published and included in the Approved Lists of Bacterial Names (Skerman *et al.*, 1980): *Flavobacterium halmephilum, Micrococcus halobius, Paracoccus halodenitrificans, Planococcus halophilus, Spirochaeta halophila* and *Vibrio costicola*. Further, the majority of species were isolated from salted foods or as laboratory culture contaminats and their taxonomic descriptions were very incomplete. The extensive studies carried out during the last years, especially those involved on the isolation and taxonomic characterization of moderately halophilic bacteria from natural hypersaline environments, permit us to have a better knowledge about the taxonomic structure of these microorganisms. Table 1 includes all the moderately halophilic bacteria validly described up to date. Besides these species, there are several other moderate halophiles that should be studied in more detail in order to determine their correct allocation and to describe them taxonomically. Examples of these moderately halophilic bacteria include: "*Pseudomonas halosaccharolytyca*" and "*Micrococcus varians* var. *halophilus*" (Ventosa, 1988).

The heterogeneity found in this physiological group is noticeable: there are several archaebacterial (Archaea) representatives, all of them methanogenic species, as well as eubacteria (Bacteria) included in many different genera. It is noteworthy that many of the moderately halophilic species are included in genera in which marine or non-halophilic species are also present. Based on phylogenetic data (16S rRNA oligonucleotide cataloging) some genera are grouped in specific families: the species of the genera *Haloanaerobium, Halobacteroides, Sporohalobacter* and the recently proposed genus *Haloincola* are included in the family Haloanaerobiaceae (Oren *et al.*, 1984a; Zhilina *et al.*, 1992), and species of the genera *Halomonas* and *Deleya* are grouped in the family Halomonadaceae (Franzmann *et al.*, 1988).

The moderately halophilic bacteria described most recently are the methanogen *Methanohalophilus portucalensis*, isolated from a Portuguese saltern (Boone *et al.*, 1993), two *Flavobacterium* species, *F. gondwanense* and *F. salegens*, isolated from Organic Lake, a hypersaline Antarctic lake (Dobson *et. al.*, 1993), and the aerobic Gram-negative rod *Arhodomonas aquaeolei*, isolated from the production fluid of a petroleum reservoir (Adkins *et al.*, 1993).

RECENT CHEMOTAXONOMIC AND MOLECULAR SYSTEMATIC STUDIES ON MODERATELY HALOPHILIC BACTERIA

Since many studies concerning the taxonomy of moderately halophilic bacteria have been published almost simultaneously in a very short period, it is evident that comparative studies of these species, based on the modern chemotaxonomic and molecular techniques currently available were needed. In fact, the chemotaxonomic studies contributed to the establishment of a more clear classification of several moderate halophiles and have permitted the elucidation of several species that were in an uncertain taxonomic position. Franzmann and Tindall (1990) analysed the respiratory quinones, polar lipid and fatty acid compositions of the species of the genera *Halomonas* and *Deleya* (grouped in the family Halomonadaceae) as well as *Pseudomonas halophila* and *Halovibrio variabilis*. These data suggest that *Halovibrio variabilis* should be transferred to a taxon within the family Halomonadaceae and, on

Table 1. Moderately halophilic bacteria described as valid species

	Species	Type strain	References
1.	**Archaea:**		
	Methanohalophilus mahii	ATCC 35705	Paterek and Smith, 1988
	Methanohalophilus zhilinae	DSM 4017	Mathrani *et al.*, 1988
	Methanohalophilus halophilus	DSM 3094	Wilharm *et al.*, 1991
	Methanohalophilus portucalensis	OCM 59	Boone *et al.*, 1993
	Halomethanococcus doii	ATCC 43619	Yu and Kawamura, 1987
2.	**Bacteria:**		
A.	**Phototrophic:**		
	Rhodospirillum salexigens	DSM 2132	Drews, 1981
	Rhodospirillum salinarum	ATCC 35394	Nissen and Dundas, 1984
	Ectothiorhodospira vacuolata	DSM 2111	Imhoff *et al.*, 1981
	Ectothiorhodospira marismortui	DSM 4180	Oren *et al.*, 1989
	Chromatium salexigens	DSM 4395	Caumette *et al.*, 1988
	Thiocapsa halophila	DSM 6210	Caumette *et al.*, 1991a
B.	**Chemoorganotrophic:**		
	Gram-negative aerobic or facultatively anaerobic:		
	Vibrio costicola	NCMB 701	García *et al.*, 1987
	Halomonas elongata	ATCC 33173	Vreeland *et al.*, 1980
	Halomonas subglaciescola	UQM 2927	Franzmann *et al.*, 1987
	Halomonas halodurans	ATCC 29696	Hebert and Vreeland, 1987
	Halomonas halmophila	ATCC 19717	Dobson *et al.*, 1990
	Deleya halophila	CCM 3662	Quesada *et al.*, 1984
	Deleya salina	ATCC 49509	Valderrama *et al.*, 1991
	Spirochaeta halophila	ATCC 29478	Greenberg and Canale-Parola, 1976
	Paracoccus halodenitrificans	ATCC 13511	Kocur, 1984
	Halovibrio variabilis	DSM 3051	Fendrich, 1988
	Pseudomonas halophila	DSM 3050	Fendrich, 1988
	Chromohalobacter marismortui	ATCC 17056	Ventosa *et al.*, 1989b
	Volcaniella eurihalina	ATCC 49336	Quesada *et al.*, 1990
	Flavobacterium gondwanense	DSM 5423	Dobson *et al.*, 1993
	Flavobacterium salegens	CCM 5424	Dobson *et al.*, 1993
	Arhodomonas aquaeolei	ATCC 49307	Adkins *et al.*, 1993
	Gram-negative anaerobic:		
	Haloanaerobium praevalens	ATCC 33744	Zeikus *et al.*, 1983
	Halobacteroides halobius	ATCC 35273	Oren *et al.*, 1984b
	Halobacteroides acetoethylicus	ATCC 43120	Rengpipat *et al.*, 1988
	Sporohalobacter lortetii	ATCC 35420	Oren, 1983; Oren *et al.*, 1987
	Sporohalobacter marismortui	ATCC 35420	Oren *et al.*, 1987
	Haloincola saccharolytica	DSM 6643	Zhilina *et al.*, 1992

Table 1. Moderately halophilic bacteria described as valid species (cont.)

Species	Type strain	References
Gram-positive:		
Micrococcus halobius	ATCC 21727	Onishi and Kamekura, 1972
Sporosarcina halophila	DSM 2266	Claus *et al.*, 1983
Marinococcus halophilus	ATCC 27964	Hao *et al.*, 1984
Marinococcus albus	CCM 3517	Hao *et al.*, 1984
Salinicoccus roseus	ATCC 49258	Ventosa *et al.*, 1990
Salinicoccus hispanicus	ATCC 49259	Márquez *et al.*, 1990; Ventosa *et al.*, 1992
Bacillus halophilus	ATCC 49085	Ventosa *et al.*, 1989a
Clostridium halophilum	DSM 5387	Fendrich *et al.*, 1990
Budding prosthecate bacteria:		
Dichotomicrobium thermohalophilum	DSM 5002	Hirsch and Hoffmann, 1989
Actinomycete:		
Actinopolyspora mortivallis	JCM 7550	Yoshida *et al.*, 1991
Sulfate-reducing bacteria:		
Desulfohalobium retbaense	DSM 5692	Ollivier *et al.*, 1991
Desulfovibrio halophilus	DSM 5663	Caumette *et al.*, 1991b

This list includes all the moderately halophilic bacteria validly published up to September 1, 1993 (included in the Approved Lists of Bacterial Names or additional Lists published in the International Journal of Systematic Bacteriology).

the other hand, the chemotaxonomic similarity among the species of the genera *Halomonas* and *Deleya* suggests that the members of these genera were chemically heterogeneous and there was no clear way of distinguishing these two genera, as currently described. Further, an analysis of the phospholipid ester-linked fatty acids of members of the family Halomonadaceae did not permit the clear distinction between members of the genera *Halomonas* and *Deleya* (Skerratt *et al.*, 1991). DNA-rRNA hybridization data showed that species of the genera *Deleya* and *Halomonas* are part of a single evolutionary lineage (Kersters, 1991). Akagawa-Matsushita *et al.* (1992) reported that isoprenoid quinone composition could be useful in differentiating species of the genera *Deleya*, *Alteromonas*, *Marinomonas* and *Shewanella*. Cellular fatty acid composition has also been used in chemotaxonomic studies of moderate halophiles (Monteoliva-Sánchez *et al.*, 1989; Vainshtein *et al.*, 1992; Ventosa *et al.*, 1992; 1993).

A recent phenotypic and chemotaxonomic study of 55 moderately halophilic Gram-positive motile cocci isolated from several hypersaline habitats showed that these isolates should be included in the species *Marinococcus halophilus* (Márquez *et al.*, 1992). This study suggests that the species *M. halophilus* is more common in hypersaline environments than expected. Similarly, a chemotaxonomic study of the moderately halophilic species *Marinococcus hispanicus* demonstrated that this organism should be transferred to the genus *Salinicoccus*. DNA-DNA homology experiments supported the placement of *Marinococcus hispanicus* and the, at that time single species of the genus *Salinicoccus*, *S. roseus*, in two different species. Thus, we proposed to transfer *Marinococcus hispanicus* to the genus *Salinicoccus*, as *Salinicoccus hispanicus* (Ventosa *et al.*, 1992). Similar studies permitted us to determine that the collection strains "*Micrococcus*" sp. CCM 168 and CCM 1405 were members of the species *Salinicoccus roseus* (Ventosa *et al.*, 1993).

PHYLOGENY OF MODERATELY HALOPHILIC BACTERIA

Moderately halophilic bacteria and halobacteria are microorganisms that inhabit hypersaline environments, in which a succession of different microbial populations can be observed at increasing salinities and in some instances both groups compete (Rodriguez-Valera, 1988). The extensive modifications of halophiles involved in the adaptation to extreme conditions permitted researchers to speculate about the origin and evolution of these procaryotes. Traditionally, halobacteria were related to *Pseudomonas* (rod-shaped cells) or to *Micrococcus-Sarcina* (cocci) (Larsen, 1962), and several opinions were expressed with respect to the possible relationships between halobacteria and other groups of procaryotes as well as about the ways in which they might be derived from other bacterial groups. In a review published in 1978, Kushner stated: "I think it is a mistake to consider the red halophiles [halobacteria] as very closely related to any other known bacterial genus. We may hope for further revelations of their taxonomic status by comparisons of specific ribosomal proteins, of RNA's, and of other macromolecules of known function".

At present, there is substantial evidence supporting the fact that halobacteria and moderately halophilic eubacteria are two independent and ancient phylogenetic groups, the first belonging to the archaebacteria (Domain Archaea) while the majority of moderately halophilic microorganisms (but not all) are eubacteria (Domain Bacteria) (Woese, 1987; Woese *et al.*, 1990). To date, the only representative species of the Archaea which are moderate halophiles are some methanogens, of which some strains of the genus *Methanohalophilus* (*M. mahii*, *M. zhilinae*, *Methanohalophilus* sp. strain SF1) have been studied with respect to their phylogenetic relationship to other members of the Archaea by comparison of the 16S rRNA sequences (Mathrani *et al.*, 1988; Olsen *et al.*, 1991). It has been suggested, on the basis of the 16S rRNA sequence analysis (Figure 1), that *M. zhilinae* should be transferred to a new genus (Boone *et al.*, 1993).

In contrast to the current knowledge on the phylogeny of the Archaea, very few studies concerning the phylogenetic situation of the moderately halophilic eubacteria have been carried out. Only very recently, and based on the 16S rRNA sequence comparison, has it been possible to determine the phylogenetic position of some representative species. Thus, by means of the old technique of comparison of the 16S rRNA oligonucleotide catalogues, the following species were studied: (i) *Spirochaeta halophila*, that is included in the spirochetes phylum (Paster *et al.*, 1984); (ii) *Halomonas elongata*, *Halomonas halmophila* and *Halomonas subglaciescola*, included in the gamma subclass of the Proteobacteria (Franzmann *et al.*, 1988; Stackebrandt *et al.*, 1988; Woese, 1987). Furthermore, this study suggested a close relationship between the genera *Halomonas* and *Deleya* (a genus that includes marine and moderately halophilic species) and it was proposed to accommodate the species of both genera in a new family, named Halomonadaceae (Franzmann *et al.*, 1988); (iii) studies on the strictly anaerobic moderately halophilic species *Haloanaerobium praevalens*, *Halobacteroides halobius*, *Sporohalobacter lortetii* and *Sporohalobacter marismortui*

(Oren *et al.*, 1984*a,b;* 1987) showed a close relationship among all these halophiles and they were placed in the family Haloanaerobiaceae. The last mentioned studies suggested that these moderate halophiles could be related to the spirochetes, but their accurate phylogenetic position should be determined by other techniques, such as the complete 16S rRNA sequence comparison of some representative strains of this family with those of the other eubacteria reported.

Very recently the 16S rRNA sequences of some moderately halophilic eubacteria have been determined and compared with sequences from other eubacterial species. The full 16S rRNA sequence analysis confirmed that *Spirochaeta halophila* is within the *Spirochaeta* group (Figure 2) of the spirochetes phylum (Paster *et al.*, 1991). *Marinococcus halophilus* (formerly *Planococcus halophilus*) was historically included in the family Micrococcaceae, and considered to be taxonomically related to the micrococci (Novitsky and Kushner, 1976; Stackebrandt and Woese, 1979). Comparative analysis of the 16S rRNA sequences revealed that *M. halophilus* formed a very distinct line of descent, not related to members of the genera *Planococcus, Bacillus* or *Sporosarcina* (Farrow *et al.*, 1992). This study also indicated that *Sporosarcina halophila* is not closely related to *Sporosarcina ureae* and may constitute a peripherical species of the *Bacillus* sensu stricto group constituted by *Bacillus subtilis* and relatives (Farrow *et al.*, 1992).

Flavobacterium gondwanense and *Flavobacterium salegens*, two recently described species isolated from a hypersaline Antarctic lake (Dobson *et al.*, 1993) are members of the "flavobacterium-bacteroides" phylum (Figure 3). The phototrophic bacteria *Rhodospirillum salexigens* and *Rhodospirillum salinarum* are included in the alpha

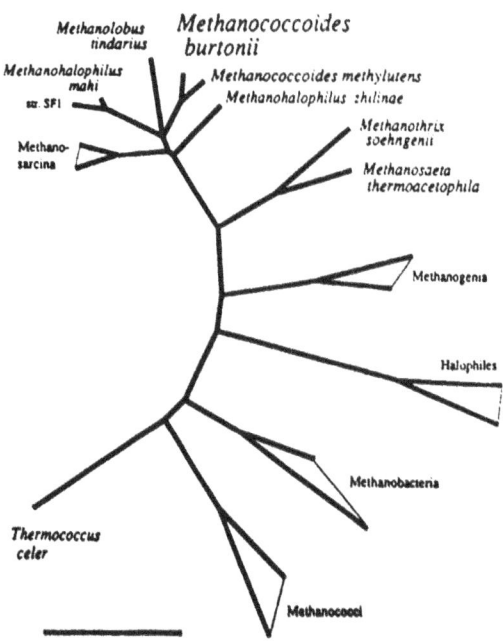

Figure 1. Phylogenetic tree showing the relationships of the moderately halophilic methanogens to other members of the Archaea. Bar equates to a Jukes-Cantor evolutionary distance of 0.05 (reproduced from Franzmann *et al.*, 1992 with permission).

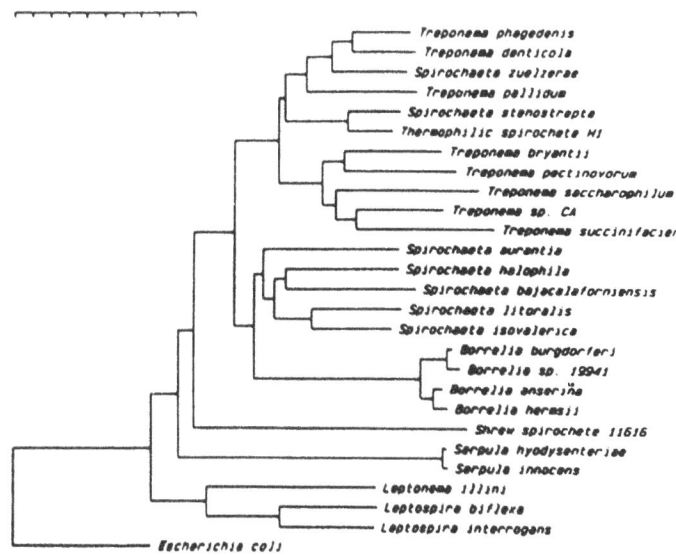

Figure 2. Phylogenetic tree showing the relationships of *Spirochaeta halophila* to other species of the spirochetes phylum. Scale bar = 10% difference in nucleotide sequence (reproduced from Paster *et al.*, 1991 with permission).

subclass of the Proteobacteria; however, in a phylogenetic study of several species of the genus *Rhodospirillum*, together with other phototrophic purple non-sulfur bacteria, an extremely diverse intra- and intergeneric relationships was observed (Kawasaki *et al.*, 1993). In fact, the moderately halophilic species *R. salexigens* and *R. salinarum* were quite distinct from other members of the alpha subclass of the Proteobacteria, and it was suggested that each species should be included in a separate genus (Kawasaki *et al.*, 1993). The recently described aerobic Gram-negative moderately halophilic species *Arhodomonas aquaeolei* is most closely related to purple sulfur bacteria (*Ectothiorhodospira* and *Chromatium*) within the gamma subclass of the Proteobacteria (Figure 4) (Adkins *et al.*, 1993). Partial 16S rRNA sequence analysis of members of the family Vibrionaceae showed that *Vibrio costicola* (that is in the gamma subclass of the Proteobacteria) is not closely related to other *Vibrio* species and perhaps should be included in a new genus (Kita-Tsukamoto *et al.*, 1993). These results agree with previous nucleic acid homology studies that showed a distant relationship of this moderate halophile to other *Vibrio* species (Brenner *et al.*, 1983). The 5S rRNA sequence comparison indicated that the strictly anaerobic saccharolytic species *Haloincola saccharolytica* should be included within the family Haloanaerobiaceae (Zhilina *et al.*, 1992).

Since quick and easy techniques of rRNA sequence analysis are currently available, it is evident that more detailed studies are required within the moderately halophilic bacteria in order to determine their phylogeny. These studies are not only necessary to establish a natural (phylogenetic) classification of moderate halophiles, but will also help to unravel the relationships among moderately halophilic eubacteria and other marine or terrestrial bacteria. Current data suggest that there are moderately halophilic representatives in four major phyla (spirochetes, Proteobacteria,

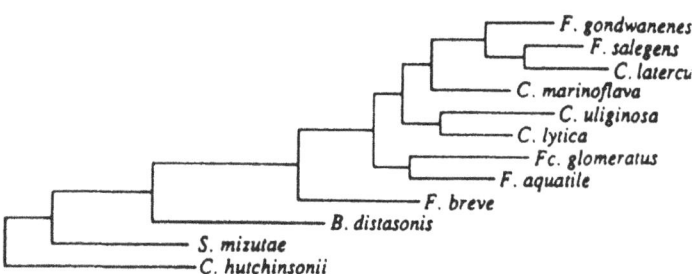

Figure 3. Tree based on the 16S rRNA sequence analysis showing the phylogenetic relationships of *Flavobacterium gondwanense* and *Flavobacterium salegens* within the flavobacterium-bacteroides phylum. Abreviations: *F., Flavobacterium*; *C., Cytophaga*; *Fc., Flectobacillus*; *B., Bacteroides*; *S., Sphingobacterium* (reproduced from Dobson *et al.*, 1993 with permission).

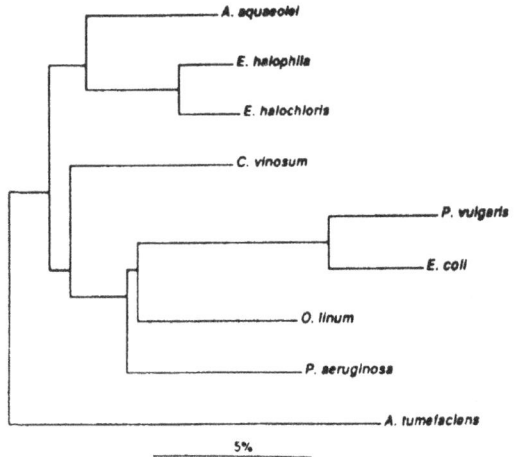

Figure 4. Phylogenetic tree showing the position of *Arhodomonas aquaeolei* within the gamma subclass of the Proteobacteria (*Ectothiorhodospira halophila* and *E. halochloris, Chromatium vinosum, Proteus vulgaris, Escherichia coli, Oceanospirillum linum, Pseudomonas aeruginosa* and *Agrobacterium tumefaciens*) (reproduced from Adkins *et al.*, 1993 with permission).

"flavobacterium-bacteroides" and Gram-positive phyla) and we may expect that there are also representatives in other phyla (moderately halophilic cyanobacteria have also been reported). Thus, moderately halophilic eubacteria should be represented at least in five of the ten major phyla of the eubacteria (Woese *et al.,* 1985; Woese, 1987), and we could speculate about the phylogenetic and evolutionary implications of this fact. It may be suggested a common phylogenetic ancestor with halophilic properties, however, since the haloadaptation mechanisms reported for moderately halophilic or halotolerant eubacteria are very different (for example, they accummulate a wide range of "compatible solutes"), it is more feasible to assume different origins of the moderate halophiles, perhaps a close relationships with marine bacteria. Future reasearch should be focused on this fascinating goal in order to elucidate the evolutionary relationship of moderately halophilic eubacteria among themselves as well as with respect to marine or non-halophilic bacteria.

ACKNOWLEDGEMENTS

I would like to thank Drs. H.G. Trüper, B.J. Tindall, J.J. Nieto and M.C. Márquez for critical reading of the manuscript and valuable suggestions. I acknowledge finantial support from the EC BIOTECH Program (BIO2-CT93-0274), Ministerio de Educación y Ciencia (PB90-0907 and PB92-0670) and Junta de Andalucía.

REFERENCES

Adkins, J.P., Madigan, M.T., Mandelco, L., Woese, C.R. and Tanner, R.S. (1993) *Arhodomonas aquaeolei* gen. nov., sp. nov., an aerobic, halophilic bacterium isolated from a subterranean brine. Int. J. Syst. Bacteriol. 43, 514-520.

Akagawa-Matsushita, M., Itoh T., Katayama, Y, Kuraishi, H. and Yamasato, K. (1992) Isoprenoid quinone composition of some marine *Alteromonas, Marinomonas, Deleya, Pseudomonas* and *Shewanella* species. J. Gen. Microbiol. 138, 2275-2281.

Boone, D.R., Mathrani, I.M., Liu, Y., Menaia, J.A.G.F., Mah, R.A. and Boone, J.E. (1993) Isolation and characterization of *Methanohalophilus portucalensis* sp. nov. and DNA reassociation study of the genus *Methanohalophilus*. Int. J. Syst. Bacteriol. 43, 430-437.

Brenner, D.J., Fanning, G.R., Hickman-Brenner, F.W., Steigerwalt, A.G., Davis, B.R. and Farmer III, J.J. (1983) DNA relatedness among *Vibrionaceae* with emphasis on the *Vibrio* species associated with human infection, in "Les baciles à Gram négatif d'intérêt médical et en Santé Publique", pp. 175-184. INSERM, Paris.

Caumette, P., Baulaigue, R. and Matheron, R. (1988) Characterization of *Chromatium salexigens* sp. nov., a halophilic Chromatiaceae isolated from Mediterranean salinas. Syst. Appl. Microbiol. 10, 284-292.

Caumette, P., Baulaigue, R. and Matheron, R. (1991a) *Thiocapsa halophila* sp. nov., a new halophilic phototrophic purple sulfur bacterium. Arch. Microbiol. 155, 170-176.

Caumette, P., Cohen, Y. and Matheron, R. (1991b) Isolation and characterization of *Desulfovibrio halophilus* sp. nov., a halophilic sulfate-reducing bacterium isolated from Solar Lake (Sinai). Syst. Appl. Microbiol. 14, 33-38.

Claus, D. Fahmy, F., Rolf, H.J. and Tosunoglu, N. (1983) *Sporosarcina halophila* sp. nov., an obligate, slightly halophilic bacterium from salt marsh soils. Syst. Appl. Microbiol. 4, 496-506.

Dobson, S.J., James, S.R., Franzmann, P.D. and McMeekin, T.A. (1990) Emended description of *Halomonas halmophila* (NCMB 1971^T). Int. J. Syst. Bacteriol. 40, 462-463.

Dobson, S.J., Colwell, R.R., McMeekin, T.A. and Franzmann, P.D. (1993) Direct sequencing of the polymerase chain reaction-amplified 16S rRNA gene of *Flavobacterium gondwanense* sp. nov. and *Flavobacterium salegens* sp. nov., two new species from a hypersaline Antarctic lake. Int. J. Syst. Bacteriol. 43, 77-83.

Drews, G. (1981) *Rhodospirillum salexigens,* spec. nov., an obligatory halophilic phototrophic bacterium. Arch. Microbiol. 130, 325-327.

Farrow, J.A.E., Ash, C., Wallbanks, S. and Collins, M.D. (1992). Phylogenetic analysis of the genera *Planococcus, Marinococcus* and *Sporosarcina* and their relationships to members of the genus *Bacillus*. FEMS Microbiol. Lett. 93, 167-172.

Fendrich, C. (1988) *Halovibrio variabilis* gen. nov., sp. nov., *Pseudomonas halophila* sp. nov. and a new halophilic aerobic coccoid eubacterium from Great Salt Lake, Utah, USA. Syst. Appl. Microbiol. 11, 36-43.

Fendrich, C., Hippe, H. and Gottschalk, G. (1990) *Clostridium halophilium* sp. nov. and *C. litorale* sp. nov., an obligate halophilic and a marine species degrading betaine in the Stickland reaction. Arch. Microbiol. 154, 127-132.

Franzmann, P.D. and Tindall, B.J. (1990) A chemotaxonomic study of members of the family *Halomonadaceae*. Syst. Appl. Microbiol. 13, 142-147.

Franzmann, P.D., Burton, H.R. and McMeekin, T.A. 1987 *Halomonas subglaciescola*, a new species of halotolerant bacteria isolated from Antarctica. Int. J. Syst. Bacteriol. 37, 27-34.

Franzmann, P.D., Wehmeyer, U. and Stackebrandt, E. (1988) Halomonadaceae fam. nov., a new family of the class Proteobacteria to accommodate the genera *Halomonas* and *Deleya*. Syst. Appl. Microbiol. 11, 16-19.

Franzmann, P.D., Springer, N., Ludwig, W., Conway de Macario, E. and Rohde, M. (1992) A methanogenic archaeon from Ace Lake, Antarctica: *Methanococcoides burtonii* sp. nov. Syst. Appl. Microbiol. 15, 573-581.

García, M.T., Ventosa, A., Ruiz-Berraquero, F. and Kocur, M. (1987) Taxonomic study and amended description of *Vibrio costicola*. Int. J. Syst. Bacteriol. 37, 251-256.

Grant, W.D. and Larsen, H. (1989) Order Halobacteriales *ord. nov.*, in "Bergey's Manual of Systematic Bacteriology", vol. 3, (Staley, J.T., Bryant, M.P., Pfennig, N. and Holt, J.G., Eds.), pp. 2216-2223. Williams & Wilkins, Baltimore.

Greenberg, E.P. and Canale-Parola, E. (1976) *Spirochaeta halophila* sp. n., a facultative anaerobe from a high-salinity pond. Arch. Microbiol. 110, 185-194.

Hao, M.V., Kocur M. and Komagata, K. (1984) *Marinococcus* gen. nov., a new genus for motile cocci with meso-diaminopimelic acid in the cell wall; and *Marinococcus albus* sp. nov. and *Marinococcus halophilus* (Novitsky and Kushner) comb. nov. J. Gen. Appl. Microbiol. 30, 449-459.

Hebert, A.M. and Vreeland, R.H. (1987) Phenotypic comparison of halotolerant bacteria: *Halomonas halodurans* sp. nov., nom. rev., comb. nov. Int. J. Syst. Bacteriol. 37, 347- 350.

Hirsch, P. and Hoffmann, B. (1989) *Dichotomicrobium thermohalophilum*, gen. nov., spec. nov., budding prosthecate bacteria from the Solar Lake (Sinai) and some related strains. Syst. Appl. Microbiol. 11, 291-301.

Imhoff, J.F., Tindall, B.J., Grant, W.D. and Trüper, H.G. (1981) *Ectothiorhodospira vacuolata* sp. nov., a new phototrophic bacterium from soda lakes. Arch. Microbiol. 130, 238-242.

Javor, B. (1989) Hypersaline environments. Microbiology and Biogeochemistry. Springer Verlag, Berlin.

Kawasaki, H., Hoshino, Y. and Yamasato, K. (1993) Phylogenetic diversity of phototrophic purple non-sulfur bacteria in the *Proteobacteria* alpha group. FEMS Microbiol. Lett. 112, 61-66.

Kersters, K. (1991). The genus *Deleya*. "The Prokaryotes", 2nd. ed. (Balows, A., Trüper, H.G., Dworkin, M., Harder, W. & Schleifer, K.H., Eds.), pp. 3189-3197. Springer-Verlag, New York.

Kita-Tsukamoto, K., Oyaizu, H., Nanba, K. and Simidu, U. (1993) Phylogenetic relationships of marine bacteria, mainly members of the family *Vibrionaceae*, determined on the basis of 16S rRNA sequences. Int. J. Syst. Bacteriol. 43, 8-19.

Kocur, M. (1984) Genus *Planococcus*, in "Bergey's Manual of Systematic Bacteriology" Vol. 1 (Krieg, N.R. and Holt, J.G., Eds.), pp. 399-402. Williams & Wilkins, Baltimore.

Kushner, D.J. (1978) Life in high salt and solute concentrations: halophilic bacteria, in "Microbial life in extreme environments" (Kushner, D.J., Ed.), 317-367. Academic Press, London.

Kushner, D.J. and Kamekura, M. (1988) Physiology of halophilic eubacteria, in "Halophilic Bacteria" vol. 1 (Rodriguez- Valera, F., Ed.), pp.109-140. CRC Press, Boca Raton.

Larsen, H. (1962) Halophilism, in "The bacteria" vol. IV, (Gunsalus, I.C. and Stanier, R.Y., Eds.), pp. 297-342. Academic Press, London.

Márquez, M.C., Ventosa, A. and Ruiz-Berraquero, F. (1990) *Marinococcus hispanicus*, a new species of moderately halophilic Gram-positive cocci. Int. J. Syst. Bacteriol. 40, 165-169.

Márquez, M.C., Ventosa, A. and Ruiz-Berraquero, F. (1992) Phenotypic and chemotaxonomic characterization of *Marinococcus halophilus*. Syst. Appl. Microbiol. 15, 63-69.

Mathrani, I.M., Boone, D.R., Mah, R.A., Fox, G.E. and Lau, P.P. (1988) *Methanohalophilus zhilinae* sp. nov., an alkaliphilic, halophilic, methylotrophic methanogen. Int. J. Syst. Bacteriol. 38, 139-142.

Monteoliva-Sanchez, M. Ventosa, A. and Ramos-Cormenzana, A. (1989) Cellular fatty acid composition of moderately halophilic cocci. Syst. Appl. Microbiol. 12, 141-144.

Nissen, H. and Dundas, I.D. (1984) *Rhodospirillum salinarum* sp. nov., a halophilic photosynthetic bacterium isolated from a portuguese saltern. Arch. Microbiol. 138, 251-256.

Novitsky, T.J. and Kushner, D.J. (1976) *Planococcus halophilus* sp. nov., a facultatively halophilic coccus. Int. J. Syst. Bacteriol. 26, 53-57.

240

Ollivier, B., Hatchikian, C.E., Prensier, G., Guezennec, J. and Garcia, J.L. (1991) *Desulfohalobium retbaense* gen. nov., sp. nov., a halophilic sulfate-reducing bacterium from sediments of a hypersaline lake in Senegal. Int. J. Syst. Bacteriol. 41, 74-81.

Olsen, G.J., Larsen, N. and Woese, C.R. (1991) The ribosomal RNA database project. Nucleic Acids Res. 19, 2017-2021.

Onishi, H. and Kamekura, M. (1972) *Micrococcus halobius* sp. n. Int. J. Syst. Bacteriol. 22, 233-236.

Oren, A. (1983) *Clostridium lortetii* sp. nov., a halophilic obligatory anaerobic bacterium producing endospores with attached gas vacuoles. Arch. Microbiol. 136, 42-48.

Oren, A., Paster, B.J. and Woese, C.R. (1984a) Haloanaerobiaceae: a new family of moderately halophilic, obligatory anaerobic bacteria. Syst. Appl. Microbiol. 5, 71-80.

Oren, A., Weisburg, W.G., Kessel, M. and Woese, C.R. (1984b) *Halobacteroides halobius* gen. nov., sp. nov., a moderately halophilic anaerobic bacterium from the bottom sediments of the Dead Sea. Syst. Appl. Microbiol. 5, 58-70.

Oren, A., Pohla, H. and Stackebrandt, E. (1987) Transfer of *Clostridium lortetii* to a new genus *Sporohalobacter* gen. nov. as *Sporohalobacter lortetii* comb. nov., and description of *Sporohalobacter marismortui* sp. nov. Syst. Appl. Microbiol. 9, 239-246.

Oren, A., Kessel, M. and Stackebrandt, E. (1989) *Ectothiorhodospira marismortui* sp. nov., an obligately anaerobic, moderately halophilic purple sulfur bacterium from a hypersaline sulfur spring on the shore of the Dead Sea. Arch. Microbiol. 151, 524-529.

Paster, B.J., Stackebrandt, E., Hespell, R.B., Hahn, C.M. and Woese, C.R. (1984) The phylogeny of the spirochetes. Syst. Appl. Microbiol. 5, 337-351.

Paster, B.J., Dewhirst, F.E., Weisburg, W.G., Tordoff, L.A., Fraser, G.J., Hespell, R.B., Stanton, T.B., Zablen, L., Mandelco, L. and Woese, C.R. (1991) Phylogenetic analysis of the spirochetes. J. Bacteriol. 173, 6101-6109.

Paterek, J.R. and Smith, P.H. (1988) *Methanohalophilus mahii* gen. nov., sp. nov., a methylotrophic halophilic methanogen. Int. J. Syst. Bacteriol. 38, 122-123.

Quesada, E., Ventosa, A., Rodriguez-Valera, F. and Ramos- Cormenzana, A. (1982) Types and properties of some bacteria isolated from hypersaline soils. J. Appl. Bacteriol. 53, 155-161.

Quesada, E., Ventosa, A., Ruiz-Berraquero, F. and Ramos- Cormenzana, A. (1984) *Deleya halophila*, a new species of moderately halophilic bacteria. Int. J. Syst. Bacteriol. 34, 287-292.

Quesada, E., Valderrama, M.J., Bejar, V., Ventosa, A., Gutierrez, M.C., Ruiz-Berraquero, F. and Ramos-Cormenzana, A. (1990) *Volcaniella eurihalina* gen. nov., sp. nov., a moderately halophilic nonmotile Gram-negative rod. Int. J. Syst. Bacteriol. 40, 261-267.

Rengpipat, S., Langworthy, T.A. and Zeikus, J.G. (1988) *Halobacteroides acetoethylicus* sp. nov., a new obligately anaerobic halophile isolated from deep subsurface hypersaline environments. Syst. Appl. Microbiol. 11, 28-35.

Rodriguez-Valera, F. (1988) Characteristics and microbial ecology of hypersaline environments, in "Halophilic Bacteria" vol. 1 (Rodriguez-Valera, F. Ed.), pp. 3-30. CRC Press, Boca Raton.

Rodriguez-Valera, F., Ventosa, A., Juez, G. and Imhoff, J.F. (1985) Variation of environmental features and microbial populations with salt concentrations in a multi-pond saltern. Microb. Ecol. 11, 107-115.

Skerman, V.B.D., McGowan, V. and Sneath, P.H.A. Eds. (1980) Approved lists of bacterial names. Int. J. Syst. Bacteriol. 30, 225-420.

Skerratt, J.H., Nichols, P.D., Mancuso, C.A., James, S.R., Dobson, S.J., McMeekin, T.A. and Burton, H. (1991) The phospholipid ester-linked fatty acid composition of members of the family *Halomonadaceae* and genus *Flavobacterium*: a chemotaxonomic guide. Syst. Appl. Microbiol. 14, 8-13.

Stackebrandt, E., Murray, R.G.E. and Trüper, H.G. (1988) *Proteobacteria* classis nov., a name for the phylogenetic taxon that includes the "purple bacteria and their relatives". Int. J. Syst. Bacteriol. 38, 321-325.

Stackebrandt, E. and Woese, C.R. (1979) A phylogenetic dissection of the family Micrococcaceae. Curr. Microbiol. 2, 317-322.

Vainshtein, M. Hippe, H. and Kroppenstedt, R.M. (1992) Cellular fatty acid composition of *Desulfovibrio* species and its use in classification of sulfate-reducing bacteria. Syst. Appl. Microbiol. 15, 554-566.

Valderrama, M.J., Quesada, E., Bejar, V., Ventosa, A., Gutierrez, M.C., Ruiz-Berraquero, F. and Ramos-Cormenzana, A. (1991) *Deleya salina* sp. nov., a moderately halophilic Gram- negative bacterium. Int. J. Syst. Bacteriol. 41, 377-384.

Ventosa, A. (1988) Taxonomy of moderately halophilic heterotrophic eubacteria, in "Halophilic Bacteria" vol. 1 (Rodriguez-Valera, F. Ed.), pp. 71-84. CRC Press, Boca Raton.

Ventosa, A., Rodriguez-Valera, F., Poindexter, J.S. and Reznikoff, W.S. (1984) Selection for moderately halophilic bacteria by gradual salinity increases. Can. J. Microbiol. 30, 1279-1282.

Ventosa, A., García, M.T., Kamekura, M., Onishi, H. and Ruiz-Berraquero, F. (1989a) *Bacillus halophilus* sp. nov., a moderately halophilic *Bacillus* species. Syst. Appl. Microbiol. 12, 162-166.

Ventosa, A., Gutierrez, M.C., García, M.T. and Ruiz- Berraquero, F. (1989b) Classification of "*Chromobacterium marismortui*" in a new genus, *Chromohalobacter* gen. nov., as *Chromohalobacter marismortui* comb. nov., nom. rev. Int. J. Syst. Bacteriol. 39, 382-386.

Ventosa, A., Márquez, M.C., Ruiz-Berraquero, F. and Kocur, M. (1990) *Salinicoccus roseus* gen. nov., sp. nov., a new moderately halophilic Gram-positive coccus. Syst. Appl. Microbiol. 13, 29-33.

Ventosa, A., Márquez, M.C., Weiss, N. and Tindall, B.J. (1992) Transfer of *Marinococcus hispanicus* to the genus *Salinicoccus* as *Salinicoccus hispanicus* comb. nov. Syst. Appl. Microbiol. 25, 530-534.

Ventosa, A., Márquez, M.C., Kocur, M. and Tindall, B.J. (1993) Comparative study of "*Micrococcus sp.*" strains CCM 168 and CCM 1405 and members of the genus *Salinicoccus*. Int. J. Syst. Bacteriol. 43, 245-248.

Vreeland, R.H., Litchfield, C.D., Martin, E.L. and Elliot, E. (1980) *Halomonas elongata*, a new genus and species of extremely salt-tolerant bacteria. Int. J. Syst. Bacteriol. 30, 485-495.

Wilharm, T., Zhilina, T.N. and Hummel, P. (1991) DNA-DNA hybridization of methylotrophic halophilic methanogenic bacteria and transfer of *Methanococcus halophilus* to the genus *Methanohalophilus* as *Methanohalophilus halophilus* comb. nov. Int. J. Syst. Bacteriol. 41, 558-562.

Woese, C.R. (1987) Bacterial evolution. Microbiol. Rev. 51, 221-271.

Woese, C.R., Stackebrandt, E., Macke, T.J. and Fox, G.E. (1985) A phylogenetic definition of the major eubacterial taxa. Syst. Appl. Microbiol. 6, 143-151.

Woese, C.R., Kandler, O. and Wheelis, M.L. (1990) Towards a natural system of organisms: proposal for the domains Archaea, Bacteria, and Eucarya. Proc. Natl. Acad. Sci. USA 87, 4576-4579.

Yoshida, M., Matsubara, K., Kudo, T. and Horikoshi, K. (1991) *Actinopolyspora mortivallis* sp. nov., a moderately halophilic actinomicete. Int. J. Syst. Bacteriol. 41, 15-20.

Yu,I.K. and Kawamura, F. (1987) *Halomethanococcus doii* gen. nov., sp. nov.: an obligately halophilic methanogenic bacterium from solar salt ponds. J. Gen. Appl. Microbiol. 33, 303-310.

Zeikus, J.G., Hegge, P.W., Thompson, T.E., Phelps, T.J. and Langworthy, T.A. (1983) Isolation and description of *Haloanaerobium praevalens* gen. nov. and sp. nov., and obligately anaerobic halophile common to Great Salt Lake sediments. Curr. Microbiol. 9, 225-234.

Zhilina, T.N., Zavarzin, G.A., Bulygina, E.S., Osipov, G.A. and Chumakov, K.M. (1992) Ecology, physiology and taxonomic studies on a new taxon of *Haloanaerobiaceae*, *Haloincola saccharolytica* gen. nov., sp. nov. Syst. Appl. Microbiol. 15, 275-284.

CHEMICAL ANALYSIS OF ARCHAEA AND BACTERIA: A CRITICAL EVALUATION OF ITS USE IN TAXONOMY AND IDENTIFICATION[1]

B.J.Tindall

DSM-Deutsche Sammlung von Mikroorganismen
und Zellkulturen GmbH   Mascheroder Weg 1b
D-38124 Braunschweig
Federal Republic of Germany

INTRODUCTION

I wonder if any task in microbiology can be so fraught with problems, prejudices, wrong information, or plain lack of scientific interest as trying to explain the full potential of the chemical analysis of prokaryotic cells? Our modern taxonomy has at its fingertips an array of powerful tools, and at no time in the history of microbial taxonomy have we been able to approach the subject in such a critical fashion. Despite this, confusion and uncertainty are rife. Much of the confusion, however, lies in very unexpected corners. I cannot help wondering how we continue to read of the "inferior" quality and interpretation of a phenetic taxonomy compared with a "phylogenetic" taxonomy. Yet, despite nearly 20 years of RNA analysis, a closer examination of these data reveals their phenetic origin, and only recently has classical cladistic methodology entered the field. Surely the power of the early analyses, based on Sab values, was its ability to illustrate a novel phenetic basis of prokaryotic taxonomy, which was the result of clearer/simpler markers of the present state of the evolution. Perhaps this is due to the general opinion that a phenetic system is confused with a methodology and philosophy pioneered and championed, among others by Sneath and Sokal (1973). It is not a phenetic data set which "fails" to be "phylogenetic" in its approach, but a particular method, given the name "phenetic taxonomy" or "numerical taxonomy" by various scientists, which did not have phylogeny as its primary objective; "a basic attitude of numerical taxonomists is the strict separation of phylogenetic speculation from taxonomic procedure" (Sneath and Sokal, 1973) . Nor can

---

[1] This chapter is dedicated to those chemists who have contributed to prokaryotic taxonomy and our understanding of evolution in a way yet to be fully appreciated.

one criticise "numerical taxonomy" for "failing" to encompass methods other than morphology and biochemical tests when its users so blatantly refrained from even attempting to incorporate other data (sequence data, chemical data) so clearly outlined in the scope of the methodology. Is it also not curious that sequence analysis so clearly has its basis in numerical principles, the evaluation of numerical data by computer? It is important to distinguish basic principle from scientific philosophy.

Molecular biology has opened many new doors and avenues in bacterial systematics. Despite the term we do not mean molecules in the chemical sense, but "the molecule", the gene. This is a far cry from the type of molecular biology so critically evaluated by van Niel (1966) nearly 30 years ago. Should we, perhaps also distinguish the work of chemotaxonomists as the study of chemical compounds? Problems with the acceptance of chemical data also appear to be linked to the fact that the 16S rDNA and 23S rDNA genes are "molecules" with lengths of about 1500 and 3000 bases, respectively, which are obviously subjected to evolutionary change. We should not forgot that the cell membrane, which includes proteins as well as lipid structures, may be under the control of about 100 genes (a conservative estimate). What fraction of the genome are we looking at if the average length of these genes is between 1000 and 1500 bases? Or do perhaps cell membranes just "happen?" The ribosome evolves, but the cell membrane is "created?"

Despite the fact that we have many of the prerequisites for developing a high resolution taxonomy, and a major revision of bacterial taxonomy was undertaken in 1980, the classical signs of chaos have begun to reappear. It is quite clear that in some cases the general overview of the present taxonomy has been lost, otherwise there would not be so many mistakes.

In view of the problems which I have outlined it is not the purpose of this article to go into detail concerning the full scope of the potential of chemical methods, but to illustrate a number of important aspects which have led to the failure to appreciate the value of this methodology. The basis of this work is a comparison of the chemical data with that from various RNA analyses, including cataloguing, reverse transcriptase and direct gene analysis, and RNA/DNA hybridization. In view of the absence of any definitions of taxa other than the species, DNA/DNA hybridisation values of above 70% are taken to be indicative of strains being in the same species.

THE DATABASE SYNDROME

One of the major problems associated with any comparison of chemotaxonomy with other data is the fact that one must place some degree of trust in published data. In recent years various databases have become available which help to simplify the work of the microbiologist in identifying strains. Unfortunately these are not entirely reliable, and I would like to briefly mention some of the past and present problems with three widely used databases, one based on biochemical testing, another using fatty acids, and perhaps the most commonly used, DNA sequence databases. In the case of the biochemical database the most interesting aspect is associated with a dendrogram used to promote the Biolog system. It is contradictory to our current knowledge to group *Halobacterium denitrificans* (which is now *Haloferax denitrificans*, and a member of the Archaea) with members of the genera *Halomonas* and *Deleya*, which are clearly members of the Bacteria. Perhaps the location of *Pseudomonas beijerinckii* outside this group is also revealing. Kersters (1991) clearly assigns this species to the *Deleya/Halomonas* complex. It is also curious that members of the genera *Sphingobacterium* and *Sphingomonas* group together.

when one considers their placement in the *Bacteroides/Flavobacterium/Cytophaga* group and α-subclass of the Proteobacteria, respectively.

Problems with the fatty acid data obviously stem from wrong strains or contamination. In one case, the strain of *Pseudomonas saccharophila* is quite clearly not a member of the β-subclass of the Proteobacteria as shown by De Ley (1991), but belongs within the *Sphingomonas* group. Other obvious mistakes include, or have included, fatty acid patterns from Gram-positive bacteria appearing under Gram-negative species names.

The various ribosomal sequence databases are also not free from problems, and one finds species such as "*Methanosarcina acidivorans*" (which is probably *Methanosarcina acetivorans*) (Eggen *et al.*, 1992), the type strain of *Flexibacter filiformis* which appears as *Flexibacter elegans* (Gherna and Woese, 1992), and *Erythrobacter longus*, which is quite distinct from *Erythrobacter* strain OCH101, the latter being the only strain, and type strain of *Erythrobacter longus* (Tsuji *et al.*, 1990). Despite the fact that the occurrence of two 16S rDNA genes have been reported in *Haloarcula marismortui* (Mylvaganam and Dennis, 1992), no-one appears to know which of the two strains in circulation has been studied (Tindall, 1991), and my attempts to obtain a subculture of the strain used have been unsuccessful. Considering the importance of the sequence data in the present trend in taxonomy it is particularly deplorable that many strains either bear no designation or are not type strains (De Ley, 1991).

Despite that fact that many extensive databases contain data on several thousand strains this is only a small part of the potential data when one considers that many people consider the total number of prokaryotic species is somewhere in the region of 100,000 to 1,000,000.

CHEMOTAXONOMY: THE DISTORTION OF FACTS

Few scientists are engaged in "classical chemotaxonomy". The majority of scientists are not familiar with it to any extent, and the impression is often given that the chemical composition of the cell is a random, chaotic assortment arrived at by chance. Unfortunately such views have their origins in the 1960s and early 1970s, when the chemical diversity of genera such as *Pseudomonas*, among others, was taken as evidence of the inherent variability in chemistry within a genus. We still find reference to such work on members of the genus *Pseudomonas* without reference to the current taxonomy, which often shows the species used now belong to different taxa. The view that members of the same genus could differ widely in chemical composition was not a view entirely shared by chemists, and one can find chemists challenging the classical taxonomy from the late 1960s onwards. Such work was not even considered in the euphoria of redefining the genus *Pseudomonas* based on 16S rRNA cataloguing (Woese *et al.*, 1984), and no mention is made of the work of Wilkinson (1967) or even the work of De Ley *et al.* (1965) nearly some 20 years previous. It is useful at this point to emphasise the work of Ikawa (1967) suggesting the use of lipids for determining "natural relationships", the critical work of Shaw (1970, 1974), which was an inspiration for some, and the pioneering work of Cummins and Harris (1956) on cell walls.

It appears that those not familiar with chemical methods in prokaryotic taxonomy are often met with a wall of misinformation. Simple statements are often the most damaging, such as "phospholipids are the most common polar lipids. They are derivatives of phosphatidic acid, and in most bacteria are quite uniform in overall pattern" (Schleifer and Stackebrandt, 1983; Trüper and Schleifer, 1991; Goodfellow and O'Donnell, 1993). Nevertheless Goodfellow and O'Donnell (1993) concede that they are of value in the actinomycetes, the group of organisms intensively studied in Newcastle, while Trüper and Schleifer (1991) acknowledge the value of phospholipid patterns in members of the Archaea. This view of the importance of phospholipid composition is not dispelled by the over emphasis of phospholipids by Lechevalier et al. (1977) and Wilkinson (1989), a point made by Shaw (1970) some twenty years ago. Despite such sweeping statements it is common to forget that, strictly speaking 1-(13-methyltetradecyl)-2-(12-methyltetradecyl)-3-phosphatidylethanolamino-glycerol (PE-iso15:0, anteiso15:0), 1-hexadecyl-2-hexadecyl-3-phosphatidylethanolamino-glycerol (PE-16:0, 16:0), and 1-hexadecyl-2-octadecyl-3-phosphatidylethanolamino-glycerol (PE-16:0, 18:0) are different compounds. This would be similar to stating that the sequences AGTATC and ATCAGT contain the same four bases, and are of no value because they are "essentially" the same.

Two points are worth noting. First, although PE may be present in *Escherichia coli* and *Bacillus stearothermophilus* the fatty acid patterns of the two species are different, and as a consequence the two species contain two modifications of the basic structure PE, i.e. two different compounds. Secondly, one only has to look at *Agrobacterium tumefaciens, Escherichia coli, Cytophaga johnsonae*, members of the genera *Deinococcus*, and *Sulfolobus* to realise that there is no one common phospholipid. Despite the extensive work on many Gram-positive organisms, certain Proteobacteria, members of the Archaea, and members of the genera *Chloroflexus, Thermus, Chlorobium*, and *Thermomicrobium*, it is often suggested that glycolipids are of limited distribution. Using the fact that *Acholeoplasma laidlawi* and streptococci have the same diglycosyl head group in the lipid (Goodfellow and O'Donnell, 1993) ignores other possible differences in the fatty acid and other lipid components. One could equally well suggest that, because a gene probe against part of the α-1 subclass of the Proteobacteria cross reacts with "*Flexistipes sinusarabicus*" or members of the spirochetes and related taxa (Manz et al., 1992), gene probes have no discriminatory power; a ridiculous suggestion. What ever happened to the aminolipids, derived from ornithine, glycine, lysine, taurine or serine which few authors, other than experts (usually chemists) even mention?

Suggestions that chemotaxonomy and numerical taxonomy do not give reliable phylogenetic classifications do not seem to be borne by the fact that chemotaxonomy very quickly provided evidence in support of the archae(o)bacteria, and that Sneath and Sokal (1973) give extensive coverage of the assumptions and problems associated with phenetic relationships, cladistic relationships, and the inference of phylogenies (Stackebrandt and Goodfellow, 1991). While Fuerst et al. (1993) have questioned the use of fatty acids as suitable compounds for inferring evolutionary relationships, it is, perhaps, more suitable to question the choice of the diunsaturated (18:2) fatty acids as characteristic of the group in question (Tindall, unpublished), and that one should place less emphasis on quantitative data.

The general opinion appears to be that the chemical composition may support groupings described by sequence analysis, but if it does not the chemistry is inferior. Unfortunately such opinions have their origin in much earlier work. While Zuckerkandl and Pauling (1965) emphasized the value of sequencing (of proteins) in determining the evolution of an organism they did not exclude the use of episematides (compounds which are neither genes nor their direct products). Although the logical disadvantages of using episematides, such as ATP or polysaccharides, are quite clear, the choice of unsuitable compounds (as they obviously are) cannot be used as evidence that the chemical composition of a cell cannot, or does not reflect evolution. It is interesting that they also convey the impression that phylogenetic interpretation using episematides requires checking, while that based on sequencing (semantides) does not. Woese et al. (1980) and Dickerson (1980) showed an interesting correlation and discordance between cytochrome sequence and 16S rRNA cataloguing data and the need has been stressed for the comparison of two or more independent data sets.

CHEMOTAXONOMY: THE MISTAKES

Irrespective of discipline, incorrect information, interpretation, or questionable results play a significant role in the development of that science. There are numerous such problems in the literature relating to the chemical composition of the prokaryotes, and they have contributed significantly to the impression that the chemical composition of an organism is a randomly arranged array of compounds in which there is no or obvious sign of evolutionary development or coherence. Unfortunately such suggestions of chaos in chemical composition do not appear to have any scientific foundation. In some cases the appearance of errors in the literature also underlines the fact that neither authors, editors, or reviewers are properly informed. It is quite surprising to read of the proposal for the development of a fatty acid data base for halobacteria (Mendala, 1990), when one considers that the predominant lipids in such organisms are ether linked (not even ester linked) polyisoprenoids. This is not an isolated example and the same error is made in discussing the affinity of the halobacteria to the Bacteria in the halobacteria/eocyte interpretation of the Archaea (Rivera and Lake, 1992).

Similar problems are to be found in the literature relating to lipoquinone analysis. One wonders how the electron transport system of *Chlorobium limicola* (Cork *et al.*, 1985) can possibly function as proposed when all members of the family Chlorobiaceae have been shown to contain menaquinones and not ubiquinones. This is not simply a case of one compound replacing another, the two compounds have different redox potentials, requiring that the other adjacent components in the respiratory chain be adjusted accordingly. Reports of the presence of ubiquinones in "*Capnocytophaga*-like" organisms (Daneshvar *et al.*, 1991) provide another example where the results indicate the "chaotic" nature of prokaryotic chemical composition. All available data indicate that members of the *Bacteroides/Flavobacterium/Cytophaga* group produce menaquinones, and that known species of the genus *Capnocytophaga* produce menaquinone 6 (Speck *et al.*, 1987). One can only conclude that either these strains do not belong within this group, which would contradict the fatty acid patterns, or that the results are in error. Considering that no mention is made of the separation into menaquinone or ubiquinone classes before HPLC, and that rapid HPLC separation leads to overlapping of ubiquinones and menaquinones, one can only conclude that these data should be checked. We also read that "menaquinones have not been found in Archaea" (Goodfellow and O'Donnell, 1993), whereas in the same book Suzuki *et al.* (1993) discuss the range of menaquinones found in members of the Archaea. One also finds an interesting bias towards ubiquinones in text books. Despite their importance in *Escherichia coli* and mitochondria the vast majority of prokaryotes which produce quinones produce menaquinones, ubiquinones are only produced by an "exotic" minority, the members of the α-, β-, and γ-subclasses of the Proteobacteria (see Collins and Jones, 1981 for a review). In some cases the conclusions reached, such as the fatty acid 11-methyl 18:1 is a specific marker of the genus *Afipia* (Moss *et al.*, 1990) is contradicted by its presence in its nearest neighbours, *Bradyrhizobium* and *Blastobacter denitrificans* (Auling *et al.*, 1988; Sittig and Hirsch, 1992), or in even more distantly related species (Sittig and Hirsch, 1992; Sittig and Schlesner, 1993; Tindall, unpublished results). The fact that one often finds reference to the fatty acid 18:1 as being characteristic of a particular genus or species is misleading when one considers its predominance in certain groups of possible class or order status.

Errors in a particular fatty acid data base include the identification of the major fatty acid of *Curtobacterium pusillum* as 18:1 rather than the ω-cyclohexyl-undecanoic acid (Suzuki *et al.*, 1981) found by mass spectrometry. This appears to be due to overlapping of the two compounds under the chromatography conditions used. However, it is not necessary to alter the run conditions in any way, and a simple, well documented experiment taking no more than 30 minutes is perfectly adequate to distinguish the two compounds. Other cases of overlap, with interesting consequences may be found in the co-elution of the methyl esters of 16:1ω7 and 2-hydroxy-*iso* 15:0 in members of the genus *Sphingobacterium* (Dees *et al.*, 1985). Considering that a variety of factors, including column loading and efficiency of the column affect whether or not the two are separated, it is interesting to see that in a commercially available database one species is listed as having both compounds, whereas others have an unresolved mixture. Again this may be solved with some simple experiments easily within reach of all laboratories. These problems point to a lack of familiarity with the literature.

Another example of such misleading data, which clearly illustrates the impression of "chemical chaos" is the publication of data on certain *Arthrobacter* species. Some of the species are clearly not even closely related to members of this Gram-positive genus. What is surprising is that data published by Amadi and Alderson (1992) and Collins (1986) are clearly contradictory. In the work of Amadi and Alderson (1992) the polar lipid patterns and lipoquinones are consistent with *Arthrobacter siderocapsulatus* NCIMB 11287 and NCIMB 11286 being Gram-negative, whereas the fatty acid data quite clearly shows that Gram-positive strains have been investigated. In contrast, two independent data sets (Collins, 1986; Tindall, unpublished) quite clearly show that the lipoquinones, polar lipid pattern, and fatty acid composition of the two strains are consistent with them being Gram-negative species. It is even possible to locate their next closest relatives (Tindall, unpublished).

It is a pity that statistical evaluation of the significance of fatty acid composition in strains of Antarctic bacteria ignores the taxonomic and evolutionary groupings examined and concentrates on the effect of temperature (Rotert *et al.*, 1993). Equally interesting are conclusions reached based on a combination of chemical and sequencing data, which place an azo dye degrading organism in the genus *Caulobacter* (Govindaswami *et al.*, 1993). The fact that all data presented (API, Biolog, fatty acids, and 16S rDNA sequence) clearly place this organism within the *Sphingomonas* group of organisms is indicative of problems in published data on the chemical and genetic properties of the reference strains.

It is not surprising, in the light of such problems that the majority of people have accepted the "chemical chaos". A logical extension of such conclusions is that the chemical composition of an organism can only be used to delineate taxa in the light of sequence analysis and biochemical testing. O'Donnell *et al.* (1993) point out that "extensive chemotaxonomic analyses have revealed the inherent variability of chemical data, the same character or group of characters being conserved in one taxon whilst varying systematically in another". They conclude "this questions the value of chemical data as the basis of a universal classification technique". Surely this illustrates the need to be flexible in one's approach and to routinely use a number of methodologies?

MOLECULAR AND EVOLUTIONARY PROBLEMS

No discussion of the role of chemotaxonomy in prokaryotic classification and identification would be complete without considering problems arising from "molecular" methods, and nucleic acid sequencing in particular. Unfortunately criticism is often looked upon negatively and I would like to point out that it is perfectly possible to accept a principle, in this case the sequencing of evolutionary conserved genes, but to evaluate the data and the end result in a scientific manner, and come to the conclusion that there are still problems to be tackled.

Problems in microbial taxonomy are also evident in nucleic acid sequencing. A cardinal mistake has been to uncouple the early cataloguing data from any other data set and provide the impression that in certain cases there was no correlation between a genetic and a phenetic taxonomy (Schleifer and Stackebrandt, 1983) a view which persists today

(Olson *et al.*, 1994). The question is what is meant by phenetic? Do we mean phenotype, a phenetic system developed by Sneath and Sokal (1973), or do we mean phenetic in the sense of being derived from the phenome? In the latter case the genome is part of the phenome, and I have great difficulty in understanding how the genome segregates from the phenome, of which it is a part.

The confusion in nucleic acid sequencing is elegantly demonstrated in the assignment of *Thiosphaera pantotropha* to the species *Paracoccus denitrificans* (Ludwig *et al.*, 1993). In an emended description of the genus *Paracoccus* one finds no mention of the recently described species *Paracoccus aminovorans*, *Paracoccus aminophilus*, *Paracoccus kocurii*, and *Paracoccus alcaliphilus* (Urakami *et al.*, 1989; 1990; Ohara *et al.*, 1990). Furthermore published chemical data indicate that there is a remarkable similarity between the lipoquinone and fatty acid patterns of *Paracoccus halodenitrificans* and those of members of the *Halomonas/Deleya* complex (Akawaga *et al.*, 1989; Franzmann and Tindall, 1990). This has been confirmed by polar lipid analyses (Marquez and Tindall, unpublished results).

Such problems are not uncommon, and the failure to check the identity of strains used, or correct nomenclature is a widespread problem. While there is the tendency to discount such problems as superficial, it makes it impossible to correlate other data with the data from 16S rRNA or rDNA sequencing results. In some cases no-one really knows which strain was examined. In the case of *Fusobacterium nucleatum* ATCC 25586 (the type strain) the cataloguing data (Pasteur *et al.*, 1985) clearly contradict the chemistry of the organism and the results obtained by reverse transcriptase analysis of a large number of *Fusobacterium* species, including the same ATCC strain (Lawson *et al.*, 1991). I have already mentioned some other problems, and these are by no means isolated examples. The strain of *Marinococcus halophilus* is not the type strain, as suggested (Farrow *et al.*, 1990). *Zooglea ramigera*, DSM 287 = ATCC 19623 = IAM 12669 is a member of the α-subclass of the Proteobacteria (Schleifer *et al.*, 1992), and not in the γ-subclass as is the type strain, ATCC 19544 = IAM 12136 (Hiraishi *et al.*, 1992). Publications confirming this view have appeared since drafting this manuscript (Shin *et al.*, 1993; Rosseló-Mora *et al.*, 1993). It is a pity that the strain of *Bacillus acidoterrestris* (this strain has now been exchanged for the correct type strain) originally examined by Ash *et al.* (1991), and since shown to be atypical by 16S rDNA sequencing (Wisotzkey *et al.*, 1992) and chemical data (Tindall, unpublished) still continues to appear in databases (Suzuki and Yamasato, 1993). Once wrong names appear in additional publications without accession numbers or strain designations it becomes impossible to locate such errors. Although the errors from cataloguing should eventually be solved by the availability of full sequence data, this does not solve problems firmly entrenched in the literature which imply that other methods are not able to recover results similar to those from 16S rRNA/rDNA analysis. In some cases strains are only shown to be members of a particular species by physiological and biochemical testing, and one can only echo the words of Cowan (1978) "do not rely on the purity of cultures received from others, or assume that the identification made by the donor (however good or senior) is correct."

Although Sneath (1989) has dealt with some of the possible problems associated with sequence analysis, it is rare to find any critical evaluation of the influence of computer programs on the analysis of 16S rRNA/rDNA data. One such exception is the review of Felsenstein (1988), and some interesting points are made. In contrast to those articles where the impression is given, either deliberate or by chance, that the topology of the dendrogram is the phylogeny of the organisms in question, discussion is made of various methods used to generate these dendrograms and the use of bootstrap, jackknife, and similar statistical methods for testing tree topology. The modern literature is littered with papers entitled "phylogeny of...", "phylogenetic diversity of.....", or "phylogenetic analysis of........" I am sure that the scientists who write such papers are not aware that they are conveying the impression that the topology of the "tree" is absolute to those not familiar with the method. While I would agree that organisms do not change their position between major groups, i.e. *Bacillus subtilis* should not group in the Proteobacteria, the effects within in groups are more critical.

In the past, 16S rRNA cataloguing and reverse transcriptase sequencing did not show much variation in the dendrograms from publication to publication, simply because only a small group of scientists, using the same methods of data evaluation, were involved in the work. The explosion in the use of DNA sequencing is a direct result of the PCR technique and the availability of automatic DNA sequencing machines, together with the development of different methods of data evaluation. While the diversification of the methodology and data evaluation should be looked on as an advantage, it is also confusing to those not familiar with the system, and judging from the many problems, I would suggest this is a significant proportion of microbiologists.

Colleagues who are involved in 16S rRNA or rDNA sequencing have pointed out that one should not place too much emphasis on certain aspects of the dendrograms, and I am grateful for their advice. However, the vast majority (scientists engaged in nucleic acid sequencing included) are influenced by designations such as α-1, α-2, α-3, and α-4 subclasses of the Proteobacteria (Woese, 1987), *Bacillus* groups 1 to 5 (Ash *et al.*, 1990), or *Flavobacterium/Cytophaga/Sphingobacterium/Bacteriodes* (Gherna and Woese, 1992) and one is directed towards dealing with dendrograms, branching order, and branching pattern. This is also where a great deal of confusion arises. Examination of some well studied groups which have been sequenced and analysed, often in parallel by different groups of scientists, are quite revealing. The α-subclass of the Proteobacteria is a case in point. The most remarkable feature is the different topologies of the dendrograms from different laboratories. Despite this, I would warn those opposed to sequencing to use this as "evidence" of inadequacies in sequencing. While Fuerst *et al.* (1993) recover the groups α-1, α-2, α-3, and α-4 subclasses, as defined by Woese (1987), Stahl *et al.* (1992) suggest an affinity between members of the α-2 and α-3 subclasses. In contrast Ludwig *et al.* (1993) consider that *Erythrobacter longus* is part of the same sub-group as *Paracoccus denitrificans*, *Rhodobacter* species, and *Thiobacillus novellus*, i.e. the α-3 and α-4 are one group. However, other studies recover the α-3 and α-4 subclasses as distinct clusters, whereas extensive fragmentation of the α-2 subclass occurs (Yanagi and Yamasato, 1993; Kawasaki *et al.*, 1993). In some cases to such an extent that the α-3 and α-4 subclasses are interspersed within the α-2 subclass, and it becomes impossible to recognise the α-2 subclass as a distinct, coherent unit. Selection of strains in the analysis is critical, otherwise mathematical problems place *Magnetospirillum magnetotacticum* (a member of the α-1 subclass) in the α-2 subclass (Willems and Collins, 1992).

One sees similar problems in the work on the bacillus group. Apart from individual strains changing position slightly (Ash *et al.*, 1990; Farrow *et al.*, 1990), there are significant changes in the topology of the dendrograms between the work of Ash *et al.* (1990) and Aquino de Muro and Priest (1993) in the group 2 bacilli and related organisms. The influence of different programs is discussed by Felsenstein (1988), and more work in this area is clearly required. The inherent instability of certain groups, the influence of the choice and range of strains, as well as the data set (i.e. masked or full sequences) is well known. Perhaps, then it is surprising that I will make no criticism of these problems, they are there and need to be solved. However, one is often given the impression that all other data should be interpreted in the light of sequence analysis. Thus, should the chemical data not fit one particular dendrogram, one concludes that chemical data do not contribute to our understanding of prokaryotic evolution. My problem is, with which version of the "phylogeny" do I compare my data in the first place? A logical consequence of this has been to adopt a system, based on chemical composition which is not based on numerical principles, and avoids such dendrograms with statistically calculated branching patterns and branching orders (Tindall, unpublished), yet is still capable of recovering groupings, and provides a new dimension in the interpretation of these "unstable dendrograms."

It is these, and a whole range of other questions which still need to be answered. What influence does the choice of a strain within a species have on the closer relationships? There is evidence that the sequence variation within a species may be as much as 1-2% in the 16S rRNA/rDNA (Fox and Brown, 1992; Fry *et al.*, 1991). Insignificant perhaps, but more important when one considers published data on members of the genus *Aeromonas* where the lowest level of 16S rRNA sequence homology between two species is about 98% (Martinez-Murcia *et al.*, 1992). Evidence for microheterogeneity of the 16S rRNA within a species also comes from data on *Leuconostoc citreum* (Martinez-Murcia and Collins, 1991; Takahashi *et al.*, 1992), the different "species" of the genera *Thermoanaerobacter* and *Thermoanaerobacterium* (Lee *et al.*, 1993; Rainey *et al.*, 1993) and *Methanobacterium thermoautotrophicum* strain THF and the *Methanobacterium thermoautotrophicum* ΔH group (Nölling *et al.*, 1993; Touzel *et al.*, 1992). Problems associated with the different methods of determining DNA/DNA hybridization also need to be taken into consideration (Grimont *et al.*, 1980; Huss *et al.*, 1983). Considering that there is 5% homology difference in the sequence homology of the two genes coding for the 16S rRNA in *Haloarcula marismortui* (Mylvaganam and Dennis, 1992) and 2.8% difference between the two 16S rRNA genes in "*Haloarcula sinaiensis*" (Kamekura and Seno, 1993) should we not ask how much variation is there in the multiple copies of the 16S rRNA gene, or are these just evolutionary peculiarities which cause confusion? I am not aware of any data on whether the PCR reaction amplifies one 16S rRNA gene, or is a composite produced? Do we not need to know about the magnitude of sequence variation within the species or genus? Are experiments demonstrating *in vitro* transfer of 16S rDNA between different strains of significance in the environment (Niebel *et al.*, 1987; Suwanto and Kaplan, 1992)? The use of additional markers would also be useful. I would not choose 23S rDNA sequences or DNA/RNA hybridisation in support of 16S rDNA data, simply because we are not studying functionally independent units. Equally well, recovering the major groups (Proteobacteria, Gram-positives, *Deinococcus*, *Planctomyces*, etc.) can also be done quite easily using chemical data, the critical evaluation needs to be done at the 16S rDNA similarity level above 90% sequence homology.

The major point is not the apparent inadequacies of the sequencing work in this context, but the fact that the question of how groups occur and which ones are correct is still in a more hypothetical stage than the literature leads us to believe. It is becoming increasingly obvious that simple evaluation of sequence homologies, whether it be similarities or evolutionary distances alone does not lead to the definition of taxa. Surely, any overemphasis of the 16S rDNA, in principle a monothetic evaluation, is as dangerous, although not to the same degree, as the overemphasis of morphology or physiological traits. In this context, trends to ignore morphology (Head *et al.*, 1993) or intracellular membrane structure (Kawasaki *et al.*, 1993) simply because the 16S rDNA homology is high (96% and 93% respectively), without providing any additional scientific data are to be viewed with extreme caution.

I see no basic need to question the value of sequencing as a method in taxonomy and identification, but there are very clear problems associated with data evaluation and interpretation. Do we, for instance, interpret the data in the light of the existing taxonomy, or do we interpret the taxonomy in the light of what we find? How important is what we find and what we expect to find?

PUTTING CHEMICAL COMPOSITION IN CONTEXT

It is not my intention to defend chemotaxonomy, I think that explained in the proper context and given a fair hearing, it has a valuable role to play. Certain points do, however, need to be made clear. Membranes are not amorphous assemblages of "fats" and proteins, they are highly organized functional units whose range of chemical components are limited genetically. They do, however, respond to environmental parameters in a fashion which is limited in scope and is genetically predetermined. Thus, phosphate limitation in a marine *Pseudomonas* species (Minnikin and Abdolrahimzadeh, 1974) results in over production of an aminolipid, whereas similar conditions in a *Bacillus* species promote synthesis of a digylcosyldiglyceride (see Minnikin and Goodfellow, 1981 for a review), simply because these are genetically predetermined responses. Biophysical studies have indicated the preference of specific head group structures for influencing the hydrophobic phase into bilayers or monolayers (Sutton *et al.*, 1991; In 't Veld *et al.*, 1993). Thus, it is not quite the roulette wheel system that picks which head groups occur in the cell membrane, there are biophysical, environmental, and genetic considerations to be taken into account. The role of phospholipids has also been suggested in sequestering ions in ion depleted environments. It is obvious that prokaryotes have a whole range of "contingency" plans for dealing with a variety of conditions. It is for these reasons, the variation in chemical composition with varying growth conditions that the majority of criticism has been made. Strict control of quantitative differences are only of importance when employing methods relying on quantitative evaluation. Considering that most laboratory media are artificial in that they are designed to promote growth, it is unlikely that adverse conditions, i.e. wrong pH, temperature, salinity or nutrient limitations will be used in taxonomic studies. Constraints are also placed on lipoquinone structure and cell wall structure. The interdependence of lipids and enzymes has also recently been emphasised (Truniger and Boos, 1993; In 't Veld *et al.*, 1993). Although differentiation to the strain level is also possible using chemical composition, quantitative variation plays a subsidiary role should one reduce the resolution to the genus level and introduce cladistic or keying methodology.

One must also be cautious in interpreting older data. In many cases the generic, or species designations are wrong, and suggestions that sulfonic acids replace phospholipids under phosphate limitation in members of the genus *Pseudomonas* (Suzuki *et al.*, 1993) must be re-evaluated in the light of the reclassification of these organisms into evolutionary distinct units, *Pseudomonas fluorescens* (γ-subclass), *Shewanella putrefaciens* (γ-subclass), and *Pseudomonas diminuta* (α-subclass). All three organisms are distinct in their fatty acid, polar lipid, lipoquinone, and polyamine patterns. It is also the comparison of data from biochemists and geneticists, using out-of-date taxonomies which can also be misleading. One finds, for instance that *Micrococcus cryophilus* is related to *Psychrobacter immobilis* (Juni and Heym, 1986), a member of the γ-subclass of the Proteobacteria, and that a *Micrococcus* species was later renamed as a member of the genus *Acinetobacter* (Pieringer, 1989). One still finds reference to the orders Pseudomonadales and Eubacteriales (Schweizer, 1989), dating from literature 20 years old! Sometimes the closer examination of superficial traits, such as isoprenoid biosynthesis has indicated that two different biosynthetic pathways may be operating (Horbach *et al.*, 1993). Is it not of relevance to our view of evolution to consider that there is 60% sequence homology between the 16S rDNA of *Escherichia coli* and *Halobacterium cutirubrum*, yet the membranes contain two fundamentally different hydrophobic regions (fatty acids and polyisoprenoids)? Furthermore the mechanisms leading to the synthesis of the fatty acids and isoprenoids are sufficiently different that there may be only a few genes in common, not to mention the absence of any homology. Is this also not worthy of our attention? Perhaps this also illustrates to the "molecular" biologist that genes are not always the answer, functionally similar units (cell membranes) do not have to be composed of the same compounds, nor do universally distributed genes have to be responsible for the synthesis of the different compounds. This also illustrates another interesting point, the analysis of genes such as the 16S rDNA indicates a high degree of homology between evolutionary diverse taxa, whereas investigating a more complex functionally conserved structure, the cell membrane, indicates a wide diversity in participating components, and probably only limited genetic homology. These data add weight to Kandler's (1993) recent hypothesis on the early diversification of life, while not denying a single genetic origin for the ribosome or the α-subunit of the ATPase.

CONCLUSIONS

It appears that the framework of a modern taxonomy is presenting a more reliable picture. It would have been impossible to locate many of the mistakes or question the interpretation of some of the data were it not for the fact that there appears to have been coherent evolutionary influence on prokaryotes. However, there are some significant problems in correlating the different data sets. The impression often given that 16S rRNA/rDNA data is "phylogenetic" does not appear to be substantiated. It may be a suitable molecule to allow us to infer the "phylogeny" of prokaryotes, however, there is more to the evolution of prokaryotes than the study of a limited number of genes. Should we not heed Felsenstein's (1988) warning of the "perils of molecular introspection?" Despite early trends to show that genetic and phenotypic taxonomies were discordant, I would predict that the study of suitable phenotypic markers (complex macromolecules etc.) will show that it is possible to harmonise phenotypic and genotypic data. Another point which also needs to be clarified is whether a phenetic or a phylogenetic system is applicable in taxonomy or identification. More importantly what do we mean by phenetic and phylogenetic? Twenty years of RNA analysis have shown the value of the method, not by being "phylogenetic", but by allowing us to determine more reliably the present state of evolutionary relationships. Adding cladistic approaches to the analysis of protein or gene sequences is not unique to this data set, and it is possible to apply such methods to chemical data with interesting results. Perhaps one should also ask "do we need to know the order of evolution (branching pattern) among species or genera in order to create a reliable taxonomy?" Are we also not distracted by the power of a method which leads us to theories which may be difficult to prove, when the requirements of taxonomy and identification are a long term stable system?

In conclusion, I can only say that we are in the process of learning in microbiology what developed in other areas within the last half a century. "For more than a century the field of biology was so extensive and growing so rapidly that no single investigator, no matter how broad might be his grasp, could keep abreast with the developments in all the numerous branches. The response of biology to this challenge was a subdivision of the general field into many disciplines, each endowed with its own material, methods, and techniques. Instead of being biologists most of us became systematists, physiologists, geneticists, embryologists, biochemists, pathologists, etc."...."This extreme compartmentalisation of biological knowledge proved fruitful in that it led to enormous accumulation of factual information; it has been deleterious in so far as it resulted in a lack of understanding between the representatives of the various disciplines and a consequent lowering of the efficiency of biological research. It stands to reason that the exigencies of the situation continue, and probably will continue, to demand that each biologist be a specialist in some small portion of the general field. During the last decade the conclusions reached by many of the specialists have begun to converge toward a set of general principles applicable to the entire realm of living matter. One can only hope that this will occur in increasing measure in future. Biology, it seems, is no longer in its childhood; as a science, it is approaching maturity" (Dobzhansky in the introduction to Mayr, 1942). One can only hope that the future developments in microbiology will see the biochemist or geneticist consulting the taxonomist concerning the identity of strains and their potential relationships to one another. That "molecular" biologists will realise that their data are part of a greater biological framework, as van Niel (1966) pointed out some 30 years ago, and that those working on nucleic acid sequencing will realise that other data have a significant contribution to make in taxonomy and understanding evolution. In my experience it is the exchange of ideas and concepts between different disciplines which often leads to fruitful discussion, the development of hypotheses and theories. Is it not this approach which Wolfe (1991) pursued in stimulating work on cell wall structures (Kandler and König, 1978) and the lipid analysis of methanogens (Tornabene *et al.*, 1978) once the distinct differences in metabolism and RNA structure became apparent in the methanogens? Without the prior work on the lipids of halobacteria, *Thermoplasma,* and *Sulfolobus, would* not the task of delineating and winning support for the Archae(o)bacteria have been more difficult? One conveniently forgets the contribution of chemical data in this area. I can only hope that the willingness of certain people to openly discuss such matters continues, the rejection of one principle by another school of thought in a fever reminiscent of the crusades or the inquisition has never been of long term benefit in science.

ACKNOWLEDGEMENTS

I would like to thank: Prof. Dr. H. G. Trüper and Dr. D. Claus for their liberal attitude towards chemotaxonomy, and thus making it possible to carry out the studies on which this article is based.
Those (few) scientists engaged in rRNA or rDNA sequence analysis, especially Dr. E. R. B. Moore, who were willing to openly and honestly discuss the topic and consider the chemical data.
Frau Elke Priemer for her invaluable technical assistance.
Finally those staff in the DSM (especially Dr. H. Hippe, Dr. N. Weiss, and Dr. P. Hoffman) without whom the task of wading through the diverse problems would have been impossible.

REFERENCES

Akagawa, M. and Yamasato, K. (1989) Synonymy of *Alcaligenes aquamarinus, Alcaligenes faecalis,* subsp. *homari,* and *Deleya aesta: Deleya aquamarina* comb. nov. as the type species of the genus *Deleya.* Int. J. System. Bacteriol. 39: 462-466

Amadi, E.N. and Alderson, G. (1992) Lipids of *Arthrobacter siderocapsulatus.* J. Appl. Bacteriol. 73: 144-147.

Aquino de Muro, M. and Priest, F.G. (1993) Phylogenetic analysis of *Bacillus sphaericus* and development of an oligonucleotide probe specific for mosquito-pathogenic strains. FEMS Microbiol. Letts. 112: 205-210.

Ash, C., Farrow, J.A.E., Wallbanks, S., and Collins, M.D. (1991) Phylogenetic heterogeneity of the genus *Bacillus* revealed by comparative analysis of small-subunit-ribosomal RNA sequences. Letts Appl. Microbiol. 13: 202-206.

Auling, G., Busse, J., Hahn, M., Hennecke, H., Kroppenstedt, R.M., Probst, A., and Stackebrandt, E. (1988) Phylogenetic heterogeneity and chemotaxonomic properties of certain Gram-negative aerobic carboxydobacteria. System. Appl. Microbiol. 10: 264-272.

Collins, M.D. and Jones, D. (1981) Distribution of isoprenoid quinone structural types in bacteria and their taxonomic implications. Microbiol. Rev., 45: 316-354

Cork, D., Mathers, J., Maka, J., and Srnak, A. (1985) Control of oxidative sulfur metabolism of *Chlorobium limicola* forma *thiosulfatophilum.* Appl. Env. Microbiol. 49: 269-272.

Cowan, S.T. (1978) "A Dictionary of Microbial Taxonomy," Cambridge University Press, Cambridge.

Cummins, C.S. and Harris, H. (1956) The chemical composition of the cell wall in some Gram-positive bacteria and its possible value as a taxonomic marker. J. Gen. Microbiol. 14: 583-600

Daneshvar, M.I., Hollis, D.G., and Moss, C.W. (1991) Chemical characterisation of clinical isolates which are similar to CDC group DF-3 bacteria. J. Clin. Microbiol. 29: 2351-2353.

De Ley, J., (1991) The Proteobacteria: ribosmal RNA cistron similarities and bacterial taxonomy in "The Prokaryotes. A Handbook on the Biology of Bacteria: Ecophysiology, Isolation, Identification, Applications," Balows, A., Trüper, H.G., Dworkin, M., Harder, W., and Schleifer, K.H., eds., Springer Verlag, New York.

De Ley, J., Park, I.W., Tijtgat, R., and Ermengen, J. (1965) DNA homology and taxonomy of *Pseudomonas* and *Xanthomonas*. J. Gen. Microbiol. 42: 43-56

Dees, S.B., Carlone, G.M., Hollis, D., and Moss, C.W. (1985) Chemical and phenotypic characteristics of *Flavobacterium thalpophilum* compared with those of other *Flavobacterium* and *Sphingobacterium* species. Int. J. System. Bacteriol. 35: 16-22.

Dickerson, R.E. (1980) Evolution and gene transfer in purple photosynthetic bacteria. Nature 283: 210-212.

Eggen, R.I.L., Geerling, A.C.M., de Groot, P.W.J., Ludwig, W., and de Vos, W.M., 1993 Methanogenic bacterium Göl: an acetoclastic methanogen that is closely related to *Methanosarcina frisia*. System. Appl. Microbiol. 15: 582-586

Farrow, J.A.E., Ash, C., Wallbanks, S., and Collins, M.D. (1992) Phylogenetic analysis of the genera *Planococcus*, *Marinococcus* and *Sporosarcina* and their relationships to members of the genus *Bacillus*. FEMS Microbiol. Letts. 93: 167-172.

Felsenstein, J. (1988) Phylogenies from molecular sequences: inference and reliability. Ann. Rev. Genetics 22: 521-565

Felsenstein, J. (1988) Perils of molecular introspection. Nature 335: 118

Fox, K.F. and Brown, A. (1993) Properties of the genus *Tatlockia*. Differentiation of *Tatlockia* (*Legionella*) *maceachernii* and *micdadei* from eachother and from other legionellas. Can. J. Microbiol. 39: 486-491.

Franzmann, P.D. and Tindall, B.J. (1990) A chemotaxonomic study of members of the family Halomonadaceae. System. Appl. Microbiol. 13: 142-147

Fuerst, J.A., Hawkins, J.A., Holmes, A. Sly, L.I., Moore, C.J., and Stackebrandt, E. (1993) *Porphyrobacter neustonensis* gen. nov., sp. nov., an aerobic bacteriochlorophyll-synthesising budding bacterium from fresh water. Int. J. System. Bacteriol. 43: 125-134.

Gherna, R. and Woese, C.R. (1992) A partial phylogenetic analysis of the "Flavobacter-Bacteroides" phylum: basis for taxonomic restructuring. System. Appl. Microbiol. 15: 513-521.

Goodfellow, M. and O'Donnell, A.G. (1993) Roots of bacterial systematics in "Handbook of New Bacterial Systematics," Goodfellow, M. and O'Donnell, A.G. eds, Academic Press, London

Govindaswami, M., Schmidt, T.M., White, D.C., and Loper, J.C. (1993) Phylogenetic analysis of a bacterial aerobic degrader of azo dyes. J. Bacteriol. 175: 6062-6066

Grimont, P.A.D., Popoff, M.Y., Grimont, F., Coynault, C., and Lemelin, M. (1980) Reproducibility and correlation of three deoxyribonucleic hybridisation procedures. Curr. Microbiol. 4: 325-330

Head, I.M., Hiorns, W.D., Embley, T.M., McCarthy, A.J., and Saunders, J.R. (1993) The phylogeny of autotrophic ammonia-oxidising bacteria as determined by analysis of 16S ribosomal gene sequences. J. Gen. Microbiol. 136: 1147-1153.

Hiraishi, A., Shin, Y.K., Sugiyama, J., and Komagata, K. (1992) Isoprenoid quinones and fatty acids of *Zoogloea*. 61: 231-236.

Horbach, S., Sahm, H., and Welle, R. (1993) Isoprenoid biosynthesis in bacteria: two different pathways? FEMS Microbiol. Letts. 111: 135-140.

Huss, V.A.R., Festl, H., and Schleifer, K.H. (1983) Studies on the spectrophotometric determination of DNA hybridisation from renaturation rates. System. Appl. Microbiol. 4: 184-192

Ikawa, M. (1967) Bacterial phosphatides and natural relationships. Bacteriol. Rev. 31: 54-64

In 't Veld, G., Driessen, A.J.M., and Konings, W.N. (1993) Bacterial solute transport proteins in their lipid environment. FEMS Microbiol. Rev. 12: 293-314

Juni, E. and Heym, G.A. (1986) *Psychrobacter immobilis* gen. nov., sp. nov.: genospecies composed of Gram-negative, aerobic, oxidase-positive coccobacilli. Int. J. System. Bacteriol. 36: 388-391

Kamekura, M. and Seno, Y. (1993) Partial sequence of the gene for a serine protease from a halophilic archaeum *Haloferax mediterranei* R4, and nucleotide sequences of 16S rRNA encoding genes from several halophilic archaea. Experientia 49: 503-512

Kandler, O. (1993) The early diversification of life, in "Early Life on Earth," Bengston, S. ed., Nobel Symposium 84, Columbia University Press, New York

Kandler, O. and König, H. (1978) Chemical composition of the peptidoglycan-free cell walls of methanogenic bacteria. Archiv. Microbiol. 118: 141-152

Kawasaki, H., Hoshino,Y., and Yamasato, K. (1992) Phylogenetic diversity of phototrophic purple non-sulfur bacteria in the Proteobacteria group. FEMS Microbiol. Letts. 112: 61-66.

Kersters, K. (1991) The genus *Deleya*, in "The Prokaryotes. A Handbook on the Biology of Bacteria: Eco-physiology, Isolation, Identification, Applications," Balows, A., Trüper, H.G., Dworkin, M., Harder, W., and Schleifer, K.H. eds., Spinger Verlag, New York.

Lawson, P.A., Gharbia, S.E., Shah, H.N., Clark, D.R., and Collins, M.D. (1991) Intrageneric relationships of members of the genus *Fusobacterium* as determined by reverse transcriptase sequencing of small-subunit rRNA. Int. J. System. Bacteriol. 41: 347-354

Lechevalier, M.P., De-Bievre, C., and Lechevalier, H.A. (1977) Chemotaxonomy of aerobic actinomycetes: phospholipid composition. Biochem. System. Ecol. 5:249-260

Lee, Y.-E., Jain, M.K., Lee, C., Lowe, S.E., and Zeikus, J.G. (1993) Taxonomic distinction of saccharolytic thermophilic anaerobes: description of *Thermoanaerobacterium xylanolyticum* gen. nov., sp. nov., and *Thermoanaerobacterium saccharolyticum* gen. nov., sp. nov.; reclassification of *Thermoanaerobium brockii*, *Clostridium thermosulfurogenes*, and *Clostridium thermohydrsulfuricum* E100-69 as *Thermoanaerobacter brockii* comb. nov., *Thermoanaerobacterium thermosulfurigenes* comb. nov., and *Thermo- anaerobacter thermohydrosufuricus* comb. nov., respectively; and transfer of *Clostridium thermohydrosulfuricum* 39E to *Thermoanaerobacter ethanolicus*. Int. J. System. Bacteriol. 43: 41-51

Ludwig, W., Mittenhuber, G., and Friedrich, C.G. (1993) Transfer of *Thiosphaera pantotropha* to *Paracoccus denitrificans*. Int. J. System. Bacteriol. 43: 363-367.

Manz, W., Amann, R., Ludwig, W., Wagner, M., and Schleifer, K.H. (1992) Phylogenetic oligonucleotide probes for the major subclasses of the Proteobacteria: problems and solutions. System. Appl. Microbiol. 15: 593-600

Martinez-Murcia, A.J., and Collins, M.D. (1991) A phylogenetic analysis of an atypical leuconostoc: description of *Leuconostoc fallax* sp. nov. FEMS Microbiol. Letts. 82: 55-60

Martinez-Murcia, A.J., Benlloch, S., and Collins, M.D. (1992) Phylogenetic interrelationships of members of the genera *Aeromonas* and *Pleisiomonas* as determined by 16S ribosomal DNA sequencing: lack of congruence with results of DNA-DNA hybridisations. Int. J. System. Bacteriol. 42: 412-421.

Mayr. E. (1942) "Systematics and the Origin of Species," Columbia University Press, New York.

Mendala, B. (1990) MIDI Technical note 103, MIDI Newark.

Minnikin, D.E. and Abdolrahimzadeh, H. (1974) The replacement of phosphatidyl ethanolamine and acidic phospholipids by an ornithine-amide lipid and a minor phosphorus-free lipid in *Pseudomonas fluorescens* NCMB 129. FEBS Letts. 43: 257-260

Minnikin, D.E. and Goodfellow, M. (1981) Lipids in the classification of *Bacillus* and related taxa, in "The Aerobic Endosporeforming Bacteria," Berkeley, R.C. and Goodfellow, M., eds Academic Press, London

Moss, C.W., Holzer, G., Wallace, P.L., and Hollis, D.G. (1990) Cellular fatty acid compositions of an unidentified organism and a bacterium associated with cat scratch disease. J. Clin. Microbiol. 28: 1071-1074.

Mylvaganam, S. and Dennis. P.P. (1992) Sequence heterogeneity between the two genes encoding 16S rRNA from the halophilic archaebacterium *Haloarcula marismortui*. Genetics 130: 399-410.

Niebel, H., Dorsch, M., and Stackebrandt, E. (1987) Cloning and expression of *Proteus vulgaris* genes for 16S ribosomal RNA. J. Gen. Microbiol. 133: 2401-2409

Nölling, J., Hahn, D., Ludwig, W., and de Vos, W.M. (1993) Phylogenetic analysis of thermophilic *Methanobacterium* sp.: evidence for a formate utilising ancestor. System. Appl. Microbiol. 16: 208-215

van Niel, C.B. (1966) Microbiology and molecular biology. Quart. Rev. Biol. 41: 105-112.

O'Donnell, A.G., Embley, T.M., and Goodfellow, M. (1993) Future bacterial systematics in "Handbook of New Bacterial Systematics," Goodfellow, M. and O'Donnell, A.G. eds, Academic Press, London

Ohara, M., Katayama, Y., Tsuzaki, M., Nakamoto, S., and Kuraishi, H. (1990) *Paracoccus kocurii* sp. nov., a tetramethylammonium-assimilating bacterium. Int. J. System. Appl. 40: 292-296

Olson, G., Woese, C.R., and Overbeek, R. (1994) The winds of (evolutionary) change: breathing new life into microbiology. J. Bacteriol. 176: 1-6

Pasteur, B.J., Ludwig, W., Weisburg, W.G., Stackebrandt, E., Hespell, R.B., Hahn, C.M., Reichenbach, H., Stetter, K.O., and Woese, C.R. (1985) A phylogenetic grouping of the bacteroides, cytophagas, and certain flavobacteria. System. Appl. Microbiol. 6: 34-42

Pieringer, R.A. (1989) Biosynthesis of non-terpenoids, in "Microbial Lipids" vol. 2, Ratledge, C. and Wilkinson, S.G., eds, Academic Press, London.

Rainey, F., A., Ward, N.L., Morgan, H.W., Taolster, R., and Stackebrandt, E. (1993) Phylogenetic analysis of anaerobic thermophilic bacteria: aid for their reclassification. J. Bacteriol. 175: 4772-4779

Rivera, M.C. and Lake, J.A. (1992) Evidence that eukaryotes and eocyte prokaryotes are immediate relatives. Science 257: 74-76

Rosselló-Mora, R.A., Ludwig, W., and Schleifer, K.H. (1993) *Zoogloea ramigera*: a phylogenetically diverse species. FEMS Microbiol. Letts. 114: 129-134

Rotert, K.R., Toste, A.P., and Steiert, J.G. (1993) Membrane fatty acid analysis of Antarctic bacteria. FEMS Microbiol. Letts. 114: 253-258

Schleifer, K.H. and Stackebrandt, E. (1983) Molecular systematics of prokaryotes. Ann. Rev. Microbiol. 37: 143-187

Schleifer, K.H., Amann, R., Ludwig, W., Rothemund. C., Springer, N., and Dorn, S. (1992) Nucleic acid probes for the identification and in situ detection of pseudomonads, in "*Pseudomonas*, molecular biology and biotechnology," Galli, E., Silver, S., and Witholt, B. eds. American Society for Microbiology, Washington.

Schweizer, E. (1989) Biochemistry of lipids, in "Microbial Lipids" vol. 2, Ratledge, C. and Wilkinson, S.G., eds, Academic Press, London.

Shaw, N. (1970) Bacterial Glycolipids. Bacteriol. Rev. 34: 365-377.

Shaw, N. (1974) Lipid composition as a guide to the classification of bacteria. Adv. Appl. Microbiol. 17: 63-108

Shin, Y.K., Hiraishi, A., and Sugiyama, J. (1993) Molecular systematics of the genus *Zoogloea* and emendation of the genus. Int. J. System. Bacteriol. 43: 826-831

Sittig, M. and Hirsch, P. (1992) Chemotaxonomic investigations of budding and/or hyphal bacteria. System. Appl. Microbiol. 15: 209-222.

Sittig, M. and Schlesner, H. (1993) Chemotaxonomic investigations of various prosthecate and/or budding bacteria. System. Appl. Microbiol. 16: 92-103.

Sneath, P.H.A. (1989) Analysis and interpretation of sequence data for bacteria systematics: the view of a numerical taxonomist. System. Appl. Microbiol. 12: 15-31.

Sneath, P.H.A. and Sokal, R.R. (1973) "Numerical Taxonomy" W.H.Freeman and Company, San Francisco.

Speck, H., Kroppenstedt, R.M., and Mannheim, W. (1987) Genomic relationships and species differentiation in the genus *Capnocytophaga*. Zentralblt. Bakteriol. Hyg. A 226: 390-402.

Stackebrandt, E. and Goodfellow, M. (1991) Introduction in "Nucleic Acid Techniques in Bacterial Systematics," Stackebrandt, E. and Goofellow, M. eds., Wiley, London

Stahl, D.A., Key, R., Flesher, B., and Smit, J. (1992) The phylogeny of marine and freshwater caulobacters reflects their habitat. J. Bacteriol. 174: 2193-2198

Sutton, G.C., Russell, N.J., and Quinn, P.J. (1991) The effect of salinity on the phase behaviour of total lipid extracts and binary mixtures of the major phospholipids isolated from a moderately halophilic bacterium. Biochim. Biophys. Acta. 1061: 235-246

Suwanto, A. and Kaplan, S. (1992) Chromosome transfer in *Rhodobacter sphaeroides*: Hfr formation and genetic evidence for two unique circular chromosomes. J. Bacteriol. 174: 1135-1145

Suzuki, K.-I., Saito, K., Kawaguchi, A., Okuda, S., and Komagata, K. (1981) Occurrence of ω-cyclohexyl fatty acids in *Curtobacterium pusillum* strains. J. Gen. Microbiol. 27: 261-266

Suzuki, K.-I., Goodfellow, M., and O'Donnell, A.G. (1993) Cell envelopes and classification in "Handbook of New Bacterial Systematics," Goodfellow, M. and O'Donnell, A.G. eds, Academic Press, London

Suzuki, T. and Yamasato, K, 1993 Phylogeny of spore-formimg lactic acid bacteria based on 16S rRNA gene sequences. FEMS Microbiol. Letts. 115: 13-18

Tindall, B.J. (1991) The family Halobacteriaceae, in "The Prokaryotes. A Handbook on the Biology of Bacteria: Ecophysiology, Isolation, Identification, Applications," Balows, A., Trüper, H.G., Dworkin, M., Harder, W., and Schleifer, K.H., eds., Springer Verlag, New York.

Tornabene, T.G., Wolfe, R.S., Balch, W.E., Holzer, G., Fox., G.E., and Oro, J. (1978) Phytanyl-glycerol ethers and squalenes in the archaebacterium *Methanobacterium thermoautotrophicum*. J. Molec. Evol. 11: 259-266

Touzel, J.P., Conway de Macario, E., Nölling, J., de Vos, W.M., Zhilina, T., and Lysosenko, A.M. (1992) DNA relatedness among some thermophilic members of the genus *Methanobacterium*: emendation of the species *Methanobacterium thermoautotrophicum* and rejection of *Methanobacterium thermoformicicum* as a synonym of *Methanobacterium thermoautotrophicum*. Int. J. System. Bacteriol. 42: 408-411

Trüper, H.G. and Schleifer, K.H. (1991) Prokaryote characterisation and identification, in "The Prokaryotes. A Handbook on the Biology of Bacteria: Ecophysiology, Isolation, Identification, Applications," Balows, A., Trüper, H.G., Dworkin, M., Harder, W., and Schleifer, K.H., eds., Springer Verlag, New York.

Truniger, V. and Boos, W. (1993) Glycerol uptake in *Escherichia coli* is sensitive to membrane lipid composition. Res. Microbiol. 144: 565-574

Tsuji, K., Tsien, H.C., Hanson, R.S., DePalma, S.R., Scholtz, R., and LaRoche, S. (1990) 16S ribosomal RNA sequence analysis for determination of phylogenetic relationship among methylotrophs. J. Gen. Microbiol. 136: 1-10.

Urakami, T., Tamaoka, J., Suzuki, K.-I., and Komagata, K. (1989) *Paracoccus alcaliphilus* sp. nov., an alkaliphilic and facultatively methylotrophic bacterium. Int. J. System. Bacteriol. 39: 116-121

Urakami, T., Araki, H., Oyanagi, H., Suzuki, K.-I., and Komagata, K., 1990 *Paracoccus aminophilus* sp. nov. and *Paracoccus aminovorans* sp. nov., which utilize N,N-dimethyloformamide. Int. J. System. Bacteriol. 39: 116-121

Willems, A. and Collins, M.D. (1992) Evidence for a close genealogical relationship between *Afipia* (the causal organism of cat sratch disease), *Bradyrhizobium japonicum* and *Blastobacter denitrificans*. FEMS Microbiol. Letts. 96:241-246.

Wilkinson, S.G. (1968) Studies on the cell walls of *Pseudomonas* species resistant to ethylenediaminetetraacetic acid. J. Gen. Microbiol. 54: 195-213.

Wilkinson, S.G. (1989) Gram-negative bacteria, in "Microbial Lipids" vol. 2, Ratledge, C. and Wilkinson, S.G., eds, Academic Press, London.

Wisotzkey, J.D., Jurshuk, P., Fox, G.E., Deinhard, G., and Poralla, K. (1992) Comparative sequence analyses on the 16S rRNA (rDNA) of *Bacillus acidocaldarius*, *Bacillus acidoterrestris*, and *Bacillus cycloheptanicus* and proposal for the creation of a new genus *Allicyclobacillus* gen. nov. Int. J. System. Bacteriol. 42: 263-269

Woese, C.R. (1987) Bacterial evolution. Microbiol. Rev. 51: 221-271.

Woese, C.R., Gibson, J., and Fox, G.E. (1980) Do genealogical patterns in purple photosynthetic bacteria reflect intraspecific gene transfer? Nature 283: 212-214.

Woese, C.R., Blanz, P., and Hahn, C.M. (1984) What isn't a pseudomonad: the importance of nomenclature in bacterial classification. System. Appl. Microbiol. 5: 179-195.

Wolfe, R.S. (1991) My kind of biology. Ann. Rev. Microbiol. 45: 1-35.

Yanagi, M. and Yamasato, K. (1993) Phylogenetic analysis of the family Rhizobiaceae and related bacteria by sequencing of 16S rRNA gene using PCR and DNA sequencer. FEMS Microbiol. Letts. 107: 115-120.

Zuckerkandl, E. and Pauling, L. (1965) Molecules as documents of evolutionary history. J. Theor. Biol. 8: 357-366

THE BIOTECHNOLOGICAL IMPORTANCE OF MOLECULAR BIODIVERSITY STUDIES FOR METAL BIOLEACHING

Brett M. Goebel[1,2] and Erko Stackebrandt[1,2]

[1]Centre for Bacterial Diversity and Identification, Department of Microbiology
The University of Queensland, St Lucia, 4072, Australia
[2]Present address: DSM-German Collection of Microorganisms and Cell
Cultures, Mascheroder Weg 1B, D-38124, Braunschweig, Germany

INTRODUCTION

Bioleaching involves the solubilization of metals from solid minerals by the direct and/or indirect metabolic activity of mixed microbial populations. Commercial-scale operations have been used extensively for the recovery of copper and uranium from low-grade ores, with an estimated 10% of the worlds copper currently being produced using dump or heap bioleach systems (Herbert, 1992). Over the past decade, the biological pretreatment of certain gold-bearing ore concentrates prior to cyanidation has also been adopted as an economically and environmentally superior option to roasting or pressure oxidation techniques (van Aswegen, 1993). Common to all of these commercial operations is the use of undefined microbial populations, usually enriched from natural acidic sites associated with the ore-body of interest. To our knowledge, there have been no reports of defined or introduced consortia being used successfully in commercial-scale bioleach operations. This is not surprising considering the incomplete understanding of sulfide biohydrometallurgy and bioleach ecology.

The use of batch or continuous flow bioreactors has long been recognised as an alternate approach to the extraction of metals from ore concentrates (Gormely et al., 1975). These tank leach systems are becoming the method-of-choice in the pretreatment of some refractory gold-bearing ores, but the potential exists for their use in the commercial extraction of higher-grade ores or other precious, semiprecious and base metals. The advantage of tank leach systems over dump, heap or in situ procedures is the ability to control critical physicochemical parameters, such as pH, temperature, aeration and dilution rate, at process-optima levels. Apart from these parameters, the most obvious variable which could be optimised is the composition of the bacterial inoculum. However, the limited knowledge of microbial diversity and dynamics in bioleach environments has only allowed intuitive manipulations. Attempts to study the ecology of bioleaching environments has been hindered by the difficulty in isolating and enumerating individual members from a mixed population. This problem is compounded by the fact that different strains, species and even genera often have similar physiological characteristics, making differentiation using classical culturable techniques onerous. As a direct consequence, only sporadic ecological studies documenting acidic bioleaching environments, both natural and commercial, are available (Harrison, 1978; Kovalenko and Malakhova, 1990; Sand *et al.*, 1992; Southam and Beveridge, 1992).

The first microorganism to be implicated in the bioleach process was *Thiobacillus ferrooxidans*, an obligately acidophilic chemolithoautotroph which oxidises iron and reduced sulfur compounds (Colmer and Hinkle, 1947). Numerous subsequent studies have shown acidic spring and mineral environments to be a rich source of autotrophic and heterotrophic microbial diversity. The bacterial taxa commonly isolated from these environments have been extensively reviewed (Brierley, 1978; Harrison, 1984; Hutchins *et al.*, 1986; Tuovinen *et al.*, 1991). More recently, novel metal-mobilising bacteria from marine and less acidic environments have been isolated (Huber and Stetter, 1989 and 1990; Drobner *et al.*, 1992), increasing the range of organisms from which an optimal, defined bioleach population could be selected.

Before "designer" consortia can be considered a viable alternative to autochthonous populations, it will be necessary to extend the present knowledge of bacterial diversity in bioleach systems. It is currently estimated that between only 1 and 10% of the existing microbial diversity has been cultivated (Bull *et al.*, 1992), although the actual percentage may be much lower than this. Recent molecular ecological studies, based on cloned 16S rDNA sequences retrieved from both terrestrial and marine environments, reinforce these estimations (Giovannoni *et al.*, 1990; Ward *et al.*, 1990; Liesack and Stackebrandt, 1992). None of the clone sequences from these studies were identical to that of cultured microorganisms, but were either distantly related to known taxa or formed new primary lines of descent within the Bacteria domain. It is therefore not unreasonable to expect that uncultivated microorganisms may represent a substantial portion of a functional bioleaching consortium. The cloning and sequencing of 16S rDNA from pure cultures and environmental samples has therefore been used in this study to evaluate the microbial diversity in a commercially-targeted bioleach process.

MATERIALS AND METHODS

Due to the commercial nature of the bioleach testing procedures, the specific features of the bioreactors and ore material used in this study are not available for publication at this time.

Sampling and Enrichment

Laboratory-scale (1-4 litre) continuous bioreactors were situated at MIM Holdings Hydrometallurgy Research Laboratory, Brisbane, Australia. These systems were used in the assessment of the bioleachability of a sulfide ore concentrate under acidic conditions. The inoculum for the bioreactors originated from the acidic runoff, pH 2.35, of a chalcocite overburden heap at the Mt Isa Mines leasehold, Mt Isa. The initial runoff sample was used to inoculate the ore concentrate being studied, and over the following year the bioreactors were reinoculated on several occasions with fresh acidic samples from the same site.

All samples for bacterial isolation were enriched in the basal mineral salts medium (MS) of Harrison (1984) to which was added single organic or inorganic growth substrates. The MS was acidified to the desired pH with 0.5 or 2 M H_2SO_4 and sterilised by autoclaving. Substrates were then added to the MS in the following concentrations (final pH of medium): 11 mM D-glucose (pH 3.5), 70 or 144 mM $FeSO_4.7H_2O$ (pH 1.5-2), 20 mM $Na_2S_2O_3.5H_2O$ (pH 4.5) or 5 g/l sublimed elemental sulfur (pH 3.5). All substrates, apart from elemental sulfur, were dissolved in distilled water and filter sterilised (0.22 µm pore size). Elemental sulfur was sterilised by steaming for 1 hr at 100°C on 3 consecutive days.

Enrichment cultures (50 ml) were inoculated with 1% (v/v) liquid sample and incubated in 250 ml shake flasks at 170 rpm at a temperature of 28, 37 or 45°C. Uninoculated controls were used concurrently with all enrichment cultures. Enrichments were incubated until growth was observed microscopically or a change occurred in the enrichment compared with the uninoculated control, i.e. decrease in pH for sulfur enrichments or ferric ion formation or precipitation in ferrous ion containing enrichments.

Isolation of Pure Cultures

Growth from enrichment cultures was streaked onto solid media and incubated at the temperature used for enrichment until colonies were visible. Solid media were identical in composition and pH to liquid media with the following exceptions: $FeSO_4.7H_2O$ was added to a final concentration of 36mM and thiosulfate-containing media were supplemented with 0.02% (w/v) yeast extract. Media were prepared as described by Harrison (1984), except an overlayer of agarose was not used. Media were solified with 0.6% (w/v) agarose (Bio-Rad Laboratories, CA., USA; cat. no. 162-0134) as a solidifying agent. Note that elemental sulfur enrichments were plated on thiosulfate-containing medium.

Isolated colonies on thiosulfate plates were subcultured into 10 ml liquid medium then restreaked repeatedly to obtain pure cultures. Pure cultures were maintained on solid media and subcultured every 1 to 2 months. Chemoautotrophic isolates were monitored for heterotrophic contaminants by streaking on glucose-containing MS agarose plates.

"*Leptospirillum ferrooxidans*" strains isolated in this study did not grow on solidified media. Pure cultures of these strains were established by continued serial dilution and subculture in liquid ferrous MS medium (pH 1.5) incubated at 37-40°C. Pure cultures were then maintained in liquid medium incubated at 28°C or 37°C and subcultured every 2 to 4 weeks.

Nucleic Acid Analysis

Cultures of iron-oxidising isolates were grown in liquid ferrous-containing MS adjusted to pH 1.6 to minimize ferric oxide precipitation (Harrison, 1986) and harvested in late logarithmic phase by centrifugation. Cell pellets were washed 3 times with MS, pH 1.6, once with MS, pH 7, and finally resuspended in 400 µL saline-EDTA. Sulfur-oxidising strains were harvested directly from agarose-gelled plates. DNA was isolated and purified from pure cultures and clones as described by Rainey et al. (1992).

The 16S rRNA gene from the pure cultures was amplified using the polymerase chain reaction (PCR) using the primers and conditions described previously by Rainey et al. (1992). Cloned DNA (see below) was amplified using the M13-20 primer and the M13 reverse primer. The 1.5-1.6 kb PCR products were purified using the Prep-A-Gene DNA purification kit (Bio-Rad Laboratories, CA., USA) and sequenced with the Sequenase Kit, Version 2.0 (United States Biochemicals, Ohio, USA). Detailed protocols for these methods have been published elsewhere (Dorsch and Stackebrandt, 1992). Note that the primers designated R1 and R8 and F2 (Dorsch and Stackebrandt, 1992), binding to positions 343 and 357 and 1100-1115, respectively on the 16S rRNA, were found to be unsuitable for sequencing thiobacilli strains.

For primer design, sequences were aligned and compared with 16S rRNA/rDNA sequences of the Ribosomal Database Project (Olsen, 1992). Based on these comparisons, a species-specific probe was designed for "*L. ferrooxidans*" and designated Lf_{176} (CGGAACCGGATACTATTC). It was not possible to define a species-specific probe for the moderately thermophilic sulfur-oxidizing isolates as their 16S rRNA sequences were virtually identical to that of the moderately thermophilic strain *T. ferrooxidans* LM2 (Marsh and Norris, 1983; Lane *et al.*, 1992). A strain-cluster specific probe was therefore designed and designated SH_{834} (CAGCACCTAAGGCGCCAA). The number associated with each primer refers to the 3' annealing position of the primer to the 16S rRNA using the *Escherichia coli* numbering system (Brosius *et al.*, 1978). All primers were synthesised according to the phosphoramidite method (Beaucage and Caruthers, 1981) and purified with OPC cartridges (Applied Biosystems, CA., USA).

Collection and Amplification of Bulk 16S rDNA

The generation of a clone library was initiated some 18 months after the original enrichment experiments. A 50 ml slurry sample was collected from a continuous reactor system which had been running at steady state conditions for 4 weeks. Ore material was allowed to settle for 1 h and a 2 ml sample of the supernatant removed. Bacterial cells from the 2 ml sample were collected and lysed in a disposable filter unit as described by Sommerville *et al.* (1989) with the following changes to the protocol. The 2 ml supernatant sample was collected in a 5 ml disposable syringe and filtered under positive pressure through the filter unit. Following filtration of the sample, the filter was washed 3 times with 20 ml MS, pH 1.6, and once with 20 ml MS, pH 7.0, prior to cell lysis with SDS and lysozyme. Extent of bacterial lysis was monitored by phase contrast microscopy.

The crude bulk DNA lysate was purified with equal volumes of phenol, phenol-chloroform and chloroform isoamylalcohol. Purified DNA was precipitated at 20°C for 30 min with an equal volume of 4M ammonium acetate and a double volume of isopropanol, washed with 70% ethanol, dried and dissolved in 50 µl sterile HPLC-grade water. The concentration of genomic DNA was determined spectrophotometrically.

PCR with $Vent_R$ DNA polymerase (New England Biolabs, MA, USA) was used to generate blunt-end, duplex products suitable for cloning into the cosmid vector pBS (+\-) which had been cut with the restriction enzyme *Sma*I. The amplification primers selected were complementary to the conserved regions flanking the 16S rRNA gene, designated fD1 (AGAGTTTGATCCTGGCTCAG) and rP2 (ACGGTTACCTTGTTACGACTT) (Weisburg *et al.*, 1991). The PCR reaction mixtures contained 0.7 µM of each primer, 10 mM KCl, 10 mM $(NH_4)_2SO_4$, 20 mM Tris-HCl (pH8.8), 4 mM $MgCl_2$, 0.1% (v/v) Triton X-100, 300 µM of each nucleotide (dATP, dCTP, dGTP, dTTP) and 400 ng bulk genomic DNA. HPLC-grade water was added to a final volume of 100 µl. The reactions were denatured at 96°C for 4 min prior to the addition of 2.5 U $Vent_R$ DNA polymerase, and finally overlayed with 80 µl light white mineral oil. Amplification was performed for 35 cycles on a Perkin Elmer Cetus 480 DNA thermal cycler using the following thermal profile: annealing at 48°C for 45 sec, extension at 73°C for 2 min, and denaturation at 93°C for 1 min. A single final extension was performed at 73°C for 10 min following reannealing at 48°C for 45 sec. PCR products were detected by electrophoresis in a 1% agarose gel supplemented with 0.5 µg/ml ethidium bromide. PCR products were precipitated with ammonium acetate and isopropanol as above, resuspended in 20 µl HPLC-grade water and visualised on a 0.7% agarose gel. The 1.5 kb band representing the 16S rDNA was excised and purified with the Prep-A-Gene DNA binding matrix (Bio-Rad Laboratories, CA., USA) following the manufactures instructions. The products were finally resuspended in 10 µL HPLC-grade water.

Cloning of PCR Products

The purified 16S rDNA products were inserted into the *Sma*I restriction site of the cosmid vector pBS(+/-) (Stratagene, Ca., USA) using blunt-end ligation. The ligation reaction contained approximately 500 ng of insert, 50 ng vector, 1.5 U T4 ligase in supplied ligation buffer (Promega, WI, USA), and HPLC-grade water to a final volume of 10 µL.

The ligation reaction was incubated overnight at 16°C. Half of the ligated product was used to transform 100 µL Epicurian coli XL1-blue supercompetent *Escherichia coli* cells (Stratagene, CA., USA), which allow blue/white colony selection. Transformants were selected on LB plates (0.5% yeast extract, 1% tryptone and 1% NaCl, final pH 7.5) containing 50 µg/ml ampicillin (Boeringer Mannheim, Germany). Eighty colonies were picked and stored on LB-Amp plates at 4°C.

For the preparation of plasmid DNA for PCR, a loopful of overnight plate culture was resuspended in 100 µl HPLC-grade water and boiled for 15 min. The samples were then centrifuged to remove cell debris and 70 µl of the supernatant removed and stored at -20°C. This method consistently yielded plasmid DNA suitable for PCR, even after storage at -20°C for 12 months.

Screening of Clones with Multi-primer PCR

Initial screening of the clone library involved partial sequencing of 17 clones. Following analysis of the sequence data, a multi-primer PCR assay was used to screen 80 clones, including those which had been partially sequenced. The amplification primers used in the assay were fD1 as a forward primer (Weisburg *et al.*, 1991) and the two reverse primers Lf_{176} and SH_{834}. PCR reaction mixtures contained 0.35 µM of each primer 50 mM KCl, 10 mM Tris-HCl (pH 8.4), 0.8 mM $MgCl_2$, 0.01% (w/v) gelatine, 20 µM of each nucleotide triphosphate, 0.3 U Taq DNA polymerase (Boeringer Mannheim, Germany) and 0.2 µl of the boiled plasmid preparation. HPLC-grade water was added to a final volume of 15 µl. Amplification was performed on a Perkin Elmer Cetus 480 DNA thermal cycler. Samples were initially placed in the preheated block and denatured at 94°C for 4 min. Products were then amplified using the following thermal profile: annealing at 54°C for 30 sec, extension at 73°C for 1 min, and denaturation at 93°C for 1 min. Products were resolved on a 1.8% agarose gel supplemented with 0.5 µg/ml ethidium bromide.

Data analysis

The sequences of strains "*L. ferrooxidans*" MIM Lf30 and *T. thiooxidans* MIM SH12 and 17 partial clone sequences were aligned to the 16S rDNA/RNA sequences of the Ribosomal Database Project (Olsen *et al.*, 1992), supplemented with sequences obtained from sulfur- and iron-oxidising bacteria (Lane *et al.*, 1992). Initially, the sequences of the two isolates were compared to those of physiologically similar organisms, omitting regions of alignment uncertainty and non-ubiquitous presence (positions 1-239, 449-679, 898-1141, 1374-3' end). Once the phylogenetically nearest neighbour of these autotrophs was determined, the position of the isolates and the clones within a wider range of organisms was determined. The position of the autotrophic species within the phyla and classes of the Bacteria domain has been proposed by Lane *et al.* (1992). For reasons indicated above, the following regions were excluded from the analysis: for strain MIM Lf30 positions 1-48, 449-487, 891-970, 999-1043, 1423-3'end, and for strain MIM SH12 positions 1-299, 507-654, 898-1141 and 1374-3'end, resulting in the analysis of about 1200 and 730 nucleotides, respectively. For some shorter clone sequences, the number had to be reduced even further. Pairwise evolutionary distances (expressed as estimated changes per 100 nucleotides) were computed from percent similarities by the correction of Jukes and Cantor (1969). Phylogenetic trees were constructed from the distance matrices by the least squares algorithm (De Soete, 1983). Sequence alignment and data comparison was performed using a SUN Sparc workstation. The sequences have been deposited in EMBL under the accession numbers X72851 (*T. thiooxidans* MIM SH12), X72852 (*L. ferrooxidans* MIM Lf30), and X72853 (clone A70).

RESULTS AND DISCUSSION

Bacterial isolation and characterisation

Four strains of "*Leptospirillum ferrooxidans*" and 5 strains of *Thiobacillus thiooxidans* were isolated from a steady-state continuous bioleach system (the genus "*Leptospirillum*" did not appear on the Approved Lists of Bacterial Names and has not been validly published since. Therefore, it has been included here in inverted commas to indicate its invalid nomenclature). Partial sequences derived from all strains of the same species were identical, and were highly related to acidophilic strains which have been sequenced previously (Lane *et al.*, 1992). Strains "*L. ferrooxidans*" MIM Lf30 and *T. thiooxidans* MIM SH12 were selected for further sequence analysis and physiological characterisation. A wide range of other chemolithotrophic and heterotrophic acidophiles, including various strains of *T. ferrooxidans*, *T. thiooxidans* (strains MIM S1 and MIM SH12, see below) and *Acidiphilium* species, were recovered from the batch culture subsequently used as inoculum for the continuous bioreactor. This would indicate that the enrichment procedures applied to the continuous reactor were suitable for the isolation of a broad diversity of acidophilic bacteria. These organisms will be discussed further in a different context (Goebel and Stackebrandt, 1994).

Strain MIM Lf30 was identified as "*L. ferrooxidans*" on the basis of its distinctive spiral or curved rod morphology (Fig. 1), rapid motility via a single sub-polar flagellum and its aerobic growth on ferrous ion at pH 1.5-2.5. Chemolithotrophic growth on reduced sulfur compounds or chemoorganotrophy was not detectable. These findings indicate that MIM Lf30 is physiologically and morphologically similar to the original "*L. ferrooxidans*" (Markosyan, 1972; Balashova *et al.*, 1974). However, the optimal growth temperature for strain MIM Lf30 was 37-40°C. The majority of "*L. ferrooxidans*" strains previously isolated have been described as mesophilic, growing optimally at 20-30°C (Balashova *et al.*, 1974; Harrison, 1984; Sand et al., 1992). A strain growing optimally at 35°C (Harrison and Norris, 1985) and the predominance of "*L. ferrooxidans*" cells in mixed cultures incubated at 40°C (Norris, 1983) have been reported.

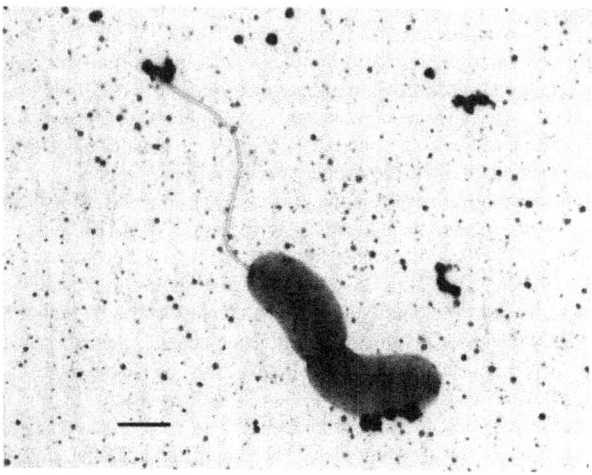

Figure 1. Electron micrograph of "*L. ferrooxidans*" MIM LF30, negatively stained with uranyl acetate. Bar = 0.5 μm.

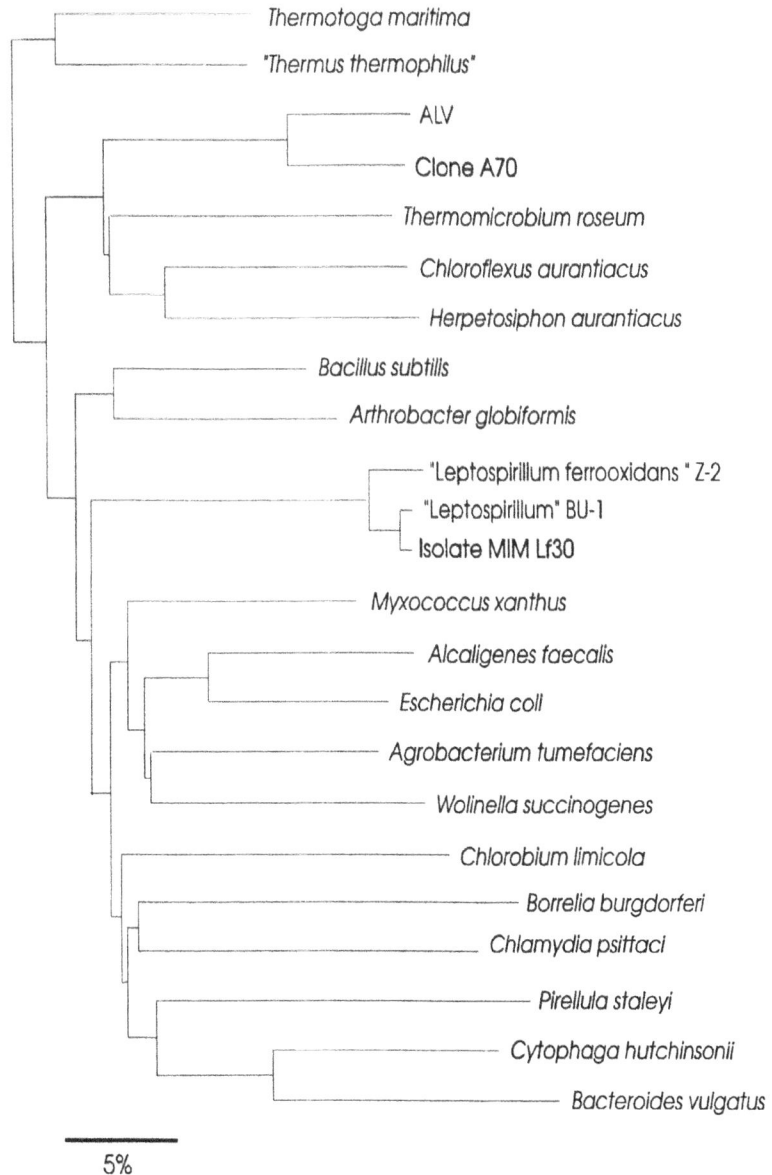

Thermotoga maritima

"Thermus thermophilus"

ALV

Clone A70

Thermomicrobium roseum

Chloroflexus aurantiacus

Herpetosiphon aurantiacus

Bacillus subtilis

Arthrobacter globiformis

"Leptospirillum ferrooxidans" Z-2

"Leptospirillum" BU-1

Isolate MIM Lf30

Myxococcus xanthus

Alcaligenes faecalis

Escherichia coli

Agrobacterium tumefaciens

Wolinella succinogenes

Chlorobium limicola

Borrelia burgdorferi

Chlamydia psittaci

Pirellula staleyi

Cytophaga hutchinsonii

Bacteroides vulgatus

5%

Figure 2. Phylogenetic position of isolate MIM Lf30 and clone A70 within the radiation of bacterial main lines of descent

Strain MIM SH12 was isolated from an elemental sulfur enrichment incubated at 37°C. Cells were Gram-negative rods which formed large spreading colonies when grown on thiosulfate agarose plates supplemented with 0.02% yeast extract.

No growth was observed on plates devoid of yeast extract or when yeast extract was added as the sole energy source. Rapid growth on elemental sulfur plus yeast extract occured in liquid cultures incubated at 37 or 45°C, with the pH of the medium decreasing from 3.5 to <1.2 after five days, but no growth occurred at these temperatures in the absence of yeast extract. growth was observed at 28°C on elemental sulfur and thiosulfate with or without the addition of yeast extract. No growth was observed with media containing ferrous ion. Although this strain is moderately thermophilic and grew most rapidly on elemental sulfur with the addition of low concentrations of organic material (slow autotrophic growth was observed on elemental sulfur at 28°C) it was presumptively identified as a strain of *T. thiooxidans*. Further studies to compare the growth mode and chemotaxonomy of this strain with authentic *T. thiooxidans* have been initiated.

Although "*L. ferrooxidans*" is routinely isolated from natural and commercial bioleaching sites, the majority of studies of microbial oxidation of metal sulphides have used *T. ferrooxidans* in pure culture or in combination with *T. thiooxidans*. The results obtained in this study agree with the findings of Helle and Onken (1988) that "*L. ferrooxidans*" was able to outcompete *T. ferrooxidans* in a continuous bioreactor system initially inoculated with both species. Neither the enrichment for *T. ferrooxidans* in the continuous culture nor the gene library generated from the same environment (see below) gave an indication of the presence of this species, usually considered to be the principle organism in bioleaching systems. It is also worth noting that selection between sulfur-oxidizing strains was also observed, since it was found that although *T. thiooxidans* strains MIM S1 and MIM SH12 were both present in the batch inoculum, MIM S1 strains could not be recovered from the steady state continuous reactor by cultural or cloning techniques.

Cloning strategy

DNA obtained from the liquid-phase bioreactor sample by lysozyme/SDS lysis and phenol/chloroform purification was sufficiently pure to be included directly into the PCR reaction. The low concentration of organic material in this environment (D. Winborne, personal communication), indicates that potential DNA interfering substances such as humic acids are absent. Phase contrast microscopy of the lysate revealed that >99.9% of cells were lysed. This is consistant with our finding that all strains we have so far isolated from this environment lyse readily with the addition of only low concentrations of SDS. Bacteria isolated from bioleaching environments are reported to adhere tightly to ore surfaces (Wakao *et al.*, 1984; Sand *et al.*, 1992). Therefore, attempts were made to isolate bulk genomic DNA directly from washed ore samples. No DNA was recovered from these samples. Subsequent electron microscopy studies, including scanning and thin-section methodologies, failed to demonstrate the presence of bacteria on the surface of the ore material. These results would indicate that direct bacterial attack of the ore concentrate was not favoured in this continuous system.

Initial cloning procedures using primers containing overhangs with restriction sites for *Sac*I and *Sal*I resulted in the formation of additional fragments to the complete 1.5 kb 16S rDNA molecule when the PCR product was cleaved with *Sal*I. The presence of an internal restriction site in clone A70 was verified at a later stage when the complete sequence was screened for the *Sal*I cleavage sequence. The presence of this recognition site around position 1450 is in accordance with the production of a digest fragment slightly smaller than the complete 16S rDNA molecule. This finding, together with the limited sensitivity of agarose gels to visualise PCR cleavage products from DNA originating from minor members of the population, led to the use of blunt-end cloning techniques. Blunt-end

PCR products were generated using Vent$_R$ DNA polymerase (Lohff and Cease, 1992) and cloned into the *Sma*I site of pBS(+/-).

Phylogeny

The almost complete16S rDNA sequences of the two isolates MIM Lf30 and MIM SH12 showed that they belonged to two different groups of sulfur and iron oxidising bacteria and they are close relatives of strains sequenced previously by Lane *et al.* (1992).

Phylogeny of the isolates. Strain MIM Lf30 shows high sequence similarity to the unnamed "*Leptospirillum*" strains BU-1 and LfLa (98.4 and 97.2% similarity, respectively), and was slightly more distantly related to "*L. ferrooxidans*" strain Z-2 (93.3 % similarity) (Fig. 2), the first "*Leptospirillum*" isolate. These latter three organisms were originally isolated from copper (Z-2, BU-1) and uranium (LfLa) mines in Armenia, Bulgaria, and Mexico, respectively. The position of the "*Leptospirillum*" cluster within the Bacteria domain is yet to be convincingly established (Lane *et al.*, 1992). We can confirm their finding that leptospirilla define a main line of descent that can not be linked to any hitherto recognised phylum. By changing the number and species of reference organisms, the leptospirilla were found to consistently branch off within the small range of internode distances that separate the main phyla. With the exception of two Gram-positive species (*Arthrobacter globiformis* and *Bacillus subtilis*), the branching order was similar to those published by Woese (1987) and the topology of the reference organisms does not changed when the iron-oxidisers are omitted from the analysis.

The second strain, the moderately thermophilic and sulfur-oxidising isolate MIM SH12, falls into the radiation of authentic *Thiobacillus ferrooxidans* and *Thiobacillus thiooxidans* strains (Fig. 3), showing quite high similarity values (between 93.3 and 96.4% between the regions compared). The main 16S rDNA sequence variation of this cluster of organisms occurs in the variable regions V3 and V5. The highest binary value of virtually 100% sequence similarity is found between strain MIM SH12 and another moderately thermophilic, elemental sulfur-preferring isolate, *T. ferrooxidans* LM2 (Fig. 3). This strain was isolated from a hot spring sediment from Iceland. The sequence of both these strains was identical in the V3 and V5 regions.

As reported previously (Lane *et al.*, 1992) the genus *Thiobacillus* is phylogenetically heterogeneous, with certain species grouping with the alpha and beta subclasses of the *Proteobacteria* (*T. acidophilus*, *T. versutus*, and *T. thioparus*, *T. perometabolis*, respectively). The affiliation of *T. neapolitanus*, *T. ferrooxidans* and *T. thiooxidans* to individual subclasses is not as clearcut. We can confirm the observation by Lane and coworkers (1992) that the branching points of these organisms are difficult to determine. The topology shown in Fig. 3 differs slightly from that of Lane *et al.* (1992), but this can be attributed to a difference in the reference strains used in the analysis and, as a consequence, the stretches of compared nucleotide sequences.

Phylogeny of the clones. In one of the early molecular biodiversity studies (Lane *et al.*, 1985) the authors recovered only two predominent 5S rRNA species from a copper leaching pond (Chino, New Mexico), which were subsequently found to be members of the "purple bacteria" (now *Proteobacteria*). While one of the 5S rRNA species was identical in sequence to *T. ferrooxidans* ATCC 19859, the other was unrelated to any *Thiobacillus* species which has been sequenced. This finding contrasts the results of recent environmental studies which indicate the presence of a high level of uncultured bacterial diversity. It was therefore intriguing to compare the culturable fraction with the total diversity which can be recovered by environmental cloning from this bioreactor system.

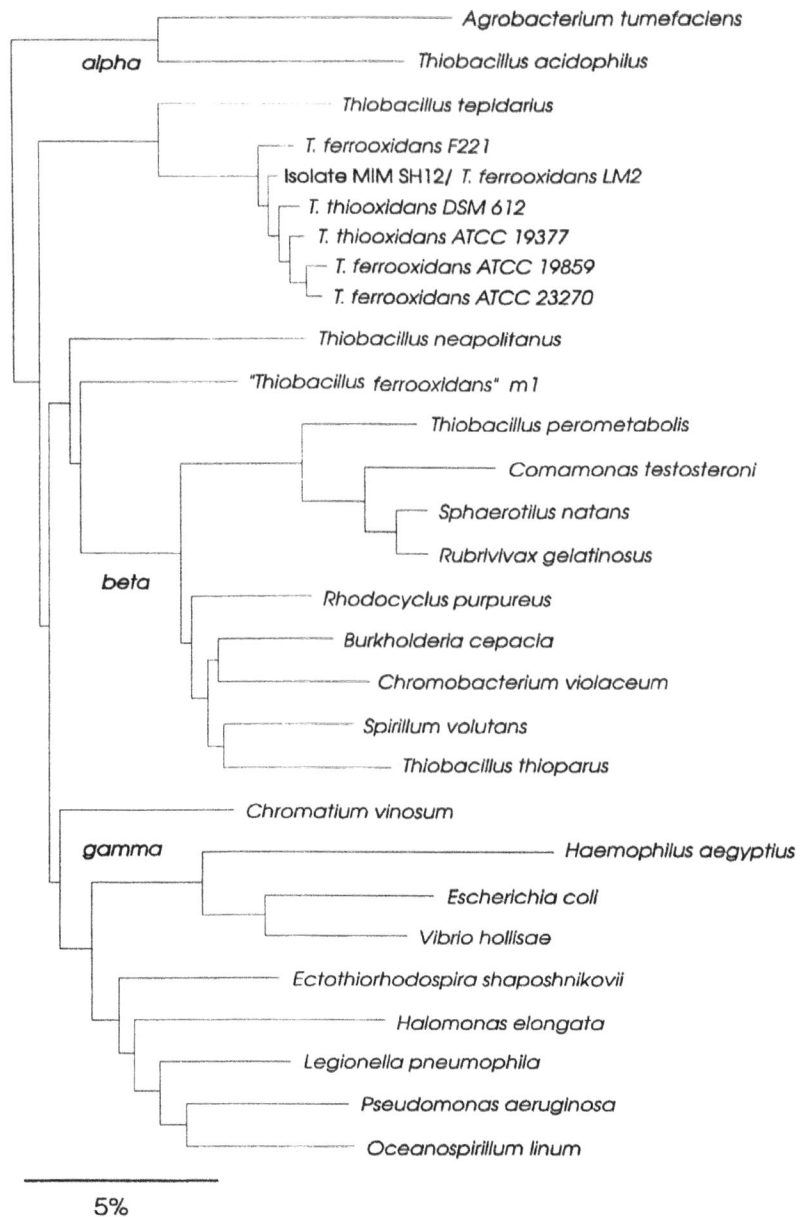

Agrobacterium tumefaciens

Thiobacillus acidophilus

alpha

Thiobacillus tepidarius

T. ferrooxidans F221

Isolate MIM SH12/ T. ferrooxidans LM2

T. thiooxidans DSM 612

T. thiooxidans ATCC 19377

T. ferrooxidans ATCC 19859

T. ferrooxidans ATCC 23270

Thiobacillus neapolitanus

"Thiobacillus ferrooxidans" m1

Thiobacillus perometabolis

Comamonas testosteroni

Sphaerotilus natans

Rubrivivax gelatinosus

beta

Rhodocyclus purpureus

Burkholderia cepacia

Chromobacterium violaceum

Spirillum volutans

Thiobacillus thioparus

Chromatium vinosum

gamma

Haemophilus aegyptius

Escherichia coli

Vibrio hollisae

Ectothiorhodospira shaposhnikovii

Halomonas elongata

Legionella pneumophila

Pseudomonas aeruginosa

Oceanospirillum linum

5%

Figure 3. Phylogenetic position of the isolate SH12 within the radiation of *Thiobacillus* and other proteobacteria

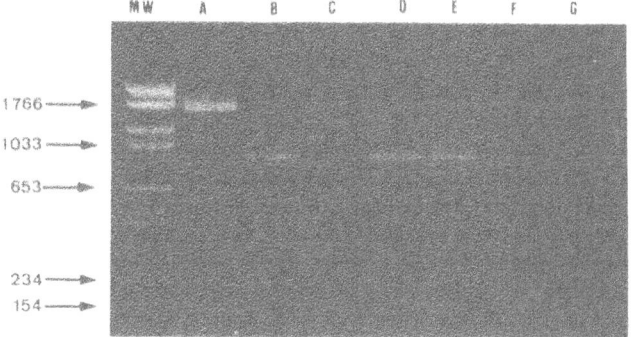

Figure 4. Multiprimer PCR assay results from cloned 16S rDNA. Lanes: MW, DNA molecular weight marker; A, complete 16S rDNA insert from clone A1 amplified using primers M13-20 and M13-reverse; B, clone A5; C, clone A1; D, clone A2; E, clone A6; F, clone A26; G, clone A79. Products from Lanes B to G were amplified using multi-primer PCR containing primers fD1, Lf_{176}, and SH_{834}.

A 16S rDNA clone library was generated and 80 clones screened; initially by partial sequence analysis and subsequently, based on this sequence information, by a multi-primer PCR assay. The multi-primer assay used is similar to multiplex PCR, in that a number of primers are included in the one PCR reaction. However, the multi-primer assay differs slightly in that the primers included in the reaction are targeted against different templates, rather than several different sites on he same template. The multi-primer assay was found to be highly specific and sensitive, with effective increases in stringency obtained most readily by decreasing the free magnesium concentration. Further details and adaptations of this method will be presented elsewhere.

Initial sequence analysis of 17 clones revealed the presence of only 3 different rDNA types, two of which (represented by A1 and A5) were identical to the two isolates from the bioreactor, while the third type, represented by A70, being phylogenetically related to moderately thermophilic iron-oxidising strains which had been sequenced previously (Lane *et al.*, 1992). The remaining clones, including those initially sequenced, were identified by multi-primer analysis (Fig. 4) and, in the case of clones showing no banding pattern, by sequence analysis. A total of 23 clones were partially or completely sequenced. The majority of initially sequenced clones (14 of 17) were identical in sequence to strain MIM SH12 over the first 500 nucleotides. Partial sequences of other regions supported this finding. This number was increased to 43 of the 80 clones after multi-primer analysis. Of the seven clones of the type A1, two which had been partially sequenced (first 500 nucleotides) were identical in sequence to the isolate MIM Lf30. Since the PCR primer was designed to specifically detect all "*Leptospirillum*" strains, it can be assumed that the five remaining clones identified by multiprimer PCR originated from strains of this species. Subsequent multi-primer analysis using a primer designed to be specific for strain MIM Lf30 confirmed that the remaining five clones were members of this strain type.

Multi-primer analysis of all clones revealed that 30 out of 80 clones did not give a banding pattern, including the two clones of the A70 type identified by sequencing. PCR products amplified using M13 primers indicated that 23 of the 30 unidentified clones did not contain any insert DNA. The remaining 7 clones were all shown to be almost identical in sequence to clone A70 over the region sequenced. The A70 clone cluster is related to the facultatively thermophilic, iron-oxidising strains ALV (93.0 % similarity) and BC (97.1 % similarity, phylogenetic position not shown) (Fig. 2). Both reference strains were isolated from coal mine drainage, Warwickshire, UK. The phylogenetic affiliation of strains ALV and BC to the *Clostridium/Bacillus* subphylum of Gram-positive bacteria is a moot point. Lane *et al.* (1992) place these organisms close to the root of this subphylum, and support their membership by the presence of certain common signature nucleotides in strain ALV. We confirm the presence of these nucleotides in clone A70. However, phylogenetic analysis performed separately with members of the two subphyla of Gram-positive bacteria, and members of other phyla, as well with a selection of representatives of all phyla, resulted in the branching of these Gram indeterminate strains ALV and BC, and clone A70, outside the confines of Gram-positive bacteria. Very consistently, these iron oxidisers clustered with members of the *Chloroflexus/Herpetosiphon/Thermomicrobium* line of descent, no matter whether one or all representatives of this phylum were included in the analysis. Of the two signature nucleotides, indicated by Lane *et al.* (1992) to support the relationship of ALV to the Gram-positives, none are exclusively found in these two taxa. The nucleotide at position 1198 (A) is also found in *Fervidobacterium*, *Bacteroides* and *Borrelia* species, while the A-residue at position 513 is found in *Chloroflexus* and its two phylogenetic neighbours, in addition to other taxa that branch deeply in the 16S rDNA tree. Position 1207 (G), suggested to reveal the phylogenetic uniqueness of strain ALV, is also found in green non-sulfur bacteria and *Bacteroides* species. More significantly, the following nucleotides indicate a specific relationship between *Chloroflexus* and related taxa and the ALV strain cluster: positions 163 (C, occurring in other deeply rooted taxa as well, and in two mycoplasmas), 223 (G, only two examples in low G+C Gram-positives), 375 (C), 484 (C), 1193 (G, occurring in other deeply rooted taxa as well). Despite the affiliation shown in Fig. 2 the branching point must still be considered provisional. The binary similarity values are low (< 85%) and the branching pattern may subsequently change with more and novel reference organisms included in the analysis.

CONCLUSIONS

The transition from a batch to continuous bioleach operation was accompanied by a resultant loss of microbial diversity. Both the enrichment procedures and the clone library gave similar results in that two strain types were recovered, one each of "*L. ferrooxidans*" and *T. thiooxidans*, although sequences from an uncultivated organism were present in the clone library. The term "strain type" is used in this context as identical 16S rDNA sequences do not necessarily indicate identity at the level of the genome (Fox *et al.*, 1992). This was highlighted by the isolation of a non iron-oxidising (*T. thiooxidans* MIM SH12) strain which was virtually identical in sequence to *T. ferrooxidans* LM2 (the small differences that occur are possibly due to the quality of the reverse transcriptase sequence technique). Also, cultures investigated for the expression of a common key phenotypic trait such as iron- or sulfur-oxidation may be quite different in other physiological properties, which, if tested for, may subsequently place this organism in a different taxon. The analysis of the clone library gave an indication of the presence of a another strain (represented by clone type A70) that was not isolated during the initial enrichment procedures. Subsequent attempts to specifically enrich for this moderately thermophilic iron-oxidising bacterium have been successful when the isolations were repeated at a later stage. This finding stresses the relevance of molecular diversity studies when trying to define mixed microbial populations. It also demonstrates that temporal fluctuations within the consortium can be monitored.

Of the 23 clones completely or partially sequenced during this study, three groups were found, with clones from 2 of the groups having exactly the same primary structure as sequences obtained from isolated strains. Considering that these clone inserts were initially obtained by PCR amplification, no detectable deviations from the original rDNA sequences were introduced over 1500 nucleotides and 35 rounds of amplification using Vent$_R$ DNA polymerase. This finding is in accordance with the results of similar studies which used *Taq* DNA polymerase. The finding also indicates that similar studies which have used Taq DNA polymerase have probably recovered real 16S rDNA sequences from the environments studied (Britschgi and Giovonnoni, 1991; Kane *et al.*, 1993). In addition, we demonstrate for the first time that the 16S rDNA sequence between certain clones and isolates were identical. The use of a multiprimer PCR in this low diversity environment appears to be a powerful alternative to probing and sequencing in that it rapidly identifies clones of taxa against which primers were designed. This system has proven to be readily adjustable for analysis of other clone libraries prepared in our laboratory.

Of the four major types of environments studied using molecular cloning techniques ie. marine, soil, hot springs and ore leaching, the latter shows the least diverse population. Both the number of species and the degree of clone sequence heterogeneity is comparatively small. This is not surprising considering the selective nature of continuous bioreactor systems toward certain physiological types. Temperature, pH, Eh, gases and concentration of dissolved metals will all influence the final composition of the bioleach consortium. Future molecular biodiversity studies on other commercially-developed continuous bioreactors will show whether similar low diversity populations are selected for. The limitation of this procedure may be that different ore compositions may select for strains with different physiological capabilities but with identical 16S rDNA primary structure. The potential of "designer" populations therefore remains promising but requires further ecological studies aimed at elucidating the mechanism of strain selection when moving from natural acidic environments to highly selective continuous commercial processes.

ACKNOWLEDGMENTS

This work was supported by a grant from MIM Holdings Hydrometallurgy Research Laboratory, Brisbane, Australia. BMG was supported by an Australian Postgraduate Research Award from the Federal Government of Australia. The excellent technical assistance of Christoph Jacobi and Rick Webb is greatly appreciated. Thanks are also extended to all the staff at HRL, Brisbane.

REFERENCES

van Aswegan, P.C. (1993) Bio-oxidation of refactory gold ores: the GENMIN experience. Proceedings of Biomine '93, Australian Mineral Foundation, Adelaide, Australia (in press).

Balashova, V.V., I.Ya. Vedenina, G.E. Markosyan, and G.A. Zavarzin (1974) The auxotrophic growth of *Leptospirillum ferrooxidans.* Microbiology 43, 491-494.

Beaucage, S.L. and M.H. Caruthers (1981) Deoxynucleoside phosporamidites: a new class of key intermediates for deoxypolynucleotide synthesis. Tetrahed. Lett. 22, 1859-1862.

Brierley, C.L. (1978) Bacterial leaching. CRC Crit. Rev. Microbiol. 6, 207-262.

Britschgi, T.B., and S.J.Giovannoni (1991) Phylogenetic analysis of a natural marine bacterioplankton population by rRNA gene cloning and sequencing. Appl. Environ. Microbiol. 57, 1707-1713.

Brosius, J., J.J. Palmer, J.P. Kennedy, and H.F. Noller (1978) Complete nucleotide sequence of a 16S ribosomal RNA gene from *Escherichia coli.* Proc. Natl. Acad. Sci. USA 75, 4801-4805.

Bull, A.T., M. Goodfellow, and J.H. Slater (1992) Biodiversity as a source of innovation in biotechnology. Ann. Rev. Microbiol. 46, 219-252.

Colmer, A.R. and M.E. Hinkle (1947) The role of microorganisms in acid mine drainage: a preliminary report. Science 106, 253-256.

De Soete, G. (1983) A least squares algorithm for fitting additive trees to proximity data. Psychometrika 48, 621-626.

Dorsch, M and E. Stackebrandt (1992) Some modifications in the procedure of direct sequencing of PCR amplified 16S rDNA. J. Microbiol. Meth. 16, 271-279.

Drobner, E., H. Huber, R. Rachel, and K.O. Stetter (1992) *Thiobacillus plumbophilus* sp. nov., a novel galena and hydrogen oxidizer. Arch. Microbiol. 157, 213-217.

Fox , E.F., J.D. Wisotzkey, and P. Jurtshuk (1992) How close is close: 16S rRNA sequence identity may not be sufficient to guarantee species identity. Int. J. Syst. Bacteriol. 42, 166-170.

Giovannoni, S.J., T.B. Britschgi, C.L. Moyer, and K.G. Field (1990) Genetic diversity in Sargasso Sea bacterioplankton. Nature (London) 345, 60-63.

Goebel, B.M., and Stackebrandt, E, Cultural and phylogenetic analysis of mixed microbial populations found in natural and commercial bioleaching environments. Appl. Environ. Microbiol. (in press)

Gormely, L.S., D.W. Duncan, R.M.R. Branion, and K.L. Pinder (1975) Continuous culture of *Thiobacillus ferrooxidans* on a zinc sulfide concentrate. Biotechnol. Bioeng. 17, 31-49.

Harrison, A.P. (1978) Microbial succession and mineral leaching in an artificial coal spoil, Appl. Environ. Microbiol. 36, 861-869.

Harrison, A.P. (1984) The acidophilic thiobacilli and other acidophilic bacteria that share their habitat. Ann. Rev. Microbiol. 38, 265-292.

Harrison, A.P. and P.R. Norris (1985) *Leptospirillum ferrooxidans* and similar bacteria: some characteristics and genomic diversity. FEMS Microbiol. Lett. 30, 99-102.

Harrison, A.P. (1986) Characterisation of *Thiobacillus ferrooxidans* and other iron-oxidizing bacteria, with emphasis on nucleic acid analysis. Biotechnol. Appl. Biochem. 8, 249-257

Helle, U and U. Onken (1988) Continuous micobial leaching of a pyritic concentrate by *Leptospirillum*-like bacteria . Appl. Microbiol. Biotechnol. 28, 553-558.

Herbert, R.A. (1992) A prospective on the biotechnological potential of extremophiles. TIBTECH 10, 395-402.

Huber, H. and K.O. Stetter (1989) *Thiobacillus prosperus* sp. nov., represents a new group of halotolerant metal-mobilizing bacteria isolated from a marine geothermal field. Arch. Microbiol. 151, 479-485.

Huber, H. and K.O. Stetter (1990) *Thiobacillus cuprinus* sp. nov., a novel facultative organotrophic metal-mobilizing bacterium. Appl. Environ. Microbiol. 56, 315-322.

Hutchins, S.R., M.S. Davidson, J.A. Brierley, and C.L. Brierley (1986) Microorganisms in reclaimation of metals. Ann. Rev. Microbiol. 40, 311-336.

Jukes, T.H. and C.R. Cantor (1969) Evolution of protein molecules, in "Mammalian Protein Metabolism" (Munro, N.H., Ed.), pp.21-132. Academic Press, New York.

Kane, M.D., Poulsen, L.K., and D.A. Stahl (1993) Monitoring the enrichment and isolation of sulfate-reducing bacteria by using oligonucleotide hybridization probes designed from environmentally derived 16S rRNA sequences. Appl. Environ. Microbiol. 59, 682-686.

Kovalenko, E.V. and P.T. Malakhova (1990) Microbial succession in compensated sulfide ores. Microbiology 59, 227-232.

Lane, D.J., D.A. Stahl, G.J. Olsen, D.J. Heller, and N.R. Pace (1985) Phylogenetic analysis of the genera *Thiobacillus* and *Thiomicrospira* by 5S rRNA sequences. J. Bacteriol. 163, 75-81.

Lane, D.J., A.P. Harrison, D.A. Stahl, B. Pace, S.J. Giovannoni, G.J. Olsen, and N.R. Pace (1992) Evolutionary relationships among sulfur- and iron-oxidizing bacteria. J. Bacteriol. 174, 269-278.

Liesack, W. and E. Stackebrandt (1992) Occurence of novel groups of the domain Bacteria as revealed by analysis of genetic material isolated from an Australian terrestrial environment. J. Bacteriol. 174, 5072-5078.

Lohff, C. J. and K. B. Cease (1992) PCR using a thermostable polymerase with 3' to 5' exonuclease activity generates blunt products suitable for direct cloning. Nucl. Acids Res. 20, 144.

Markosyan, G.E. (1972) A new acidophilic iron bacterium *Leptospirillum ferrooxidans*. Biol. Zh. Armenii. 25, 26-29.

Marsh, R.M. and P.R. Norris (1983) The isolation of some thermophilic, autotrophic, iron- and sulfur-oxidizing bacteria. FEMS Microbiol. Lett. 17, 311-315.

Norris, P.R. (1983) Iron and mineral oxidation with *Leptospirillum ferrooxidans*, in "Progress in Biohydrometallurgy 1983, Cagliari," (G. Rossi and A.E. Torma, Eds.), pp. 83-96. Associazione Mineraria Sarda, Italy.

Olsen, G.J., R. Overbeek, N. Larsen, T.L. Marsh, M.J. McCaughey, M.A. Maciukenas, W.K. Kuan, T.J. Macke and C.R. Woese (1992). The ribosomal database project. Nucl. Acids Res. 20 (Supplement), 2199-2200.

Pronk, J.T., R. Meulenberg, W. Hazeu, P. Bos, J.G. Kuenen (1990) Oxidation of reduced inorganic sulphur compounds by acidophilic thiobacilli. FEMS Microbiol. Lett. 75:293-306.

Rainey, F.A., M. Dorsch, H.W. Morgan, and E, Stackebrandt (1992) 16S rDNA analysis of *Spirochaeta thermophilia*: its phylogenetic position and implications for the systematics of the order *Spirochaetales*. System. Appl. Microbiol. 15, 197-202.

Sand, W., K. Rohde, B. Sobotke, and C. Zenneck (1992) Evaluation of *Leptospirillum ferrooxidans* for leaching. Appl. Environ. Microbiol. 58, 85-92.

Sommerville, C.C., I.T. Knight, W.L. Straube and R.R. Colwell (1989) Simple, rapid method for direct isolation of nucleic acids from aquatic environments. Appl. Environ. Microbiol. 55, 548-554.

Southam, G. and J.J. Beveridge (1992) Enumeration of thiobacilli within pH-neutral and acidic mine tailings and their role in the development of secondary mineral soil. Appl. Environ. Microbiol. 58, 1904-1912.

Tuovinen, O.H., B.C. Kelley, and S.N. Groudev (1991) Mixed cultures in biological leaching processes and mineral biotechnology, in "Mixed Cultures in Biotechnology" (J.G. Zeikus and E.A. Johnson, Eds.), pp. 373-427. McGraw-Hill, New York.

Wakao, N., M. Mishina, Y. Sakurai, and H. Shiota (1984) Bacterial pyrite oxidation III. Adsorption of *Thiobacillus ferrooxidans* cells on solid surfaces and its effect on iron release from pyrite. J. Gen. Appl. Microbiol. 30, 63-77.

Ward, D.M., R. Weller, and M.M. Bateson (1990) 16S rRNA sequences reveal uncultured inhabitants of a well studied thermal community. FEMS Microbiol. Rev. 75, 105-116.

Weisberg, W.G., S.M.Barns, D.A. Pelletier, and D.J. Lane (1991) 16S ribosomal DNA amplification for phylogenetic study. J. Bacteriol. 173, 697-703.

Woese, C.R. (1987) Bacterial evolution. Microbiol. Rev. 51, 221-271.

SYSTEMATICS OF INSECT PATHOGENIC BACILLI: USES IN STRAIN IDENTIFICATION AND ISOLATION OF NOVEL PATHOGENS

Fergus G. Priest, Marilena Aquino de Muro and Denise A. Kaji

Department of Biological Sciences
Heriot Watt University
Edinburgh EH14 4AS
Scotland, UK

INTRODUCTION

Microbiological insecticides comprise micro-organisms which cause disease and ultimately death in insects. They are being used increasingly for control of agricultural pests and insect vectors of disease, either alone or in combination with chemicals in "integrated pest management" programmes. The growing acceptance of microbial insecticides partly arises from disillusionment with chemical insecticides due to increasing resisitance among target insects. Currently more than 500 insect and mite species have acquired resistance resulting in major agricultural and health problems. For example, mosquitoes resistant to most of the commonly used chemical insecticides have now emerged in many areas of the world (reviewed by Rawlins, 1989). One particularly well studied case involves resistance to organophosphates in the mosquito *Culex quinquefasciatus* which resulted from amplification of an esterase gene. The same mutation was detected in resistant mosquitoes from USA, India, Africa and south east Asia showing that a single mutation had arisen and the resistant population had rapidly become disseminated and established in different continents, presumably as a result of airline traffic (Raymond *et al.,* 1991). Resistance to microbial insecticides, on the other hand, is currently very limited (McGaughey and Whalon, 1992; Hougard and Back, 1992) but unless strategic measures to avoid the problem are taken, it is anticipated that resistance will develop rapidly in target populations as they are used more intensively (McGaughey and Whalon; 1992; May, 1993).

An additional advantage of biological control agents over chemical insecticides is high target specificity. Bacterial toxins are exquisitely specific and can be used to control target insects while leaving predators, other insects and small animals unaffected. The lack of mammalian toxicity also benefits handling and spraying of the materials.

Microbiological insecticides can be brought on to the market more quickly and less expensively than synthetic chemicals and can be markedly improved using modern biotechnological techniques. Indeed this has led to a new generation of genetically engineered plants such as cotton, tobacco and tomato expressing insect toxin genes in their foliage (Ely, 1993).

Finally, growing concern about the environmental impact of chemical spraying is focusing attention on ecologically acceptable alternatives.

Despite their advantages, microbial products command a low share of the total insecticide market (about 1%) although some industry observers speculate that this could rise to 10-12% by the year 2000. This has promoted extensive screening programmes with the intention of isolating bacteria with novel toxicities or increased activity (Feitelson *et al.*, 1992). Microbial screening programmes are most effective if based on a sound taxonomic framework (Bull *et al.*, 1992). This allows for the development of effective selective isolation strategies and the accurate recognition of the required organism or of novel microbes. Without the taxonomic considerations, screening programmes often result in continuous re-isolation and examination of numerous strains previously characterized and discarded. With this in mind, we have embarked on a taxonomic study of some insect pathogenic, endospore-forming bacteria used in biological control programmes with a view to developing improved isolation and screening procedures for novel pathogens.

INSECT PATHOGENIC BACTERIA OF THE GENUS *BACILLUS*

The genus *Bacillus* encompasses the rod-shaped bacteria which differentiate into endospores under aerobic conditions. Extensive phenetic (Priest *et al.*, 1988) and phylogenetic (Ash *et al.*, 1991a) classifications have resulted in the description of five major phylogenetic lineages within the genus. Interestingly, insect pathogenicity has developed, presumably independently, in four of these; the only exception is the branch containing the thermophilic species. Moreover, toxicity to mosquitoes has also been noted in the anaerobic endospore-forming bacterium, *Clostridium bifermentans* (Nicolas *et al.*, 1993). Since these bacterial phyla diverged before the emergence of insects, it is predictable that the molecular basis of insect pathogenicity in these different lineages would be unrelated. Although current knowledge of the the mechanisms of pathogenicity is limited outside *B. thuringiensis*, it does seem that the *B. thuringiensis* model (reviewed by Aronson, 1993a) is not necessarily applicable to other entomopathogenic bacilli.

Within rRNA group 1, which is based on *B. subtilis*, the notable entomopathogens are *B. thuringiensis*, which is treated in detail below, and two beetle (Coleopteran) pathogens; *B. lentimorbus* and *B. popilliae*. The latter are the etiological agents of "milky diseases" in certain beetle larvae, such as those of the Japanese beetle, an important pest of pastures and grasslands in the USA. The name of the disease derives from the massive growth and sporulation of bacteria in the haemolymph of the infected larvae which gives the almost translucent grubs an opaque appearance. The disease is therefore an infection rather than a toxicity and although strains of *B. popilliae* produce a refractile parasporal body, its role in pathogenicity is unclear (reviewed by Stahly *et al.*, 1992). These bacteria are nutritionally fastidious, catalase negative and produce a large oval spore in a swollen sporangium, so phenetically they are unlike most other members of rRNA group 1 which produce oval endospores which do not distend the mother cell.

RNA group 2 encompasses several insect pathogens, or suspected pathogens, notably *B. larvae*, the etiologic agent of American foulbrood in honeybees and *B. pulvifaciens* which is associated with "powdery scale" in these insects. These two bacteria are close phylogenetically (Ash *et al.*, 1991a) and can be distinguished phenetically by the less fastidious growth requirements of *B. pulvifaciens*. In common with other members of this group, these bacteria are facultative anaerobes which ferment a variety of sugars. They differentiate into oval endospores which distinctly swell the mother cell or sporangium.

Bacilli which differentiate into spherical endospores comprise a separate phylogenetic lineage (RNA group 3 of Ash *et al.*, 1991a). The lack of *meso*-diaminopimelic acid in the cell walls of these bacteria is characteristic, as is their strictly aerobic metabolism. Some strains of *B. sphaericus,* which is the key member of this group, are highly toxic to mosquito larvae and are used for the biological control of this important vector of tropical diseases (see below).

Bacillus laterosporus which, together with *B. brevis* comprise a separate phyletic lineage (RNA group 4 of Ash *et al.*, 1991a), may be pathogenic to mosquito larvae (Favret and Yousten, 1985). Sixteen of 29 strains examined displayed toxicity to mosquito larvae but not to lepidopteran larvae. Although strains of *B. laterosporus* produce parasporal bodies, the toxicity was associated with the cell rather than the spore.

Of these insect pathogens, three have been commercialized as biocontrol agents. *Bacillus popilliae* has been used since the 1940's for control of Japanese beetle in the USA but has not gained widespread application. *Bacillus sphaericus* has great potential as a larvicide for control of mosquito populations in the tropics but, because of the relatively poor market has not been developed as extensively as it might have been. *Bacillus thuringiensis*, on the other hand, is the most successfull and widely used biocontrol agent. Indeed, the annual market for *B. thuringiensis* products for crop protection is in excess of $75 million (Coombs, 1993).

In this review, we shall consider the systematics of *B. sphaericus* and *B. thuringiensis* with emphasis on the use of a taxonomic approach for the selective isolation and characterization of novel strains for biological control applications.

BACILLUS SPHAERICUS

Systematics of *Bacillus sphaericus*

Virtually all mesophilic bacilli which differentiate into spherical endospores are currently placed in *B. sphaericus.* These bacteria have an oxidative metabolism and use organic acids rather than sugars as a source of carbon and energy (Russell *et al.*, 1989; Alexander and Priest, 1990). As a result they are negative in most of the traditional tests used for the classification of *Bacillus*, such as acid formation from sugars, starch hydrolysis, Voges Proskauer reaction etc. (Gordon *et al.*, 1973), and strains are automatically allocated to *B. sphaericus* on the basis of these negative physiological reactions and sporal morphology. It is therefore not suprising that there is considerable diversity within *B. sphaericus sensu lato.*

The foundation of *B. sphaericus* classification was established by Krych *et al.* (1980) who assigned 50 strains to five distinct DNA homology groups with 12 strains ungrouped. These DNA homology groups were subsequently ratified by numerical analysis of phenotyopic features using assimilation tests and other characters of diagnostic value rather than sugar fermentations (Alexander and Priest, 1990). Moreover, the numerical classification highlighted additional taxa of probable species status. Although probabilistic identification of strains to the DNA homology groups on the basis of phenotypic characteristics is effective (Alexander and Priest, 1990), the scarcity of sound diagnostic features (De Barjac *et al.*, 1980; Krych *et al.*, 1980; Nagel & Andreesen, 1992) has delayed their introduction of these groups as new species.

DNA homology group I of *B. sphaericus* contains the type strain but group II is of particular importance because some strains in this group are pathogenic to certain mosquito larvae, particularly those of the genus *Culex* which is an important vector of encephalitis and filariasis in tropical countries. In the original DNA homology study, group II strains formed two subgroups at 60-70% relatedness; IIA and IIB (Krych *et al.* 1980). The stability of hybrid duplex formation (Δ*T*m) between DNA from strains of the two subgroups was significantly less than that of the homologous duplexes indicating that the hybrids were poorly matched. Indeed these features of 60-70% sequence homology and Δ*T*m about 6°C are compatible with the strains being allocated to separate subspecies (Wayne *et al.*, 1987). Interestingly, all mosquito pathogens were recovered in group IIA (Krych *et al.*, 1980), indicating that pathogenicity was associated with this discrete taxon or subspecies and the group IIB strains were non-pathogenic. Although, group IIB has been named as *B. fusiformis* (Priest *et al.*, 1988), group IIA has yet to be named formally.

We have analysed strains of *B. sphaericus sensu lato* using rRNA gene restriction fragment length polymorphisms (RFLP) or ribotyping. In this procedure, chromosomal DNA is cleaved with restriction endonuclease(s), separated by agarose gel electrophoresis, blotted onto a nylon or nitrocellulose membrane and hybridized with a 16S rRNA gene probe. We routinely use a probe which is PCR-amplified from *B. subtilis* chromosomal DNA using primers to the extremities of the gene. The cDNA is labelled with digoxigenin by including the labelled nucleotide in the amplification reaction.

Strains of *Bacillus* are thought to contain 10 rRNA operons (based on the *B. subtilis* model; Jarvis *et al.*, 1988) each comprising the same pattern of 16S-23S-5S rRNA genes. The regions between the operons vary from less than 1 kb (*rrnI, rrnH, rrnG*) to almost one quarter of the entire chromosome (Green and Vold, 1993). The spacer regions between the 16S and 23S genes within the operons, and the spaces between the operons are less highly conserved than the genes themselves, because they are not subject to the same functional constraints, although in two cases (*rrnA* and *rrnO*) the intra-operon spacer regions contain tRNA genes (reviewed by Green and Vold, 1993). The RFLP patterns therefore derive from the distribution of restriction sites immediately outside the operons, in the spacer regions between the 16S and 23S genes and in the rRNA genes themselves.

Ribotyping of *B. sphaericus* strains has revealed that all mosquitocidal strains and all members of DNA homology group IIA examined to date (which is not exactly the same thing since some members of group IIA may not be pathogenic) have a common rRNA RFLP when using either *Hind*III or *Eco*RI digested chromosomal DNA, although the former generally gives more readily discernible patterns (Figure 1). Indeed, we have now examined more than 30 mosquito pathogenic strains from a wide geographical distribution and all give identical ribotype patterns. Moreover, ribotyping succesfully distinguishes strains from DNA homology groups I (*B. sphaericus sensu stricto*), and group IIB (*B. fusiformis*) from the mosquito pathogens (Aquino de Muro *et al.* 1992). Too few strains from groups III, IV and V have been studied for any firm conclusions to be made about these taxa. The full implications of these results are discussed later in this chapter.

More recently, we determined the full 16S rRNA gene sequences for representative strains of the DNA homology groups of *B. sphaericus* (Aquino de Muro and Priest, 1993). The phylogenetic tree derived from these sequences and selected sequences obtained from the EMBL database is shown in Figure 2. The sequences were aligned using programs in the Wisconsin Molecular Biogy package (Devereux *et al.*, 1984). A distance matrix was prepared using Kimura's two parameter model contained in DNADIST. The tree was constructed using FITCH which estimates phylogenies under the "additive tree model" using the Fitch-Margoliash criterion and some related least squares criteria All programs were obtained from PHYLIP (Felsenstein, 1993). The pattern of branching is different from that previously published by us (Aquino de Muro and Priest, 1993) and we believe this to be

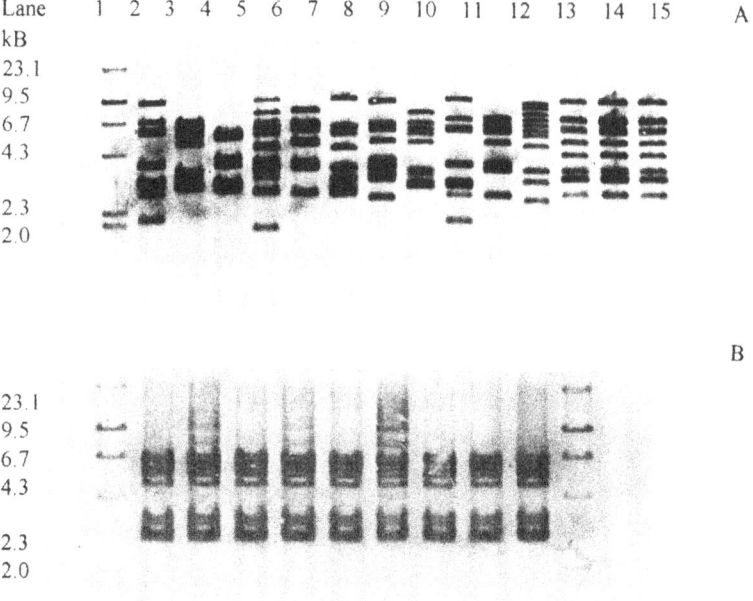

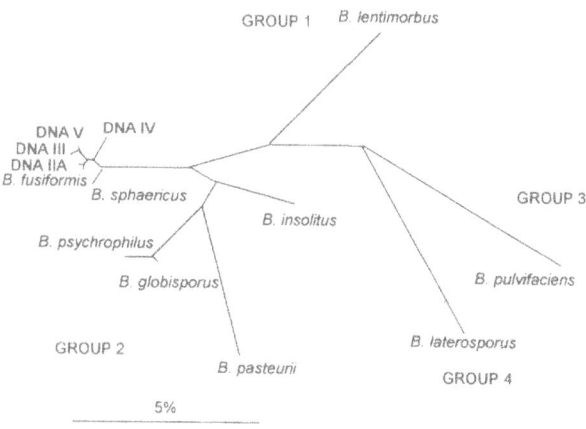

Figure 1. Ribotype patterns of some strains of *Bacillus sphaericus*.

A. Lanes 2 to 15 strains ATCC 14577 [I], 2362 [IIA], ATCC 7055 [IIB], NRS 592 [III] NRS 400 [IV], NRS 1198 [V] , 033, 152, 626, 657, LP4-b, 3, 4, and 5. DNA homology groups shown in [ ]; for details of strains see Table 2. Lane 1, size markers, λ *Hind*III digest.
B. Lanes 2 to 10, DNA homology group IIA strains 2362, 235, 261, LP12-as, LP7-a, LP1-g LP20-e LP14-8 and 2297 (for details of strains see Table 2). Lanes 1 and 11, λ *Hind*III markers.

Figure 2. Phylogenetic tree of some strains of *Bacillus sphaericus* and relatives derived from 16S rRNA sequences. The tree was prepared from a distance matrix of the alignment using FITCH (see text).

a better representation of the relationships. The tree shows the mosquito pathogenic strains to be closely related to group IIB (Figure 2) and *B. sphaericus sensu stricto* to be more distantly related to these bacteria. As in our previous tree and in other studies (Ash *et al.*, 1991a; Farrow *et al.*, 1992), the round spore formers appear to comprise two lineages, one containing *B. sphaericus* and the mosquito pathogens and the second the psychrophilic species such as *B. insolitus* and *B. psychrophilus*.

Strain Typing

As with bacteria of clinical importance, there is a need for typing strains of *B. sphaericus* at the sub-species level. Two systems have been employed which give largely congruent results. A serotyping scheme based on antibody reactions to flagellar antigens was developed by de Barjac *et al.* (1980). Currently, pathogenic strains have been assigned to 10 serotypes (Table 1). A phage typing scheme (Yousten, 1984) is used to allocate pathogenic strains to 8 phage types (Table 1), although some isolates cannot be typed using the existing battery of phages.

The only major discrepancy between the typing systems is in phage type 3 which encompasses serotypes 5a5b and 6. The basis of the specificity of the phage typing scheme is unclear but is presumably partly based on restriction and modification and the surface layer proteins of these bacteria (Lewis *et al.*, 1987).

There is a correlation between the typing schemes and pathogenicity. The early strains isolated by Kellen in the USA (phage type 1; serotype H1a) have low toxicity as do serotype H2a2b (phage type 2) strains. These organisms (where examined) contain a single toxin, the 100 kDa product of the *mtx* gene (see below). Phage type 3 strains (serotype 5a5b) are highly toxic to *Culex* larvae. These comprise the largest group of isolates held in the "Collection of Bacillus thuringiensis and Bacillus sphaericus" at the Institut Pasteur, presumably because they have been selectively isolated and collected on the basis of high larvicidal activity. In all cases examined, these bacteria contain two toxins, the 100 kDa-toxin and a crystal protein. These bacteria are not only highly active against *Culex pipiens* but also toxic to *Anopheles stephensii*, *Aedes aegypti* (Thiery and de Barjac, 1989) and some *Mansonia* larvae (Petcharat, 1991). Other serotypes and phage types are less well represented, although strain 2297 (phage type 4; serotype H25) is a highly toxic strain which has received considerable attention as a promising biocontrol agent and serotype H26a26b strains might be interesting since they appear to have no genes which hybridize with the crystal protein and 100 kDa-toxin gene probes.

DNA fingerprinting of *B. sphaericus* strains using M13 DNA as a probe can be used to type strains (Abajadeva *et al.* 1992) and preliminary studies using RAPDS have indicated that serotype specific patterns can be generated following judicous choice of primers and amplification conditions (Woodburn and Yousten, personal communication). The correlation of RAPDS with serotyping and phage typing suggests a clonal structure to group IIA *B. sphaericus* in which strains of the same serotype/RAPD group are members of the same clone and have a common recent ancestor (see Selander *et al.*, this volume, and Maiden, 1993 for discussion of clonal population genetics of bacteria).

Pathogenicity of *Bacillus sphaericus*

The first mosquitocidal strains of *B. sphaericus* which were isolated by Kellen in California (Kellen *et al.*, 1965) were weakly toxic to *Culex* larvae. Following an intensive screening programme organized by the WHO, several more highly toxic strains were isolated (reviewed by Singer, 1988) and these subsequent strains form the basis of current biocontrol programmes; particularly strains 2362 and 1593 (Table 1). Most of the the low

Table 1. Characteristics of some mosquito-pathogenic strains of *Bacillus sphaericus*

Strain	Origin	Serotype[1]	Phage type	Toxin genes Binary	100 kD	Reference
K	USA	1a (3)	1	-	+	2,6
Q	USA	1a	1	ND[2]	+	5
SSII-1	India	2a2b(25)	2	-	+	2,4,6
1889	Israel	2a2b	2	ND	+	6
1883	Israel	2a2b	2	-	ND	4
SL 42	USA	3 (6)	ND	ND	ND	
1593	India	5a5b (180)	3	+	+	2,4,6
1691	El Salvador	5a5b	3	ND	+	6
2013.6	Roumania	5a5b	3	+	ND	4
2317.3	Thailand	5a5b	3	+	+	3,6
2362	Nigeria	5a5b	3	+	+	1,2,6
2500	Thailand	5a5b	3	+	ND	4
BSE18	Scotland	5a5b	3	+	+	1,4,6
IAB59	Ghana	6 (38)	3	+	+	3,6
COK31	Turkey	9 (2)	8	ND	+	6
2297	Sri Lanka	25 (28)	4	+	+	1,2,6
2173	India	26a26b (12)	NT[3]	-	-	4,6
2315	Thailand	26a26b	NT	-	-	4,6
2377	Indonesia	26a26b	ND	-	ND	4
2115	Philippines	27 (1)	6	-	ND	4
IAB872	Ghana	48 (3)	3	ND	ND	
LP-1G	Singapore	ND	NT	+	-	1,5
LP-7A	Singapore	ND	NT	+	-	1,5

[1] Numbers in brackets are the numbers of strains of the appropriate serotype held in the "Collection of Bacillus thuringiensis and Bacillus sphaericus", Institut Pasteur.
[2,3] ND and NT represent not done and not typable respectively
Detection of toxin genes was by hybridization (or sequencing, reference 3)as reported in the references;1, Aquino de Muro and Priest (1993); 2, Baumann *et al.* (1988); 3, Berry *et al.* (1989); 4, Guerineau *et al.* (1991); 5, Liu *et al.* (1993); 6, Thanabalu *et al.* (1991).

toxicity and high toxicity strains synthesize an ADP-ribosylating-type toxin (Mtx; mosquitocidal toxin) of about 100 kDa (Thanabalu *et al.*, 1991). This protein is synthesized at low level during exponential phase growth and is therefore associated with the vegetative cell (Myers and Yousten, 1980). When ingested by mosquito larvae, Mtx protein is processed by gut proteases into 27 and 70 kDa peptides (Thanabalu *et al.*, 1992). The former, derived from the N-terminal region of the native protein, ADP-ribosylates two proteins of 38 and 42 kDa in extracts of *Culex* cells in addition to ADP-ribosylating itself (Thanabalu *et al.*, 1993). Indeed it behaves very much like other ribosylating toxins such as cholera and pertussis toxins. The larger moiety is responsible for morphological changes in mosquito cells in culture and both peptides are needed for larval toxicity (Thanabalu *et al.*, 1993).

High toxicity strains synthesize a crystal protein in addition to, or instead of the Mtx protein. This small parasporal body is synthesized during the early stages of sporulation and toxicity to larvae parallels the appearance and growth of the crystal. The crystal is composed of equimolar amounts of two proteins, the 42 and 51 kDa-toxins which act synergistically in their toxicity to larvae (reviewed by Baumann *et al.*, 1991; Porter *et al.*, 1993). When the crystal is ingested by susceptible larvae, the proteins dissolve in the alkaline conditions of the larval gut and are processed by larval proteases into smaller, highly toxic proteins of 39 and 43 kDa. This processing step activates the toxins resulting in about a 50-fold increase in toxicity. The 42 kDa protein is the major toxin (Nicolas *et al.*, 1993) and is bound to susceptible cells in conjunction with the 51 kDa protein (Oei *et al.*, 1992). Recent experiments with cloned binary toxin genes indicate that the 42 kDa protein is responsible for host range, and that minor alterations in the amino acids around position 100 can have marked effects on toxicity to Aedes larvae (Berry *et al.*, 1993).

Binary toxin genes are located on the chromosome of group IIA strains (Aquino de Muro *et al.*, 1992) and have been cloned and sequenced from numerous organisms including representatives of serotype H5a5b (strains 2362, 1593, 2317.3 and BSE 18), serotype H6 (strain IAB59) and serotype H25 (strain 2297) (Berry *et al.*, 1989, unpublished results). Genes from organisms in serotype H5a5b have identical sequences and there are minor differences in binary toxin genes between serotypes. This reinforces the concept of clonality in these bacteria although too few strains have been studied for firm conclusions to be drawn.

DNA Probes for *Bacillus sphaericus*

Bacillus sphaericus has important advantages over *B. thuringiensis* serovar *israelensis* as a mosquito control agent. In particular, persistence in the environment, ability to retain activity in polluted waters and evidence for recycling in target insects combine to give rise to prolonged larval control. However, one of the drawbacks to its fuller implementation is the lack of activity against certain *Aedes* species, the vectors of dengue, yellow fever and some forms of filariasis. Given that subtle differences in crystal protein stucture may have important implications in host range (Berry *et al.*, 1993), we have instigated a programme for the isolation of new strains of *B. sphaericus* DNA homology group IIA which might have different spectra of toxicities.

Even with the most successful selective media (Yousten *et al.*, 1985; Guerineau *et al.*, 1991), fewer than 1% of colonies arising from pasteurized samples of mud or insects are group IIA pathogens. Characterization of these colonies as mosquito pathogens requires removal of the colony to mosquito larvae and examination for lethality over a 2-day period. This process is very time-consuming and requires large numbers of mosquito larvae since at best one in one hundred colonies will be toxic. We have therefore developed a series of DNA probes with the purpose of rapid identification of pathogens on isolation plates so that the number of colonies subjected to toxicity testing can be reduced to those few with high potential for larval toxicity.

Of course it is impossible to design a hybridization probe for the truly novel toxin since by its very nature, the gene responsible is unknown. In our attempt to detect novel pathogens, we therefore conceived a programme in which a "species"-specific probe is used to identify group IIA colonies (i.e. potential pathogens). Hybridizing colonies are then examined with probes for the 100 kDa- and binary toxin genes. Strains which hybridize to the species-specific probe but NOT to the toxin gene probes may represent novel pathogens (or non-pathogens!). Since only group IIA strains are examined for toxicity against

Table 2. Strains used for slot-blot hybridization

Strain B. sphaericus	DNA homology group(R)	Hybridization to chromosomal DNA from strain 2362	Hybridization to oligonucleotide probe	Hybridization to 100 kDa toxin gene probe	Hybridization to 51.4 and 41.9 kDa toxin gene probe	Phage type	Position in Figure 3	Source
ATCC 14577*	I	-	-	-	-	NT	A1	1
2362	IIA	+	+	+	+	3	B1	1
2297	IIA	+	+	+	+	NT	C1	1
SSII-1	IIA	+	+	+	-	2	C5, A8	1
ATCC 7055*	IIB	+	-	-	-	2	A2	1
NRS 592*	III	-	-	-	-	NT	B2	1
NRS 400*	IV	-	-	-	-	NT	C2	1
NRS 1198*	V	-	-	-	-	NT	A3	1
LP12-as	IIA	+	+	+	+	8	A4	2
LP7-a	IIA	+	+	+	+	8	B4	2
LP6-b	IIA	+	+	+	-	2	C4	2
LP1-g	IIA	+	+	-	+	NT	B5	2
LP20-e	IIA	+	+	-	+	8	B6	2
LP14-8	IIA	+	+	-	-	8	C6	2
3**	UG	-	-	-	-	NT	A9	3
4	UG	-	-	-	-	NT	B9	3
5	UG	-	-	-	-	NT	C9	3
10	UG	+	-	-	-	ND	A11	3
17	UG	+	-	-	-	ND	C12	3
20/1	UG	+	-	-	-	NT	A13	3
26	UG	+	-	-	-	ND	A14	3
37	UG	+	-	-	-	ND	A15	3
033	UG	-	-	-	-	NT	A16	4
152	IIA	+	+	+	-	2	B16	4
235	IIA	+	+	+	+	3	C16	4
261	IIA	+	-	+	+	3	A17	4
626	UG	-	-	-	-	NT	B17	4
657	UG	-	-	-	-	8	C17	4

UG = ungrouped; (*) = reference strains; NT = not typed; ND = not determined;

(R) DNA homology groups determined by rRNA gene restriction patterns (Aquino de Muro et al., 1991);

(**) Additional 11 UG strains did not hybridize to chromosomal DNA from strain 2362 or either oligonucleotide probe, positions A10-A15 in Figure 3;

(1) Laboratory collection;
(2) Dr. A. Porter, Institute of Molecular and Cell Biology, National University of Singapore;
(3) Dr. A. Mahasneh, Faculty of Science, University of Jordan, Amman;
(4) Dr. L. Rabinovitch, Laboratorio de Fisiologia Bacteriana, Departamento de Bacteriologia, Instituto Oswaldo Cruz, FIOCRUZ, Rio de Janeiro, Brazil.

appropriate mosquito larvae, the toxicity-testing workload is reduced to less than 1% of similar programmes which do not use this screening approach. This in turn allows many more pathogens to be examined.

The group IIA-specific probe comprises an oligonucleotide targeted at a position centered at about nucleotide 190 of the 16S rRNA gene (Aquino de Muro and Priest, 1993). This 19-mer probe hybridizes specifically with members of DNA homology group IIA and not with bacteria from any of the other homology groups of *B. sphaericus* or with representatives of the other rRNA groups of the genus (Figure 3a). This probe is much more discriminatory than randomly-labelled chromosomal DNA from a group IIA strain (Figure 3b) for the identification of group IIA strains. For example in Figure 3, the group IIB strain at position A2 hybridizes with chromosomal DNA but not the oligonucleotide, as does a non-pathogenic soil isolate at position A13. These probes can also be used in colony blots to identify potentially pathogenic strains in screening programmes (Aquino de Muro and Priest, 1994). We also use probes derived from the toxin genes of strains 2362 and SSII-1. The former is the cloned, 3.5 kb *Hind*III fragment encompassing the 42 and 51 kDa-toxin genes. This hybridizes with all crystal protein toxin genes studied to date since the sequence conservation of these genes, even between serotypes is high (Berry *et al.*, 1989). A second probe is the *mtx* gene (100 kDa-toxin gene) which we PCR amplify from chromosomal DNA from strain SSII1 in the presence of digoxigenin-labelled dUTP to produce a labelled probe (Aquino de Muro and Priest, 1994). Using these probes we can determine the identity of group IIA strains and the occurrence of toxin genes in their chromosomes (Figure 3c, d; Table 2). For example, in Figure 3c, the occurrence of both toxin genes in phage type 3 strains (e.g. positions A17, B1, C16) and the single 100 kDa-toxin gene in phage type 2 strains is evident (e.g. positions A8, B16, C4, C5). Using this

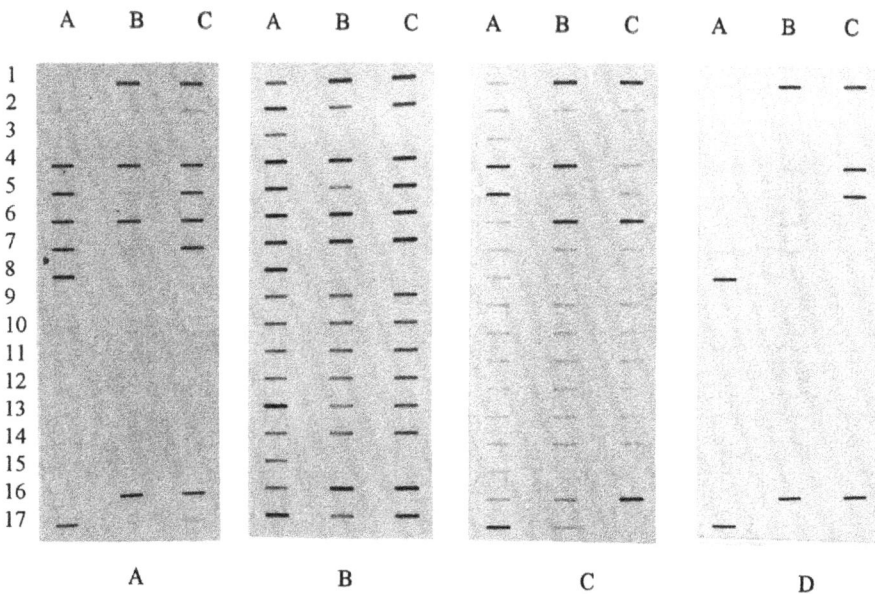

Figure 3. Slot-blot hybridizations of chromosomal DNA from some strains of *Bacillus sphaericus* with; A. Oligonucleotide probe for group IIa strains, B. Chromosomal DNA from strain 2362, C. Binary toxin gene probe and D. 100 kDa-toxin gene probe. For details of strains see Table 2.

approach, we recently isolated a novel type of group IIA pathogen from soil and mud samples in Singapore in which the crystal protein gene is present but the 100 kDa-toxin gene is absent (Liu *et al.*, 1993) (e.g. positions A4, B4, B6, C6 in Figure 3).

BACILLUS THURINGIENSIS

Systematics of the *Bacillus thuringiensis* group

Bacillus thuringiensis is a taxonomic enigma. It was originaly isolated from diseased silkworms in Japan by Ishiwata in 1901 and subsequently, and apparently independently, in Germany from the diseaed larvae of the Mediterranean mealmoth by Berliner (reviewed by Stahly *et al.*, 1992). The close phenetic relationship of *B. thuringiensis* with *B. anthracis, B. cereus* and *B. mycoides* was emphasized by Smith *et al.* (1946) in their studies of *Bacillus*, who recommended that *B. cereus* should be recognized as the "parent species" or stable species and that *B. anthracis, B. mycoides* and *B. thuringiensis* should be reduced to varieties of *B. cereus*. This caused understandable consternation among microbiologists concerned with diseases of man, animals and insects and although later editions of the taxonomic monograph (Smith *et al.*, 1952; Gordon *et al.*, 1973) maintained this classification, separate species status was provided for *B. anthracis, B. cereus B. thuringiensis* and *B. mycoides* in the "Approved Lists of Bacterial Names" and in the first edition of "Bergey's Manual of Systematic Bacteriology" (Claus and Berkeley, 1986) in recognition of the pathogenic properties of strains of these taxa.

The reasons for the confusion in the classification of the *B. cereus* complex are plainly evident. Morphologically the four species are virtually identical with the exceptions that strains of *B. mycoides* produce a highly rhizoid colony type, but this feature is often unstable on repeated subculture. *Bacillus thuringiensis* strains produce insecticidal crystal proteins or parasporal bodies, but again these may be lost on laboratory culture. *Bacillus anthracis* grows in longer chains than does *B. cereus* or *B. thuringiensis*, is capsulated and non-motile. However, many of these features are plasmid encoded including crystal protein synthesis in *B. thuringiensis* (reviewed by Aronson, 1993b) and capsule synthesis in *B. anthracis* (Thorne, 1993) and when the plasmids are lost, diagnosis of the latter species may be difficult. Physiologically, the four species are very similar. *Bacillus anthracis* can be distinguished from *B. cereus* by its sensitivity to penicillin, absence of haemolysin activity and failure to degrade tryrosine (Thorne, 1993) but there are no such features for differentiating *B. cereus* from *B. thuringiensis* (de Barjac and Frachon, 1990).

Nucleic acid analyses are beginning to unravel the systematics of the the *B. cereus* group. Both 16S and 23S rRNA sequences are highly homologous for the four species and show the variation expected within a single species (Ash *et al.*, 1991b; Ash and Collins, 1992). The 16S rRNA gene sequences from *B. anthracis* and *B. cereus* are identical for a continuous stretch of 1446 bases and different from the sequence of a second strain of *B. cereus* by only one base. The 16S rRNA sequences of *B. mycoides* and *B. thuringiensis* differ from each other and from the sequences of *B. cereus* and *B. anthracis* by four to nine nucleotides. This high conservation within 16S rRNA genes has also been noticed in other *Bacillus* strains, where organisms from distinct species, (*B. globisporus* and *B psychrophilus*) as judged by DNA sequence homology, have identical 16S rRNA genes (Fox *et al.*, 1992). Moreover, the 23S rRNA sequences from *B. anthracis* and *B. cereus* show only two differences in almost 2900 nucleotides (Ash and Collins, 1992). A tree showing the evolutionary relationships of *B. thuringiensis*, some relatives from rRNA group 1 and representatives from the other rRNA groups is shown in Figure 4. This was constructed in the same way as the tree shown in Figure 2.

DNA reassociation studies are more helpful than rRNA sequence comparisons within recently diverged taxa because sequence divergence is greater throughout the chromosome than within the confines of the conserved stable RNA genes. *Bacillus anthracis* strains consistently show less than 60% homology with DNA from *B. cereus* and *B. thuringiensis* (Ash, C. personal communication) confirming the separate identity of these species. This is particularly important from the safety viewpoint of using *B. thuringiensis* in the environment and confirms that *B. thuringiensis* cannot "become *B. anthracis*" through aquisition of the necessary plasmids. By currently accepted criteria (Wayne *et al.*, 1987), *B. anthracis* is a separate and independent species.

Moreover, DNA reassociation between strains of *B. cereus* and *B. thuringiensis* is consistently less than 70% establishing these organisms as independent species (Nakamura, 1994) although their physical genetic maps are very similar (Carlson and Kolsto, 1993). There is developing therefore a firm molecular basis for the separate speciation of *B. anthracis*, *B. cereus* and *B. thuringiensis*, and despite many of their distinctive features being plasmid borne, the evolutionary association of plasmid with host is presumably sufficiently strong to prevent the natural association of *B. anthracis* plasmids with *B. thuringiensis* strains and *vice versa*. Although these plasmids can be introduced into the "wrong" hosts in the laboratory (Gonzales *et al.*, 1982), should this event happen in the environment, the hybrid presumably has no competitive advantage and is lost.

Strain Typing

Bacillus thuringiensis strains are traditionally typed serologically using H (flagellar) antisera (de Barjac and Frachon, 1990). Thirty four serotypes are listed in the latest (1992) edition of the catalogue of the the the "Collection of Bacillus thuringiensis and Bacillus sphaericus" issued by the Insitiut Pasteur. Serotypes are given varietal names and correlate to an extent with toxicity to larvae of certain Lepidopera and Diptera (Jaquet *et al.*, 1987; Lecadet and Mantouret, 1987). For example, serotype 14 strains are almost invariably toxic to mosquito and blackfly larvae (or may be non-toxic; Ohba *et al.*, 1988). In other serotypes

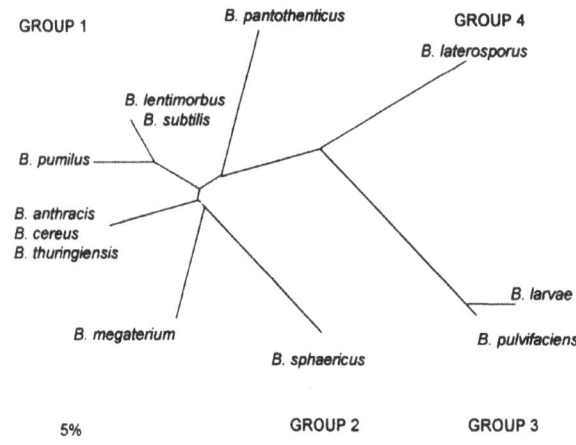

Figure 4. A tree showing the relationships of *Bacillus thuringiensis* with some other *Bacillus* species. The tree was calculated from a distance matrix of 16S rRNA gene sequences using the FITCH algorithm (Felsenstein, 1993).

286

Table 3. Distribution of toxin genes in some serovars of *Bacillus thuringiensis*

Serovar (biovar)	Strain	Serotype	*cry* genes	Principal toxicities
thuringiensis	1715 HD-2	1	*cryIA (b)* *cryIB*	Several lepidoptera Cabbage white butterfly
alesti	HD-4	3a	*cryIA (b)*	
kurstaki	HD-1	3a3b	*cryIA (a)* *cryIA (b)* *cryIA (c)* *cryIIA* *cryIIB*	Silk moth Tobacco budworm Lepidoptera & Diptera Tobacco hornworm
dendrolimus	HD-37	4a4b	*cryIA (a)*	
kenyae	-	4a4c	*cryIC*	Cotton leafworm
galleriae	-	5a5b	*cryIG*	Greater waxmoth
entomocidus	601	6	*cryIC*	
aizawai	HD-137 IC-1	7 7	*cryIC* *cryIA (b)* *cryIF*	
morrisoni	HD-12	8a8b	*cryIA*	
(tenebrionis)		8a8b	*cryIIIA*	Colorado potato beetle
morrisoni	PG14	8a8b	*cryIVA* *cryIVB* *cryIVC* *cryIVD*	Diptera
tolworthi	EG2838	9	*cryIE* *cryIIIB*	Beet armyworm Colorado potato beetle
israelensis		14	*cryIVA* *cryIVB* *cryIVC* *cryIVD*	Diptera

the relationship with toxicity breaks down and serotype 8a8c strains, for example, may be toxic to Lepidopetra, Coleoptera or Diptera (deBarjac and Frachon, 1990). Since toxin genes are plasmid borne and strains often synthesize more than one toxin type, any correlation between toxicity and serotype might, at first sight, be considered suprising.

The implications of the serotyping scheme of *B. thuringiensis* are significant beyond the nature of the flagella. DNA reassociation studies (Nakamura, 1994) of eight to ten strains belonging to 13 common serovars have shown that intragroup DNA relatedness among most serovars (*aizawai, alesti, darmstadiensis, dendrolimus, galleriae, israelensis, kurstaki, morrisoni, sotto,* and *thuringiensis*) is 90-100% but two serotypes, *canadiensis* and *kenyae* each comprised two, well defined subgroups of strains. Each showed intragroup relatedness of 80% or more and intergroup relatedness of 60 to 70%. However, when strains from serotypes were compared with each other, *alesti* and *dendrolimus* strains showed about 90% homology and strains of *aizawai, galleriae, kurstaki* and *morrisoni* clustered together. On the other hand, strains of *entomocidus, israelensis, sotto, thuringiensis, canadiensis* (both groups separately) and *kenyae* grouped separately at the 85% DNA homology level. Nakamura concludes that *B. thuringiensis* serovars are genetically distinct but closely related.

Carlson and Kolsto (1993) have examined several strains of *B. thuringiensis* by pulsed field gel electrophoresis of *Not*I-digested chromosomal DNA. Strikingly different *Not*I patterns were seen for most of the strains, although two strains of serovar *israelensis* gave the same banding patterns as did two strains of serovar *kurstaki*. Unfortunately, too few strains have been examined to fully confirm the DNA reassociation data of Nakamura (1994), but in support of his findings, strains of serovars *entomocidus, dendrolimus* and *subtoxicus* were different but clearly related while serovar *kenya* was remarkably different. Further evidence for the heterogeneity of *B. thuringiensis* has been obtained from a preliminary study of whole cell protein patterns which revealed continuity among strains of serovar *israelensis* and the separateness of these strains from other serovars (Kaji *et al.*, 1994).

There is a strong correlation between ribotyping and serotyping among strains of several serovars of *B. thuringiensis*. In particular, serovar *israelensis* strains conform to unique *Hin*dIII (Priest *et al.*, 1994) and *Eco*RI ribotype patterns (Figure 5) and representatives from several other serovars such as *aizawai, alesti, galleriae, kurstaki, morrisoni* and *thuringiensis* reveal one or a few ribotype patterns within each serotype (Priest *et al.*, 1994).

An alternative, chromosmal DNA typing system based on M13 DNA as the probe produces complex banding patterns which are serotype specific in several instances (Miteva *et al.*, 1991). In particular, serovars *kenyae, kurstaki* and *thuringiensis* were genotypically homogeneous by M13 typing. Moreover, RAPDS revealed that strains belonging to the same serovar produced closely related or identical banding patterns (Brousseau *et al.*, 1993). On the other hand, multiplex PCR reactions can distinguish individual strains if necessary (Bourque *et al.*, 1993).

A small study of *B. thuringiensis* strains using multilocus enzyme electrophoresis lends some support to the nucleic acid typing (Zahner *et al.*, 1989). Strains of serovar *israelensis* produced identical zymograms and strains of serovar *kurstaki* were highly conserved or identical. Although there is a need for comprehensive population genetic studies of *B. thuringiensis* before firm conclusions can be drawn, it would seem that the species is clonal in structure.

Pathogenicity of *Bacillus thuringiensis*

The first strains of *B. thuringiensis* to be isolated were invariably pathogens of lepidopteran larvae. Indeed for many years it was thought that pathogenicity in *B.*

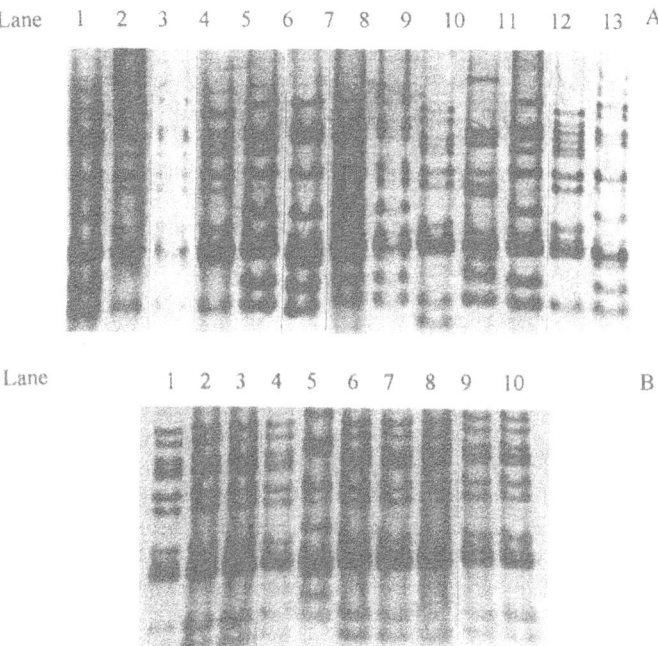

Lane 1 2 3 4 5 6 7 8 9 10 11 12 13 A

Lane 1 2 3 4 5 6 7 8 9 10 B

Figure 5. Ribotype patterns of some strains of *Bacillus thuringiensis*.

A. Lanes 1 to 13; strains *B.t. tolworthi* S478, *B.t. thuringiensis* S461, *B.t thuringiensis* S331, *B.t. tenebrionis, B.t. sotto* S468, *B.t. kurstaki* CCT0434, *B.t. kurstaki* HD-73, *B.t. kurstaki* HD-1, *B.t. israelensis* IPS82, *B.t. entomocidus* S474, *B.t. dendrolimus* S469, *B.t. alesti* S332 and *B.t. aizawai* S477.
B. Lanes 1 to 4 and 6 to 10; strains of *B.t. israelensis;* CCT2266, CCT2340, CCT2339, CCT2259, CCT2258, CCT2342, CCT2257, CCT2341, and IPS82. Lane 5 unassigned strain CCT2338.

thuringiensis was limited to the butterflies and moths until the isolation of strains in Israel from dead mosquito larvae found in a dried pool (Margalit and Dean, 1985). These serotype 14 strains were highly toxic to *Aedes* larvae and had low activity against *Culex* larvae. Unlike *B. sphaericus*, serovar *israelensis* strains are toxic to blackfly larvae, which has proved immensely useful for the control of blackflies, and hence river blindness, in Africa (Walsh, 1986). More recently, strains with toxicity against Coleptera, in particular Colorado potato beetle (serovar *tenebrionis*), have been discovered and with more extensive screening programmes strains with toxicity to a variety of small animals including nematodes and even protozoa have been discovered (Feitelson *et al.*, 1992).

The principal toxic agent *B. thuringiensis* is the crystal protein or parasporal body. When ingested, the protein dissolves in the alkaline gut of the larva and is processed by larval proteases. The toxins bind with high affinity and considerable specifity to receptors (large proteins or glycoproteins) in the larval midgut epithelium. Interestingly, resistant larvae often have altered receptors to which the toxin cannot bind (Ferre *et al.*, 1991). After binding, there is probably a conformational change in the toxin followed by its insertion into the membrane. Insertion leads to permeability of the membrane to selective ions, uptake of potassium, uptake of water to balance the osmotic pressure, lysis of the cells; a process termed colloid osmotic lysis. The larvae stop feeding and death ensues, often as a result of starvation (Knowles and Ellar, 1987).

Toxin genes from more than 40 strains of *B. thuringiensis* have been cloned and sequenced (reviewed by Feitelson *et al.*, 1992; Aronson, 1993b). Five major classes (labelled *cryI* to *V*) with various subclasses are currently recognized on the basis of sequence similarities and toxicities to certain insects. However, the distribution of these toxins to serovars follows little rationale (Table 3). The principal reasons for this lie in the plasmid location of these genes with the associated opportunity for genes to spread throughout bacterial populations and for recombination between two toxin genes resident in the same host. Two examples demonstrate the complexity of the situation.

Bacillus thuringiensis serovar *kurstaki* is the principal organism used for biological control of lepidopteran larvae. The HD-1 strain of this taxon produces as many as five distinct toxins (Table 3) and contains numerous size classes of plasmid (see Aronson, 1993a,b for reviews). Four toxin genes are simultaneously lost when the strain is cured of a 110 mDa plasmid and the fifth, the *cryIA(b)* gene is located on an unstable 44 Mda plasmid. The opportunities for recombination between toxin genes are therefore considerable and it has been suggested that the *cryIA(b)* gene has arisen from recombination between the *cryIA(a)* and *cryIA(c)* genes (Geiser *et al.*, 1986). To add to the opportunities for gene transfer and recombination, some of these toxin genes are located on transposons or flanked by insertion sequences (Lereclus *et al.*, 1993).

A different situation is presented by serovar *morrisoni*. Serotype 8a8c strains are generally toxic to lepidoptera but strain PG-14 contains an inclusion identical to those of strains of serovar *israelensis* comprising the four dipteran specific gene products CryIVA, B, C and D. It may be that this represents an atypical strain of serovar *israelensis* in which flagellin genes have been transferred laterally from a serovar *morrisoni* strain. Alternatively, it is an atypical strain of the *morrisoni* clone containing the plasmids of serovar *israelensis*. Moreover, some strains with H8a8c antigens are toxic to Coleoptera and given the name *tenebrionis*. These are different from serovar *morrisoni* strains by ribotyping and whole cell protein patterns (Kaji *et al.*, 1994) and may represent a separate clone within this taxon, perhaps again arising by lateral transfer of flagellar antigen genes.

CONCLUDING REMARKS

Bacillus sphaericus and *B. thuringiensis* are both insect toxigenic, endospore-forming bacteria but present remarkably different taxonomic and ecologic scenarios. Mosquitocidal strains of *B. sphaericus* comprise a well defined taxon of species or subspecies (of *B. fusiformis*) status. Bacteria in this taxon possess a clonal population structure, that is the species comprises numerous independent clones of strains. These clones are widely distributed geographicaly and each is evolving independently. The correlation of phage type, serotype and RAPD pattern portrays the lack of gene exchange with subsequent recombination between strains and the consequent strong linkage disequilibrium within clones. If gene exchange and recombination were common between strains of this taxon, as it is in the naturally transformable species *B. subtilis* (Cohan *et al.*, 1991), linkage equilibrium would result and serotype, phage type and RAPD determinants would become intermingled. Since the toxin genes are located on the chromosome in this bacterium, it would be expected that these would conform to the clonal structure. Unfortunately limited data are available, but the sequence studies completed to date support the clonal structure with identical binary toxin genes from serotype 5a5b strains whatever their origin (for example India, Nigeria, Scotland and Thailand for strains 1593, 2362, BSE18 and 2317.3) and small variations in other serotypes. Thus it seems likely that variation in toxin genes will conform to the variation seen in the genetic stucture of the species. How does this affect the quest for new strains with novel toxicities? First, it seems likely that the restriction of toxin genes to the chromosome will hamper the introduction of new toxin gene structures by recombination. Consequently, novelty in toxin genes will only be derived from small changes associated with mutagenic drift. Such changes can have significant effects on insect target range and can dramatically change specific activities against individual insects. However, the target range is unlikely to be modified beyond the genus or order, and it seems unlikely that toxins active on insects other than the Diptera are likely to be generated.

The second implication of the clonal structure of group IIA of *B. sphaericus* is that screens aimed at new phage types or serotypes are the most likely to result in new toxins. Phage against common clones (e.g. phage type 3) could be incorporated in isolation protocols to selectively isolate new phage types and by implication new toxin types. It seems likely that with intensified screening programmes new useful organisms could be isolated.

Finally, the clonal structure of mosquitocidal strains of *B. sphaericus* has implications for the release of genetically engineered strains. So long as recombinant toxin genes are integrated into the chromosome, it is unlikely that transfer into other organisms will ensue.

Bacillus thuringiensis presents a fascinating taxonomic problem. The high sequence identity in 16S rRNA genes within the four species in the *B. thuringiensis* complex presents particular taxonomic problems concerning the definition of the species which have been discussed by Fox *et al.* (1992). It would seem that *B. anthracis, B. cereus, B. mycoides* and *B. thuringiensis* are currently speciating. This is evident from DNA reassociation experiments which confirm the individual identity of *B cereus* and *B. thuringiensis* (Nakamura, 1994). But there are additional species emerging in this complex. The emetic variety of *B. cereus*, which causes food poisoning in man, *B. thuringiensis* serovar *israelensis* and *B. thuringiensis* serovar *kenyae* are taxa which can be distinguished by DNA hybridization and, in the case of serovar *israelensis* also by whole cell protein patterns (Kaji *et al.*, 1994) and on these criteria could be afforded species status.

The taxonomy of the *B. thuringiensis* group may be clarified from the population genetic approach. The correlation of ribotyping with serotyping has confirmed an extensive clonal population structure to *B. thuringiensis* and it would be interesting to extend these studies into *B. cereus* and *B. anthracis*. The exceptionally large number of clones in *B. thuringiensis* is seen in few other bacterial species and, although plasmids may transfer readily between *B. thuringiensis* strains, the clonal structure indicates that chromosomal gene transfer is rare.

The remarkable diversity of toxin genes in *B. thuringiensis* presumably results from their location on transmissible plasmids and the prevalence of bordering transposable sequences. Recombinational events when two toxin genes are in the same host will give rise to considerable variation in gene structure and on occasion the resultant hybrid toxin will have novel properties against a particular insect. This could allow the parental strain to compete more effectively through death of the insect and growth of the bacterium in the cadaver. Such a plasmid/toxin/bacterium combination must be ecologically competitive for the association to become established and stabilized and will finally result in a taxon such as *B. thuringiensis* serovar *israelensis* in which plasmid complement, toxin genes and clonal background are almost invariably uniform. Thus, although these plasmids are transmissible, the general correlation of toxicity with serovar suggests that there are constraints in the environment on the competitiveness of random combinations and that most plasmid transfer events result in a non-competitive transconjugant and the plasmid is lost. Nevertheless, toxigenic plasmids are a driving force in the evolution of *B. thuringiensis* and could be responsible for the extensive number of distinct clones. New strains of *B. thuringiensis* with novel toxicities are continually being reported. It is to be hoped that the ingenuity of the microbiologist in screening programmes coupled with the technological excellence of the genetic engineer will result in new toxins for the continual battle against insect vectors of disease and crop pests.

ACKNOWLEDGEMENTS

Work from the authors laboratory was supported by scholarships from CNPq and CAPES for D. Kaji and M Aquino de Muro respectively and funding from the the the World Health Organization/UNDP/World Bank/WHO Special Programme for Research and Training in Tropical Diseases (TDR). We are grateful to Preben Nielsen for assistance with computing phylogenetic trees.

REFERENCES

Abajadeva, A., Miteva, V. and Grigorova, R. (1992) Genomic variations in mosquitocidal strains of *Bacillus sphaericus* detected by M13 DNA fingerprinting. J. Invert. Pathol. 60, 5-9.

Alexander, B. and Priest F.G. (1990) Numerical classification and identification of *Bacillus sphaericus* including some strains pathogenic for mosquito larvae J. Gen. Microbiol. 136, 367-376.

Aquino de Muro, M., Mitchell, W.J. and Priest, F.G. (1992) Differentiation of mosquito-pathogenic strains of *Bacillus sphaericus* from non-toxic varieties by ribosomal RNA gene restriction patterns J. Gen. Microbiol. 138, 1159-1166.

Aquino de Muro, M and Priest, F.G. (1993) Phylogenetic analysis of *Bacillus sphaericus* and development of an oligenucleotide probe specific for mosquito-pathogenic strains. FEMS Microbiol. Letts. 112, 205-210.

Aquino de Muro, M and Priest, F.G. (1994) A colony hybridization procedure for the identification of of mosquitocidal strains of *Bacillus sphaericus* on isolation plates. J. Invert. Pathol. In press.

Aronson, A.I. (1993a) Insecticial toxins, in "*Bacillus subtilis* and Other Gram-Positive Bacteria : Biochemistry Physiology and Molecular Genetics" (Sonenshein, A.L., Hoch, J.A. and Losick, R. Eds.) pp 953-963. American Society for Microbiology, Washington D.C.

Aronson, A.I. (1993b) The two faces of *Bacillus thuringiensis*; insecticial proteins and post-exponential survival. Molec. Microbiol. 7, 489-496.

Ash, C., Farrow, J.A.E., Wallbanks, S. and Collins, M.D. (1991a) Phylogenetic heterogeneity of the genus *Bacillus* revealed by small-subunit-ribosomal RNA sequences. Letts. Appl. Microbiol. 13, 202-206.

Ash, C., Farrow, J.A.E., Dorsche, M. Stackebrandt, E. and Collins, M.D. (1991b) Comparative analysis of *Bacillus anthracis*, *Bacillus cereus* and related species on the basis of reverse transcriptase sequencing of 16S rRNA. Int. J. Syst. Bacteriol. 41, 343-346.

Ash, C. and Collins, M.D. (1992) Comparative analysis of the 23S ribosomal RNA gene sequences of *Bacillus anthracis* and emetic *Bacillus cereus* determined by PCR-direct sequencing. FEMS Microbiol. Lett. 94, 25-80.

Baumann, L., Broadwell, A. H. and Baumann, P. (1988). Sequence analysis of the mosquitocidal toxin genes encoding 51.4- and 41.9- kilodalton proteins from *Bacillus sphaericus* 2362 and 2297. J Bacteriol. 170, 2045-2050.

Baumann, P., Clark, M.A., Baumann, L. and Broadwell, A.H. (1991) *Bacillus sphaericus* as a mosquito pathogen: properties of the organism and its toxins. Microbiol. Rev. 55, 425-436.

Berry, C., Jackson-Jap, J., Oei, C. and Hindley, J. (1989) Nucleotide sequence of two toxin genes form *Bacillus sphaericus* IAB59: sequence comparisons between five highly toxigenic strains. Nucleic Acids Res. 17, 7516.

Berry, C., Hindley, J., Erhardt, A.F., Grounds, T., de Souza, I. and Davidson, E.W. (1993) Genetic determinants of the host range of the *Bacillus sphaericus* mosquitocidal toxins. J. Bacteriol. 175, 510-518.

Bourque, S.E., Valero, J.R., Mercier, J. Lavoie, M.C. and Lavesque, R. (1993) Multiplex polymerase chain reaction for detection and identification of the microbial insecticide *Bacillus thuringiensis*. Appl. Environ. Microbiol. 59, 523-527.

Brosseau, R., Saint-Onge, A. Prefontaine, G., Masson, L. and Cabana, J. (1993) Arbitrary primed polymerase chain reaction is a powerful method to identify *Bacillus thuringiensis* serovars and strains. Appl. Environ. Microbiol. 59, 114-119.

Bull, A.T., Goodfellow, M. and Slater, H. (1992) Biodiversity as a source of innovation in biotechnology. Annu Rev. Microbiol. 46, 219-252.

Carlsen, C.R. and Kolst, A.B. (1993) A complete physical map of a *Bacillus thuringiensis chromosome*. J. Bacteriol. 175, 1053-1060.

Claus, D. and Berkeley, R.C.W. (1986) Genus *Bacillus* Cohn 1872, in "Bergey's Manual of Systematic Bacteriology", Vol. 2. (Sneath, P.H.A., Mair, N.S., Sharple, M.E. and Holt, J.G., Eds.) pp 529-550. Williams and Wilkins, Baltimore.

Cohan, F.M., Roberts, M.S. and King, E.C. (1991) The potential for genetic exchange by transformation within a natural population of *Bacillus subtilis*. Evolution 45, 1383-1421.

Coombs, R. (1993) *Bacillus thuringiensis* the success story continues. Agro. Food Ind. Hi-Tech 6, 7-9.

de Barjac, H., Véron, M. and Cosmao Dumanoir, V. (1980) Characterisation biochemique et sérologique de souches de *Bacillus sphaericus* pathogéns ou non pour les moustiques. Ann. Microbiol. (Inst. Past.) 131B, 191-201.

de Barjac, H. and Frachon, E. (1990) Classification of *Bacillus thuringiensis*. Entomophaga 35, 233-240.

Devereux, J. P., Haeberli, P and Smithies, D. (1984) A comprehensive set of sequence programs for the VAX. Nucleic Acids Res. 12, 387-395.

Ely, S. (1993) The engineering of plants to express *Bacillus thuringiensis* δ-endotoxin, in "*Bacillus thuringiensis*, an Environmental Pesticide" (Entwistle, P.F., Cory, J.,S., Bailey, M.J. and Higgs, S., Eds.), pp. 105-124. Wiley, Chichester.

Farrow, J.A.E., Ash, C., Wallbanks, S. and Collins, M.D. (1992) Phylogenetic analysis of the genera *Planococcus*, *Marinococcus* and *Sporosarcina* and their relationships to members of the genus *Bacillus*. FEMS Microbiol. Letts. 93, 167-172.

Favret, M.E. and Yousten, A.A. (1985) Insecticidal activity of *Bacillus laterosporus*. J. Invert. Pathol. 45, 195-203.

Feitelson, J.S., Payne, J. and Kim, L. (1992) *Bacillus thuringiensis*: insects and beyond. Biotechnology 10, 271-275.

Felsenstein, J. (1993) PHYLIP (Phylogeny Inference Package) version 3.5c. Department of Genetics, University of Washington, Seattle.

Ferré, J., Real, M.D., Van Rie, J., Jansens, S. and Peferoen, M. (1991) Resistance to the *Bacillus thuringiensis* bioinsecticide in a field population of *Plutella xylostella* is due to a change in a midgut membrane receptor. Proc. Natl. Acad. Sci., USA 88, 5119-5123.

Fox, G.E., Wisotzkey, J.D. and Jurtshuk, Jr., P. (1992) How close is close: 16S rRNA sequence identity may not be sufficient to guarantee species identity. Int. J. Syst. Bacteriol. 42, 166-170.

Geiser, M., Schweitzer, S. and Grimm, C. (1986) The hypervariable region in the genes coding for entomopathogenic crystal proteins of *Bacillus thuringiensis*: nucleotide sequence of kurdh1 gene of subspecies *kurstaki* HD1. Gene 48, 109-118.

Gonzales, J. M., Jr., Brown, B.S. and Carlton, B.C. (1982) Transfer of *Bacillus thuringiensis* plasmids coding for δ-endotoxin among strains of B. *thuringiensis* and B. *cereus*. Proc. Natl. Acad. Sci. USA 86, 4037-4041.

Gordon, R.E., Haynes, W.C. and Pang, C.H. (1973) "The Genus *Bacillus*". Agricultural Handbook No. 427. US Department of Agriculture, Washington, D.C.

Green, C.J. and Vold, B.S. (1993) tRNA, tRNA processing and aminoacyl-tRNA synthetases. In "*Bacillus subtilis* and Other Gram Positive Bacteria. Biochemistry, Physiology and Molecular Genetics" (Sonenshein, A.L., Hoch, J.A. and Losick, R., Eds.) pp. 683-698. American Society for Microbiology, Washington D.C.

Guerineau, M., Alexander B. and Priest, F.G. (1991) Isolation and identification of *Bacillus sphaericus* strains pathogenic for mosquito larvae. J. Invert. Pathol. 57, 325-333.

Hougard, J.M. and Back, C. (1992) Perspectives on the bacterial control of vectors in the tropics. Parisitol. Today 8, 364-367.

Jaquet, F., Hütter, R. and Lüthy, P. (1987) Specificity of *Bacillus thuringiensis* delta-endotoxin. Appl. Environ. Microbiol. 53, 500-504.

Jarvis, E.D., Widom, R.L., La Fanci, G., Setoguchi Y., Richter, I.R. and Rudner, R. (1988) Chromosomal organization of rRNA operons in *Bacillus subtilis*. Genetics 120, 625-635.

Kaji, D.A., Rosato, Y.B., Canhos, V.P. and Priest, F.G. (1994) Characterisation by polyacrylamide gel electrophoresis of whole-cell proteins of some *Bacillus thuringiensis* subsp. *israelensis* strains isolated in Brazil. Syst. Appl. Microbiol. In Press.

Kellen, W.R., Clark, T.B., Lindegren, J.E., Ho, B.C., Rogoff, M.H. and Singer, S. (1965) *Bacillus sphaericus* Neide as a pathogen of mosquitoes. J. Invert. Pathol. 7, 442-448.

Knowles, B.H. and Ellar, D.J. (1987) Colloid osmotic lysis is a general feature of the mechanism of action of *Bacillus thuringiensis* δ-endotoxins with different insect specialities. Biochim. Biophys. Acta 924, 509-518.

Krych, V., Johnson, J.L. and Yousten, A.A. (1980) Deoxyribonucleic and homologies among strains of *Bacillus sphaericus*. Int. J. Syst. Bacteriol. 30, 476-484.

Lecadet, M.M. and Mantouret, D. (1987) Host specificity of the *Bacillus thuringiensis* δ-endotoxins toward lepidopteran species: *Spodoptera littoralis* Bdv. and *Pieris brassicae*. J. Invert. Pathol. 49, 37-48.

Lereclus, D., Delécluse, A. and Lecadet, M.M. (1993) Diversity of *Bacillus thuringiensis* toxins and genes, in "*Bacillus thuringiensis*, an Environmental Biopesticide : Theory and Practice (Entwistle, P.F., Cory, J.S., Bailey, M.J. and Higgs, S. Eds.) pp. 37-70. Wiley, Chichester.

Lewis, L.O., Yousten, A.A. and Murray, R.G.E. (1987) Characterisation of the surface protein layers of the mosquito-pathogenic stains of *Bacillus sphaericus* . J. Bacteriol. 169, 72-79.

Liu, J.W., Hindley, J., Porter, A.G. and Priest, F.G. (1993) New high-toxicity mosquitocidal strains of *Bacillus sphaericus* lacking a 100-kilodalton-toxin gene. Appl. Environ. Microbiol. 59.

Maiden, M.C.J. (1993) Population genetics of a transformable bacterium : the influence of horizontal genetic exchange on the biology of *Neisseria meningitidis*. FEMS Microbiol. Letts. 112, 243-250.

May, R.M. (1993) Resisting resistance. Nature 361, 593-594.

Margalit, J. and Dean, D. (1985) The story of *Bacillus thuringiensis* var. *israelensis* (B.t.i.). J. Am. Mosq. Cont. Assoc. 1, 1-7.

McGaughey, W.H. and Whalon, M.E. (1992) Managing resistance to *Bacillus thuringiensis* toxins. Science 258, 1451-1455.

Miteva, V., Abadjieva, A. and Grigorova, R. (1991) Differentiation among strains and serotypes of *Bacillus thrungiensis* by M13 DNA fingerprinting. J. Gen. Microbiol. 137, 593-600.

Nagel, M. and Anderson, J.R. (1992) Utilisation of organic acids and amino acid by species of the genus *Bacillus*: a useful means in taxonomy. J. Basic Microbiol. 32, 91-98.

Nakamura, L.K. (1994) DNA relatedness among *Bacillus thuringiensis* serovars. Int. J. Syst. Bacteriol. 44, 125-129.

Nicholas, L., Haman, S., Frachon, E., Sibald, M. and de Barjac, H. (1990) Partial inactivation of the mosquitocidal activity of *Clostridium bifermentans* serovar. malaysia by extracellular proteases. Appl. Microbiol. Biotech. 34, 36-41.

Nicolas, L.C., Nielsen-Leroux, C., Charles, J.F. and Delécluse, A. (1993a) Respective roles of the 42- and 51- kDa components of the *Bacillus sphaericus* toxin over-expressed in *Bacillus thuringiensis*. FEMS Microbiol. Lett. 106, 275-280.

Nicolas, L., Charles, J.F. and de Barjac, H. (1993b) *Clostridium bifermentons* serovar. malaysia: characterisation of putative larvicidal proteins. FEMS Microbiology Lett. 113, 23-28.

Oei, C., Hindley, J. and Berry, C. (1992) Binding of purified *Bacillus sphaericus* binary toxin and its deletion derivatives to *Culex quinquefasciatus* gut: elucidation of functional binding domains. J. Gen. Microbiol. 138, 1515-1526.

Ohba, M., Yu, Y.M. and Aizowa, K. (1988) Occurrence of non-insecticidal *Bacillus thuringiensis* flagellar serotype 14 in the soil of Japan. Syst. Appl. Microbiol. 11, 85-89.

Petcharat, J. (1991) Toxicity of *Bacillus sphaericus* strain 2362 on *Mansonia* spp. larvae. Southeast Asian J. Trop. Med. Public Health 22, 429-435.

Porter, A.G., Davidson, E.W. and Liu, J.W. (1993) Mosquitocidal toxins of *bacilli* and their genetic manipulation for effective biological control of mosquitoes. Microbiol. Rev. 57, 838-861.

Priest, F.G., Goodfellow, M., Todd, C. (1988) A numerical classificaiton of the genus *Bacillus*. J. Gen. Microbiol. 134, 1847-1882.

Priest, F.G., Kaji, D.A., Rosato, Y.B. and Canhos, V.P. (1994) Characterisation of *Bacillus thuringiensis* and related bacteria by ribosomal RNA gene restriction fragment length polymorphisms. Microbiology. In Press.

Rawlins, S.C. (1989) Biological control of insect pests affecting man and animals in the tropics. CRC Crit. Rev. Microbiol. 16, 235-249.

Raymond, M., Callaghan, A., Fort, P. and Pasteur, N. (1991) Worldwide migration of amplified insecticide resistance genes in mosquitoes. Nature 350, 151-153.

Russell, B.L., Jelley, S.A. and Yousten, A.A. (1989) Carbohydrate metabolism in the mosquito pathogen *Bacillus sphaericus*. Appl. Environ. Microbiol. 55, 294-297.

Singer, S. (1988) Clonal populations with special reference to *Bacillus sphaericus*. Adv. Appl. Microbiol. 33, 47-74.

Smith, N.R., Gordon, R.E. and Clark, F.E. (1946) "Aerobic Sporeforming Bacteria". US Misc. Publ. No. 559. US Department of Agriculture, Washington D.C.

Smith, N.R., Gordon, R.E. and Clark, F.E. (1952) "Aerobic Spore-Forming Bacteria". Agriculture Monograph No. 16. United States Department of Agriculture, Washington D.C.

Stahly, D.P., Andrews, R.E. and Yousten, A.A. (1992) The genus *Bacillus* - insect pathogens, in "The Prokaryotes", 2nd edition (Balows, A., Trüper, H.G., Dworkin, M., Harder, W. and Schleifer, K.H., Eds.) pp. 1697-1745. Springer Verlag, New York.

Thanabalu, T., Hindley, J., Jackson-Jap, J. and Berry, C. (1991) Cloning sequencing, and expression of a gene encoding a 100-kilodalton mosquitocidal toxin from *Bacillus sphaericus*. J. Bacteriol. 173, 2776-2785.

Thanabalu, T., Hindley, J., and Berry, C. (1992) Proteolytic processing of the mosquitocidal toxin from *Bacillus sphaericus* SSII-I. J. Bacteriol. 174, 5051-5056.

Thanabalu, T., Hindley, J., and Berry, C. (1993) Cytotoxicity and ADP-ribosylating activity of the mosquitocidal toxin from *Bacillus sphaericus* SSII-1: possible roles of the 27- and 70- kilodalton peptides. J. Bacteriol. 175, 2314-2320.

Thiery, I. and deBarjac, H. (1989) Selection of the most potent *Bacillus sphaericus* strains based on activity ratios determined on three mosquito species. Appl. Microbiol. Biotechnol. 31, 577-581.

Thorne, C.B. (1993) *Bacillus anthracis*. In "*Bacillus subtilis* and Other Gram-positive Bacteria: Biochemistry, Physiology and Molecular Genetics" (Sonenshein, A.L., Hoch, J.A. and Losick, R., Eds.) pp. 113-124. American Society for Microbiology, Washington D.C.

Walsh, J. (1986). River blindness, a gamble pays off. Science 232, 922-925.

Wayne, L.G., Brenner, D.J., Colwell, R.R., Grimont, P.A.D., Kandler, P., Krichevsky, M.I., Moore, L.H., Moore, W.E.C., Murray, R.G.E., Stackebrandt, E., Starr, M.P., and Trüper, H.G. (1987) Report of the *ad hoc* committee on reconciliation of approaches to bacterial systematics. Int. J. Syst. Bacteriol. 37, 463-464.

Yousten, A.A. (1984) Bacteriophage typing of mosquito pathogenic strains of *Bacillus sphaericus*. J. Invert. Pathol. 43, 124-125.

Yousten, A.A., Fretz, S.B. and Jelley, S.A. (1985) Selective medium for mosquito-pathogenic strains of *Bacillus sphaericus*. Appl. Environ. Microbiol. 49, 1532-1523.

Zahner, V., Momen, H., Salles, C.A. and Rabinovitch, L. (1988) A comparative study of enzyme variation in *Bacillus thuringiensis*. J. Appl. Bacteriol. 67, 275-282.

INDUSTRIES REQUIREMENTS WITH REGARD TO IDENTIFICATION
OF BACTERIA

Hanne Gürtler and Lisbeth Anker

Microbiology
Novo Nordisk A/S
Novo Allé
DK-2880 Bagsvaerd
Denmark

INTRODUCTION

Microorganisms constitute a rich source of diverse chemical compounds with versatile activities, of which many have found commercial application in the health care sector, agriculture and food/feed sector and in the chemical sector.

Although microbial taxonomy plays an important role in the industrial exploitation of microbial products, the importance to classify and identify microorganisms has often been a neglected task in industrial biotechnology.

The existing taxonomic systems seem inadequate for a thorough classification due to the large number of apparently new microbial taxa being discovered. Furthermore they often include methods which are very time-consuming and not very reliable. These systems therefore do not meet the demands from legal authorities and industry requiring unambiguous and fast-working systems. Useful identification systems are a prerequisite for exploring the phylogenetic and metabolic diversity of microorganisms in nature and for the development of new biotechnological applications therefrom.

The aim of this article is to give an overview of industries' requirements for identification and characterization of bacterial cultures. In this context the need for reliable and standardized identification methods is discussed. The use of taxonomic tools in industrial isolation and screening programmes and the need for taxonomists in future are also touched upon.

REQUIREMENTS FOR IDENTIFICATION

As illustrated in Figure 1, identification and characterization of microorganisms are important both in the discovery, the development and manufacturing of microbial products.

A reliable identification is required by the authorities for safety evaluation of the producing organism and for the approval of the final product.

Furthermore identification and characterization of microorganisms play an important role in many phases during the discovery, development, and manufactoring processes of a microbial product e.g. industrial isolation and screening programmes, patent applications, quality control and process validation. In countries with strict quarantine regulations a reliable identification of microorganisms also becomes important in connection with import of cultures.

The demands in the individual phases may vary depending on the type of organism, the type of product and the application of the final product.

The requirements for identification in the discovery, development, and manufacturing of bacterial products respectively are summarized below.

Characterization of Cultures for Patent Purposes

When a new product has been discovered a patent has to be procured to protect the intellectual property. The patent has to cover the source of the product as well as the description of the process. A characterization of the producing organism is necessary. This description has to be an enabling one, meaning that when the patent is issued and the culture becomes available the recipient of the culture has to be able to cultivate the culture and observe the same characteristics as are cited in the patent. The microorganism has to produce the product for which the claims were made. Therefore it is obvious to include a thorough description of the microorganism in the patent. The culture has to be deposited in a culture collection recognised by the patent office.

Industry is very concerned with accurate descriptions of cultures for patent applications as an accurate description always will strengthen the final patent.

It is, however, difficult to make such an accurate identification, because there are no standard procedures for identification of microorganisms including bacteria. There exists

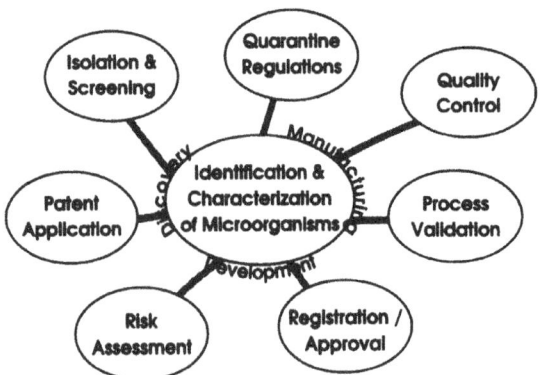

Figure 1. Importance of taxonomy in the exploitation of microbial products

no agreement among industrial taxonomists on what procedures should be followed. The differences in the view of the methods in biosystematics are probably even greater between the academic and industrial taxonomists. This inconsistency in methods and viewpoints makes it extremely difficult for industry to carry out proper and reliable searches in literature or to compare results obtained in different laboratories.

It is quite clear that a consistent use of a common standard in the identification and characterization of microorganisms would be of great help in evaluating the novelty of an invention and in ensuring stronger and more clear patent claims.

Safety of Producing Organism

The vast majority of microorganisms is harmless. Some microorganisms, however, are known as pathogens to man, animals, or plants.

From a safety point of view, it is extremely important to exclude candidates from industrial development and application which could be harmful to man or the environment.

The European Federation of Biotechnology (EFB) has accepted a classification system of aetiological agents including 4 risk groups of medical concern as shown in Table 1. (Küenzi, 1985). It is recommended that no microorganisms belonging to a risk group above 1 should be candidates for large scale production (Frommer *et al.*, 1989, Collins, 1992)

It is therefore extremely important to have an identification sufficient to exclude pathogens but to allow commercial and valuable bacteria to be developed even when they are very much alike. Knowledge about the relationships of different organisms is also very important as this influences on the demands for characterization. *Bacillus anthracis/ Bacillus cereus/Bacillus thuringiensis* represents a group of very close related bacteria (Ash *et al.*, 1991) which includes both highly pathogenic species and nonpathogenic species of considerable commercial value (Frommer, 1989). In such a case reliable methods are required to distinguish between these species

When authors of scientific papers are describing microorganisms as pathogenic strains, it is of major importance that the organisms are described in a proper way both using reliable methods and proof of the pathogenicity of the particular organisms. A wrong conclusion of either the identity of the organism or the pathogenicity may unreasonably result in exclusion of an organism and phylogenetically related species from commercial exploitation.

A major part of the risk evaluation in biotechnology relies on a proper identification of a microorganism. The risk evaluation of the microorganism includes safe handling in the laboratory, pilot plants or production facilities. Furthermore, unintended environmental release or outdoor application are also included in the risk evaluation. The approval of the final product is dependent on a regulatory acceptable organism.

Quality Control

For the quality control of industrial products, especially in the pharmaceutical industry, bacterial identification plays an important role. The industry has to guarantee that its products do not contain any pathogens and have a high hygienic standard. To ensure this, microbial monitoring of raw materials, process intermediates and the final product is necessary.

Normally the procedures involve the monitoring of pathogens such as *Salmonella, Listeria, Staphylococcus, Bacillus cereus* and enteropathogenic *E. coli.*

For the different steps in quality control, several types of both pathogenic and non pathogenic bacteria are monitored.

Table 1. Relationships of EFB-classification of microorganisms according to risk and OECD safety precautions

EFB-Class		Safety Precautions
1.	Harmless	GILSP
2.	Low	Containment Category 1
3.	Medium	Containment Category 2
4.	High	Containment Category 3

Source: Frommer *et al.* (1989)

Detection of organisms such as streptococci, *E. coli,* and other coliform bacteria are part of the evaluation of the hygienic standard. The organisms and the methods prescribed by the authorities, however, vary from country to country. Practically this means that the industry has to use several different methods to monitor one particular organism causing a costly extra effort for the industry to fulfil the different regulations. Furthermore the methods prescribed by the authorities are not regularly updated and are therefore not necessarily the most appropriate ones. Classical methods based on morphological and biochemical properties (API, Biolog etc.) are often used (FDA, 1984). However, immunological methods e.g. enzyme-linked immunosorbent assays, ELISA, are being increasingly used today.

For the industrial microbiologists involved in quality control work it would be desirable to have simple, rapid and reliable methods. The sensitivity of the methods also needs to be high, as it should be possible to detect for example, one *Salmonella* cell per 25 g of product. To achieve this, reliable fingerprinting methods would be very useful.

From an environmental point of view, it is in some cases important to check the manufacturing surroundings for an unintentional release of the producing organisms.

Process Validation

Process validation is yet another area where detection, and to a certain extent identification, of microorganisms play an important role. The cleanliness and sterility of the process have to be checked for instance by proving absence of specific organisms for example *Salmonella.* Moreover, a process validation procedure often prescribes requirement for securing that the identity of the producing organism is exactly the same from one process run to the next.

Quarantine Regulations

Some countries have strict regulations on the import of microbial cultures or materials for isolation.

An import permission has to be obtained from the authorities. Organisms that are not indigenous to the country in question and represents a threat to plant production, man or animals can not be imported without special permission. Industry often needs to exchange cultures with subsidiaries or collaborating partners around the world. Rapid identification methods to exclude organisms that are considered to represent a risk would be not only very useful, but probably are a necessity in the future.

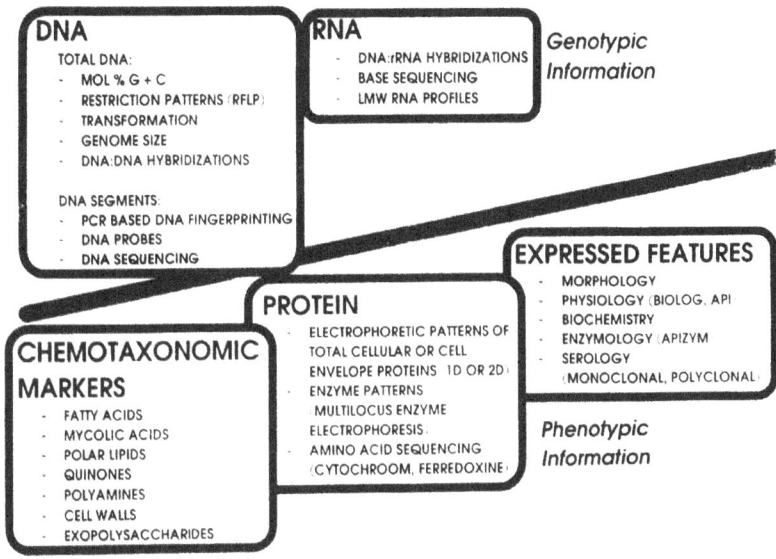

Figure 2 Genotypic and phenotypic information on bacterial taxa
Source: J. Swings: IUBS, IUMS, Scope Workshop, August 1993

NEED FOR RELIABLE AND STANDARDIZED IDENTIFICATION METHODS

As indicated above, a reliable identification is important to the industry as the identity has great impact on the evaluation and exploitation of a microbial product. However, there is no consensus on the criteria for identification and there are no minimum descriptive characteristics for use in patent applications or for publications in scientific journals. As a consequence there is no consensus in the requirements and prescriptions by the authorities.

It would be of great help both, for the industry and the authorities, if more standardized methods were used by all industries and academia in identifying bacteria.

Increasing amounts of genotypic and phenotypic information are available on bacterial taxa (Figure 2).

Today there exist many methods for characterizing bacteria based on either genotypic information on DNA or RNA (e.g. DNA base composition, restriction patterns, DNA hybridization, sequencing and DNA probes) or phenotypic information on proteins (electrophoretic protein patterns and electrophoretic enzyme patterns), chemotaxonomic markers (e.g. fatty acids, mycolic acids and quinones) and on expressed features (morphology and metabolism). As new techniques are being developed and applied, new information for characterization is obtained. Indeed, the final aim of characterization is never reached because continuing progress in scientific and technological methods allows the study of an even increasing number of characters or properties of a species. This continuous progress results in a higher reliability in identification.

Today the classification of bacteria reflects phylogeny. In the second edition of the Prokaryotes the old determinative classification is giving way to a phylogenetic system (Woese, 1992).

Phylogenetic relations have been assessed at all levels by analysis of rRNA or rRNA genes. A significant finding of the rRNA work and DNA-DNA reassociation studies was to point out the genetic heterogeneity of many phenotypically defined genera (Stacke-brandt, 1992). There is no consensus, however, on which system to use for taxonomic purposes including identification. The taxonomists are split into three groups (Stackebrandt, 1992).

The first group prefers the existence of two systems, the traditional one for practical purposes and another purely phylogenetic one for other than taxonomic purposes.

A second group of taxonomists would like to establish a system based on molecular grounds, visualizing the definition of taxa based exclusively on signature oligonucleotides. (Woese *et al.*, 1985).

The third group of taxonomists is searching for an integrated system in which the ranks of phylogenetically defined groups are circumscribed by phenotypic data (Murray *et al.*, 1990).

A polyphasic approach including phenotypic and phylogenetic data has been recommended for characterization and identification of bacteria as differences in evolutionary rates in various groups of organisms and other considerations prevent the use of phylogenetic parameters alone (Trüper and Schleifer, 1992). The taxonomic resolution of currently used techniques is summarized in Figure 3. The use of the different techniques in identification of bacteria has been summarized by Trüper and Schleifer (1992).

The first step in the identification of bacteria is to try to assign organisms to genera. Subsequently the organisms have to be assigned to the species level.

Speciation in many bacterial genera has become extremely difficult because of the lack of standardized methods and availability of reliable keys. Thus from an industrial

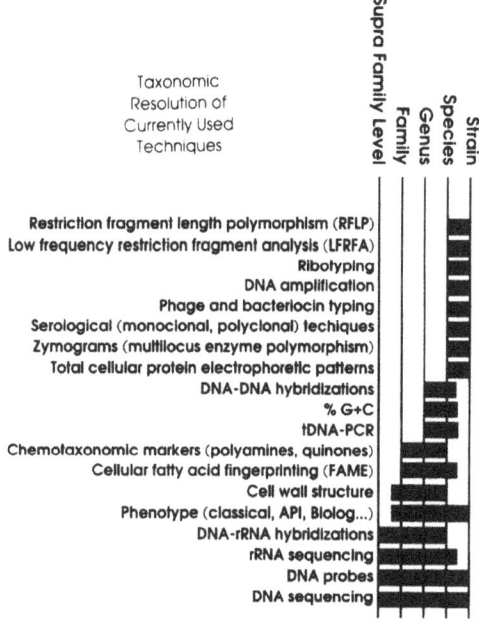

Figure 3. Taxonomic resolution of currently used techniques.
Source: J. Swings: IUBS, IUMS, Scope Workshop, August 1993.

point of view genus and species are those ranks for which proper descriptions are needed most urgently.

For some groups of organisms used by industry, however, it is also necessary to identify the organisms to the subspecies level and in some cases even to the infra-subspecific level, for example serovar and pathovar for strains of *Bacillus thuringiensis*.

With so many different methods available today and no consensus on the criteria used for identification it is very difficult to meet industries' requirements for identification. These can only be met if more standardized methods for identification and characterization of bacteria are developed. The methods should be reliable, reproducible and rapid. From an industrial point of view, as indicated previously, the identity of an organism has great impact on the evaluation and exploitation of a microbial product. It is therefore important that the selected methods are reliable. It is also important that the results are reproducible. Standardized procedures need to be developed so that results obtained in different laboratories can easily be compared.

Rapid methods would appeal to the industrial laboratories that perform their own identification. The advantage of having data available in a short time and needing fewer people to run the tests would support the cost of buying kits or other materials and instruments.

Some companies do not have in-house taxonomists and rely on the experts at the culture collections or in academia to identify their cultures. The identification service conducted by the culture collections or other institutions could be improved by applying faster identification schemes.

Methods are necessarily going to be dependent on the microorganisms in question. The scope of identification is so broad that the issue has to be approached within related groups of microorganisms. For every group it would be helpful to have minimum descriptive characteristics for publication. This would allow everyone to run the same tests and be able to compare results with those published for other strains.

Some of these suggestions are lengthy research projects but might be accomplished by a cooperative endeavour.

A coordination committee could be set up with members from universities and industries.

The aims of the coordination committee could be the development of

1) Consensus on the criteria for identification.
2) Rapid and reliable identification methods.
3) Standardized identification procedures.
4) Minimum descriptive characteristics for publication.
5) Recommendations to the authorities.

A workshop for industrial microbiologists should be arranged to define the problems and needs of the industry for identification and characterization of microorganisms including bacteria. A workshop could subsequently be arranged for taxonomists from both academia and industry to analyse the potential of the various methods and to prepare a plan for the development of more standardized methods.

It is important to ensure increased funding of research projects that enable the expertise and resources of the leading laboratories in the area of microbial identification to focus on the development of standardized methods. One way to ensure this is by visualizing the needs to the public.

In order to promote transfer of technology between academia and industry there could be established Associated Industrial Research Groups for example in connection with the formation of Industrial platforms. Industrial platforms are suggested by the commission of the European Communities in connection with, for example the Bridge T-project on

"High Resolution Automated Microbial Identification". The Industrial platform is to provide a specialized forum where companies will meet on a regular basis with the research groups responsible for the implementation of the project to discuss common problems and formulate view points and actions in relation to R. & D.

One of the responsibilities of the subcommittees of the International Committee on Systematic Bacteriology, ICSB, is to produce minimal standards for description of the taxa assigned to them (Wayne *et al.*, 1991).

It is recommended that the ICSB prepare guidelines for preparation of these standards to ensure a uniform classification system for bacteria. Although technology and taxonomic concepts are changing rapidly it is hoped that such standards could be authorized. We encourage the ICSB and its subcommittees to play a more active role in advising authorities and industry in the area of identification and characterization of bacterial cultures. It would also be desirable if the ICSB could play a more active role in the development of standardized methods for the identification and characterization of bacteria. This needs to be done on an international basis.

APPLICATION OF TAXONOMIC TOOLS IN INDUSTRIAL ISOLATION AND SCREENING PROGRAMMES

Today microorganisms are used in many different industrial applications as producers of e.g. drugs, enzymes, vitamins, food colourings, flavourings, fuels and acids. Furthermore the manufacturing of many food products such as cheese and bread also involves exploitation of microorganisms.

It seems that the discovery of new potential industrial applications of microorganisms to a large extent depends upon the discovery of new types of microorganisms and exploitation taken of their primary or secondary metabolism. Nature can be considered as an almost inexhaustible source of new microorganisms. The application of modern techniques of molecular biology is enabling the detection of hitherto completely unknown groups of microbes and also is revealing the extent of genetic diversity within microbial taxa. The latest estimates for fungi and bacteria add up to around 1.5 million species for fungi and 3 millions species of bacteria (Hawksworth, 1992, Hawksworth and Calwell, 1992).

Only a minor part of the total microbial population, less that 5% of the probable number of species has been revealed until now (Nisbet and Fox, 1991) and the potential for identification of new biotechnological uses of microorganisms must be considered as enormous.

The success of industrial screening programmes depends on both the number and the diversity of the organisms examined. Today many companies are focusing on new and rare bacteria in order to increase the possibilities of discovering new metabolites. Concurrently with this development rapid taxonomic tools have become more important in industrial isolation and screening programmes (Figure 4)

The selection of soil samples and other materials for isolation is normally based on ecological parameters. The development of methods for direct detection *in situ* of particular groups of microorganisms gives the microbiologists an additional tool to select samples for isolation.

The use of probes based on 5S, 16S, and 23S rRNA together with the isolation, separation and sequencing of RNA molecules have led to the establishment of new phylogenetic relationships. By choosing the appropriate probe for a particular rRNA

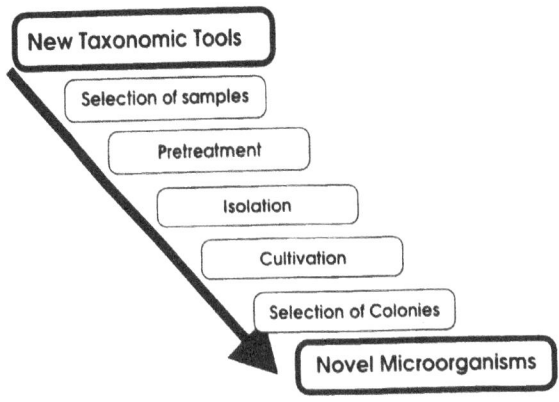

Figure 4. Important steps in the isolation of new microorganisms.

sequence, one can analyse the members of communities at different levels of resolution from kingdom to species (Stahl and Amann, 1991).

The direct extraction, purification and probing for 16S rRNA from soil allows the detection of specific target groups without prior cultivation or isolation. This approach may be of value to help overcome the difficulties associated with the poor culture ability of many organisms.

The sensitivity of the nucleic acid probe technique has been improved by amplification of target nucleic acids with the polymerase chain reaction (PCR). However, improved DNA/RNA extraction procedures which permit the removal of contaminants like humic acid are needed. An efficient protocol to avoid PCR false positives has to be developed at the same time.

DNA fingerprinting by restriction fragment length polymorphism (RFLP) analysis is another accepted technique for evaluating relationships between organisms, especially if they are closely related. The power of the method for revealing the diversity and distribution of particular organisms and particular properties has been demonstrated by several studies (Bull and Hardman, 1991).

Taxonomic information is also important in the development of selective isolation procedures for rare and novel microbes. Bull *et al.* (1992) suggested the generation of high quality databases that can be used to improve existing systems for microbial classification and identification and to formulate new media for the selective isolation of rare and novel microbes. The information in taxonomic databases can be used to formulate media select-ive for members of target taxospecies and for the search and discovery of additional active organisms following identification and the formulation of new selective media. By combin-ing the needs of industrial screening programmes and advances in microbial classification and identification, totally new selective media can be formulated to promote the growth of rare, uncommon and novel microorganisms.

Serological studies have previously been used to establish inter and intrageneric relationships amongst a range of bacterial genera. Polyclonal and monoclonal antibodies to cell surface epitopes can be used in combination with immunomagnetic beads e.g. Dynabeads® to selectively isolate targeted groups of bacteria (Christensen *et al.*, 1992).

But maybe even more important is the availability of rapid taxonomic methods for distinguishing between different bacteria and for selection of interesting groups and deselection of undesirable ones. It is rarely appreciated that the choice of microorganisms for industrial screening programmes (especially those with a low throughput) is primarily a problem of distinguishing between known organisms and of recognising novel ones. Methods allowing rapid detection, circumscription, and identification of novel and target organisms on primary isolation plates and the recognition of colonies that have arisen from identical propagules are increasingly important.

Novel microorganisms can be regarded either as those that can not be assigned to validly described taxa or as known organisms that have not been examined using in-house screening systems. These definitions of novelty presuppose the availability of reliable taxonomies and accurate identification procedures.

Different techniques have been used by industry to evaluate the diversity of the isolated cultures or to select targeted groups. Some companies have used expressed features such as morphology, API, Biolog, and metabolite profiles, while others have used fatty acid analysis or protein electrophoresis. Furthermore, nucleic acid probes have been used by some companies for rapid identification of bacteria. The feasibility of 16S rRNA specific probes or primers for selection of targeted groups of bacteria and for deselection of pathogenic isolates is being investigated in some industrial laboratories. However, for routine use in industrial isolation and screening programmes, a direct colony hybridization technique to select organisms of interest on isolation plates would be useful. DNA-DNA hybridizations have also been used by some companies for identification of bacteria.

Whole organism fingerprinting by pyrolysis mass spectrometry (PyMS) has been recommended as a fast method for the selection and rapid identification of large numbers of actinomycete strains in industrial screening programmes (Saddler *et al.*, 1988).

Rapid methods for detection and identification of novel and targeted organisms directly on isolation plates would improve the industrial screening programmes aiming at the discovery of new microbial products.

NEED FOR GOOD TAXONOMISTS ALSO IN THE FUTURE

At present there is a tendency for funding and training in biosystematics to decline, a tendency that is a major obstacle to both the assessment of biodiversity and industrial exploitation of the diversity. The reason for this decline is probably caused by the fact that the usefulness and contribution of taxonomy as a tool for successful exploitation of micro-organisms have not been acutely stressed (Boussienguet, 1991).

In the academic world, biosystematics has almost lost credibility and depends to a great extent on scientists having preference for specific groups of organisms. As a consequence the taxonomic skills are splendid in relatively few classes of microorganisms and for other classes totally missing.

From an industrial point of view the lack of skilled taxonomists and the decline in funding and training of new taxonomists must be considered next to a catastrophic. The need for such skills is very much in demand and increasingly soughts through the different phases in the development of new microbial products.

Hopefully, with the intensified global focus on biodiversity, joint efforts by academia and industry will have the necessary impact and importance to foster renewed interest in microbial systematics.

ACKNOWLEDGEMENT

We wish to thank Dr. J.J. Sanglier, Sandoz Pharma Ltd., Dr. P. Jeannin, Rhône-Poulenc Rorer, Dr. M. Jackson, Abbott Laboratories, Miss A. Dietz, Microtax, Dr. J.C. Hunter-Cevera, Blue Sky Research Service, Dr. L.H. Huang, Pfizer Inc., Dr. P.B. Fernandes, Bristol-Myers Squibb., for their contribution to this presentation.

REFERENCES

Ash, C., Farrow, J.A.E., Dorsch, M., Stackebrandt, E., Collins, M.D (1991) Comparative analysis of *Bacillus anthracis, Bacillus cereus,* and related species on the basis of reverse transcriptase sequencing of 16S rRNA. Int. J. Syst. Bacteriol. 41, 343-346.

Boussienguet, J. (1991) Problems of assessment of biodiversity, in "The Biodiversity of Microorganisms and Invertebrates: Its Role in Sustainable Agriculture" (Hawksworth, D.L., Ed.), pp. 31-35. CAB International, Surrey, UK

Bull, A.T. and Hardman D.J. (1991) Microbial diversity. Curr. Opin. Biotechnol. 2, 421-28.

Bull, A.T., Goodfellow, M. and Slater, J.H. (1992) Biodiversity as a source of innovation in Biotechnology. Ann. Rev. Microbiol. 46, 219-52.

Christensen, B., Torsvik, T., and Lien, T. (1992) Immunomagnetically captured thermophilic sulfate-reducing bacteria from North Sea oil field waters. Appl. Environ. Microbiol. 58, 1244-1248.

Collins, C.H. (1992) Hazard groups and containment categories in microbiology and biotechnology, in "Safety in Industrial Microbiology and Biotechnology", 2nd Edition. (Collins, C.H. and Beale, A.J., Eds.), pp. 23-33. European Pharmacopeia, Maisonneuve,

FDA Bacteriological Analytical Manual of the Division of Microbiology (1984) Center for Food Safety and Applied Nutrition, 6th Edition.

Frommer, W., Ager, B., Archer, L., Brunius, G., Collins, C.H., Donikian, R., Frontali, C., Hamp, S., Houwink, E.H., Küenzi, M.T., Krämer, P., Lagast, H., Lund, S., Mahler, J.L., Normand-Plessier, F, Sargeant, K., Tuijenburg Muijs, G., Vranch, S.P. and Werner, R.G. (1989) Safe biotechnology. III. Safety precautions for handling microorganisms of different risk classes. Appl. Microbiol. Biotechnol. 30, 541-552.

Goodfellow, M. (1977) Numerical taxonomy, in "CRC handbook of Microbiology, 2nd ed., Vol 1." (Laskin A.I. and Lechevalier H.A. Eds.), pp. 579-596. CRC Press, Cleveland,

Hawksworth, D.L. (1992) Microorganisms in global biodiversity. Status on the Earth's living resources (Groombridge B. Ed.), pp. 47-54. Chapman & Hall, London.

Hawksworth, D.L. and Colwell, R.R. (1992) Microbial diversity 21: Biodiversity Amongst Microorganisms and its Relevance, Biodiversity and Conservation 1, 221-226.

Küenzi, M., Assi, F., Cumiel, A., Collins, C.H., Donikia, M., Dominguez, J.B., Financsek, L., Fagarty, L.M., Frommer, F., Hasko, F., Hovland, J., Houwink, E.H., Mahler, J.L., Sandkvist, A., Sargeant, K., Sloover, C., Tuijenburg Muijs, G.(1985) Safe biotechnology-General considerations. Appl. Microbiol. Biotechnol. 21, 1-6

Murray, R.G.E., Brenner, D.J., Colwell, R.R., Vos, P. De, Goodfellow, M., Grimont, P.A.D., Pfennig, N., Stackebrandt, E. and Zavarzin, G.A. (1990) Report of the ad hoc committee on approaches to taxonomy within the proteobacteria. Int. J. Syst. Bacteriol. 40, 213-215.

Nisbet, L.J. and Fox, F.M. (1991) The importance of microbial biodiversity to biotechnology, in "The Biodiversity of Microorganisms and Invertebrates: Its Role in Sustainable Agriculture" (Hawksworth, D.L. Ed.), pp. 229-244. CAB International, Surrey, UK.

Saddler, G.C., Falconer, C., Sanglier, J.J. (1988) Preliminary experiments for the selection and identification of actinomycetes by pyrolysis mass spectrometry. Actinomycetologica 2, 53-54.

Stackebrandt, E. (1992) Unifying phylogeny and phenotypic diversity, in "The Procaryotes" 2nd Ed., (Balows, A., Trüper, H.G., Dworkin, M., Harder, W., Schleifer, K.H., Eds.), pp. 19-47. Springer-Verlag, New York.

Stahl, D.A., Amann. R. (1991) Development and application of nucleic acid probes, in "Nucleic Acid Techniques in Bacterial Systematics" (Stackebrandt E., Goodfellow, M., Eds.), pp. 205-48. Wiley, Chichester.

Trüper, H.G., Schleifer, K.H. (1992) Prokaryote characterization and identification, in "The Prokaryotes" 2nd Ed., (Balows, A., Trüper, H.G., Dworkin, M., Harder, W., Schleifer, K.H., Eds.), pp. 126-148. Springer-Verlag, New York.

Wayne, L.G. et al. (1991) Judicial commission of the International Committee on Systematic Bacteriology. Minutes of Meeting 14 September 1990, Osaka, Japan. Int. J. Syst. Bacteriol. 41, 185-187.

Woese, C.R. (1992) Prokaryote systematics: The Evolution of a science, in "The Prokaryotes" 2nd Ed. (Balows, A., Trüper, H.G., Dworkin, M., Harder, W., Schleifer, K.H., Eds.), pp. 3-18. Springer-Verlag, New York.

Woese, C.R., Stackebrandt E., Macke T.J., Fox G.E. (1985) A phylogenetic definition of the major eubacterial taxa. System. Appl. Microbiol. 6, 143-151.

PRESENT TRENDS AND FUTURE PROSPECTS FOR RAPID METHODS AND AUTOMATION IN THE CLINICAL LABORATORY

Kevin A. Feltham[1], and Michael Stevens[2]

[1]Feltham Associates Limited
Carlton House
Kibworth Hall Park
Kibworth Harcourt
Leicester LE8 0PE, U.K.
[2]Department of Health Evaluation Unit
Public Health Laboratory
Leicester Royal Infirmary
Leicester LE1 5WW, U.K.

INTRODUCTION

Clinical microbiology laboratories handle a wide range of specimen types from the human body, excretory products and the environment. These require detailed, local handling protocols for the initial detection, isolation and subsequent analysis of any potential pathogens or pathogenic products. Clinical microbiology laboratories have a responsibility to the patient for ensuring an accurate diagnosis of any microbiological effect and guidance towards the optimum treatment. The majority of microbes causing human illness are not immediately life threatening and time has always been on the clinical microbiologist's side. However, socioeconomic pressures for more cost effective techniques are increasing, as are pressures from preventative and epidemiological lobbies to identify the presence of specific pathogens at earlier stages and provide guidance for their optimum treatment.

The level of automation in the clinical laboratory has gradually increased over the past 20 years. Wholescale automation throughout the clinical microbiology laboratory, however, has not occurred in the United Kingdom (U.K.), indeed instrumentation has been partially successful in only a few specific areas such as immunology and serology where liquid handling equipment has been introduced. Clinical applications for mechanized robotic sample handling devices and serological optical readers have increased in popularity, particularly as the availability and quality of reagents have improved. The majority of serological applications are based around multi-well plastic plate technology. Instruments for automating sample screening, antibiotic sensitivity testing and bacterial identification became available for diagnostic medical microbiology during the 1970s but never really fulfilled their initial promise and advantage to the clinical laboratory except for blood culture screening.

The advent of the microprocessor age during the early 1980s introduced a wider variety of automation with many early models being upgraded with additional facilities (e.g. Bactec-NR range of non-radiometric equipment). The handling of the raw microbiological data has also been partially automated through the widescale use of laboratory computers. The principal uses of laboratory computers have included complex daybooks for routine reporting and fast enquiry of sample results, but sophisticated epidemiological and graphical models for reporting incidents of infection are now being described and reported. Direct electronic interfaces to automated sample analyzers greatly enhance the efficiency and effectiveness of the information disseminating role of the laboratory.

Despite a concerted effort by both industry and academia and the recent advances in molecular biology, U.K. clinical laboratories are still relying on traditional cultural methods in bacteriology. This paper discusses some of the attempts to make the work of U.K. microbiologists become more automated and some of the reasons for their apparent reluctance to utilize the new techniques.

Although a plethora of automated equipment has been introduced into the U.K. clinical laboratory market over the past 20 years, very few commercial systems have been successful and remained in widespread use over an extended period of time. The exceptions have been in the fields of immunology, serology, blood culture and antibiotic assay.

We have limited this review to two clinical applications which epitomize the advances in rapid methods: (1) sample screening for potential pathogens and (2) bacterial identification. These two fields of endeavour have seen considerable change over the past 10 years and are likely to demonstrate a continuation of this trend over the next 10 years. They epitomize the struggle by system developers to demonstrate accuracy and quality to the U.K. clinical microbiologist and the changes now being brought about by economic pressure rather than clinical pressure. We also discuss the impact of molecular biology and offer some views on the likely changes the new technology will bring to clinical microbiology.

SAMPLE SCREENING

In order to illustrate the key changes which have led to a renewed interest in automated sample screening in clinical microbiology, we have chosen two important body fluids commonly used to screen for potential pathogens: urine and blood. Both of these body fluids are usually sterile in healthy individuals and so the presence of micro-organisms, particularly in a pure state, is indicative of disease. In addition, urine provides a range of microscopical indicators of disease or infection of the urinary tract or kidney including red and white blood cells, epithelial cells and organic crystals. Microscopic examination of blood can provide evidence of parasitic infestation.

Urine Screening

Early attempts at reducing technician workload and making the service more cost-effective focused on urine analysis. U.K. clinical laboratories analyzing in excess of 300 urine samples a day are not uncommon. In recent years, several attempts have been made to automate and rapidly detect viable bacterial and somatic cells in urine samples. The instruments have employed a variety of techniques to indicate problems, including automated microscopy using acridine orange (Autotrak), particle counting (Orbec RAMUS), colorimetric filtration (Bac-T-Screen), the detection of adenosine triphosphate (ATP) by bioluminescence (Lumac and UTIscreen) and the limulus amoebocyte lysate (LAL) test (LeukoBact). It is interesting to note than no system has met with general acceptance in routine clinical U.K. laboratories and therefore, there still remains a need for an accurate, rapid, cost-effective system.

Questor (Difco Laboratories) is a new, automated system which has been designed to categorize and enumerate bacteria and other cells found in urine using novel particle counting technology. The major component of the system is a glass probe with a small orifice (≤ 100 μm) and two electrodes. An electrical field is established between the electrodes via the electrolyte in the orifice. The sample is drawn through the orifice, enters the electrical field where it displaces its own volume of electrolyte, causing a change in the impedance of the electrical circuit. The changes are detected as pulses, the number of which indicate the quantity of particles present in the sample. The height and area of the pulse is proportional to the mean spherical volume of the particle. A histogram of the particles and their sizes is formed and analyzed by the computer software to categorize the particles into different groups. This principle has been used to great effect in automating blood cell counting in clinical haematology laboratories. The novel feature of **Questor** is that the volume of the orifice can be varied by moving a tapered spike into and out of the orifice. Thus the characteristics of the orifice can be optimized for the size of the particles of interest (from 0.5 to 40 μm). The analyser records the volume distributions for four different particle size ranges corresponding to organisms, erythrocytes, leucocytes and epithelial cells.

The system has recently been clinically evaluated in the U.K. (Stevens *et al.*, 1993) and it was concluded that the performance of the system, in terms of sensitivity and specificity, was an improvement over that obtained from other automated bacteriuria screening systems evaluated to the same protocol. The system was easy to use, was not intrinsically hazardous and was described as a useful new method for cost-effectively screening for bacteriuria. The system could have many applications in wider microbiological fields.

Blood Culture Screening

The processing of blood cultures is one of the few areas in clinical microbiology where innovative methods, particularly involving some degree of automation, have become widely established in U.K. laboratories. Spencer (1988) reported a 50% increase in the prevalence of septicaemia in hospital patients between 1978 and 1988, i.e. from 10 to 15 cases per 1000 admissions. Automated systems have helped laboratories examine the associated increase in numbers of blood cultures submitted for investigation. Traditionally, blood cultures involve the incubation and regular examination of patients' blood samples which have been introduced into rich broth cultures. The media in the broth culture bottles can be modified to enhance the growth of aerobic and/or anaerobic organisms. Manual methods are extremely time consuming and are susceptible to contamination because of the need for regular monitoring for the presence of pathogens in the broth. Automated systems have helped U.K. laboratories handle more blood culture samples without relative staff increases. Once positive samples have been identified, traditional cultural techniques are used to characterize any pathogens and corresponding antibiotic resistance patterns.

The **Bactec** 460, using radiometric detection methods, and the NR-660 and NR-730 non-radiometric (NR) instruments, using infra-red spectroscopic detection of carbon dioxide have been the mainstay of automated blood culture technology since the early 1970s. Although all employ a significant degree of automation, they require continual manual attention and the specimens can only be tested at periodic intervals during the day. The **Bactec** NR-860 instrument has a greater degree of automation, but the blood culture vials or bottles must still be examined sequentially.

Many older systems, particularly the **Bactec** 460, suffered from being labelled as potentially hazardous because of the low levels of radioactivity used in the commercially-manufactured blood culture media. Laboratories who are now contemplating the purchase of their second, or even third, generation of blood culture screening equipment are able to choose from a growing variety of non-hazardous equipment and consumables which use continuous blood culture monitoring techniques.

Continuous Monitoring Systems. The new generation of blood culture systems have the advantage that the sample is continually monitored for indications of bacterial growth. The following systems are currently being sold in the U.K.: **Sentinel** (Difco, UK), **BacT/Alert** (Organon Teknika), **Bactec 9000** series - 9120 and 9240 (Becton Dickinson), **Vital** (bioMérieux), **OASIS** (Unipath), **ESP** (Difco USA), **RBC** [Rapid Blood Culture] (Baxter), and **Bio Argos** (Sanofi: Diagnostics Pasteur).

All systems include an integral incubator, usually employing agitation to enhance microbial growth, a continuously operational detection system which in turn provides signals to a computer, which uses algorithms to determine that growth has occurred. Any indication of microbial growth is transmitted to the operator using a number of aural and visual indicators and alarms.

The blood culture bottles are analysed for a period defined by the laboratory (normally five to seven days) and there is no requirement for subculturing negative samples at the end of the incubation period unlike manual or non-continuous systems. Many of the instruments have a stand-by capability in case of computer failure and have battery back-up for critical functions in case of complete power failure.

Although all these continuous systems have had a remarkably similar gestation period, the detection methods used vary considerably. A major change from manual or early mechanized systems is that almost all the continuous systems use a bacterial growth detection system which is an integral part of the blood culture bottle. **Sentinel** uses a novel bacterial detection method: a voltage is generated continuously in the bottle using a combination of dissimilar gold and aluminium metal electrodes. When organisms grow, the oxygen tension is reduced, the voltage falls, and the potential difference can be detected by the system. This event is reported to the instrument's computer and a printed report is produced. With some of the systems an alarm can be set off to attract the attention of a microbiologist who may choose to inform the patient's own clinician immediately with the preliminary information.

With the exception of **Sentinel**, all the other commonly available systems rely on the production of CO_2 as an indicator of microbial growth, but the methods used to detect this gas vary. The methods employed include: colorimetric sensors (**BacT/Alert**), increases in fluorescence (**Bactec 9000** series and **RBC**), decreases in fluorescence (**Vital**), monitoring gas pressure changes (**OASIS**), measurement of gas production and consumption (**ESP**) and detection of CO_2 by infra-red spectroscopy (**Bio Argos**).

Media. A range of commercially-produced culture media are available from the supplier of each system, including broths containing antibiotic-adsorbent resins, anti-microbial neutralizing agents such as sodium polyanetholsulphonate (SPS), bottles for paediatric patients and media specially formulated for the detection of fungi or yeasts.

Advantages. All the described systems make more efficient use of laboratory effort, enabling resources to be concentrated on the positive samples, while ignoring the negative ones. A feature common to most systems is a continuous "at a glance" facility for checking positive samples. A greatly reduced likelihood of contamination is ensured by minimizing the bottle handling. Once bottles have been entered into the system, there is no manual intervention by the operator until they are reported as positive or the incubation period (normally 5, 7 or 10 days) has been completed and the bottle discarded as negative, i.e. the original blood sample was considered sterile. Terminal subculture of negative bottles is considered unnecessary using these instruments. Sample identification and control are greatly enhanced using bar coding of bottles. This facility is available with several of the systems. All the systems produce printed reports and an electronic interface for communicating results to hospital or laboratory data handling computers.

Bactec instruments prior to the latest 9000 series used needles to pierce the septum in order to sample the CO_2 inside the bottle. It was also necessary to replace the gas samples, and therefore a bottled supply of gas was required. All the continuously monitored systems have non-invasive methods of detection and supplies of bottled gas are not required. The non-invasive methods also limit any potential intra-laboratory contamination.

Disadvantages. The capacity of systems can be a disadvantage because they all use modular incubation units which can be combined together in various configurations and controlled by the same computer. The smallest system is usually a single incubation drawer holding some 72 to 80 bottles, whereas the largest systems have capacities for 1200 bottles. For laboratories choosing a five day incubation period this size would give a maximum of 240 samples per day. Ten day incubation protocols would however only be able to handle 120 samples a day and for large U.K. hospital laboratories this would be insufficient and require more than one system.

Conclusions. The performance of the systems, in terms of detecting those organisms which cause septicaemia and bacteraemia has been investigated in comparative evaluations. Stevens *et al.* (1992) reported that the **Sentinel** continuous system had an overall performance comparable to the **Bactec** radiometric and non-radiometric mechanized methods. The sensitivity of the technology was considered equivalent to manual techniques.

All of the systems will run around the clock, 365 days per year but it is questionable whether laboratory staff are available at all times in U.K. laboratories to maximize the effectiveness of the information as soon as the system detects growth.

It is interesting to note that none of the continuous detection methods described above was predicted 10 years ago at a symposium dealing with the current state and future prospects of blood culture (Gould and Duerden, 1983).

BACTERIAL IDENTIFICATION

Traditional diagnostic methods are still widely used especially in developing countries. However, diagnostic kits for the identification of pathogenic microbes are continuing to be developed and some have become a part of the microbiologists' standard toolbox. Barrow and Feltham (1993), in collaboration with key workers in selected fields, have extensively revised and updated one of the standard reference works for U.K. medical bacteriologists: Cowan & Steel's *Manual for the Identification of Medical Bacteria*. This not only includes traditional diagnostic tables to aid technologists identify unknown pathogens but also chapters reviewing the technical aspects of kits and automated identification systems.

Manual Kits

The pioneering work of Buissière and Nardon (1968) on miniaturized, multi-test methods, from which the API system evolved, has led to the development of several off-the-shelf kits with specialized plastic strips and reagents for bacterial identification. The API strip is now as much a part of our vocabulary as the Gram's stain. Many other techniques have been devised and more recently several kits have been developed to enable automated interpretation of the results after incubation of the plastic strips. It is these systems of automated bacterial identification that we will discuss here.

Automated Identification Systems: Past

The first generation of automated identification systems included the **Autobac**, **Abbott MS-2** (latterly called the **Avantage**) and Abbott **Quantum II**. All had a degree of automation, in particular reading and interpretation of results was automatic. These systems can no longer be purchased but have been used in many European and North American laboratories

Automated Identification Systems: Present

In their review on automated systems, Stager and Davis (1992), concentrated on **Vitek**, **Sensititre**, **Walkaway (autoSCAN)**, **Aladin**, **Biolog** and the **MIDI** gas chromatography system. For this paper, we shall use some of these instruments to highlight the prospects for the future and underline the lessons learnt from the past. These instruments have been designed to handle pure cultures. Most of these systems use conventional tests such as increases in cell density or colour changes due to variations of pH. The exception is the **MIDI** system which analyzes cellular fatty acids using gas chromatography.

The **Vitek** modular instrument was introduced in the early 1970s following McDonnell Douglas' contract with the National Aeronautics and Space Administration (NASA) to build a system to automatically allow the monitoring of pathogens in astronauts' urine. The **Vitek** AutoMicrobic System (AMS) provides screening, identification and antimicrobial assay by the use of small playing-card sized plastic cards. Three standard identification cards have proved popular: the Gram-Negative Identification (GNI), Gram-Positive Identification (GPI) and Yeast Biochemical (YBC). Cards are automatically dosed, and moved manually to a separate reader-incubator. Readings are carried out automatically at 20 minute intervals and confirmed identifications are usually available within 4 to 18 hours.

Sensititre is a modular system incorporating a computer and automated reader. The system uses fluorescently-labelled substrates to enable both Gram-negative identification and antibiotic sensitivities for Gram-positive and Gram-negative organisms. Test plates are incubated off-line and may be read after 5 or 18 hours of incubation.

Both the **Walkaway and autoSCAN** instruments, from Baxter Diagnostics, are computer controlled and enable incubation of multi-well test plates and automatic interpretation of biochemical or susceptibility tests. Fluorogenic panels enable results to be obtained after 3.5 to 7 hours for antibiotic sensitivities and 15 to 42 hours for the identification of Gram-negative rods and Gram-positive cocci. Additional panels enable identification of *Neisseria*, *Haemophilus*, *Gardnerella*, anaerobes and yeasts.

The **Biolog** system was introduced in 1989 for identifying aerobic, Gram-negative bacteria by determination of 95 substrate carbon source utilization profiles. In 1992, Gram-positive profiles were added. Preliminary identifications can be obtained after 4 hours or after overnight incubation (16 to 24 hours) for interpretation of the ability of the bacteria to utilize the carbon source. The system is being used extensively as a rapid means of biotyping.

The Automated Tests for Bacteriology (**ATB**) instrument, available in Europe from bioMérieux sa, includes modular equipment for dispensing standardized inocula, reading and interpretation of results. Gram-negative rods, staphylococci and yeasts can be identified in 24 hours or using the rapid ID kits, 73 species of enterobacteria and 76 species of anaerobic bacteria can be identified in 4 hours. The system also reads and interprets antibiotic sensitivity panels.

Gas chromatography of cellular fatty acids is a rapid and reliable means of identifying organisms encountered in the clinical laboratory. The **MIDI** Microbial Identification System (MIS) is fully automated and can analyze more than 300 fatty acid methyl esters. The system has a large database containing profiles of yeats, anaerobic bacteria, and aerobic bacteria, including mycobacteria. The system was originally developed for agricultural laboratories for identifying environmentally important bacteria causing crop spoilage but additions to include a clinical database are now available and preliminary evaluations of the system show very promising results.

Some of these automated identification systems have been evaluated comparatively for Gram-negative rod (i.e. a combination of results for both *Enterobacteriaceae* and non-fermenting aerobes) identifications. The aggregated results from several evaluations and taken from the review by Stager and Davis (1992) are shown in Table 1.

Table 1. Performance of automated systems for the identification of Gram-negative aerobes

System	Number of		% of identifications that were:		
	Evaluations	Correct	Unclear	Not made	Incorrect
Vitek GNI	10	91.7	3.4	3.2	1.9
Sensititre	4	84.8	2.8	2.2	9.8
autoSCAN-4	3	86.2	10.4	1.3	2.1
Walkaway-96	7	77.9	14.3	0	7.8
Biolog GN	2 4h	46.6	6.2	36.1	11.1
	16h	58.2	11.1	21.0	9.7

Correct identifications for members of the *Enterobacteriaceae* were Vitek (95.8%), Sensititre (88.5%), autoSCAN-4 (90.2%), Walkaway-96 (84.8%) and Biolog (52% for 4 h and 60% for 16 h).

All these systems have proven capabilities but there is still room for improvement, including more rapid results and expanded ranges of micro-organisms identified. A major shortcoming of all these systems is that the organism must be cultured before the identification process can be performed.

Other approaches to the identification of micro-organisms include flow cytometry (Beckmann and Connolly, 1990), polyacrylamide gel electrophoresis (Pot *et al.*, 1993) and pyrolysis mass spectrometry (Larsson, 1991). These methods have shown potential for accurately identifying and differentiating between strains of pathogenic organisms, but are outside the scope or resources of routine U.K. clinical laboratories save for specialized reference investigations such as epidemiology or biotyping. Immunology and serology are utilizing the advances made in monocolonal and polyclonal antibody techniques (Payne *et al.*, 1988).

Considerable research effort has been expended on the development of procedures, both manual and automated, that can directly detect and even identify organisms in specimens using molecular biological techniques, such as monoclonal antibodies or nucleic acid probes. For some applications, such as brucellosis or meningitis in serum or cerebro-spinal fluids, the consequences of the infection and the target range of pathogens is sufficiently limited to make such a direct approach desirable and feasible. In other conditions, such as diarrhoea, where there are naturally high numbers of different commensals mixed with possible pathogens, it is not currently economically viable to utilize molecular techniques.

Nucleic Acid Probes

Probes for microbes causing sexually transmitted diseases and respiratory pathogens are used in many laboratories. Nucleic acid probes have been shown to be frequently less sensitive than traditional culture methods (Ossewaarde *et al.*, 1992; Hall, 1993) and the multi-stage molecular procedures themselves have tended to be labour-intensive and complex. There is a natural reluctance in routine U.K. clinical laboratories to use radioactivity because of the potential hazards and short shelf life. Where small numbers of bacteria are suspected, amplification techniques such as polymerase chain reaction (see below) appear to be more popular and effective in developed countries. Commercially available DNA probes for detecting of organisms in clinical samples include those for *Chlamydia trachomatis, Gardnerella vaginalis*, Group A streptococci, *Legionella pneumophila, Neisseria gonorrhoeae, Trichomonas vaginalis* and human papillomaviruses. Commercial culture confirmation assays have been developed for slow growing mycobacteria, thermophilic campylobacters, enterococci, *Listeria monocytogenes, Haemophilus influenzae* and various clinically important respiratory fungi.

Antimicrobial resistance probes covering beta-lactam, aminoglycoside, tetracycline, chloramphenicol, trimethoprim and heavy-metal resistance genes have been described. An exciting development has been described by Archer and Pennell (1990) for the *mec* resistance determinant, which encodes the protein responsible for most methicillin resistance in staphylococci. Problems have been found with chromosomally mediated ampicillin determinants that are inducible or become activated when patients receive therapy. A problem has been found when attempting to distinguish between clinically susceptible and resistant organisms. This negative probe attribute has prevented the widespread use of this technology as a replacement for clinical susceptibility testing but the methods will continue to be feasible for specific applications where organisms are present in normally sterile body sites or fluids and have relatively few modes of resistance.

Polymerase Chain Reaction (PCR)

Polymerase Chain Reaction (PCR) technology was introduced into clinical microbiology in the late 1980s. PCR has been deemed the 'diagnostic tool of the 1990s' and should, within a few years, be in use in most clinical microbiology laboratories in the developed world (Pallen and Butcher, 1991). The techniques have been discussed by others in these proceedings and so the detailed methodology has not been included here.

The application of PCR into clinical laboratories (Tables 2 and 3) has concentrated on those areas which traditionally presented a poor return on time and resources. PCR has immense potential in the rapid, sensitive and specific diagnosis of infection (Eisenstein, 1990). The whole process can be completed in a few hours, making same-day diagnosis possible and the sensitivity of the technique is comparable to or better than conventional microscopic or cultural methods. It is however, relatively expensive and dependent on the availability and quality of commercial primers. PCR kits have been developed for a broad range of organisms and these include methods for *Neisseria, Helicobacter, Chlamydia trachomatis, Mycobacterium tuberculosis* and the *Mycobacterium avium-Mycobacterium intracellulare* complex (MAC), hepatitis B, herpes simplex (HSV), HIV, and cytomegalovirus (CMV) (Schulz, 1993).

Chlamydia *Chlamydia* causes major problems in women, and *C. trachomatis* is a major contributor to causing the debilitating pelvic inflammation disease (PID). Moreover, chlamydial infection is a major cause of female infertility with over 120,000 cases per year throughout the European Community. Male patients presenting with non-gonococcal urethritis (NGU) show a high rate of co-infection with *C. trachomatis*. The majority of chlamydial infections in men (75%) stay asymptomatic. Effective screening and treatment of genito-urinary clinic attendees, diagnosed as infected with *Chlamydia*, would reduce the spread of the disease and save the higher costs of treatment later. Traditional cell culture methods are laborious and offer a sensitivity of between 70 and 90% in routine laboratories, with experienced laboratories rarely achieving a level of even 95% (Stamm, 1990). PCR demonstrates a sensitivity of over 96% in routine use. The critical population of infected patients not diagnosed correctly and subsequently inadequately treated is therefore considerably reduced. The effectiveness of early antibiotic treatment for *Chlamydia* is some 95%. The advantages of early diagnosis and effective treatment are clearly evident.

Table 2. Example of PCR diagnosis of bacterial infectious disease

Pathogen	Bacterial infections
	Respiratory organs
Bordetella pertussis	Whooping cough
Corynebacterium diphtheriae	Diphtheria
Legionella pneumophila	Legionnaire's disease
Mycobacterium tuberculosis	Tuberculosis
Mycoplasma pneumoniae	Atypical pneumonia
	Genital area
Chlamydia trachomatis	Inflammatory processes
Neisseria gonorrhoeae	Gonorrhoea
Treponema pallidum	Syphilis
	Nervous system
Haemophilus influenzae	Meningitis
Neisseria meningitidis	Meningitis
	Gastrointestinal tract
Helicobacter pylori	Gastritis
Campylobacter jejuni	Gastritis
Salmonella typhi	Typhoid fever
Shigella dysenteriae	Dysentery
	Systemic
Bacillus anthracis	Anthrax
Borrelia burgdorferi	Lyme disease (inflamed joints)
Leptospira sp.	Leptospirosis
Mycobacterium leprae	Leprosy
Yersinia pestis	Plague

Mycobacterial infections. Infections with mycobacteria cause three to four million deaths per year throughout the world. Tuberculosis is fast re-appearing as a major concern in the U.K. and throughout the world and can be an opportunistic pathogen in immuno-compromised patients. Traditional cultural and microscopic methods for the diagnosis of tuberculosis in sputum, pleural effusion, urine and aspirates can take 6 to 10 weeks. The immunological detection of mycobacterial antigens and the use of nucleic acid probes lack the sensitivity for testing clinical specimens. PCR, however, has been used to detect *M. tuberculosis* DNA rapidly in samples, found to be negative by microscopic and cultural techniques, from patients receiving treatment for tuberculosis over a 2 month period (Folgueira *et al.*, 1993). Tubercular patients may be hospitalized for extended periods awaiting a confirmed diagnosis. PCR methods can yield a clear result from a wide range of sample types within 6 hours. Infections caused by the *M. avium -M. intracellulare* complex (MAC), associated with AIDS, have been on the increase for several years (Pitchenik, 1990). Rapid and accurate identification and discrimination between *M. tuberculosis* and MAC are clinically important, since the treatment of each is significantly different. PCR has been used as a rapid and reliable method for the detection of potentially viable *Mycobacterium leprae* in human biopsies (Woods and Cole, 1989).

Borrelia burgdorferi. The agent of Lyme's disease, *Borrelia burgdorferi*, is transmitted by ticks, and can be carried for months before symptoms appear. If left untreated, chronic arthritis, heart disease and neurological abnormalities can occur. Antibody tests at an early stage of the infection are inconclusive yet, at this early stage, antibiotic therapy would be decisive in preventing long-lasting damage. PCR detection of *Borrelia* (Rosa and Schwan, 1989; Malloy *et al.*, 1990) allows the organism to be positively identified in urine (Goodman *et al.*, 1991) early in the course of the disease.

Hazardous organisms. An important application of PCR has been the detection of hazardous organisms, such as *Bacillus anthracis* (Turnbull, 1992), *Francisella tularensis* (Long *et al.*, 1993), *Salmonella typhi* (Song *et al.*, 1993) and *Yersinia pestis* (Hinnesbusch and Schwan, 1993), where routine cultural methods in the clinical laboratory are ruled out on safety grounds.

Organisms difficult to identify. PCR is being used extensively for the specific identification and discrimination of virulent organisms, including *Bartonella bacilliformis* (Maass *et al.*, 1992) and *Tropheryma whippellii* (Relman *et al.*, 1992) and between toxigenic and non-toxigenic strains of *Clostridium difficile* (Wren *et al.*, 1990) and *Corynebacterium diphtheriae* (Pallen, 1991).

Viruses. Herpes simplex virus (HSV) has an estimated total prevalence worldwide of 600 million with 90 million requiring treatment. Detection of *in vitro* human immunodeficiency virus type 1 (HIV1) was described by Kwoh *et al.* (1989) and Pachi *et al.* (1992) have recently described a technique for the direct detection of HIV RNA in plasma. HIV diagnosis research is costing £200 million of public funds in the U.K. alone and with a rapidly increasing worldwide infection rate, techniques such as PCR are vital. PCR has been used in a number of direct detection applications such as adenoviruses from faeces (Allard *et al.*, 1992).

318

Table 3. Diagnosis of viral infectious disease

Pathogen	Viral infection
Adenoviruses	Diarrhoea
Rotaviruses	Diarrahoea
Hepatitis B/C virus (HBV, HCV)	Liver inflammation and carcinoma
Human Immunodeficiency Virus (HIV1,2)	AIDS
Human T-cell leukaemia virus (HTLVI, II)	Acute T-cell leukaemia
Papilloma viruses	Certain cervical carcinomas
Herpes simplex virus (HSV1,2)	Lesions on genitalia, oral cavity etc.

The socioeconomic relevance of molecular biology to clinical microbiology cannot be doubted. The importance of PCR as a clinical tool for microbiologists is still only in its infancy but huge savings estimated at $400 million for Europe alone (Schulz, 1990) cannot be overlooked. The arguments against the technique have always been attributed to its high sensitivity. False positive results were a major problem when the methods were being developed but the introduction of modified techniques including dilution and further enzymic treatment, such as with N-uracyl-glycolase to remove unwanted uracil, have significantly reduced this effect.

As with most diagnostic methods, PCR is appropriate for the detection of some disease causing organisms especially those which have baffled traditional methodologies (Meier *et al.*, 1993). Epidemiologically, the technique is expected to prove invaluable during outbreaks to screen environmental samples. The presence of non-culturable organisms, e.g. *Campylobacter*, in water has been difficult to prove with traditional techniques but PCR can now be used to provide the necessary epidemiological proof (Birkenhead *et al.*, 1993).

Current techniques for product detection have been cumbersome and unsuited to the routine clinical laboratory but the commercial colorimetric assays (Rumpianesi *et al.*, 1993) now appearing will ensure the emergence of rapid and automated PCR methods. PCR has the potential for handling large numbers of samples and should radically alter our knowledge of the clinical and environmental microbiological world about us provided the cost per test is not prohibitive and kits are developed in line with clinical microbiologists' requirements.

PROSPECTS FOR THE FUTURE

Almost twenty years ago, Sherris (1974) wrote that "clinical microbiologists looked with some embarrassment at the developments in technology and performance control which were occurring in clinical chemistry, but consoled themselves with the thought that their own discipline was much more difficult (which it is), required the continued application of informed judgement (which it does) and was, therefore, perhaps inappropriate for the application of statistical standards of performance and the use of automated procedures (which it does not)". The aims of clinical microbiology in terms of patient care and the difficulties of the discipline remain the same.

Instruments available in clinical chemistry in the 1960s were capable of testing for single analytes and required significant operator involvement. The modern chemistry analyzer is capable of performing many analyses with multiple reagents, requires less sample volume and preparation and operator involvement is minimal. We speculate that this is a glimpse of what the future holds for clinical microbiology. Stager and Davis, in 1992, stated that we should expect these instruments of the future generations to be highly automated, cost-effective, accurate, reliable, flexible and able to provide rapid turn-round times.

The sample screening procedures outlined in this paper indicate considerable resource has been expended over the past two decades to ease the clinical microbiologist's burden without compromising accuracy or clinical specificity. The most economic, but sadly less than accurate or specific form of urine screening, has traditionally been to hold the fresh sample up to a bright light source. An indication of infection, either by pus cells or bacterial growth, can readily be seen as turbidity. Unfortunately a wide number of other factors can also cause cloudiness of the sample. The various ingenious attempts at automating urine screening have frequently been frustrated by the same wide range of factors involved with this complex body fluid. The **Questor** system does at last seem to hold some hope for cost-effectively automating the preliminary examination and screening of urine samples.

The screening of blood cultures was an early candidate for automation and the **Bactec** series of instruments has been widely used for over a decade. The novel continuous monitoring systems now being marketed indicate the considerable importance of this technique to clinical microbiology. The improved safety features and lack of manual intervention of the modern instruments are welcome advantages.

Culture techniques are time consuming and labour intensive but currently provide the most cost-effective means of aiding the detection and identification of bacteria for clinical diagnosis. The current limited success of nucleic acid probe-based assays indicates their continued use in clinical laboratories particularly for *in situ* hybridization assays. In immunosuppressed patients, methods used to demonstrate unusual microorganisms include non-cultural nucleic acid techniques (Kiehn *et al.*, 1989). These new technologies are enabling clinical microbiologists to gather important diagnostic information more readily. Commercial PCR assays are now available but at more than 150% of the present culture costs are unlikely to supplant the traditional methods save for those organisms where culture techniques are unreliable or unavailable or where early diagnosis is clinically important such as with *Chlamydia* and *Mycobacterium tuberculosis*. The high sensitivity and specificity of PCR tests together with enhanced automation will hopefully make these methods more cost-effective with time. PCR promises to add an extra practical dimension to diagnostic bacteriology (Eisenstein, 1990; Wayne, 1993) and virology (Smith *et al.*, 1992).

Despite the major improvements in techniques that are continuing to be made, the major breakthrough allowing laboratories to analyze all routine specimens directly and accurately with results available within minutes, has yet to be made. It would appear that the considerable advances in molecular biology technology, especially PCR, may yet prove to be the greatest addition to the microbiologist's toolbox since the loop!

REFERENCES

Allard, A., Albinsson, B. and Wadell, G. (1992) Detection of adenoviruses in stools from healthy persons and patients with diarrhoea by two-step polymerase chain reaction. J. Med. Virol. 37, 149-157.

Barrow, G.I. & Feltham, R.K.A. (1993) (Eds.) "Cowan & Steel's Manual for the Identification of Medical Bacteria". Cambridge University Press.

Beckmann, E. and Connolly, P. (1990) Flow cytometry: introduction and microbiological applications. Clin. Microbiol. Newsl. 12, 105-112.

Birkenhead, D., Hawkey, P.M., Heritage, J., Gascoyne-Binzi, D.M. and Kite, P. (1993) PCR for the detection and typing of campylobacters. Letts. Appl. Microbiol. 17, 235-237.

Buissière, J. and Nardon, P. (1968) Microméthode d'identification des bactéries. I. Intérêt de la quantification des caractères biochimiques. Annal Inst. Pasteur, (Paris), 115, 218-231.

Eisenstein, B.I. (1990) New molecular techniques for microbial epidemiology and diagnosis of infectious diseases. J. Infect. Dis. 161, 595-602.

Folgueira, L., Delgado, R., Palenque, E. and Noriega, A.R. (1993) Detection of *Mycobacterium tuberculosis* DNA in clinical samples by using a simple lysis method and polymerase chain reaction. J. Clin. Microbiol. 31, 1019-1021.

Goodman, J.L., Jurkovich, P., Kramber, J.M. and Johnson, R.C. (1991) Molecular detection of persistent *Borrelia burgdorferi* in the urine of patients with active lyme disease. Infect. Immun. 59, 269-278.

Gould J.C. and Duerden B.I. (1983) Blood culture - current state and future prospects. J. Clin. Pathol. 36, 963-977.

Hall, G.S. (1993) Probe technology for the clinical microbiology laboratory. Arch. Path. Lab. Med. 117, 678-583.

Hinnesbusch, J. and Schwan, T.G. (1993) New method for plague surveillance using polymerase chain reaction to detect *Yersinia pestis* in fleas. J. Clin. Microbiol. 31, 151 1-1514.

Kiehn, T.E., Ellner, P.D. and Budzko, D. (1989) Role of the microbiology laboratory in the care of the immunosuppressed patient. Rev. Infect. Dis. 11 (Supplement 7), 1706-1710.

Kwoh, D.Y., Davis, G.R., Whitfield, K.M., Chappelle, H.L., DiMichele, L.J. and Gingeras, T.R. (1989) Transcription-based amplification system and detection of amplified human immunodeficiency virus type 1 with a bead-based sandwich hydridization format. Proc. Natl. Acad. Sci. USA 86, 1173-1177.

Larsson, L. (1991) Gas chromatography and mass spectrometry, in "Automation in Clinical Microbiology" (Jorgensen, J.H., Ed.), pp. 153-166. CRC Press, Florida.

Long, G.W., Oprandy, J.J., Narayanan, R.B., Fortier, A.H., Porter, K.R. and Nacy, C.A. (1993) Detection of *Francisella tularensis* in blood by polymerase chain reaction. J. Clin. Microbiol. 31, 152-154.

Maass, M., Schreiber, M. and Knobloch, J. (1992) Detection of *Bartonella bacilliformis* in cultures, blood and formalin preserved skin biopsies by use of the polymerase chain reaction. Trop. Med. Parasit. 43, 191-194.

Malloy, D.C., Nauman, R.K. and Paxton, H. (1990) Detection of *Borrelia burgdorferi* using the polymerase chain reaction. J. Clin. Microbiol. 28, 1089-1093.

Meier, A., Persing, D.H., Finken, M. and Bottger, E.C. (1993) Elimination of contaminating DNA within polymerase chain reaction reagents: implications for a general approach to detection of unusual pathogens. J. Clin. Microbiol. 31, 646-652.

Ossewaarde, J.M., Rieffe, M., Rozenberg-Arska, M., Ossenkoppele, P.M., Nawrocki, R.P. and van Loon, A.M. (1992) Development and clinical evaluation of a polymerase chain reaction test for detection of *Chlamydia trachomatis*. J. Clin. Microbiol. 30, 2122-2128.

Pachi, C.A., Kern, D.G., Sheridan, P.J., Stempien, M.M., Todd, J.A., Zhu, Y.S., Gong, Y., Cimino, G.D., Wilber, J.C., Urdea, M.S. and Neuwald, P.D. (1992) Quantitative dtection of HIV RNA in plasma using a signal amplification probe assay. Program Abstr. 32nd Intersci. Conf. Antimicrobial Agents Chemother., abstr. 1247.

Pallen, M.J. (1991) Rapid screening for toxigenic *Corynebacterium diphtheriae* by the polymerase chain reaction. J. Clin. Pathol. 44, 1025-1026.

Pallen, M.J. and Butcher, P.D. (1991) New strategies in microbiological diagnosis. J. Hosp. Infect. 18 (Supplement A), 147-158.

Payne, W.J., Marshall, D.L., Shockley, R.K. and Martin, W.J. (1988) Clinical laboratory applications of monoclonal antibodies. Clin. Microbiol. Rev. 1, 313-329.

Pitchenik, A.E. (1990) Tuberculosis control and the AIDS epidemic in developing countries. Ann. Intern. Med. 113, 89-91.

Pot, B., Vandamme, P. and Kersters, K. (1993) Analysis of electrophoretic whole-organism protein fingerprints, in "Chemical Methods in Bacterial Systematics" (Goodfellow, M. and O'Donnell, A.G., Eds.) (in press). J. Wiley & Sons, Chichester, UK.

Relman, D.A., Schmidt, T.M., MacDermott, R.P. and Falkow, S. (1992) Identification of the uncultured bacillus of Whipple's disease. New Eng. J. Med. 327, 293-301.

Rosa, P.A. and Schwan, T.G. (1989) A specific and sensitive assay for the Lyme disease spirochaete *Borrelia burgdorferi* using the polymerase chain reaction. J. Infect. Dis. 160, 1019-1029.

Rumpianesi, F., LaPlaca Jr., M., D'Antuono, A., Negosanti, M. and Pavan, G. (1993) Assessment of the "Amplicor" PCR test in the diagnosis of *Chlamydia trachomatis* infection. Microbiologica 16, 293-295.

Schulz, U. (1993) Value added diagnostics. Medical Lab. World, April, 25-26.

Sherris, J.C. (1974) Recent and future changes in the clinical microbiology laboratory, in "Modern Methods in Medical Microbiology: systems and trends" (Prier, J.E., Bartola, J.T. and Friedman, H., Eds.), pp. 89-106. University Park Press, Baltimore.

Smith, T.F., Wold, A.D. and Epsy, M.J. (1992) Diagnostic virology - then and now. Adv. Exp. Med. Biol. 312, 191-199.

Song, J.H., Cho, H., Park, M.Y., Na, D.S., Moon, H.B. and Pai, C.H. (1993) Detection of *Salmonella typhi* in the blood of patients with typhoid fever by polymerase chain reaction. J. Clin. Microbiol. 31, 1439-1443.

Spencer R.C. (1988) Blood cultures: where do we stand? J. Clin. Pathol. 41, 668-670.

Stager, C.E. and Davis, J.R. (1992) Automated systems for identification of microorganisms. Clin. Microbiol. Rev. 5, 302-327.

Stamm, W.E. (1990) Laboratory diagnosis of chlamydial infections, in "Chlamydial Infections" (Bowie, W.R., Caldwell, H.D., Jones, R.P., Mårdh, P.-A., Ridgway, G.L., Schacter, J., Stamm, W.E. and Ward, M.E., Eds.), pp. 459-470. Cambridge University Press.

Stevens M., Patel H., Walters A., Burch K., Jay A., Dowling N., Mitchell C.J., Swann R.A., Willis A.T., Shanson D.C. and MacDonald C.A. (1992) Comparison of Sentinel and Bactec blood culture systems. J. Clin. Pathol. 45, 815-818.

Stevens M., Mitchell C.J., Livsey S.A., MacDonald C.A. (1993) Evaluation of the Questor urine screening system for bacteriuria and pyuria. J. Clin. Pathol. 46, 817-821.

Turnbull, P.C., Hutson, R.A., Ward, M.J., Jones, M.N., Quinn, C.P., Finnie, N.J., Duggleby, C.J., Kramer, J.M. and Melling, J. (1992) *Bacillus anthracis* but not always anthrax. J. Appl. Bact. 72, 21-28.

Wayne, L.G. (1993) The impact of new technology on the laboratory's contribution to the diagnosis and management of mycobacterial disease. Kekkaku 68, 113-129.

Woods, S.A. and Cole, S.T. (1989) A rapid method for the detection of potentially viable *Mycobacterium leprae* in human biopsies: a novel application of PCR. FEMS Microbiol. Lett. 53, 305-309.

Wren, B.W., Clayton, C.L. and Tabaqchali, S. (1990) Rapid identification of toxigenic *Clostridium difficile* by polymerase chain reaction. Lancet 335, 423.

322

Contributors

Amann, Rudolf
Lehrstuhl für Mikrobiologie, Technische Universitat München, 8029 München, Germany.

Anker, Lisbeth
Microbiology, Novo Nordisk A/S, Novo Allé, DK-2880 Bagsvaerd, Denmark.

Aquino de Muro, Marilena
Department of Biological Sciences. Heriot-Watt University, Riccarton, Edinburgh, EH14 4AS, UK

Atalan, Ekrem
Department of Microbiology, Medical School, University of Newcastle upon Tyne, Newcastle upon Tyne, NE2 4HH, UK.

Boyd, E. Fidelma
Institute of Molecular Evolutionary Genetics, Pennsylvania State University, University Park, Pennsylvania 16802, USA.

Chun, Jongsik
Department of Microbiology, Medical School, University of Newcastle upon Tyne, Newcastle upon Tyne, NE2 4HH, UK.

Collins, Nadine C.
Department of Microbiology, University of Leicester, Leicester, LE1 9HN, UK.

Conway de Macario, Everly
Wadsworth Center for Laboratories and Research, New York State Department of Health, P.O. Box 509, Albany, New York, 12201-0509, USA.

Dewettinck, Dirk
Laboratorium voor Microbiologie, Universiteit Gent, K.L. Ledeganckstraat 35, B-9000 Gent, Belgium.

da Costa, Milton S.
Departmento de Bioquímica, Universidade de Coimbra, 3049 Coimbra Codex, Portugal

Embley, T. Martin
Microbiology Group, The National History Museum, Cromwell Road, London, SW7 5BD, UK.

Feltham, Kevin, A.
Feltham Associates Ltd., Carlton House, Kibworth Hall Park, Kibworth Harcourt, Leicester, LE8 0SE, UK.

Finlay, Bland J.
Institute of Freshwater Ecology, Windermere Laboratory, Far Sawrey, Ambleside, LA22 0LP, UK.

Gicquel, Brigitte
Unité-Génétique Mycobactérienne, Institut Pasteur, Rue du Dr. Roux, 75 724 Paris, Cadex 15, France.

Goebel, Brett M.
Centre for Bacterial Diversity and Identification, Department of Microbiology, University of Queensland, St. Lucia, Queensland 4072, Australia.

Goméz-Lus, Rafael
Unidad Microbiología, Faculdad de Medicina, Universidad Zaragoza, C/Domingo Miral sn, 50009 Zaragoza, Spain.

Goodfellow, Michael
Department of Microbiology, Medical School, University of Newcastle upon Tyne, Newcastle upon Tyne, NE2 4HH, UK.

Grant, William D.
Department of Microbiology, University of Leicester, Leicester, LE1 9HN, UK.

Gürtler, Hanne
Microbiology, Novo Nordisk A/S, Novo Allé DK-2880 Bagsvaerd, Denmark.

Helm, Dieter
Abteilung für Cytologie, Robert Koch-Institut, des Bundesgesundheitsamtes, Nordufer 20, D I3353, Berlin, Germany.

Huddleston, Annaliesa S.
Department of Biological Sciences, University of Warwick, Coventry, CV4 7AL., UK.

Jones, Brian E.
Gist-Brocades B.V. R & D, Postbus 1, 2600 MA Delft, The Netherlands.

Kaji, Denise A.
Department of Biological Sciences. Heriot-Watt University, Riccarton, Edinburgh, EH14 4AS, UK.

Kersters, Karel
Laboratorium voor Microbiologie, Universiteit Gent, K.L. Ledeganckstraat 35, B-9000 Gent, Belgium.

Li, Jia
Institute of Molecular Evolutionary Genetics, Pennsylvania State University, University Park, Pennsylvania 16802, USA.

Ludwig, Wolfgang
Lehrstuhl für Mikrobiologie, Technische Universitat München, 8029 München, Germany.

Macario, Alberto J.L.
Department of Biomedical Sciences, School of Public Health, The University of Albany, Albany, New York, 12201-0509, USA.

Marsh, Peter
Department of Biological Sciences, University of Warwick, Coventry, CV4 7AL., UK.

Martín, Carlos
Unidad Microbiología, Faculdad de Medicina, Universidad Zaragoza, C/Domingo Miral sn, 50009 Zaragoza, Spain.

Mwatha, Wanjiru E.
Department of Botany, Kenyatta University, Nairobi, Kenya.

Naumann, Dieter
Abteilung für Cytologie, Robert Koch-Institut, des Bundesgesundheitsamtes, Nordufer 20, D 13353, Berlin, Germany.

Nelson, Kimberlyn
Institute of Molecular Evolutionary Genetics, Pennsylvania State University, University Park, Pennsylvania 16802, USA.

Nobre, M. Fernanda
Departmento de Zoologia, Universidade de Coimbra, 3049 Coimbra Codex, Portugal.

Otal, Isabel
Unidad Microbiologiá, Faculdad de Medicina, Universidad Zaragoza, C/Domingo Miral sn, 50009 Zaragoza, Spain.

Pot, Bruno
Laboratorium voor Microbiologie, Universiteit Gent, K.L. Ledeganckstraat 35, B-9000 Gent, Belgium.

Priest, Fergus G.
Department of Biological Sciences. Heriot-Watt University, Riccarton, Edinburgh, EH14 4AS, UK.

Samper, Sofía
Unidad Microbiología, Faculdad de Medicina, Universidad Zaragoza, C/Domingo Miral sn, 50009 Zaragoza, Spain.

Sanglier, Jean-Jacques
Preclinical Research, Sandoz Pharma Ltd., CH-4002 Basle, Switzerland.

Schleifer, Karl H.
Lehrstuhl für Mikrobiologie, Technische Universitat München, 8029 München, Germany.

Schultz, Christian
Abteilung für Cytologie, Robert Koch-Institut, des Bundesgesundheitsamtes, Nordufer 20, D
13353, Berlin, Germany.

Selander, Robert K.
Institute of Molecular Evolutionary Genetics, Pennsylvania State University, University Park,
Pennsylvania 16802, USA.

Stackebrandt, Erko
DSM-German Collection of Microorganisms and Cell Cultures, Mascheroder Weg 1b, D-
38124 Braunschweig, Germany.

Stevens, Michael
Department of Health Evaluation Unit, Public Health Laboratory, Leicester Royal Infirmary,
Leicester, LE1 5WW, UK.

Tindall, Brian J.
DSM-German Collection of Microorganisms and Cell Cultures, Mascheroder Weg 1b, D-
38124 Braunschweig, Germany.

Torck, Urbain
Laboratorium voor Microbiologie, Universiteit Gent, K.L. Ledeganckstraat 35, B-9000 Gent,
Belgium.

Torrea, Gabriella
Unité-Génétique Mycobactérienne, Institut Pasteur, Rue du Dr. Roux, 75 724 Paris, Cadex 15,
France.

Vancanneyt, Marc
Laboratorium voor Microbiologie, Universiteit Gent, K.L. Ledeganckstraat 35, B-9000 Gent,
Belgium.

Vandamme, Peter
Laboratorium voor Microbiologie, Universiteit Gent, K.L. Ledeganckstraat 35, B-9000 Gent,
Belgium.

Vauterin, Luc
Laboratorium voor Microbiologie, Universiteit Gent, K.L. Ledeganckstraat 35, B-9000 Gent,
Belgium.

Ventosa, Antonio
Department of Microbiology and Parasitology, Faculty of Pharmacy, University of Sevilla,
41012 Sevilla, Spain.

Wang, Fu-Sheng
Institute of Molecular Evolutionary Genetics, Pennsylvania State University, University Park,
Pennsylvania 16802, USA.

Wellington, Elizabeth M.H.
Department of Biological Sciences, University of Warwick, Coventry, CV4 7AL, UK.

INDEX

Actinobacillus, 200
Alicyclobacillus, 177
Alignment of sequences, 4, 119
Alkaliphiles, 195-200
    definitions, 195
Alkylglycerol diethers, 177
Anacystis, 5, 6, 7
Antigens
    flagellar, 21-23, 280, 287-289
    fingerprinting, 163
Aquifex pyrophilus, 175, 177
Archaea, 5, 6, 9, 120, 157, 173, 243-258
    DNA probes for, 120
Arcobacter species, 55
Arcobacter butzleri, 59-62
Arcobacter cryaerophilus, 59-62
Arcobacter skirrowi, 59-62
Arcobacter nitrofigilis, 59-62
Arhodomonas aquaeoli, 237
ATPase, 3
Azoarcus species, 55

Bacillus, 8, 19, 177-178, 200, 221
    insect pathogens, 275-295
Bacillus alcalophilus, 196
Bacillus anthracis, 285, 317
Bacillus cereus, 285
Bacillus cohnii, 196
Bacillus flavothermus, 177
Bacillus macerans, 121
Bacillus polymxa, 121
Bacillus sphaericus, 277-285
Bacillus subtilis, 69
Bacillus thermocloacae, 177
Bacillus thermoruber, 177
Bacillus thuringiensis, 285-290
Bacteria, 173
    DNA probes for, 120
Bacteroides forsythus, 121
Bacteroides ureolyticus, 94
Bioleaching, 259-273
BIOLOG, 211
Bioreactors, 162-164
Blood culture screening, 311

Bordetella, 19, 317
Bordetella avium, 55-56
Borrelia burgdorferi, 318
Bradyrhizobium japonicum, 148
Butyrivibrio, 175-176

Caedibacter coryophila, 128
Campylobacter cinaedi, 55
Campylobacter concisus, 55
Campylobacter hyointestinalis, 55
Campylobacter jejuni, 317
Campylobacter jejuni subsp. "doylei", 55
Campylobacter species, 55
    analysis by SDS PAGE, 58
Caryophanon, 8
CHARSEP program, 220
Chemotaxonomy, 245-248
Chemosystematics, 173-193
Chlamydia, 317
Chlorflexux auriantiacus, 176
Citrobacter freundii, 69
Cladistic analyses, 243
Classification
    guiding concepts, 1
Clonal populations, 18, 19, 290-291
Clostridium, 178
Clostridium botulinum, 55
Clostridium thermocellum, 178
Clostridium thermohydrosulfuricum, 178
Clostridium thermosulfurogenes, 178
Colony hybridization, 282
Communities, 129-131, 161-171
Comomonas species, 55
Comomonas terrigena, 214
Comomonas testosteroni, 121
Computer-assisted
    analysis of PAGE, 53-54
    classification, 214-217, 220-223
    identification, 220
Consortia for bioleaching, 260
Corynebacterium jeikeium, 94
Corynebacterium novyi, 69
Corynebacterium septicum, 69
Curie point MS, *see* Mass Spectrometry

Cytophaga, 5,7

Databases, 244-245
  for identification, 51
Deinococcus, 5-7
Deleya, 233, 234, 244
Deleya marina, 217, 220
Desulfobacter, 120
Desulfohalobium retbaense, 206
Desulfovibrio halophilus, 206
DIACHAR program, 220
Digoxigenin, *see* Label digoxigenin
DNA
  fingerprinting, 109-111
  isolation from soil, 138-141
  plasmid, 290
DNA homology groups, 277-278
DNA hybridization, 51
DNA probes, *see* Nucleic acid probes
DNA reassociation, 285-286
DNA sequencing. 21-23, 27-34
  analysis of, 17-49

Ectothiorhodospiraceae , 206, 211
Electrophoretic types, 19
Endosymbionts, 126-128, 153-160
Enzyme electrophoresis, 19
Escherichia coli, 17, 40, 90
  clonal structure, 20
Eubacteria 169
Eucarya, 6, 7, 117

Fatty acids
  of alkaliphiles, 222
  cycloheptyl, 177
  cyclohexyl, 177
  for identification, 315
  of *Thermus*, 183-187
Fervidobacterium islandicum, 175
Fervidobacterium nodosum, 175
Fibrobacter intestinalis, 121
Fibrobacter succinogenes, 121
Filibacter, 8
Fingerprinting techniques, 62, 109, 305
Flavobacterium , 232
Flavobacterium godwanense, 236
Flavobacterium salegens, 236
Flow cytometry, 125
Fourier transform IR (FT-IR) microscopy,
  75-77
  of microcolonies, 75-76
Fourier transform IR (FT-IR) spectroscopy,
  67-85
  analysis of cells, 79-80
  experimental procedures, 68
  of staphylococci, 68-72
Frankia, 121, 140-141

Gene
  analysis, 11-23
  detection of in soil, 143-148
  evolution, 23-42
  trees, 6, 7, 30, 31, 33, 38, 39, *see also*
    Phylogenetic trees
Genotypic data
  integration with phenotypic data, 8-9,
    243-258
GENSTAT, 97
Geotoga petraea, 175
Geotoga subterranea, 175
Gluconobacter oxydans, 57
Glyceraldehyde-3-phosphate dehydrogenase,
  32
Green non-sulphur bacteria, 176-177

Haemophilus , 19
Haemophilus influenzae, 19, 40, 55
  DNA probes for, 316
Haloanaerobium , 232
Haloarcula, 231, 235
Halobacteriales, 231-242
Halobacterium denitrificans, 244
Halobacteroides, 232
Halococcus, 231
Haloferax, 231
Halomonas, 219, 234,
Halomonas elongata, 217, 235, 244
Halomonas halmophilum, 217, 220
Halophilic bacteria, 231-242
  definitions, 231
  diversity, 233-234
Herpetosiphon auriantiacus, 176
Holospora elegans, 127
Holospora obtusa, 127
Horizontal gene exchange, 17, 18
Human immunodeficiency virus (HIV), 318-
  319
Hybridization, *see* Nucleic acid probes
Hydrogenobacter thermophilus, 175
Hydrogenophaga, 51-52, 55

Identification
  automated, 314-316
  chemotaxonomy and, 173-193
  computer-assisted, 215, 223
  in industry, 298-301
  kits, 314
  matrix, 215, 223
  probes, 120-121, 316; *see also* Nucleic
    acid probes
  pyrolysis, 87-104
Igatibacter hanningtonii, 220
Infrared microscopy, 75-77
Infrared spectroscopy, 67-85; *see also*
    Fourier transform IR
Infrared spectroscopy, sample preparation 413

Insertion elements, 107-113
Insertion sequences, 107-113
    and fingerprinting, 109
Isoprenoid quinones, 176, 177, 178, 216,
    219, 221, 222

Kauffman-White scheme, 21, 26

Label
    digoxigenin, 12, 124, 278-279
    fluorescent, 123
Lactic acid bacteria, 12
Lactobacillus, 8-9
Lactobacillus acidophilus, 55
Lactobacillus casei, 55
LacZY as marker genes, 138, 148
Leuconostoc, 55
Legionella, 19
Legionella pneumophila, 69, 80-82, 94,
    317
Leptospirillum ferrooxidans, 261, 264-270
Ligation-mediated PCR (LMPCR), 111-112
Lipopolysaccharide (LPS), 72-74
Listeria monocytogenes, 94
Luciferase, 107-108
Lux genes, 138, 148

Magatibacter afermentans, 217
Magnetospirillum, 120, 128-129, 250
Magnetotactic bacteria, 128-129
Malate dehydrogenase, 27, 29-30
Marinococcus halophilus, 234
Marinococcus hispanicus, 234
Mass spectrometry, 87-104
MATIDEN program, 220
Menaquinones, 176, 177, 178, 216, 219,
    of alkaliphiles, 222
    of thermophiles, 177, 178
Methanogens, 153-160, 162-171
Methanobacterium alcaliphilum, 210
Methanobacterium bryantii, 163
Methanobacterium formicum, 163
Methanobacterium thermoautotrophicum,
    163-164, 167-168, 210, 251
Methanobacterium thermoalcalophilum, 210
Methanobrevibacter arboriphilus, 163-164,
    168
Methanobrevibacter ruminantium, 163
Methanobrevibacter smithii, 163-164, 168
Methanococcus vanielli, 163
Methanogenium cariaci, 164
Methanohalophilus mahii, 235
Methanohalophilus origonense, 207-208
Methanohalophilus portucalensis, 232
Methanohalophilus zhilinae, 207-208, 235
Methanomicrobium mobile, 162
Methanosarcina barkeri, 163, 167
Methanosarcina mazei, 165

Methanosarcina thermophilus, 163-164, 168
Methanospirillum hungatei, 164
Metopus contortus, 155-157
Metopus palaeformis, 155-157
Metopus striatus, 155-157
Microcolonies
    analysis by FT-IR, 75-76
MIDI system, 314
Molecular chronometers, 2-3
Molecular taxonomy, 1-15
Moraxella canis, 55
Moraxella lincolnii, 55
Morganella morganii, 55
Multilocus enzyme electrophoresis, 17
Mycobacterium, 19, 105-114, 318
Mycobacterium africanum, 105
Mycobacterium avium, 9, 106, 108, 317-318
Mycobacterium bovis, 96, 106
Mycobacterium fortuitum, 106
Mycobacterium intracellulaire, 9, 317-318
Mycobacterium mycoti, 96
Mycobacterium partuberculosis, 9
Mycobacterium tuberculosis, 55, 90, 96, 105-
    113, 317-318
    fingerprinting by RFLP, 109-110
    PCR diagnosis, 317
Mycoplasma mycoides, 55

Natronobacterium, 220, 231
Natronobacterium gregori, 231
Natronobacterium magadii, 213
Natronobacterium pharaonis, 213
Natronococcus occultus, 220, 213
Natronococcus vacuolatum, 213
Neisseria, 19
Neisseria gonnhoreae, 40
Neisseria meningitidis, 19, 40
Neural networks, 91-93, 95-96
    applications, 95
Non-culturable bacteria, 12, 17, 304
Nucleic acid probes, 10-12, 106, 115-135,
    147, 155, 282-284, 304, 316
    clinical diagnosis, 316
    colony hybridization, 282
    design, 11, 116-121
    digoxigenin labelling, *see* label,
        digoxigenin
    in situ hybridization, 115-135,
    sensitivity, 122-123
    16S and 23S genes, 11, 119-121, 304,
        316
    specificity, 121-122
    target sites, 117-118
    whole cell hybridization, 11, 119-122,
        125-132
Nucleic acids
    amplification, *see* Polymerase chain
        reaction

Nucleic acids (*cont'd*)
   from soil, 139-143
   hybridization, 11
   sequence analysis, 17-49
Numerical taxonomy, 214-224

Oligonucleotide probes, *see* Nucleic acid probes

Paracoccus alcalophilus, 249
Paracoccus aminophilus, 249
Paracoccus aminovorans, 249
Paracoccus halodenitrificans, 232, 249
Paracoccus kocurii, 249
Paramecium caudatum, 127
Peptidoglycan types, 9
Patents, 298-299
Petrotoga miotherma, 175
Phenetic taxonomy, 243-244
Phospholipids, 177, 246
Phylogenetic
   relationships, 2-9, 243-244, 248-251
   trees, 5,6,7,117, 156, 236, 237, 238, 250-251, 279, 286
Polar lipids
   of alkaliphiles, 222
   of *Bacillus*, 177
   of thermophiles, 175-186
Polyacrylamide gel electrophoresis (PAGE), 51-66
Poly-ß-hydroxybutyrate
   analysis by FT-IR, 80-81
Polymerase chain reaction (PCR), 144, 154, 262, 316
   for detection of bacteria, 12-13
   for diagnosis, 105-113, 316-317
   of probes, 284
   from soil, 148
Populations, genetic structure, 17-49
Porphyomonas gingivalis, 120
Proline permease, 30-31
Propionibarterium species, 55
Protein patterns, 51-66
Proteobacteria, 5-7, 9, 247
Proteus mirabilis, 55
Proteus penneri, 55
Proteus vulgaris, 55, 69
Protozoa, 153-160
Providencia alcalifaciens, 55
Pseudomonas, 55
Pseudomonas aeruginosa, 20, 69, 94
Pseudomonas cepacia, 120
Pseudomonas putida, 120, 214
Pseudomonas stutzeri, 214
Pyrolysis, 87-104
   apparatus, 88-90
   data analysis, 90-93
   use in screening programmes, 96-100, 306

Quality control, 299-300
Quarantine regulations, 300

Radiometric detection, 311
Random amplified polymorphic DNA (RAPDS)
   of *Arcobacter*, 61-62
   of *Bacillus sphaericus*, 280
Recombination
   evidence for, 24, 34-39
   and evolution, 21
   rates of, 41
Restriction fragment length polymorphism (RFLP), 105-113, 278, 305,
   of mycobacteria, 109, 11
Rhizobium, 19, 55
Rhizobium meliloti, 20
Rhodospirillum salexigens, 237
Rhodospirillum salinarum, 237
Rhodothermus marinus, 178
Ribosomal RNA
   determination of relationships by sequence comparison of, 4-9, 105, 154, 244, 248-251
   from soil, 138, 140-143
   structure, 116-119
Ribotyping, 279, 287-289, 301, 302
Rothia dentocariosa, 55

Salmonella, 17-49
   antigenic formulae, 22
   serovars, 26
Salmonella enterica, 17, 21-23, 27-39
Salmonella enteritidis, 24, 94
Salmonella minnesota, 69, 72-74
Salmonella typhi, 317
Salmonella typhimurium, 21, 69, 149
Sarcobium lyticum, 128
SDS-PAGE, 51-66
Serological typing, 26, 169
Shigella, 19, 38-39
Slot-blot hybridization, 284
Sporohalobacter, 232, 235
Sporosarcina halophila, 236
Sporosarcina ureae, 236
Staphylococcus
   classification by FT-IR 69-71
Staphylococcus arlettae 69-71
Staphylococcus aureus, 55, 69-71, 78
Staphylococcus capitis, 70-71
Staphylococcus caprae, 70
Staphylococcus cohnii, 70
Staphylococcus epidermidis, 70
Staphylococcus equorum, 70
Staphylococcus gallinarum, 70
Staphylococcus kloosii, 70
Staphylococcus saprophyticus, 69- 70
Staphylococcus xylosus, 70

Streptococcus agalactiae, 69
Streptoccus pneumoniae, 94
Streptococcus pyogenes, 19, 40, 69, 94
Streptomyces, 55, 96
Streptomyces albidoflavus, 96-98
Streptomyces anulatus, 96-98
Streptomyces griseus, 147
Streptomyces lividans, 146-147
Streptoverticillium, 55
Sulphate-reducing bacteria, 120, 176
Symbionts, 120, 153-160

Taxonomy, *see* Classification
Teichoic acids, 9
Thermophilic bacteria, 173-193
    definition, 174
Thermotoga maritima, 175
Thermotoga neapolitana, 175
Thermotoga thermarum, 175
Thermotogales, 175-176
Thermus, 179-187
    chemotaxonomy, 181-187
Thermus aquaticus, 179
Thermus filiformis, 179-187
Thermus ruber, 180-187
Thermus scotoductus, 180-187
Thermus thermophilus, 180-187
Thiobacilus
    and metal leaching, 259-273
    physiology of, 267-268

Thiobacillus ferrooxidans, 260
Transposable genetic elements, *see* Insertion
sequences
Trimyena, 157
Tuberculosis, 105-113; *see also*
Mycobacterium tuberculosis
Typing
    by FT-IR, 67-85
    with insertion sequences, 107-113
    by protein electrophoresis, 51-66
    by PyMS, 94
    by RAPDS, 61-62
    by RFLP, 105-113
    by ribotyping, *see* Ribotyping
Ubiquinones, 214, 247
Uncultured microorganisms, 126, 137
Urine screening, 310-311

Viruses, detection and identification, 318-320

Wadi Natrum, 213
Ward's method, 70, 71, 73, 74
Whole cell hybridization, 11, 119-122, 125-
132

Xanthomonas, 55
Xanthomonas maltophilia, 94
XylE, 138, 145

The manufacturer's authorised representative in the EU is Springer
Nature Customer Service Centre GmbH, Europaplatz 3, 69115 Heidelberg,
Germany. If you have any concerns regarding our products, please
contact ProductSafety@springernature.com

Printed and bound by CPI Group (UK) Ltd, Croydon, CR0 4YY

23/04/2026

02095629-0011